国家自然科学基金项目（编号：51278416）
陕西省社会发展攻关计划项目（编号：2012K12-01-05）

中国历史城市的更新与社会资本
Social Capital and Urban Regeneration in Chinese Historic Cities

翟斌庆　著
Zhai Binqing

中国建筑工业出版社

图书在版编目（CIP）数据

中国历史城市的更新与社会资本/翟斌庆著. —北京：中国建筑工业出版社，2014.10
ISBN 978-7-112-16973-3

Ⅰ.①中… Ⅱ.①翟… Ⅲ.①城市史-建筑史-研究-中国-古代 Ⅳ.①TU-098.12

中国版本图书馆CIP数据核字（2014）第124915号

责任编辑：李　鸽
责任校对：姜小莲　刘梦然

中国历史城市的更新与社会资本
翟斌庆　著

*

中国建筑工业出版社出版、发行（北京西郊百万庄）
各地新华书店、建筑书店经销
北京嘉泰利德公司制版
北京中科印刷有限公司印刷

*

开本：787×1092毫米　1/16　印张：23　字数：475千字
2014年11月第一版　2014年11月第一次印刷
定价：73.80元
ISBN 978-7-112-16973-3
（25748）

版权所有　翻印必究
如有印装质量问题，可寄本社退换
（邮政编码　100037）

前言 / PREFACE

自 20 世纪 90 年代以来,中国的快速城市化进程对许多历史城市的发展提出了双重挑战。一方面,历史城市需要更新改造其破败的旧城区;另一方面,当地经济需要进一步发展。在这种情形下,许多地方往往通过大规模城市再开发的更新政策来达到上述双重目的。不过,这种更新策略在为当地带来巨额经济收入的同时,也对历史城市的固有特征造成了许多无法挽回的破坏。

已有文献中关于城市更新的研究表明,城市更新涉及的主要利益相关者除了包括地方政府和私营部门,还包括地方社区居民。事实上,地方社区居民在城市更新过程中,对保护历史住区原住民的生活内容发挥着重要作用。社会资本使得社区居民以社区为基础紧密团结在一起,并会为实现历史地段更新和保护的综合目标而积极参与其中。基于上述理论,本书探讨了社会资本在中国历史城市的更新过程中将发挥怎样的作用。

本书旨在对中国历史城市的更新过程,作出一种特定历史条件和语境下的分析与解读。同时,针对当前转型期中国历史城市更新过程中的诸多问题,探讨了社会资本在城市更新过程中的作用。在当前中国城市更新管治模式中,地方政府发挥着决定性的作用:一方面作为城市再开发项目的倡导者与参与

With rapid urbanization since the 1990s, many Chinese historic cities have faced the dual challenges of regenerating dilapidated historic inner urban areas and promoting local economic development. Rampant urban redevelopment-oriented planning and practices have been undertaken in many places. While bringing enormous economic returns, this trend also poses many threats to the character of the historic cities.

Literature on urban regeneration shows that local communities should play an active role in regenerating a place in addition to other key actors such as the government and the private sectors. Local communities also play a critical role in conserving the indigenous lives of historic residential districts. Social capital enables local communities to act together to pursue shared objectives in the community-based regeneration processes. Based on this theoretical premise, the study seeks to examine the role of social capital in the regeneration of Chinese historic cities.

This book aims to provide a historical and contextual understanding of the evolution of urban regeneration in Chinese historic cities. The book also explores the role of social capital in the current controversies surrounding urban regeneration in transitional China. In the current mode of urban regeneration

者；另一方面也作为再开发项目的管理者。私人开发商往往会积极参与到城市更新项目当中；而地方社区居民往往还被排斥于整个更新项目之外。中国历史城区的更新活动通常是通过对少数文物建筑的保护和大规模的房地产再开发项目来进行，其目的是达到刺激地方经济发展和改善物质环境。由于城市保护政策侧重于对物质环境的保护，许多原住民的生活并不在保护范畴之内，他们常常在当地实施房地产再开发项目时，需要被重新安置，因此，本书认为缺乏原住民参与的城市更新过程，可能很大程度上会妨碍历史城市更新和保护综合目标的实现。

通过对西安历史文化名城中两个历史地段更新保护实践的案例分析，本书得出如下主要结论：在西安鼓楼回民区的更新保护案例中，传统的回族社区居民生活内容与商业活动是当地经济繁荣的重要基础。基于对当地社区居民间具有较强的社区凝聚性这一事实的考虑，研究认为，该回民区实际上可以保持一种自我更新的过程。在这当中，地方政府可能需要对当地破败的城市基础设施进行适时维修。因而，下一步的问题将是如何建立一套制度化的更新机制来协助完成整个历史街区的更新。与回民区案例形成鲜明对照的是三学街历史街区的更新保护工程。相

governance in China, local governments often play a dominant role as both redevelopment advocates and project managers. Private developers are actively involved while local indigenous residents are often excluded from the regeneration processes. Urban regeneration practices in Chinese historic cities are often conducted through urban conservation-cum-redevelopment strategies to spur local economic growth and improve the physical environment. Since the focus is on the conservation of the physical environment, many local indigenous lives have to be excluded and relocated to give way to local redevelopment projects. The lack of community involvement in local regeneration processes severely undermines the goals of comprehensive urban regeneration and integrated urban conservation plans.

Xi'an, a typical Chinese historic city, has been chosen as the study site. Two solid local case studies have produced the following major findings. In the Drum Tower Muslim District, traditional Muslim lives and businesses have sustained a vibrant local economy. Together with cohesive community relationships, this book argues that the historic Muslim district can actually sustain a self-regeneration process, given proper maintenance of the dilapidated urban infrastructures by local governments. The question is an institutionalized mechanism

比较而言，三学街历史街区居民间的邻里关系比较薄弱，当地社区居民间的社会资本水平也较低。不过，值得注意的是，这里城市更新实践的"效率"则比回民区明显要高。然而，这种高"效率"的更新实践却显示出该城市更新策略的不够综合化，同时当地原住民的邻里生活方式和生活内容也未能得到很好保护。

为使转型期的中国历史城市在更新过程中达到综合化的城市更新目标，本书提出如下观点：(1) 对当地原住民的生活内容和地方社会文化元素的保护，应当在历史城市的更新策略中占据重要位置；(2) 较高的社会资本水平有助于协助社区居民更好地融入城市更新过程中，从而也有助于对当地原住民生活方式和生活内容的保护；(3) 制度化的社区参与式的城市更新管治模式，是保障城市更新的综合目标在中国历史城市的社区层面得以实现的重要途径。

to facilitate this kind of regeneration. The implementation of the government-led regeneration projects within the Sanxuejie Historic District, where community relationships are weak with low level of social capital, was more "efficient". However, it also means that regeneration efforts were less comprehensive and indigenous lives of the original neighborhoods were not conserved.

To achieve the goals of comprehensive urban regeneration in historic cities, this book maintains the following: (1) the conservation of indigenous lives and local socio-cultural elements is important for local regeneration plans; (2) a strong social capital contributes to the conservation of local indigenous lives by facilitating community involvement in local urban regeneration processes; (3) an institutionalized community participatory mode of urban governance is essential for a comprehensive regeneration plan at the local level.

目录 / CONTENTS

第一章　绪论———— 001 ————CHAPTER 1　Introduction
  1.1　概述———— 002 ————1.1　Introduction
  1.2　研究问题———— 005 ————1.2　Research Questions
  1.3　研究目的———— 007 ————1.3　Goals and Objectives
  1.4　研究意义———— 007 ————1.4　Significance of the Study
  1.5　研究范围———— 009 ————1.5　Delimitation
  1.6　方法论———— 010 ————1.6　Research Methodology
  1.7　本书架构———— 030 ————1.7　Outline of the Book

第二章　城市更新与社会资本———— 035 ————CHAPTER 2　Urban Regeneration and Social Capital
  2.1　城市更新议程———— 037 ————2.1　The Urban Regeneration Agenda
  2.2　历史城市的更新———— 058 ————2.2　Urban Regeneration in Historic Cities
  2.3　社会资本和城市更新———— 074 ————2.3　Social Capital and Urban Regeneration
  2.4　总结———— 091 ————2.4　Conclusion

第三章　转型期中国历史城市的更新———— 093 ————CHAPTER 3　Urban Regeneration and Social Capital in Historic Cities of Transitional China: Towards an Analytical Framework
  与社会资本：研究架构的建立

  3.1　自1949年以来中国历史城市的———— 095 ————3.1　Urban Development and Regeneration in China from 1949 to Present
    发展与更新
  3.2　影响中国历史城市更新———— 118 ————3.2　Key Factors in Urban Regeneration in Chinese Historic Cities
    的关键因素
  3.3　中国历史城市更新中的社会———— 149 ————3.3　Social Capital, Urban Conservation, and Urban Governance in the Regeneration of Chinese Historic Cities
    资本、城市保护与城市管治

第四章　西安的社会资本与城市更新―― 153 ――CHAPTER 4　Social Capital and Urban Regeneration in Xi'an

    4.1　西安：地理介绍―― 155 ――4.1　The City of Xi'an: A Geographical Introduction
    4.2　西安的城市发展与更新策略―― 157 ――4.2　Urban Development and the Implications on Regeneration in Xi'an
    4.3　西安城市更新中的社会资本―― 190 ――4.3　Social Capital in Xi'an's Urban Regeneration
    4.4　总结―― 206 ――4.4　Conclusion

第五章　具有较强社会资本的历史―― 209 ――CHAPTER 5　Urban Regeneration in Historic District with Social Capital based on Stronger Community Relationships
　　　　街区及其更新实践

    5.1　西安鼓楼回民区简介―― 211 ――5.1　Introduction to the Xi'an Drum Tower Muslim District
    5.2　鼓楼回民区的更新实践―― 221 ――5.2　Urban Regeneration Practices in Xi'an DTMD
    5.3　鼓楼回民区更新实践中的社会资本―― 237 ――5.3　Social Capital in the Regeneration of Xi'an DTMD
    5.4　总结―― 252 ――5.4　Conclusion

第六章　具有转型期社会资本典型特征―― 255 ――CHAPTER 6　Urban Regeneration in a Historic District with Social Capital based on the Transformation of Community Relationships
　　　　的历史街区及其更新实践

    6.1　西安三学街历史街区简介―― 257 ――6.1　Introduction to the Sanxuejie Historic District
    6.2　三学街区的更新政策与更新实践―― 265 ――6.2　Transitional Urban Regeneration Policies and Practices in Xi'an SHD
    6.3　三学街历史街区更新中的社会资本―― 292 ――6.3　Social Capital in the Regeneration of Xi'an SHD

6.4　总结——— 307 ———6.4　Conclusion

第七章　结论——— 309 ———**CHAPTER 7　Conclusion**

7.1　综述——— 310 ———7.1　Summary
7.2　主要结论与讨论——— 312 ———7.2　Major Findings and Discussions

参考文献——— 332 ———**BIBLIOGRAPHY**
后记——— 359 ———**POSTSCRIPT**

第一章 绪论
CHAPTER 1 Introduction

1.1 概述

2008年联合国人居署发布的报告显示(UN-HABITAT，2008:4)，全球的城市居民总数已超过全球总人口数量的一半。随着城市化的快速进程，不少挑战也接踵而来，尤其对城市化速度还较低的发展中国家而言，挑战将更为严峻。世界银行发布的研究报告表明，到2020年，城市居民总人口将达到41亿人（约占世界总人口的55%），其中增加的城市居民中有将近94%的人口来自于发展中国家（Léautier，2006:viii）。城市化在为发展中国家带来很多积极变化的同时，还给这些国家带来不少挑战，包括能源消耗和气候变化，同时也会因社会阶层的分离与排斥，而引发社会不安。

在亚洲，尤其是一些历史古都城市，重点关注的往往是在实现经济快速发展的同时，如何进行城市保护（Stovel，2002:107）。亚洲很多国家，都是通过资本投资来促进城市改造和旅游业发展，并使当地城市经济进入了空前的繁盛发展期。不过，有关发展政策的提出，不仅导致环境和文化遗产资源的大量消耗，而且也产生很多社会排斥的问题，造成"社会—文化资本"存量的惊人损耗（Engelhardt，2002:33）。2003年在国际学术界，

1.1 Introduction

According to the UN-HABITAT (2008: 4), in 2008, more than half of the world's population lived in cities. This rapid urbanization poses many challenges especially to developing countries, where urbanization speed is usually much faster than in developed countries. A World Bank research reports that by 2020, 4.1 billion people (about 55 percent of the world's population) will live in urban areas and that almost 94 percent of the increase will occur in developing countries (Léautier, 2006: viii). The forces of urbanization promise to reshape the developing world, posing not only challenges such as energy consumption and climate changes but also worrisome trends in terms of social deprivation and exclusion.

In Asia, particularly in some historic cities, the focus is on how to deal with its urban conservation during rapid development (Stovel, 2002: 107). Many Asian countries have experienced unprecedented economic growth through capital investment to promote urban renewal and tourism industry. On the one hand, these development policies result in a heavy cost of the environmental and cultural heritage resources. On the other hand, they give birth to many problems of social exclusion, leading to an "alarming depletion of the common stock of [socio-]'cultural capital'" (Engelhardt, 2002: 33). In 2003, the notions of "living heritage" and "indigenous lives" were introduced into the arena of urban conservation internationally. Since then, the

"活态遗产"和"原住民"的概念，被引入城市保护的讨论范畴。自此，在城市发展和城市更新的过程中，人们开始更多地关注对原住民的保护以及当地居民的参与；尤其在发展中国家，由于人们对弱势群体参与城市保护过程尚存较大争议，从而上述议题显得尤为突出（Filippi, 2005; Imon, 2006; Miura, 2005）。

20世纪80年代，随着"可持续发展"议程的提出（WCED, 1987），人们开始关注与城市的综合发展目标相关的研究（Carley & Kirk, 1998; Filippi, 2005），提出城市发展不仅要考虑城市空间和物质环境的变化，还应当考虑社会、文化和政策的发展目标。相应地，社会和文化因素也被认为是，历史城市保护整合内容中的一部分（Cohen, 2001）。从20世纪90年代开始，人们对城市更新的综合化目标进行了许多研究，并基于西欧国家的实践情况，提出了城市更新的目标（Kleinhans, Priemus, & Engbersen, 2007: 1070）。很多研究者在研究当中，强调了城市更新过程的整体性特点（Ginsburg, 1999; May, 1997; Roberts, 2000），正如梅（May）所提到的：

> "城市更新的目的是改善城市物业的物质状况，促进地方的经济发展，提高就业率，关注包括医疗保健在内的社会和社区问题，通过构建当地的社区能力来实现可持续发展，并促进社会和经济问题之间的交流与互动（May, 1997: 12）。"

随着城市更新综合目标的提出，近年来，人们越来越关注城市更新中的社会资本问题。很多城市规划决策者指出，城市更新不仅要改

conservation of indigenous lives and the participation of local residents in local urban development or urban regeneration processes have obtained much attention, particularly in developing countries where the participation of local disadvantaged communities in the practices of urban conservation has been very contentious (Filippi, 2005; Imon, 2006; Miura, 2005).

With the proposal of "sustainable development" in the late 1980s (WCED, 1987), comprehensive urban development objectives have been discussed by many studies (Carley & Kirk, 1998; Filippi, 2005), which generally contend that urban development should consider not only spatial transformation and physical environment but also social, cultural, and political goals. Accordingly, social and cultural aspects are likewise deemed integral in the conservation of historic urban areas (Cohen, 2001). Since the 1990s, many studies have examined the comprehensive aims of urban regeneration, which have been proposed based on the practices in western European countries (Kleinhans, Priemus, & Engbersen, 2007: 1070). The holistic features of urban regeneration processes have been emphasized by many researchers (Ginsburg, 1999; May, 1997; Roberts, 2000). As May put it,

> The aims of urban regeneration are to improve the physical condition of estates; to stimulate the local economy; to increase employment; to pay attention to social and community issues, including health care; to provide sustainable development through capacity building at the local level; and to consider the interaction of social and economic issues (May, 1997: 12).

With the proposal of the comprehensive aims of urban regeneration, there have been a growing importance and concern for social capital in the discourse on urban regeneration

善城市居住区的物质环境状况，也要提高城市居民的社会幸福度（Imrie & Raco, 2003；Kearns, 2003；Kleinhans et al., 2007；Lees, 2003）。在最近的一些研究报告里，学者们开始在城市更新的讨论话题中，引入社会资本问题（Kleinhans et al., 2007；Middleton, Murie, & Groves, 2005）。密顿（Middleton）等人认为，社会资本在城市社区发展过程中，具有如下重要作用：

"社会资本被认为是维护社会稳定和构建社区自助能力的基石。社会资本的缺失被认为是导致社区衰败的关键因素（Middleton et al., 2005：1711）。"

结合上述城市更新理论的发展趋势与综合式城市保护的目标，一个常常被人们忽略的重要研究课题是，针对历史城市的更新过程中社会资本所具有的作用的探讨。这些探究不仅将有利于保护当地的非物质文化遗产——即原住民的生活内容，和促进社区居民在城市更新过程中的参与，而且将有利于整个历史城区的可持续发展。同时，多数已有的城市更新方面的研究和探讨，主要以发达国家为研究背景，如英国（Healey, 1991；Roberts, 2000；Turok, 1992）和美国（Gans, 1962, 1967；McKIE, 1974）。而以亚洲的国家城市为背景，以可持续发展和城市更新为研究框架下的探讨尚显较少（Ng & Tang, 2002a, 2002b）。到目前为止，甚至还没有发现以城市更新、社会资本和整合式的城市保护为理论框架，并以发展中国家的历史城市为研究背景的系统研究工作。

in recent years. Many decision-makers in urban planning point out that urban regeneration should improve not only the physical quality of urban neighborhoods but also their social well-being (Imrie & Raco, 2003; Kearns, 2003; Kleinhans et al., 2007; Lees, 2003). Recently, some studies have begun to integrate the idea of social capital in the discussion on urban regeneration (Kleinhans et al., 2007; Middleton, Murie, & Groves, 2005). Middleton accurately emphasized the substantial contribution of social capital to community development as the following,

Social capital is seen as the foundation on which social stability and a community's ability to help itself are built; and its absence is thought to be a key factor in neighborhood decline (Middleton et al., 2005: 1711).

Considering this trend of urban regeneration theory with the integrated aims of urban conservation mentioned above, a very significant but often overlooked research area is the exploration into the significant role of social capital in the regeneration practices of historic urban areas. These practices will contribute not only to the conservation of local intangible cultural heritage in terms of indigenous lives and the involvement of local communities in local urban regeneration processes but also to the overall sustainable development of the historic urban areas. Moreover, most of the available studies and discussions on urban regeneration have been carried out mainly in developed countries, for example, the U.K. (Healey, 1991; Roberts, 2000; Turok, 1992) and U.S. (Gans, 1962, 1967; McKIE, 1974). Very few studies on the Asian context are found within the framework of sustainable development and urban regeneration (Ng & Tang, 2002a, 2002b). And no studies are found so far on the context of historic cities of developing countries within the framework of urban regeneration, social capital, and integrated urban conservation.

1.2 研究问题

基于上述讨论,本书提出对中国历史城市中的社会资本和城市更新问题进行探索。尽管城市更新的概念涵盖了物质环境、经济环境、社会和自然环境在内的许多基础课题(Roberts, 2000: 3),但在中国的城市发展实际情况下,这些内容仍属于棘手问题。虽然在2001年提出的第十个"五年计划"中,中央政策就提出在地方城市发展或更新规划中,要注重对社会文化因素的改善(Hu, 2001: 4),但在中国目前的政策环境下,这点往往还没引起人们的足够重视。针对中国城市中社会文化和经济发展速度不匹配的现象,赛池(Saich)认为,"中国和很多处于转型期的国家一样,还未能较好地解决社会转型问题,而社会政策的发展也明显滞后于国家的经济发展水平"(Saich, 2001: xv)。有文献显示,在诸多因素当中,社会资本在促进地方弱势群体之间的交往及其在当地城市发展过程中的参与方面发挥着重大作用,同时也有助于形成多方参与和基于社区的城市更新过程(Forrest & Kearns, 2001; Kearns, 2003; Kleinhans et al., 2007; Middleton et al., 2005)。根据对中国历史城市中的城市更新和社会资本问题的分析,本书旨在探讨社会资本在中国历史城市更新中发挥的作用。因此,本书的主要研究问题为:

社会资本在中国历史城市更新中发挥着怎样的作用?

为了解答该问题,本研究还提出若干如下子问题:

1)在中国,城市更新涉及哪些问题?中

1.2 Research Questions

Based on the discussion above, this book then proposes to explore social capital and urban regeneration in Chinese historic cities. Although the notion of urban regeneration covers many fundamental topics such as the physical, economic, social, and environmental dimensions (Roberts, 2000: 3), it still is a very tricky issue in China. In China's current political context, social improvement in the local development or regeneration plan is often marginalized, although it has already been emphasized in national development agenda since the 10th Five-year plan in 2001 (Hu, 2001: 4). The mismatches between social and economic development in China were examined by Saich who said, "China, like many transitional economies, has not been able to deal well with the social transition and the development of social policy has lagged badly behind economic development" (Saich, 2001: xv). Accordingly, literature shows that social capital, among other various factors, plays a significant role in enabling the connection and involvement of local disadvantaged communities in urban development process, which promotes multi-participatory and community-based regeneration practices (Forrest & Kearns, 2001; Kearns, 2003; Kleinhans et al., 2007; Middleton et al., 2005). Based on the analysis of urban regeneration and social capital in Chinese historic cities, this book aims to investigate the role of social capital in the regeneration of Chinese historic cities. Accordingly, the main research question of the study is:

What is the role of social capital in the regeneration of Chinese historic cities?

To answer the main problem, the following sub-questions should be investigated:

1) What is urban regeneration in China? What are the historic cities in China?

2) Why is there a need for urban regeneration in Chinese historic cities? Who identifies these needs?

国历史城市的内容是什么？

2）为何要对中国历史城市进行更新？由谁界定更新的内容？

3）对于转型期的中国而言，影响历史城市更新的因素有哪些？

4）城市更新过程涉及的主要利益相关者，即地方政府、地方社区和私营部门，在城市更新过程中扮演着怎样的角色？

5）社会资本指的是什么？影响社会资本的因素有哪些？

6）社会资本是如何影响中国历史城市的更新过程的？

7）本研究对中国历史城市的更新有何普适意义和建议？

本研究中的许多关键概念，即城市更新和社会资本，都是基于西方国家的城市背景下的实践经验发展而来。所以，本书首先探讨了转型期中国历史城市的更新问题，并分析了中国历史城市更新的实际需求，然后对影响城市更新政策、策略和综合式城市更新结果的主要因素展开调查研究。这些因素包括，转型期中国的城市保护法律法规、城市更新机制和更新管治模式。在研究中国历史城市更新的管制体系时，我们发现，社会资本的存在可以让人们更好地理解社区居民在城市更新过程中的参与情况。然而，本研究也注意到，中国城市中的社会资本与西方国家背景下的社会资本概念存在较大差异。因此，本研究会围绕社会资本概念以及影响转型期中国地方社会资本的关键因素展开探讨。通过讨论关键变量和影响因素之间的关系，最终总结出社会资本在中国历史城市更新过程中发挥的作用及其贡献。

3) What factors affect the urban regeneration policies of historic cities in transitional China?

4) What roles do the various stakeholders, that is, local government, local communities, and private sectors, play in this process?

5) What is social capital? What factors affect the social capital of particular communities?

6) How can social capital affect the regeneration processes in Chinese historic cities?

7) What implications and recommendations can this book provide to conduct better urban regeneration practices in Chinese historic cities?

Many key concepts in the study, that is, urban regeneration and social capital, have been developed mainly based on the practices and experiences in Western context. Therefore, this book initially explores the issue of urban regeneration in the context of historic cities in transitional China. Following this, the actual or genuine need for urban regeneration in Chinese historic cities will be investigated. Various factors affecting urban regeneration policies, strategies, and comprehensive regeneration outcomes will then be examined, including urban conservation laws and regulations, urban regeneration mechanisms, and mode of regeneration governance in transitional China. In examining the system of regeneration governance in Chinese historic cities, social capital facilitates the understanding of community involvement in the urban regeneration process. However, it is observed that the social capital in urban China greatly differs from that in the Western context where this concept originated. Therefore, this book will examine the concept of social capital and analyze some corresponding key factors affecting the generation of local social capital in transitional China. By discussing the relationships among the key variables and influencing factors, this book will finally explore the impact and contribution of social capital on the urban regeneration process of Chinese historic cities.

1.3 研究目的

本研究的主要目的是揭示中国历史城市中城市更新与社会资本两者间的关系。这个目标还可以具体划分为如下几个分目标：

1）构建一个理论分析框架，以了解中国历史城市中的城市更新和社会资本。

2）了解几个关键因素（即城市更新和社会资本）的现状及其关系，以及转型期中国历史城市中的其他因素。

3）了解转型期中国城市更新的真正需求、地方城市更新的政策以及历史城市的实践活动。

4）了解各种利益相关者在转型城市更新管治模式中所扮演的不同角色。

5）了解社会资本是如何促进社区发展的，以及社会资本在地方城市更新过程中的作用。

6）从社会资本的角度出发，来了解社区居民参与对地方城市更新活动所起的作用。

7）基于对社会资本在中国历史城市综合式更新中的作用的理解，探讨本研究的启示，并提出相关的政策建议。

1.4 研究意义

自20世纪90年代以来，中国的快速城市化进程和倾向于经济增长的城市发展政策[1]，对中国历史城市保护与发展提出了诸多挑战。例如，不少珍贵的历史物质环境的破坏和大量原住民的重新安置，其目的都是优

---

[1] 2005年中国城市的城市化率是43%（DED，2006：224），到2020年有望达到60%（Shan，2006：9）。

1.3 Goals and Objectives

With the main goal of providing a better understanding of the relationships between urban regeneration and social capital in Chinese historic cities, this research has the following specific objectives:

1) To build a theoretical framework to understand urban regeneration and social capital in Chinese historic cities.

2) To understand the state and relationship of the key factors, that is, urban regeneration and social capital, as well as the other related factors in historic cities of transitional China.

3) To understand the genuine need for urban regeneration, local regeneration policies and practices in the historic cities of transitional China.

4) To understand the roles of the various stakeholders in the transitional mode of regeneration governance.

5) To understand the role of social capital to community development and involvement in the local urban regeneration process.

6) To understand the role of community involvement based on the level of social capital to the outcomes of local urban regeneration practices.

7) To develop research implications and policy recommendations on the role of social capital in the comprehensive urban regeneration of Chinese historic cities.

1.4 Significance of the Study

Rapid urbanization and economic growth-biased urban development policies in China since the 1990s[1] have posed many challenges to urban conservation and the development of Chinese historic cities, for example, the demolition of valuable historic physical environments and the

---

[1] Urbanization proportion of Chinese cities was 43% in 2005 (DED, 2006: 224), and this figure is expected to reach 60% in 2020 (Shan, 2006: 9).

先考虑了房地产再开发或城市营销进程[1](Ruan & Sun, 2001：28)。但是，重新安置原住民，不仅可能破坏社区中原住民的邻里关系与交往，而且不利于城市保护的整体内容，尤其是不利于保护原住民的生活内容和提高社区的凝聚力。有学者认为，中国历史城市的更新甚至陷入了一种两难境地，就是一方面要促进私人发展商主导的再开发活动，以保持当地经济的大发展；另一方面要增进社会凝聚力和社区建设(Shan, 2006：98)。基于社区的发展规划有助于改善社区关系，增加对原住民的保护，完善城市可持续更新的综合目标。然而，如果对社会资本和城市更新策略认识不足，也会严重影响这一目标的实现。鉴于此，本研究致力于探讨转型期中国历史城市中的社会资本和城市更新。希望通过这一研究，让人们更好地理解历史城市中的社会资本以及城市更新政策和实践问题。同时，人们也可以更好地理解社会资本、城市更新和其他相关的关键性因素之间的关系。

至于本书在实践层面的贡献，研究中最重要的问题是，如何理解并将原住民的生活内容，融入西安历史城市的整个城市保护或更新进程中。毫无疑问，本研究将促进当地历史城市的更新结果，不仅有利于改善城市的物质环境，而且有利于保护当地的非物质社会文化环境。此外，研究通过总结其政策启示和提出策略性建议，将有助于改善西安历史城市更新规划的决策环节。最后，作者

---

[1] 2002年房屋拆迁面积约为1.2亿平方米，占当年商品房面积3.2亿平方米的37.5%。2003年，房屋拆拆面积总共是1.61亿平方米，占当年商品房面积3.9亿平方米的41.3%（Shan, 2006：98）。

resettlement of many indigenous residents to give way to redevelopment-led or city marketing processes[1] (Ruan & Sun, 2001: 28). The resettlement of original residents not only disrupts the original relationships and connections of neighborhoods but also undermines the quality of urban conservation in terms of indigenous lives and the improvement of social cohesion. Some academic argued, the dilemma in the regeneration of Chinese historic cities lies in the rapid economic growth, caused by the redevelopment activities of private developers, and the improvement of social cohesion and community development (Shan, 2006: 98). A community-based development plan contributes to the improvement of community relationships, the conservation of indigenous lives, and the comprehensive objectives of sustainable urban regeneration. However, the lack of understanding of the relationships between social capital and urban regeneration strategies has remained a major obstacle in achieving this objective. Given this background, this book endeavors to explore social capital and urban regeneration in transitional Chinese historic cities. It is expected that through this study, the issues of social capital and urban regeneration policies and practices in transitional Chinese historic cities will be understood better. Moreover, the relationships among social capital, urban regeneration, and related key factors will be understood better as well.

As regards the practical contributions of this book, the most critical issue in the current regeneration practices of the historic city of Xi'an is how to understand and incorporate local

---

[1] Housing removal areas in 2002 were about 120 million square meters, which was 37.5 percent of 320 million square meters, the accomplished commodity housing areas in that year. In 2003, the housing removal areas totally were 161 million square meters, which was 41.3% of 390 million square meters accomplished commodity housing areas in that year (Shan, 2006: 98).

希望，本书中的研究结论还可以对中国其他历史城市的更新过程有实践性的启发。

residents' indigenous lives into the integrated urban conservation or regeneration processes. This book, without a doubt, will contribute to the regeneration outcomes of local historic urban areas not only in terms of helping improve the physical urban environments but also in terms of conserving the local intangible social-cultural environment. Moreover, this book will contribute to the decision-making processes in the regeneration plan of Xi'an city through its policy implications and recommendations. Finally, it is hoped that the findings of this book will provide empirical implications for the regeneration practices in other Chinese historic cities.

1.5 研究范围

本书以西安历史城区中的居住区的保护与更新为研究重点。相比较下，西安历史城区中的商业区或企业单位的更新则可能涉及非常不同的工作机制，例如，这些企事业单位可能需要与地方政府建立更多的联系，并且这些企事业单位可能对当地经济发展具有很重要的作用等。这些迥异的工作机制，可能会使历史城区的更新过程趋于复杂化，因此本研究并未涉及西安历史城区的商业或企业单位的更新内容。本研究中的社会资本也限于研究当地社区邻里间的社会资本，从而有助于研究居民在当地更新过程中的参与。由于本研究对社会资本的认识基于社区关系和联系，所以该研究不包括以其他社会网络形式存在的、特定社区之外的地方居民的社会资本，但却会涉及居民特定的工作单位和社会地位等其他动态因素。同时，由于有限的时间和人力原因，本研究没有深入调研原住民在被重置后社会资本的变化，这个问题没有被置为本研究的重点。但是，这种社会资本的变化对今后的研究工作仍具有意义。

1.5 Delimitation

This book focuses on examining the regeneration of historic residential areas in Xi'an city. Comparatively, the regeneration of the business or entrepreneurial units within the historic urban core of Xi'an city is not included because doing so requires very different mechanisms, such as their relationships with the local governments and their roles in local economic development, which can complicate the regeneration picture of the historic city. The study is also confined to the social capital of concerned residents at the level of local neighborhoods, hence contributing to the investigation of the involvement of local residents in local regeneration processes. The understanding of social capital in the book is based on community relationships and connections. Therefore, the social capital of local residents outside particular communities in other forms of social networks is not included in the study; instead, it is concerned with other dynamic factors such as their specific work units and social status among others. Due to limited time and manpower, this research does not delve into the social capital changes of the original residents after being relocated because they are not the focus of the study. However, they are deemed significant for future research.

1.6 方法论

本书在探讨提出的研究问题时，采用了质性研究方法。按照克雷斯韦尔（Creswell）的理论，质性研究方法是"调查者常常从构建主义的角度（如在社会上或历史上构建的、以发展理论或模式为目的的个人经验的多种方式）、倡议或者参与的角度（如政策性的、问题为导向或以协作、变化为导向）或者从两种角度同时出发，来提出知识主张的一种方法"（Creswell, 2003：18）。麦克斯韦尔（Maxwell）指出了质性研究方法的特点，即"没有固定的出发点或过程，不需遵循特定的步骤，认识到相互联系和相互作用在不同设计元素中的重要性"（Maxwell, 1996：3）。本研究在城市更新和社会资本的研究框架下探讨中国历史城市的更新，这与当地城市更新过程中基于社区的参与程度密切相关。所以，质性研究方法可以较好地实现该研究目的。而在各种质性研究方法中，即传记方法、现象学方法、扎根于地方社区以及案例研究的方法当中（Creswell, 1998：47），本研究采用案例研究的方法。根据斯塔克（Stake）的观点，案例研究是"对单一案例的特殊性和复杂性进行的研究，目的是了解案例在很多重要情形下的情况"（Stake, 1995：xi）。斯塔克进一步阐述到，当研究是要探讨事物与其所处具体环境之间的相互关系的细节时，就更适合采用案例研究的方法（Stake, 1995：xi）。同时，殷（Yin）认为，如果案例本身与研究现象之间非常吻合，而研究本身也需要涉及具体的研究背景时，案例研究的方法就是较好的选择（Yin, 1994：13）。鉴于上述分析，

1.6 Research Methodology

To explore the research question, this book adopts a qualitative research approach. According to Creswel (2003), it is "one in which the inquirer often makes knowledge claims based primarily on constructivist perspectives (i.e., the multiple meanings of individual experiences, meanings socially and historically constructed, with an intent of developing a theory or pattern) or advocacy/participatory perspectives (i.e., political, issue-oriented, collaborative, or change oriented) or both" (Creswell, 2003: 18). Maxwell (1996) pointed out the features of qualitative research, which "does not begin from a fixed starting point or proceed through a determinate sequence of steps, and it recognizes the importance of interconnection and interaction among the different design components" (Maxwell, 1996: 3). This study explores the regeneration of Chinese historic cities within the framework of urban regeneration and social capital, which has much to do with community-based participation in the local urban regeneration process. Therefore, a qualitative research method fits this study purpose well. In addition, among the various qualitative approaches, that is, biography, phenomenology, grounded theory, ethnography, and case study (Creswell, 1998: 47), this research applies the case study method. According to Stake (1995: xi), a case study is "the study of the particularity and complexity of a single case, coming to understand its activity within important circumstances" (Stake, 1995: xi). Stake further explained that it is used when the research is looking for details of interaction with its contexts (Stake, 1995: xi). Yin affirmed that the case study method is preferable when the research wants to "cover contextual conditions," where the case is seen as highly pertinent to the phenomenon of study (Yin, 1994: 13). Considering the analysis above, this book chose the case study as its research

本研究将选择案例研究作为本研究设计方法，其他具体原因如下：

1) 本研究旨在探讨社会资本在中国历史城区更新中的作用，所以在给定环境中研究相互作用的因素可以帮助人们更好地理解有关研究目的。

2) 由于本研究的时间有限，而且研究内容的界限，即城市更新中的社会资本与研究背景的界限，并不明确，所以开展详细的案例研究方法将更适合本研究项目。

本书的研究方法，从三方面进行阐述（有关内容会在后面部分详述）：(1) 大量查阅有关历史城市中的社会资本、城市更新、城市保护政策和相关法规的文献资料；(2) 开展实地调研，包括借阅地方政府可以对外公开的档案资料、问卷调查、实地观察，以及采访当地社区居民代表、社区组织、地方政府（各级政府部门）和当地的专业工作人员（包括学术机构和社会机构）；(3) 与相关领域中的其他专家学者进行讨论。通过以上三种方式获取的信息和知识，有助于在研究初期初步制定一个研究框架。结合当地的具体特点，将使研究框架进一步的地方化，从而使人们能够更好地理解本研究所选案例。

1.6.1 理论架构

基于目前已有的文献资料，本书首先提出一个初步的理论分析架构来探讨所提的研究问题。该理论架构主要基于历史城市中的城市更新和社会资本等几个关键变量。在分析架构中，城市更新的概念是基于罗伯茨所给出的定义（Roberts，2000：17）以及中国城市的实践特征，归纳如下：

design approach mainly for the following reasons:

1) The research intends to explore the role of social capital in the regeneration of Chinese historic urban areas; therefore, studying interactive components in a given context helps us to understand the research aims better.

2) Considering the limited time span, a detailed case study is appropriate for the research because the boundaries between the phenomenon, that is, social capital in urban regeneration and study contexts, are not evident.

Accordingly, the research methods of this book are developed in three ways, which will be elaborated in later sections: (1) extensive review of the literature on social capital, urban regeneration and urban conservation policies, and legislations in historic cities; (2) field studies including the local governments' documentaries, questionnaire surveys, field observations, and interviews with local community representatives, organizations, local authorities (various levels of governments), and local professionals (academic institutes and societal institutions); and (3) discussions with other experts in related fields. The information and knowledge obtained from these three perspectives help to frame a tentative research framework at the beginning. Based on some local particular features, the initial framework is then refined to provide a better understanding of the specific situations in the selected case, which is Xi'an city.

1.6.1 An Analytical Framework

Based on current literature studies, the research sets out with a tentative analytical framework hinging on several key variables, namely, urban regeneration and social capital in historic cities, to explore the research question. In the analytical framework, the concept of urban regeneration, based on the definition by Roberts (2000: 17) and the practices in urban China, is defined as:

"[城市更新是指]综合性与整合性的理念和行动,以解决中国历史城市中存在的问题;一般是通过保护历史文化遗产和原住民的生活内容,以及理性地再开发或复兴那些正面临严峻的社会、经济与环境等问题的地区;其目的是使历史城市或地段在多层次内容上可以达到持续的提高与改善。"

我们注意到,罗伯茨是在研究英国城市更新问题的基础上,提出了城市更新的定义,但是这一定义与中国历史城市的实际情况还存在较大差异。由于中国目前处于转型期,其城市更新的要求、政策和策略与英国的情况也不同。所以要想更好地理解这一概念,还需要结合中国历史城市的具体情况。后面的章节将会对这一点进行更多的阐述。同样地,在社区居民参与城市更新过程方面,社会资本的研究范围主要强调了地方居民之间,以及某些特殊群体和政府机构之间的关系。以帕特南提出的定义(Putnam, 1993: 35)和中国国情为基础,本研究界定社会资本的定义如下:

"地方居民之间或其与政府机构之间的社会关系所具有的某种特点,如关系网、社会规范与社会信任;这些内容往往以政治、宗教和种族为基础,并有助于增强社区居民之间的联系,以及促进社区居民在城市更新过程中的参与。"

如上文所述,从西方国家背景下发展而来的城市更新和社会资本的概念并不完全适应转型期中国的实际情况(Zhai & Ng, 2009)。根

Comprehensive and integrated vision and action leading to the resolution of urban problems in Chinese historic cities, by way of conserving historical heritages and indigenous lives, and reasonably redeveloping or revitalizing the places where the social, economic and environmental problems are outstanding, with the aim of bringing about a lasting improvement in the multi-dimensional conditions.

It is noticed that Roberts put forth his definition when he was researching the regeneration issues in England, which are very different from the problems laid before many Chinese historic cities. Given the context of transitional China, the regeneration needs, policies and strategies are very different from the British practices. Therefore, a better understanding of this concept cannot be separated from the specific context of Chinese historic cities. This will be discussed further in the later chapters. Similarly, the study dimensions of social capital mainly accentuate the relationships among local residents and between particular communities and governmental agencies, in terms of community participation in urban regeneration process. The definition of social capital in the book, rooted on the definition by Putnam (1993: 35) and the realities in China, is proposed as the following:

Features of social relationships among local residents or between the residents and governmental agencies, such as networks, norms and social trust, based on politics, religion and ethnicity issues, which facilitate community connections and involvement in urban regeneration process at the local level.

As mentioned above, the concepts of urban regeneration and social capital from the Western context do not fully agree with the realities in transitional China (Zhai & Ng, 2009). Given China's transitional political economy, social capital in urban China has

据转型期中国的政治经济状况，中国城市中的社会资本有很多独一无二的特点。比如，1978年之前，社会主义时期的公共伦理和政治意识形态深深地植入到人们的日常生活中（Vogel，1965：59）。而自1978年中国逐步实行市场化倾向的经济改革之后，人们之间的关系明显开始具有物质化的特点（Gold，1985；Yue，Wang，& Wang，2002）。同时，在中国许多城市更新过程中，有多种因素影响着当地的社会资本水平。所以，本书在中国城市更新和社会资本的基础上，将进一步探索理论架构，以便从历史和城市情景的角度来理解本书所涉及的诸多关键变量。

1982年，国务院提出了中国历史城市或历史文化名城的概念，以保护珍贵的历史文化遗产免受城市快速发展的破坏。截至2007年，中国共有108座城市被国务院审批为中国历史文化名城（Jia，2007：3）。根据《中华人民共和国文物保护法》，历史文化名城指的是：

"文化遗产特别丰富、具有重大历史价值和革命意义的城市，经国务院审批为具有历史和文化价值的名城（SCNPC，2007：第14条）。"

many unique characteristics. For instance, in socialist China before 1978, public ethics and government ideologies significantly penetrated into individual lives (Vogel, 1965: 59). However, in a market-oriented economy after 1978, human relationships in China were largely characterized by commoditization (Gold, 1985; Yue, Wang, & Wang, 2002). Moreover, during the urban regeneration of many Chinese cities, various factors affecting local social capital emerged. Therefore, this book will further develop an analytical framework based on the situations of evolving urban regeneration and social capital in urban China for a historical and contextual understanding of these key variables in the book.

The concept of historic cities or famous cities of historical and cultural value in China was put forth by the State Council in 1982 to protect valuable historical heritage from destruction in rapid urban development. By 2007, 108 cities in China had been designated by the State Council as historic cities (Jia, 2007: 3). As laid out in the Law of the People's Republic of China on the Protection of Cultural Relics, historic cities in China are defined as:

[the] cities with an unusual wealth of cultural relics of important historical value or high revolutionary memorial significance, [which] shall be verified and announced by the State Council as famous cities of historical and cultural value (SCNPC, 2007: Article 14).

The relationships among the key variables in the study are illustrated in Figure 1-1.

In China's transitional economy and politics, the various key stakeholders concerned in the regeneration process mainly include the state (central, municipal, and district governments), public and private developers, and the local community. Generally, during market-oriented economic development, local development policies strongly promote economic growth. In addition, the close coalition between local governments and private entrepreneurs

图1-1：研究中关键变量之间的关系
Figure 1-1: Relationships among the key variables of the study

图 1-1 表明了本研究中关键变量之间的关系。

在中国转型期的政治经济环境下，与城市更新有关的利益相关者主要涉及国家机构（中央政府、市政府和区政府）、公共或私人发展商和地方社区居民。在以市场为导向的经济发展中，地方发展政策的提出通常会极大地促进经济繁荣。此外，地方政府和私营企业者的紧密合作，也有助于发展以再开发为主导的城市更新政策，同时，这也是增加地方政府税收的重要渠道（Zhu，1999：539）。在地方更新过程中，有学者认为，地方社区居民的利益往往会遭受损失（Yu，2007：2）。此外，那些侧重于经济发展的地方更新政策和策略，往往对历史城区会带来更多影响。随着强制拆迁和居民重置工作的开展，历史街区中的原有邻里会随之消失。除了对原住民的影响，当地传统的生活方式与内容也会相应减少，而不利于当地非物质文化遗产的保护。同时需要指出的是，中国现有的有关历史城市保护的法律法规，非常关注对城市物质环境的保护。地方政府的工作议程，还没有足够重视对社会文化等非物质因素的保护。其他研究背景下的实证研究和文献告诉人们，以社区为基础的城市更新过程，常常更有助于当地社会环境因素的保护与改善；在这过程中，社会资本非常有助于促进该更新过程，并对当地城市更新规划的结果产生重要影响（Forrest & Kearns，2001；Kearns，2003；Kleinhans et al.，2007；Middleton et al.，2005）。

1.6.2 案例选择
以西安为例

作为中国六大历史古都之一，西安始建

encourage the redevelopment-led urban regeneration policies, which also turn out to be the ways for local governments to generate revenues (Zhu, 1999: 539). During local regeneration processes, some scholar maintains that local communities often suffer heavy losses (Yu, 2007: 2). Furthermore, local economic-biased regeneration policies and strategies have more influence on historic urban areas. With enforced eviction and resettlement, original neighborhoods of historic residential areas are demolished. Aside from the original residents, local traditional living styles and activities also diminish. This undermines the conservation of local intangible cultural heritage in terms of indigenous lives. At the same time, it should be noted that in China today, the existing laws and regulations on urban conservation in Chinese historic cities emphasize much on the dimension of urban physical environments. On the contrary, the conservation of local intangible aspects, for example, social and cultural aspects, is not on the local governments' agenda. Empirical studies and literature in other contexts show that the community-based regeneration process often promotes the conservation and improvement of local social aspects, during which social capital strongly facilitates community-based regeneration processes and has a significant influence on the outcomes of local regeneration plans (Forrest & Kearns, 2001; Kearns, 2003; Kleinhans et al., 2007; Middleton et al., 2005).

1.6.2 Selection of Case Studies
Xi'an city as the research focus

Xi'an city is one of China's six great historic capitals. In Chinese history, since this city was established during the Zhou dynasty (B.C.1046-B.C.256), it has connected China's northwest region with the southwest region in the aspects of transportation and social and economic development (Ma, 1985: 149). After the Tang dynasty, Xi'an was no longer the capital of China, and it entered into the period called the "Post-Capital Period" (*hou du cheng*

于周朝（公元前 1046～前 256 年），这期间，西安交通发达，经济和社会蓬勃发展，成为连接中国西北部与西南部的重要纽带（Ma，1985：149）。自唐朝之后，西安便不再是国都，从而进入了"后都城时代"（Cheng & Liu，2003）。直到 1949 年中华人民共和国成立之后，西安破败的历史城区才得以进一步发展和振兴。随着中国政治与经济翻天覆地的变化，西安的历史城区也发生了剧变，并展现出中国历史城市发展过程中所共有的典型特征与许多问题。自 20 世纪 80 年代以来，中国进入了市场为导向的城市发展时代，这期间颁布的城市政策也比较倾向于发展地方经济，对西安历史城市的发展与建设产生了重要影响。无论是 1980 年西安提出的第二次总体规划，还是 2008 年提出的第四次总体规划，针对破败的历史城区制定的城市保护和更新政策与策略都占据着重要地位。根据第四次总体规划，西安市政府也提出从 2004 年开始重建整个城区，包括对中心历史城区的更新。由于西安属于历史都城，城市保护在当地规划中占据重要意义。同时，西安城市保护规划的侧重点也发生了变化，从重点保护文物古迹扩展到保护整个历史城区。不过，当地城市保护政策法规的内容，还是非常强调对城市物质环境的保护，而在当地城市保护或更新过程中尚未重视对非物质文化遗产内容的保护，尤其是对历史居住区中的传统生活方式与内容的保护。

西安城市建设与发展的显著特点之一在于，多年来综合利用了其历史城市中心区，从而使该历史中心区也成为今日西安的政治、经济、文化、交通运输和流通中心（Chen,

shi dai) (Cheng & Liu, 2003). Large-scale urban development and revitalization of the derelict historic inner urban areas in Xi'an did not take place until the establishment of the People's Republic of China (henceforth, China) in 1949. Since then, with the tremendous political and economic changes in China, Xi'an's historic urban areas have changed rapidly and displayed many typical characteristics and problems of urban development in Chinese historic cities. During China's market-oriented urban development period that began in the 1980s, pro-growth urban policies had a significant impact on the development trajectory of the historic city of Xi'an. From the city's second master plan in 1980 to the latest fourth master plan in 2008, urban conservation and renewal policies and strategies for the derelict historic urban core have always been crucial in these plans. According to the fourth master plan, the Xi'an municipal government proposed to restructure the whole city area including the regeneration of its historic inner urban areas beginning 2004. As Xi'an is a historic capital city, urban conservation plays a significant part in local plans. In Xi'an, the conservation plans of the various periods turn from being focused on heritage monuments to covering all the historic urban areas. Moreover, local urban conservation policies and regulations mainly stress on the urban physical environments, and the intangible cultural heritage, especially the traditional lives in the historic residential districts, is often ignored in local urban conservation or regeneration processes.

One feature of the urban development in Xi'an is that for many years, the urban development and growth of Xi'an city have been based on the entire utilization of the historic urban core, hence making the historic urban core the urban political, economic, cultural, transportation, and circulation center (Chen, Wang, & Wen, 2003a: 22). This situation has caused much burden to the area as a historic urban area, leading to many urban problems.

Wang, & Wen, 2003a：22）。这种情况致使该历史中心区承受着巨大压力，并引发很多城市问题。该地段的综合式城市保护策略也面临很大挑战。随着快速的城市化进程和许多有助于城市发展的激励因素的提出，如20世纪90年代中央政府提出的西部大开发战略，西安市政府花费了很大气力来重组整个城市空间。早在第四次总体规划实施之前，西安为了招商引资，就形成了目前的城市空间格局（He，2005）。按照这一规划，除了需要保护好个别历史文物古迹，还将对历史城区中心的很多居住区进行再开发，将其发展成旅游和商贸中心，从而为当地繁荣的旅游业服务（He，2004）。

根据西安最新颁布的总体规划，对历史街区实施城市再开发与保护政策，被看作是唐皇城"复兴"规划的一部分。值得注意的是，该"复兴"规划具有以下特点：建造传统城市景观，改善当地的物质环境，以及促进经济增长（XCPB，2005）。为了满足当地倾向于经济发展的城市更新需求，尽管历史居民区中的原住民会被重置，但还是有不少地方居民借助高水平的社会资本而自行组织起来，参与到城市更新活动中。而高水平的社会资本对当地城市更新规划的过程与结果影响重大。所以，西安城市的更新规划和实践活动，非常有助于人们理解本研究之前所提的主要研究问题。

西安市旧城区的历史街区

根据最新颁布的《西安历史文化名城保护规划》（XCPB，XACH，& UPDI，2005），历史街区被定义为物质遗产比较集中，能较完整地体现出某一特定历史时期的西安传统城

It has also challenged the comprehensive urban conservation policies in this area. With the rapid urbanization and many incentives for urban growth, for example, the Western China Development Program promoted by the central government since the late 1990s, the Xi'an municipal government has exerted great efforts to restructure the city's overall urban spaces. These urban spaces have been formed in the previous periods through the implementation of the fourth master plan in order to attract more investors (He, 2005). According to this plan, many historic residential areas within the historic urban core would be redeveloped into tourist and commercial centers to cater to the exuberant local tourism industry in addition to the conservation of several historic monument buildings (He, 2004).

As laid out in Xi'an's latest master plan, the urban redevelopment-cum-conservation policies in the historic areas are regarded as parts of the "regeneration" plan of the Tang imperial city. What is noteworthy about this "regeneration" plan is that it has the outstanding features of the re-fabrication of traditional urban scenes, the improvement of local physical environments, and substantial economic growth (XCPB, 2005). Although disadvantaged residents in the historic residential districts need to be relocated to give way to the local pro-growth urban regeneration plans, some local residents are autonomously organized and involved in the regeneration processes based on the high level of local social capital. However, this high level of social capital has a significant impact on the processes and outcomes of local urban regeneration plans. Therefore, urban regeneration plans and practices in Xi'an provide the perfect context in which to explore the major problem of this book.

Historic districts within the Xi'an historic urban core

According to the latest Conservation Master Plan for the Historic and Cultural City of Xi'an (*Xi'anshi lishi wenhua mingcheng baohu guihua*) (XCPB, XACH, & XUPDI, 2005), historic districts/areas *(lishi jiequ)* are defined as urban areas

市风貌、居住形式、生活方式、传统生活文化、民族习俗以及西安古老的商业特色的街区。在历史街区中，社区居民过着群体生活，并展示着丰富的物质、精神与实践活动（XCPB et al.，2005：37）。根据《西安历史文化名城保护条例》，西安市政府将历史城区中的几个历史居住区划定为历史街区，包括鼓楼回民区（或北院门历史街区）、三学街区、竹笆市、德福巷、湘子庙和七贤庄（SCXPC，2002）。图1-2表明了这些历史街区在西安历史城区的位置。

本研究在西安选择案例的基础工作，是通过一系列的前期实地调研来进行，主要包括现场考察、采访当地居民和社区代表以及对历史街区的历史背景开展有关文献研究（Chen，Wang，& Wen，2003b；Chen，1994；Chen，2001；Dong，1996；Gao & Jiang，2005；Hoyem，1989；SMDPI, XUAT, UTPRI, RERI, & RUDRI, 1997；Xiao，2004；XMHDPPO，2003）。为了在案例选择中，能够获取关于西安历史街区最具代表性的信息，本书在以下几个方面进行着重衡量。

where physical heritage is converged to reflect integrally Xi'an's traditional urban features at a specified period and where residential forms, living styles, traditional living culture, nationality custom, and ancient Xi'an commercial features are revealed. These historic districts/areas are the places where urban residents live in groups and where physical, spiritual, and practical activities are performed (XCPB et al., 2005: 37). With reference to the Conservation Regulations for the Historic and Cultural City of Xi'an (*Xi'an lishi wenhua mingcheng baohu tiaoli*), the Xi'an municipal government classified several historic residential areas within the Xi'an historic urban core as historic areas, including the Drum-tower Muslim historic district (or Beiyuanmen district) (*Beiyuanmen*), Sanxuejie historic district (*Sanxuejie*), Zhubashi historic district (*Zhubashi*), Defuxiang historic district (*Defuxiang*), Xiangzimiaojie historic district (*Xiangzimiaojie*), and Qixianzhuang historic district (*Qixianzhuang*) (SCXPC, 2002). Figure 1-2 shows the locations of these historic districts within the Xi'an historic urban core.

The bases for selecting the case studies in Xi'an city were developed through a pilot field study involving site visits, interviews with local residents and community representatives, and extensive literature studies on the historical and contextual background of these historic districts

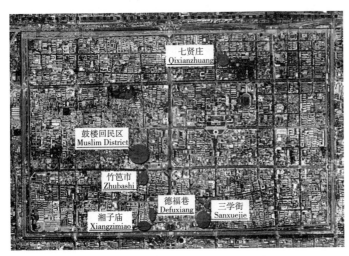

图1-2：西安历史城区中的历史街区
Figure 1-2: Historic districts within Xi'an historic urban core

物质环境方面

● 无论是物质环境方面还是社会文化环境方面，作为西安历史城区中重要的一部分，所选案例应当体现西安这座历史文化名城的某些重要特征。换句话说，研究案例应反映西安历史城市的背景，并反映当地政府主导的城市保护—更新规划和实践的特色。

● 研究案例应包含诸多历史元素。同时，案例现有的历史城市环境有进行城市更新的迫切需求。

实践方面

● 认识所选历史街区里的有关人士对必要资料的获取非常重要。在西安，由于倾向于经济发展的城市再开发政策和城市保护目标之间存在较大的矛盾和冲突，所以，城市更新和历史街区的发展，长期以来都是一个颇有争议的问题。因此，关于当地城市更新的规划和策略的资料有很多，既有政府和专业工作者级别的，也有当地社区老百姓级别的。但是，由于地理位置和所受教育程度的局限，很多西安历史街区的居民都较保守。在这种情况下，认识所研究区域中的某些人士，对后期资料搜集和开展大量的调查研究，显得极为重要。

另一方面，长期以来，一些当地的专家学者一直比较关注西安历史城区的保护和再发展。他们之前的研究成果，有助于本研究对所选案例进行特定历史时期和一定语境下的理解。因此，亲自找到这些以前的研究成果，或者亲自见到当地的学者，将对这项调查研究起到很大的推动作用。

● 现有的及可用的文献资料（例如，地图、之前的研究成果、报告、文献、官方文件等

(Chen, Wang, & Wen, 2003; Chen, 1994; Chen, 2001; Dong, 1996; Gao & Jiang, 2005; Hoyem, 1989; Ma, 2005; SMDPI, XUAT, UTPRI, RERI, & RUDRI, 1997; Xiao, 2004; XMHDPPO, 2003). To obtain the most representative information on Xi'an historic districts in the selection of the case studies, the following aspects have been intensively examined.

Physical aspects

● The case study areas constitute a significant part in Xi'an historic urban core in terms of the historic urban identity of Xi'an either from the physical or socio-cultural perspectives. In other words, the case studies represent Xi'an's historic urban contexts and reflect the many features of urban conservation and regeneration plans and practices led by the local governments.

● The case studies contain many historical elements. In addition, the existing historic urban setting faces an urgent need for urban regeneration.

Practical aspects

● Knowing someone from the selected historic districts is important for the collection of necessary data. The urban regeneration and development of historic districts has long been a controversial issue in Xi'an given the conflicts between pro-growth urban redevelopment policies and urban conservation intentions. Therefore, there is much information available on local urban regeneration plans and strategies not only at the government and professional levels but also at the local community level. However, because of the geographic limitations and their limited education, many local community members in Xi'an's historic districts are introversive. Thus, knowing someone from the study areas will contribute significantly to the easy access of information and intensive investigation.

On the other hand, some local scholars have paid long time attention to the issue of urban conservation and redevelopment of Xi'an historic core. Their previous studies provide

资料)。就如上述所提到的,许多研究者对西安历史核心区的保护与更新工作也产生了很久的兴趣,尽管他们的研究方法和侧重点与本研究有所不同。现有的及可用的文献资料,例如,政府报告、报纸、时事通信、档案以及其他资料,可为本研究提供基本的初始数据,即人口统计数据、当地的更新政策和规划,这可以为本研究节省大量的时间和人力。另外,之前的研究及结论也可以作为进一步分析工作的基础信息和知识。

基于上述衡量考察因素,本研究开展了大量的实地试点调研,并从诸多历史街区案例中选择了鼓楼回民历史街区和三学街历史街区,以这两个历史街区作为本研究的案例研究地区。研究中的案例选择标准及有关内容见表1-1;最后所选历史街区案例的位置图见图1-3。

a historical and contextual understanding of the study areas. Therefore, access to these previous studies or local scholars in person will contribute to this research substantially.

● The presence and availability of documentary resources (e.g., maps, previous studies, reports, archives, official documents, and others). As mentioned above, the issue on urban conservation and regeneration of Xi'an's historic core has drawn the interest of many researchers despite the fact that the available study approaches and foci are different from those of this research. The presence and availability of documentary resources, for example, government reports, gazettes, newsletters, archives, and others, on the selected sites can save this study considerable time and energy because these resources can provide basic preliminary data, namely, demographic statistics, local regeneration policies, and plans. In addition, previous studies and findings can provide basic information and knowledge for further analysis.

选定的研究案例
Study sites selection matrix

表1-1
Table 1-1

地点 Sites		标准 Crieria	历史特色 Historic characteristics	更新的必要性 Necessity of regeneration	当地联系人 Available local contacts	资料文献 Available local documentation
TR&RC	鼓楼回民历史街区 Muslim historic district		●●●	●●●	●●	●●●
RC&WRC	三学街历史街区 Sanxuejie historic district		●●●	●●	●●●	●●
C&WRC	竹笆市历史街区 Zhubashi historic district		●●	●●		
C&WRC	德福巷历史街区 Defuxiang historic district		●●			
C&WRC	湘子庙历史街区 Xiangzimiaojie historic district		●●			
E	七贤庄历史街区 Qixianzhuang historic area		●●●	●		●●●

注释:TR—传统居住区;RC—住宅和商业;C—商业区;E—展览建筑;WRC—家属院
●—低;●●—中;●●●—高;
阴影部代表最后选定的研究案例

Note: TR—traditional residential area; RC—residential-cum-commercial; C—commercial; E—exhibit rooms; WRC—work-units residential community (*jia shu yuan*);
●—little presence; ●●—moderate presence; ●●●—high presence;
The shading areas indicate the selected cases.

图 1-3：西安历史城区中选定的历史街区案例
Figure 1-3: Locations of the selected historic districts in Xi'an urban areas

表 1-1 表明所选的历史街区符合本次研究目标的设定标准。现对所选案例简介如下：

1）鼓楼回民历史街区（简称回民区，也称为北院门历史街区）

地处西安历史城区的核心区或称明城区的最中央，回民区是西安为数不多的被完全划定为历史街区的其中一处（SCXPC，2002；XCPB et al.，2005）。根据《西安历史文化名城保护条例》（下文简称《西安保护条例》），整个回民区东起社会路、西至早慈巷、南至西大街、北至红埠街。全区占地面积约54公顷，拥有大约6万人,其中3万多人是穆斯林居民，其余的为汉民（SCXPC，2002；XMHDPPO，2003）。该街区范围在图1-3中用红色标注出来；蓝线代表西安城墙和护城河的位置，表明了西安明朝城区的范围。

回民区的更新议题早在20世纪90年代就被提出了。相应地，为了开展城市更新，当地政府主导的再开发规划要求项目涉及的居民进行搬迁。不过，由于当地特殊的社会经济条件，鉴于居民当中存在非常亲密的社

Through intensive pilot field studies based on the examining factors as discussed above, the two historic districts of the Muslim District and the Sanxuejie Historic District are selected as the case study areas from among the various historic areas. The study sites' selection matrix is shown in Table 1-1, while the locations of the selected historic districts are illustrated in Figure 1-3.

The matrix above shows that the selected historic districts meet the criteria set for the goals of the study. A general description of the selected cases is given as follows.

1) Drum Tower Muslim District (DTMD; also known as the Beiyuanmen District)

Located at the very centre of Xi'an's historic urban core or the city wall area (thereafter, CWA), the DTMD is one of the few areas entirely declared as a historic district in Xi'an (SCXPC, 2002; XCPB et al., 2005). According to the Conservation Regulations on Xi'an Historic and Cultural City (thereafter, Xi'an *Conservation Regulations*), the entire DTMD begins from the Shehui Road in the east to Zaoci Lane in the west, and from West Avenue in the south to Hongbu Street in the north. The entire area is about 54 hectares, with a population of almost 60,000. More than 30,000 of the population are Muslims, and the other half are Han Chinese (SCXPC, 2002; XMHDPPO, 2003). The boundary of this district is shown in the red square in Figure 1-3, with the blue ring representing the city wall and the moat, which delineate the boundaries of Xi'an's CWA.

The issue of urban regeneration in the DTMD has been proposed since the late 1990s. Accordingly, the local government-led redevelopment plan requires the residents to be resettled to give way to it. However, considering the local specific socio-economic situations, local residents in the DTMD have a high level of social capital given their very close community relationships and interactions. Based on this kind of strong community tie or bonding social capital, local residents have a significant influence on the processes and outcomes of

区关系和交流，回民区居民之间有着高水平的社会资本。基于这种较强的社区关系或聚合型社会资本，地方居民对当地城市更新计划的进程和结果都产生了很大影响。

2) 三学街历史街区（简称三学街）

《西安保护条例》指出，三学街历史街区环绕碑林博物馆布局，东至开通巷、西至南大街、南至城墙、北至东木头市。街区占地面积约36.5公顷（XUPDI，2001：1），居民为汉族人口。2008年该街区的常住居民人口为8742人，其中不包括当地的流动人口[1]。明清年间，该街区是西安的文化和教育中心。今天，这里仍留有一些建筑遗址和遗迹，例如，碑林博物馆、关中书院、卧龙寺、奎星阁等。图1-3的红色区域标出了它在西安旧城区的位置。

从清朝末年至今，三学街区发生了巨大变化，由古代的文化教育中心演变为今天西安的历史居住区。自20世纪80年代以来，该街区在不同阶段历经了城市修复和以再开发为中心的发展过程。自2005年西安颁布第四次总体规划以来，该区的更新实践一直被认为是西安历史城市核心区中所有城市更新规划的试点工程。面对当地破败的居住环境，尤其是随着大量外来务工者的迁入，使该街区确实具有开展城市更新的必要。在以政府为主导的城市更新进程中，当地居民的实际需求却很少被兼顾到。由于当地较低的社会资本水平（可能是由于较松散的社区关系和脆弱的地方归属感所造成），很多居民也把政府主导的再开发工程，看作是改善当地居住

---

[1] 2008年的人口资料由碑林区公安局提供。

local urban regeneration plans.

2) Sanxuejie Historic District (SHD)

As indicated in Xi'an *Conservation Regulations* (SCXPC, 2002), the SHD encompasses the Steles Forest Museum (*Beilin*) and begins from Kaitong Lane in the east to Southern Avenue in the west, and from the City Wall in the south to East Wood Street in the north. This area is about 36.5 hectares (XUPDI, 2001: 1) and is inhabited by the Han Chinese. The permanent residents in 2008 were 8,742; this number does not include local mobile populations[1]. During the Ming and Qing dynasties, this area was the cultural and educational center of Xi'an. Today, a number of heritage buildings and physical remains still exist here, for example, the Steles Forest Museum (*Beilin*), Guanzhong School (*Guanzhong Shuyuan*), Dragon Crouching Temple (*Wolong Si*), and Literature God Hall (*Kuixing Ge*), among others. The red square in Figure 1-3 shows its location within Xi'an's CWA.

Since the late Qing dynasty, this area has changed dramatically from being the ancient cultural and educational center to a historic residential district in Xi'an. Furthermore, it has gone through urban rehabilitation and redevelopment-oriented processes at different stages since the late 1980s. Given Xi'an's fourth master plan in 2005, the regeneration practices in this area have been regarded as the pilot projects in the overall urban regeneration plan within Xi'an's historic center. There is a genuine need for urban regeneration in this area, for example, in the local dilapidated living environments and the massive number of migrant workers. In the local government-led urban regeneration process, local residents are largely excluded. Due to the low level of local social capital (as may be caused by loose community relationships and weak place attachment), many residents see the

---

[1] The population data in 2008 is obtained from local police station.

条件的唯一出路。很多人都非常支持政府主导的再开发计划，许多原住居民被重新安置。显然，随着原住居民的再安置，当地许多原有的集体生活方式与内容也都逐渐消失了。这与回民区的情况可以形成鲜明对比。

1.6.3 资料来源

本研究中所用资料主要来源于如下途径：

1）采访当地社区居民代表、政府工作人员、当地专家学者以及当地社会团体的成员（如回民区的清真寺工作人员）；

2）实地调查和观察；

3）国内外的研究文献、当地图书馆的有关资料以及不同级别政府部门（市政府、区政府和街道办事处）的报告与文件等；

4）当地社会团体所提供的资料、主要报纸和网络上的文献资料（包括，对这两个历史街区已作过调研的学生论文）。

针对不同的研究目标和目的，本研究采用不同的文献资料。例如，为"了解几个关键因素（即城市更新和社会资本）的现状及其关系，以及转型期中国历史城市中的其他因素"，本研究就需要对中国历史城市的城市发展、更新规划以及实践进行大量的文献阅读。另外，还应该对中国历史城市的相关保护法律法规进行查阅。

同样，为"了解社会资本是如何促进社区发展的，以及社会资本在地方城市更新过程中的作用"，本研究需要对当地的社区居民和社会团体进行实地调查和研究。通过问卷调查的方式可以获取地方居民对社区在更新过程中参与情况的看法。同时，研究现有的文献资料，对理解地方社会资本和社区参与

government-led redevelopment projects as their only choice to improve their living conditions. Therefore, many people strongly support the local government-led redevelopment plans, through which most original residents have been resettled outside the CWA area. Notably, with the relocation of these residents, many local traditional group activities have also disappeared. This forms a sharp contrast with what has happened in the DTMD case.

1.6.3 Data sources

The relevant data for this study are obtained from various sources including the following:

1) Interviews with local residents and community representatives, government officials, local professionals, and members of local organizations (e.g., the mosques in the DTMD);

2) Field surveys and observations;

3) Reports and documents from abroad and from local libraries and government departments at different levels (municipal and district governments and street offices);

4) Other documents and resources from local organizations, newspapers, and the Internet (e.g., several local research students' dissertations on these two historic districts).

Different types of data sources are closely related to the manner of achieving specific research goals and objectives. For instance, to "*understand the state and relationships of key variables, that is, urban regeneration and social capital, as well as the related factors in historic cities of transitional China*," a study may require an extensive documentary and archive studies of urban development, regeneration plans, and practices in Chinese historic cities. In addition, related urban conservation laws and regulations in Chinese historic cities should also be reviewed.

In the same way, to "*understand the role of social capital in community development and involvement in local urban regeneration process*," field surveys and interviews in local

两者之间的状态及其关系,可以提供进一步的认识。当然,大多数情况下,一种文献资料能够适用于多个研究目的。通过搜集各种综合性的文献资料,本研究最终将"构建一个理论分析框架,以了解中国历史城市中的城市更新和社会资本"。

在针对选定的案例采集资料数据时,采用的主要方法是实地调研和问卷调查相结合的方法。这两种方法主要应用在二手文献资料缺失的情况下。

采访和问卷调查

在本研究中,除了采用对已有文献研究这一重要的研究方法之外,还进行了采访和问卷调查的方式,以完善本研究中所需的重要数据。案例研究中涉及的主要人员分为下列四类:

(1)政府工作人员;(2)当地专家学者;(3)开发公司工作人员;(4)社区居民(包括,当地居民、商人和社会团体成员)。

受访的当地专家学者,以前都曾在不同时期直接参与过本文选择的案例项目。通常,这些专业人士在当地占据着较高的社会地位,如西安城市规划设计研究院的前任院长、在当地大学从事城市规划和更新项目的教授等。许多受访人士都表达出了批判性的观点,并提供了珍贵的第一手官方数据,如人口统计和经济数据,这些为本研究奠定了坚实的基础。作者曾分别在2006年6月20日~7月4日,2007年5月1日~5月14日,以及2008年2月18日~2月29日对这些专家学者进行了

communities and organizations are necessary. Moreover, conducting questionnaire surveys is a way to obtain the opinion of the local communities on local community involvement in regeneration processes. At the same time, available literature and documentary studies help complement the understanding of the state and relationships between local social capital and community participation. Of course, in most cases, a specific data source can often yield information needed by more than one aim. Through various combined sources of data, the study eventually helps to "*build a theoretical framework to understand urban regeneration and social capital in Chinese historic cities.*"

With regard to data collection in the selected case studies, field studies and questionnaire surveys serve as the primary data collection methods especially when second-hand literature or documentary data are not available.

Interviews and questionnaire surveys

Aside from the literature and documentary studies, which are the main research methods in the study, the interviews and questionnaire surveys complete the necessary data collection in the study. The key actors in the case studies are divided into the following four types:

(1) local authorities; (2) local professionals; (3) development companies; and (4) local communities (including local residents, businessmen, and local organizations)

Local professional interviewees were mainly academics who directly participated in studying the selected cases in this book at different times. Generally, these professionals had senior status in local studies, such as the former dean of the Xi'an City Planning and Design Institute, and professors in urban planning and regeneration programs in local universities. They provided many critical points to the study and supplied valuable first-hand official data, such as information on demographics and economics, which were fundamental to this study. These

面访。

本研究采用了访问和问卷调查的形式对当地社区居民进行了调查。在 2006 年 6 月 20 日～7 月 4 日,2007 年 5 月 1 日～5 月 14 日,以及 2008 年 2 月 18 日～2 月 29 日期间,作者进行了实地调研,并对社区的居民代表进行了采访;在 2008 年 2 月 18 日～2 月 29 日期间,开展了问卷调查,采集到了当地居民对于地方政府更新规划和实践的宝贵意见。同时也收集到了有关社区居民之间的联系与交流的意见。关于社区居民和地方政府之间的关系,主要是通过问卷调查和采访的形式进行资料采集。不过,由于作者有限的人力资源和时间限制,最终的受访者和参与问卷调查的人数也比较有限。

根据最近的统计数据,2006 年,回民区中的穆斯林居民约有 3 万人,占到回民区总人口的一半左右(Yang, 2007:377)。作者于 2008 年 2 月 18 日至 2 月 29 日期间,对当地 60 名居民进行了面访和问卷调查。在回民区,回族和汉族居民生活在不同的聚居地。例如,由于宗教传统的关系,穆斯林居民每天都会定时去清真寺做礼拜。图 1–3 标注出的红色区域是大多数穆斯林居民的居住地,而大部分汉民通常住在该区的西部和北部。本研究采访了当地 60 名回族居民,主要考虑到如下四个因素。第一,由于穆斯林宗教生活的关系,该区拥有两套行政管理体系。行政管理体系中,其一是地方政府部门或社区居民委员会;其二是清真寺管理委员会(简称寺管会)。每个管理体系的职责明确,各具特点,这也使得穆斯林居民的生活方式与汉族居民的生活方式非常不同。例如,以

professionals were interviewed separately on June 20-July 4, 2006, on May 1-14, 2007, and on Feb 18-29, 2008.

Local residents in this study were investigated through both interviews and questionnaire surveys. The interview was conducted separately during the author's field observations on June 20-July 4, 2006, on May 1-14, 2007, and on Feb 18-29, 2008. The questionnaire survey, conducted on February 18-29, 2008, collected the necessary data on local residents' opinions regarding local regeneration plans and practices. It also gathered their views on the relationships and interactions among local community members. The data concerning the relationship between the local communities and the local governments in this study were mainly obtained through questionnaire surveys and interviews. Nevertheless, considering the author's limited manpower and time, the total number of interviewees and questionnaire respondents was restricted.

According to the latest available statistical data, approximately 30,000 Muslim residents were in the DTMD in 2006, accounting for half of the local total population (Yang, 2007: 377). The author successfully surveyed 60 local residents on February 18-29, 2008. Within the DTMD, Muslim and Han residents live in separate geographic groups. For instance, related to their religious tradition, Muslim residents schedule their day around their regular mosque visits. Majority of Muslim residents live within the area delineated in the red square in Figure 1-3, while majority of Han residents live to the west and to the north of this area. The study interviewed 60 Muslim indigenous residents for the next four reasons. First, this area comprises two sets of administration systems related to the Muslim's religious lives. One administration system is the local government organization or the Community Residents' Committee, and the other is the Mosque Management Committee. Each administration system has distinctive responsibilities and makes the lifestyle of

种族和宗教组织为基础，穆斯林居民对周围发生的城市更新过程表现得更加关注，这与回民区中的汉民情况有所不同。第二，回民区的大部分汉族居民在回民区以外的地方工作，而不像很多回民那样，在本地拥有家族商业。第三，大部分汉族居民邻里之间不像穆斯林居民那样，拥有世代的亲属关系。值得一提的是，这种亲属关系也是促使穆斯林社群积极参与到当地城市更新过程中，并使居民之间保持一种密切社区联系的重要因素。这种亲属关系使得与穆斯林社群生活相关的城市更新活动显得很与众不同，也要比汉民社群中的城市更新活动显得更复杂。第四，在所选研究案例的更新活动中，受影响的家庭和私营企业主要是穆斯林居民。因此，参与到城市更新活动中的绝大部分人也是穆斯林居民。

值得一提的是，回民区中60名穆斯林受访者，几乎都是自出生就生活在这里。他们中的一多半（51.6%）是当地的个体工商户，36.7%的人是在西安其他地方工作的私营工商户，11.7%的受访士人没有固定工作。作为即是穆斯林居民，同时也是商户的双重身份，很多受访者的生活都不同程度地受到了城市更新活动的影响。由于穆斯林居民的文化和宗教传统，本研究较难对女性穆斯林居民进行采访。因此，在回民区，接受采访的受访者绝大部分（88.3%）都是男性穆斯林。尽管有的受访者对作者的提问比较警觉，并要求匿名回答问题，但总体而言，穆斯林受访者在本研究的调研过程中，都非常积极热情。根据他们的要求，调查中的所有受访者都保持匿名。相比较而言，本研究中，汉族

Muslims distinctive from that of Han Chinese residents. For instance, based on the ethnic and religious organizations, Muslim residents are actively involved in the local regeneration process. This involvement is less visible among the Han Chinese residents. Second, most Han Chinese residents in the DTMD take jobs in other workplaces in the city, unlike most Muslim residents who have in-situ family-based businesses. Third, most Han residents do not have kinship relationships as Muslims do. Kinship ties are essential among Muslim groups to support their involvement in the local regeneration process and retain their intimate community connections. These ties make regeneration related to Muslim's lives in the DTMD unique and more complicated than that among the Han Chinese groups. Four, affected households and private businesses in the local regeneration process are mainly Muslim residents. Accordingly, most participants in the regeneration process are Muslims.

All 60 Muslim respondents in the DTMD have lived in the area since birth. Among them, more than half (51.6%) are local private businessmen, 36.7% are private businessmen working in the city, and 11.7% are local unemployed residents. As both Muslim residents and indigenous businessmen, the lives of many respondents have been affected, to varying degrees, by local regeneration practices. Given their cultural and religious traditions, the author found access to female Muslim interviewees difficult. Therefore, majority of the respondents (88.3%) in the DTMD are male Muslims. Generally, the Muslim respondents were enthusiastic and cooperative during the author's survey, though some respondents were vigilant and requested anonymity in answering the questionnaires. As indicated during their brief, all respondents in the survey remained anonymous. Comparatively, the issues of the Han Chinese residents during the urban regeneration process in this study are addressed more in the SHD case.

居民在城市更新进程中的特点主要是通过三学街历史街区的案例来展现。

本研究于2008年2月18日至2月29日期间，在三学街区成功完成了70份问卷调查，其中40份在三学街历史核心区，30份在大吉昌巷区。根据最近的统计数据，2001年三学街历史街区的总人口是7600人（XUPDI, 2001：2）；其中，约有1200位原住居民在2005年为了配合地方政府主导的城市更新项目而进行重置（EBXY, 2006：337）。在作者的问卷调查中，三学街的70位受访者全是汉族居民。他们中的绝大部分（88.6%）已经在这个地方生活了30多年，受访者中多数（60%）也是老年人。自20世纪90年代以来，受访者中不少人要么退休、要么下岗。所有受访者中，18.6%的人是在西安其他地方工作的私营工商户；21.4%的人在各种政府事业单位工作。所以，虽然40%的受访者居住在三学街，但他们却在街区以外的其他地方工作。

三学街受访的居民中大部分人都是老年人。由于当地的年轻人基本上都在市区的其他地方工作，所以常常只有老年人留在该区的家中。另外，很多当地家庭已经搬到其他地方居住，他们通常会把这里的房子出租给外来居民。当地调研结果显示，2002年，三学街中大约42%的住户为外来务工者（Ju, 2005：40; Zhu, 2006：54）；另外，在2005年，大约81%的家庭仅仅是居住在这里，而在西安其他地方从事私营商业活动（Ju, 2005：40; Liu, 2006a：82; Zhang, 2002c：31）。因此，在三学街，白天能够受访的绝大部分都是老年人。面对作者的调研，大部分受访者也都比较敏感，并要求在面访和问卷调查中保持

In the SHD, 70 questionnaire surveys were distributed on February 18-29, 2008, that is, 40 in the historic core and 30 in the Dajichang-alley site. According to the latest available data, the total population in the SHD was 7,600 in 2001 (XUPDI, 2001: 2), and about 1,200 original residents had been relocated with the completion of local government-led regeneration projects in 2005 (EBXY, 2006: 337). In the author's questionnaire survey, all the 70 respondents in the SHD are Han Chinese and most of them (88.6%) had lived in that area for more than 30 years. More than half of them (60%) were senior residents. These residents were either retired or had been laid off since the end of the 1990s. 18.6% of the total respondents were private businessmen in other workplaces in the city, and 21.4% of the respondents took jobs in various government departments. Hence, around 40% of the total respondents were employed in the workplaces outside of the historic district, though they resided in the SHD.

Most surveyed residents in the SHD were senior residents because most indigenous young residents have taken jobs in other workplaces of the city, and generally, only senior residents in the family live in the area. In addition, many indigenous families have moved out to live in other places and have rented their original houses to mobile populations in this area. According to local studies, approximately 42% of the residents in the SHD in 2002 were migrant workers (Ju, 2005: 40; Zhu, 2006: 54). In addition, approximately 81% of the total households simply resided there while working as private businessmen in other places in Xi'an city in 2005 (Ju, 2005: 40; Liu, 2006a: 82; Zhang, 2002c: 31). Therefore, most available indigenous interviewees during the daytime in the SHD were senior residents. Most respondents were vigilant regarding the author's study and requested anonymity during the interview and questionnaire survey. Accordingly, all respondents in the SHD remained anonymous in the author's field survey.

匿名。

为了减少在三学街案例中因年龄差距而造成的局限性，以及在回民区案例中因性别而造成的局限性，本研究也综合考虑了当地政府的文件以及其他研究工作者所提供的一系列二手资料。表1-2罗列出了两个历史街区中受访者的基本资料。

在针对具体研究问题搜集资料期间，尽管在两个历史街区分发的调查问卷相同，但是本研究发现，一些问题在两个街区中所起的作用并不相同。例如，在回答"你会怀念以前的老邻居吗？如果答案肯定，那会是什么呢？"时，在三学街区，不同的受访者给出了不同的答案。但是在回民区，几乎所有的受访者都表示他们世代生活在这里，从没离开过这儿。因此，对他们而言，新旧邻居没有什么变化和可比性。总体来讲，这两个历史街区有以下几个方面的主要差别：

1）两个街区中，社会资本的性质或当地居民间的社区关系有较大差别。在回民区，社区邻里关系基本建立在亲密的亲情关系、家属关系以及血缘关系之上。对很多居民而言，他们的邻居也是他们的亲戚、朋友和同事。相比较下，三学街区以汉民为主的社区，就没有这个特点。

2）除了当地比较亲近的社区邻里关系之外，两个历史街区的住户对当地的依赖程度也不尽相同。在回民区，70%以上的居民以家庭私营商业为生，这与当地繁荣的旅游业市场也有密切关系（SMDPI et al.，1997：70）。不过，这种情况在三学街就不太一样。

3）回民区的保护与更新问题已引起当地专家学者很久的关注，因此，本研究可以获

To reduce the limitations caused by age bias in the SHD case, and the gender bias of Muslim interviewees in the DTMD case, the study strove to combine a variety of secondary data provided by local government documents and other research groups in the analysis. The profiles of the respondents as representatives of indigenous residents in the two historic districts are illustrated in Table 1-2.

During the data collection for the specific research objectives, although the same questionnaires were given in these two historic districts, the study found that some questions did not work equally well in the two districts. For instance, for the question *"Do you miss anything about your old neighborhood? If YES, what is it?"* the situation varies among the different interviewees in the SHD. However, in the DTMD, all the interviewees stated that they had lived in this district for generations, and they had never left since then. Hence, there was nothing for them to compare between the old and new neighborhoods. Generally, there are several major differences between the DTMD and the SHD as shown in the following:

1) The nature of social capital or community relationships among local residents is very different between the two districts. In the DTMD, community relationships are based on intimate relative relationships, kinships, and blood ties. To many local residents, their neighbors are also their relatives, friends, and business colleagues. However, this situation is not observed in the SHD because it is a Han Chinese community.

2) In addition to local intimate community ties, the levels of place attachment to the two historic districts are different. In the DTMD, more than 70% of the local residents have lived on family-based commercial businesses, which closely relate to local flourishing tourism industry (SMDPI et al., 1997: 70). Nevertheless, this situation does not apply to the residents in the SHD.

3) The issue of urban conservation and regeneration in the DTMD has attracted much of the local professionals' attention.

调研中两个历史街区的受访者的基本资料　　表 1–2
Respondents' profiles in questionnaire surveys in the two historic districts　　Table 1–2

项目 Items	内容 Contents	分类 Categories	回民区人数 (%) DTMD Number (%)	三学街人数 (%) SHD Number (%)
性别 Gender	男性 Male		53 (88.3%)	29 (41.4%)
	女性 Female		7 (11.7%)	41 (58.6%)
	总数 Sub-total		60 (100%)	70 (100%)
年龄 Age	10 岁以下 Below 10		0	0
	11–20		0	0
	21–40		15 (25.0%)	0
	41–60		38 (63.3%)	28 (40.0%)
	60 岁以上 Above 60		7 (11.7%)	42 (60.0%)
	总数 Sub-total		60 (100%)	70 (100%)
种族 Ethnicity	汉族 Han Chinese		0	70 (100%)
	回族 Muslim residents		60 (100%)	0
	其他 Others		0	0
	总数 Sub-total		60 (100%)	70 (100%)
职业 Employment	当地私营工商户 Local private businessmen		31 (51.6%)	0
	其他地方工作的私营工商户 Private businessmen in other places		22 (36.7%)	13 (18.6%)
	政府事业单位的员工 Employees in government departments		0	15 (21.4%)
	其他 Others		7 (11.7% 无业) 7 (11.7% unemployed)	42 (60.0% 退休或解雇) 42 (60.0% retired or laid off)
	总数 Sub-total		60 (100%)	70 (100%)
教育水平 Levels of education	非正式教育 Informal schooling		0	19 (27.1%)
	小学 Primary school		0	4 (5.7%)
	初中 Middle school		22 (36.7%)	21 (30.0%)
	高中 High school		23 (38.3%)	18 (25.7%)
	技校 Technical school		7 (11.7%)	8 (11.4%)
	大专或以上 College and above		8 (13.3%)	0
	总数 Sub-total		60 (100%)	70 (100%)
月收入 (人民币¥) Monthly income (RMB yuan)	¥500 以下 Below ¥500		7 (11.7%)	0
	¥500–1000		0	48 (68.6%)
	¥1000–1500		15 (25.0%)	0
	¥1500–2000		23 (38.3%)	14 (20.0%)
	¥2000–3000		15 (25.0%)	0
	¥3000–5000		0	8 (11.4%)
	¥5000 以上 Above ¥5,000		0	0
	总数 Sub-total		60 (100%)	70 (100%)
居住情况 Residential situation	少于一年 Less than one year		0	0
	1–5		0	0
	6–15		0	0
	16–30		0	8 (11.4%)
	30 年以上或自出生起 Above 30 or since birth		60 (100%)	62 (88.6%)
	总数 Sub-total		60 (100%)	70 (100%)

来源：2008 年 2 月 18 日~2 月 29 日作者进行的调研
Source: Conducted by author on Feb 18–29, 2008

得很多二手的数据和资料。相比之下，在三学街区，这方面的资料就很少，尤其是针对当地社会资本的研究方面更显缺乏。为了探讨当地社区居民关系和联系，本研究就需要通过对受访者的面访和问卷调查，来获得足够的第一手资料。

在探讨当地的社会资本或社区居民关系时，鉴于两个历史街区存在的较大差异，本书综合了多方资料来源。对三学街的分析主要依赖于第一手资料，以及其他研究者为数不多的调研结果。例如，白宁曾在三学街历史街区成功进行了约50份问卷调查（Bai，2000）。在对回民区的研究中，其他研究者早期进行的调查结果对本研究也有重要意义。例如，霍伊姆（Hoyem）曾在回民区成功进行了193份问卷调查（Hoyem，1989）。从多方面来看，早期开展的研究结果为本研究提供了大量的可比性资源，也有助于帮助人们更好地理解当地物质环境的转变和地方居民之间关系的变化。

实地调研与观测

在进行问卷调查和采访当地社区居民时，实地调研与观测很有必要。这可以补充通过面访主要利益相关者所收集到的资料。本研究中，作者参加了若干次当地居民组织的社区会议。在会议中，居民对当地的再开发和重置问题进行了讨论。另外，为了研究地方社区居民之间的关系，作者对当地传统的经济活动和历史建筑物也做了调查。这些都有助于更好地了解社区居民之间的关系，以及他们对城市物质环境的依附感，而这些都与当地社区的社会资本水平有很大关系。

Accordingly, there are many available data and resources for further analysis. Comparatively, studies on the SHD have been scarce, especially those on social capital in this area. To explore the issue of community relationships and connections, this book particularly needs to obtain sufficient primary data from the interviews and questionnaires' surveys.

Given the major differences between DTMD and SHD in exploring the local social capital or community relationships in these two districts, various data sources are used. For SHD, the analyses mainly rely on primary data in addition to the limited surveys done by other researchers. For example, Bai carried out 50 questionnaire surveys at the historic core of the SHD area (Bai, 2000). In a study on the DTMD, the survey results conducted by other researchers in an earlier stage significantly contributed to this research. For instance, Hoyem conducted 193 questionnaire surveys in the DTMD (Hoyem, 1989). From many perspectives, the results of the previously conducted surveys provide significant comparative resources that help us understand the local transforming physical environments and the evolving relationships among local residents.

Field studies and observations

Field studies and observations are helpful when conducting questionnaire survey and interviewing local communities. They complement the data gathered by interviewing some key stakeholders. For this research, observations were conducted by participating in local community assemblies when the residents discussed the local redevelopment and resettlement issues. In addition to observing the relationships among local community members, the relationships between local traditional economic activities and the historic urban fabric were also observed. All these help us understand better the relationships between the community members and their economic attachment to the physical environments in relation to the level of local social capital.

1.7 本书架构

第一章简要介绍了本书的总体大纲，概括了研究的主要观点；分析了以往或相关的研究文献，并提出了研究空白。同时，本章还分解了主要研究问题，并辨析了子问题间的逻辑关系。探讨了本研究的目标和目的、意义和界定。最后，第一章中很重要的内容是关于研究方法的讨论，建立了一个理论分析架构，并介绍了案例的选择方法和数据的采集方法。

第二章和第三章致力于建立一个能说明转型期中国历史城市的更新与社会资本关系的研究理论架构，并讨论了本研究所涉及的几个关键性因素之间的关系。

第二章首先回顾了西方文献中有关城市更新的政策和实践。它探讨了两个亚洲城市，即京都和吴哥遗址的以保护为导向的城市更新实践，揭示了亚洲城市的城市保护政策和实践特点。在保护主导的城市更新框架下，本章系统回顾了城市保护的理论以及联合国教科文组织议程中的相关议题。在基于社区的城市更新中，社会资本或社区联系有助于社区参与到城市更新规划中，并保护原住民的生活内容。本章还回顾了社会资本的概念、类型和测量方法。此外，还对城市更新过程中，影响社会资本的主要因素进行了分析。

第三章首先回顾了中国城市建设和更新的历史演变过程，以及各时期的主要特点。其次，本章着重探讨了两种典型的城市更新策略：一种是再开发主导的更新策略，另一种是保护为主导。接下来，讨论了以再发展为主导的，侧重于经济增长的城市更新过程

1.7 Outline of the Book

Chapter 1 provides the overall outline of the book. It describes the general study context of the book, reviews the main arguments, discusses the reviewed or related studies and literature, and identifies the research gap. In the same chapter, the main research question is proposed, and the logic in the various sub-questions is discussed. The research goals and objectives, significance, and delimitation of the study are explored. Finally, the research methodology is examined, wherein a tentative analytical framework is set up, and the case study areas and data collection methods are introduced.

Chapters 2 and 3 focus on the development of a theoretical framework that can address social capital and urban regeneration in the context of transitional Chinese historic cities. Related key issues in this research are also explored.

Chapter 2 initially reviews the evolution of urban regeneration policies and practices through related literature from the West. It explores the conservation-led urban regeneration practices in two Asian cities, Kyoto and the Angkor heritage site, which reveal the urban conservation policies and practices in Asian cities. In the framework of conservation-led urban regeneration, this chapter reviews the theory of urban conservation and related issues included in the UNESCO's agenda. With regard to the community-based urban regeneration, social capital or community connections promote community involvement in urban regeneration plans and in the conservation of indigenous lives. Furthermore, this chapter reviews the social capital surrounding the concept, its types, and measurement. In addition, some major factors affecting social capital in the regeneration process are explored. In conclusion, the chapter discusses the relationships among these key variables.

First, Chapter 3 reviews briefly the urban development and the regeneration evolution in urban China, providing a historical and contextual understanding of Chinese cities' evolving urban

中，城市管治模式的转变。最后，为了促进社区参与到历史城市的保护和更新过程中，本章分析了社会资本因素，及其在促进社区参与到城市更新过程中的积极作用。基于第二、三章的内容，最后建立了一个理论分析框架，以便更好地理解中国历史城市中的社会资本和城市更新问题。

第四章以西安为例，开展案例研究。首先，本章简要介绍了西安的历史和地理，并回顾了自1949年以来，西安城市建设与更新政策的演变。其次，探讨了西安的社会资本问题，及其在社区居民关系和社区参与方面的作用。最后，本章讨论了影响社会资本生成的关键性因素，及其在城市更新当中的作用。

第五章以西安鼓楼回民历史街区为例，探讨了拥有较高水平社会资本的历史街区，对城市更新的影响。首先，本章介绍了西安回民区，并以两个历史地段为例，介绍了西安当地政府主导的倾向于经济发展的城市更新政策。其次，本章分析了拥有较高水平社会资本的社区特点，探讨了影响社区高水平社会资本的主要因素。最后，基于对当地高水平的社会资本的理解，本章在社区参与问题上探讨了当地居民的自建活动。

第六章以西安三学街历史街区为例，探讨了拥有较低水平社会资本的历史街区更新与保护的主要特点。首先，本章简要介绍了西安三学街，分析了该区转型的城市发展政策和策略。在对2001年以来该区的更新政策和实践进行介绍时，本章着重介绍了两个具体案例：2001年以来，三学街历史核心区的保护与再发展；2004年以来大吉昌巷的再发展。此外，本章分析了拥有较低水平社会资

policies on urban renewal and regeneration. Second, this chapter probes into two outstanding urban approaches, redevelopment-led approach and urban conservation, both of which border on the urban regeneration policies and strategies in Chinese historic cities. Third, the transforming mode of urban governance in Chinese cities in relation to the redevelopment-led and pro-growth urban regeneration policy is explored. Finally, to promote community participation in the conservation and regeneration processes of Chinese historic cities, this chapter investigates the issue of social capital and its contributions to community participation in the urban regeneration process. Based on the discussions in Chapters 2 and 3, an analytical framework to understand better social capital and urban regeneration in Chinese historic cities is developed.

Based on the analytical framework, Chapter 4 explores these issues in relation to a specific Chinese historic city, Xi'an. Moreover, this chapter gives a brief historical and geographical introduction on Xi'an City and reviews its transforming urban development and regeneration policies since 1949. In light of the more comprehensive concerns in the regeneration of Xi'an's historic urban core, this chapter also explores the issue of social capital in terms of community relationships and involvement in urban regeneration processes. At the same time, some key factors affecting the generation of social capital are studied. Finally, this chapter ends with an emphasis on the role of social capital in the urban regeneration program of Xi'an's historic urban core.

Following the discussions on Xi'an's transitional urban regeneration policies, urban management mechanism, and various forms of social capital in urban development, Chapter 5 investigates the urban regeneration practices and social capital in a specific historic district, namely, the Xi'an Drum Tower Muslim District (DTMD). The chapter introduces the DTMD and explores Xi'an's local government-led pro-growth urban regeneration policies in two

本的社区特点，探讨了影响当地社会资本的主要因素。本章在结尾重申，当地的城市更新政策和策略与脆弱的社区关系也有关系，低水平的社会资本进一步导致城市更新过程中社区参与的缺乏，而这也影响到了整合的城市更新和综合的城市保护结果。

第七章对本研究进行了总结，讨论了研究结果，并对在中国历史城市中，开展城市更新的实践提出了策略性建议。具体来说，本章总结并讨论了西安案例研究的主要结论；探讨了促进西安进行真正的城市更新实践的意义以及策略性建议。同时，本章研究了中国历史城市更新实践的理论意义，并讨论了本研究对已有文献的理论贡献。最后，本章指出了本研究的不足之处以及将来研究的方向（图1-4）。

historic quarters. Moreover, probes into the social capital in the DTMD and considers the key factors affecting the community social capital at the same time. Finally, based on the understanding of the high level of local social capital, the issue of community involvement in terms of self-constructed activities is explored.

Based on another historic district, the Xi'an Sanxuejie Historic District (SHD), Chapter 6 further explores the role of social capital in local urban regeneration processes. This chapter briefly introduces the historic district of Xi'an SHD and examines the transitional urban development policies and strategies in this area. In examining the area's regeneration policies and practices since 2001, this chapter focuses on two specific cases: the conservation and redevelopment of the historic core of SHD since 2001 and the redevelopment of the Dajichang alley site since 2004. Moreover, this chapter probes into the issue of the low level of social capital within SHD. Several key factors affecting this low level of social capital are studied as well. This chapter concludes by reiterating that local urban regeneration policies and strategies are related to weak community relationships, and that with the lack of community involvement in local regeneration processes due to the low level of social capital, comprehensive urban regeneration and integrated urban conservation outcomes are undermined.

Chapter 7 concludes the study by discussing the findings and providing recommendations and policy implications for genuine urban regeneration practices in Chinese historic cities. Specifically, this chapter summarizes and discusses the major findings of the case studies in Xi'an. The implications and policy recommendations in promoting genuine urban regeneration practices in Xi'an are examined. This chapter also probes into the theoretical implications for the urban regeneration practices in Chinese historic cities and discusses the contribution of the study to the current literature. In conclusion, the limitations of the study and topics for further research are outlined (Figure 1-4).

转型期中国城市面临的变化与挑战	URBAN CHANGES AND CHALLENGES IN TRANSITIONAL CHINESE CITIES
• 自20世纪90年代以来，随着快速城市化进程，许多中国历史城市面临着巨大挑战： 　促进当地经济的发展 　修复破败的历史城区 　保护历史城市环境 • 以再发展为核心的城市规划和实践在很多地方开展： 　巨额经济回报 　对历史城市特色的威胁	• With rapid urbanization since the 1990s, many Chinese historic cities have faced the challenges: 　Promoting local economic development 　Regenerating dilapidated historic inner urban areas 　And protecting historic urban scenes • Rampant urban redevelopment-oriented planning and practices have been undertaken in many places: 　Massive economic returns 　And many threats to the character of the historic cities

↓

文献资料和理论背景	LITERATURE AND THEORETICAL BACKGROUND
• 历史性回顾西方城市背景下的城市更新 • 联合国教科文组织议程中的城市保护 • 西方城市背景下的社会资本及其与城市更新的关系 　关键因素： 　　• 社区参与和参与式城市管治模式	• A historical overview of urban regeneration in the Western context • Urban conservation on the UNESCO's agenda • Social capital and the relations with urban regeneration in the Western context 　Key factors: 　　• Community involvement and mode of participatory urban governance
• 转型期中国城市的发展与更新 　主要参与者： 　　• 政府部门（即中央政府、市政府和区政府） 　　• 公共和私人开发商 　　• 当地社区 • 转型期中国城市的社会资本及其与城市更新的关系 　关键因素： 　　• 转型期中国的城市保护法律法规 　　• 企业化的城市管治模式	• Urban development and regeneration in transitional China 　Key actors: 　　• the state (i.e., central, municipal, and district governments) 　　• public and private developers 　　• local communities Social capital and the relations with urban regeneration in transitional China 　Key factors: 　　• Conservation laws and regulations in transitional China 　　• Mode of entrepreneurial governance in China

↓

理论架构：以便更好地理解中国历史城市中的社会资本和城市更新
Theoretical framework for the better understanding of social capital and urban regeneration in Chinese historic cities

↓

案例研究	CASE STUDIES
• 西安城市发展和城市更新政策及实践 • 社会资本的特点 　• 聚合型社会资本 　• 联合型社会资本 • 城市更新、社会资本以及相关要素之间的关系	• Urban development and regeneration policies and practices in Xi'an • Characteristics of social capital 　• The bonding social capital 　• The bridging social capital • Relationships between urban regeneration and social capital and related factors

| • 两个选定案例的城市更新规划、政策与实践：回民区和三学街历史街区
• 当地社区级别社会资本的特点
• 社会资本、城市更新以及相关要素的关系
　• （回民区）基于高水平社会资本的社区在当地城市更新过程中的参与
　• （三学街）低水平社会资本对当地城市更新过程的影响
• 不同水平的社会资本对地方城市更新规划的影响及不同结果 | • Urban regeneration plans, policies, and practices in two selected cases: DTMD and SHD
• Characteristics of social capital at the local community level
• Relationships between social capital, urban regeneration, and related factors
　• (in DTMD) Community involvement in local regeneration practices based on a stronger local social capital
　• (in SHD) Impacts of lower level social capital on local urban regeneration processes
• Distinct outcomes of local regeneration plans based on different levels of social capital |

↓

结论	CONCLUSIONS
• 讨论社会资本在西安历史城市更新中的作用 • 对促进西安城市更新实践的意义和策略性建议 • 对中国历史城市更新实践的理论意义及启示 • 本研究对已有文献的理论贡献	• Discussing the role of social capital in the regeneration of the historic city of Xi'an • Implications and policy recommendations to promote genuine urban regeneration practices in Xi'an • Theoretical implications contributing to the urban regeneration practices in Chinese historic cities • Contribution of the book to the current literature

图 1-4：研究流程图

Figure 1-4: Flowchart of the study

第二章　城市更新与社会资本
CHAPTER 2　Urban Regeneration and Social Capital

建立一个能够说明处于经济转型期的中国历史城市的理论框架，涉及的内容非常宽泛，因此本研究将通过两个章节的内容来构建该理论框架。本章将主要讨论一些西方文献，并引出目前关于主要影响因素的讨论；第三章将探讨转型期中国历史城市发展过程中的一些关键影响因素及其相互关系。

在探讨中国历史城市的城市更新和社会资本中，将涉及三个主要变量，即城市更新、社会资本和历史城市。在这当中，城市保护也是一个考虑内容。首先，本章回顾了西方文献中城市更新政策与实践的演变。西方的城市更新政策和实践在二战之后大概分为4个阶段："推土机"式的清除阶段、街区修复阶段、经济复苏和公私合营阶段、运用多维方法进行全面更新的阶段。然而，鉴于西方历史的具体的政治经济体系，本研究发现这种城市更新的发展阶段不适用于亚洲，因此，关于第二个变量，本书在两个亚洲城市，即京都和吴哥遗址，探讨了以城市保护为主导的城市更新实践，揭示了亚洲城市的城市更新政策和实践。其次，这些地方的更新实践已经暴露了一些问题，彰显了以社区为基础的城市保护或更新实践的重要性。本章也同样回顾了这一点。再次，在以城市保护为主导的城市更新框架中，本章回顾了城市保护

Owing to the complexity of developing a framework that can address the specific context of historic cities in China's transitional economy, the book will develop the theoretical framework in the next two chapters. This chapter focuses on the Western literature, canvassing current discussions on key variables central to the book, while Chapter 3 will explore the corresponding variables and their relations in the context of historic cities in transitional China.

In exploring urban regeneration and social capital in Chinese historic cities, three main variables cover urban regeneration, social capital, and historic cities, where urban conservation is a concern. First, this chapter reviews the evolution of urban regeneration policies and practices in Western literature. Urban regeneration policies and practices in the Western world have been roughly divided into four periods since World War II: the bulldozing era, neighborhood rehabilitation, economic revitalization and public-private partnership, and comprehensive regeneration concerns through multi-dimensional approaches. Nevertheless, given the specific political economy and institutions in the Western context, the study finds that this periodization does not apply to Asia and for that matter, China. Hence, the second variable is concerned with the conservation-led urban regeneration practices in two Asian cities, that is, Kyoto and the Angkor heritage site, which revealed the urban conservation policies and practices in Asian cities. In addition, these places have uncovered some problems that

理论以及联合国教科文组织议程的相关问题，例如，综合城市保护、"活态遗址"、"本土生活"及其他问题。这些问题在很多亚洲国家的城市保护和更新中非常突出。因此，本章探讨了在城市更新中以社区为基础的城市管治问题，这和地方居民的参与紧密相关。在以社区为基础的城市更新过程中，社会资本或社区联系有助于促进社区居民参与到城市更新规划中，并有效保护原住居民的生活。因此，本章回顾了社会资本的概念、类型和测量方法。另外，本章还探索了在更新过程中影响社会资本的主要因素，例如，管治模式、民族和宗教问题。最后，本章讨论了这些关键变量之间的关系。

2.1 城市更新议程

自 20 世纪 90 年代以来，很多学者已经讨论了城市更新的概念（Carmon，1997；Ng，2005b；Ng & Tang，2002b；Roberts，2000）。正如罗伯茨（Roberts）所说，城市更新是城市发展中众多因素相互作用的产物，包括物质、社会、环境和经济方面的因素，城市更新通常被看作是"机遇和挑战的产物"，在特定的城市中体现为城市退化（Roberts，2000：9）。与城市改造相比，有学者认为城市改造的特征是"清除破败地区，建设美好家园，

spotlight the significance of community-based urban conservation or regeneration policies. This is likewise reviewed in this chapter. Third, in the framework of conservation-led urban regeneration, this chapter reviews the theory of urban conservation and related issues on the UNESCO's agenda, for example, integrated urban conservation, "living heritage site" and "indigenous lives," and other issues. These issues are very prominent in urban conservation and in the regeneration of many Asian developing countries. This chapter then explores the issue of community-based urban governance in urban regeneration, which is closely related to the involvement of local communities. With regard to the community-based urban regeneration, social capital or community connections promote community involvement in urban regeneration plans and in the conservation of indigenous lives. Therefore, this chapter reviews social capital surrounding the concept, its types, and measurement. In addition, some major factors affecting social capital in the regeneration process are explored, for example, mode of governance and ethnic and religious issues. Finally, this chapter discusses the relationships among these key variables.

2.1 The Urban Regeneration Agenda

The concept of urban regeneration has been discussed by many academics since the 1990s (Carmon, 1997; Ng, 2005b; Ng & Tang, 2002b; Roberts, 2000). As Roberts put it, as an outcome of the interplay of many factors in urban development concerning the physical, social, environmental, and economic aspects, urban regeneration is commonly seen as "a response to the opportunities and challenges" presented by urban degradation in particular urban places (Roberts, 2000: 9). Compared with urban renewal, some academics argue that the approach of urban renewal is often featured by "the eradication of blighted areas

提供舒适居住条件"(Western et al., 1973：589)，而城市更新过程的特征是让城市生活水平有一个更加综合和整体的提高。

本书对于城市更新的理解是基于罗伯茨(Roberts)所给出的定义：

"城市更新是指用一种综合的、整体性的观念和行为来解决各种城市问题，应致力于在经济、社会、物质环境等各方面对处于变化中的城区作出长远的、持续性的改善和提高（Roberts, 2000：17）。"

在这个概念中，罗伯茨强调了城市更新项目的综合性结果，而这在理论上需要各种因素之间的互相作用。通过输入各种主题，产生了多种不同的综合结果，这样形成了一

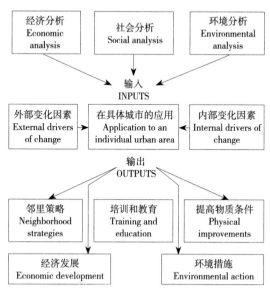

图2-1：西方文献中的城市更新过程
Figure 2-1: The urban regeneration process in Western literature
资料来源：(Roberts, 2000：20)
Source: revised from (Roberts, 2000: 20)

and the provision of decent homes and suitable living conditions" (Western et al., 1973: 589), while the process of urban regeneration is characterized by a more comprehensive and incorporative improvement of urban life.

In the book, the provisional understanding of urban regeneration is based on the definition by Roberts:

Comprehensive and integrated vision and action which leads to the resolution of urban problems and which seeks to bring about a lasting improvement in the economic, physical, social and environmental condition of an area that has been subject to change (Roberts, 2000: 17).

By this definition, Roberts emphasizes the comprehensive outcomes of the urban regeneration program, which theoretically requires the interaction among various factors. The comprehensive urban regeneration process from the inputs of various themes to the multiplicity of integrated outputs is illustrated in Figure 2-1 (Roberts, 2000: 20).

Although the above theoretical discussion shows a rosy picture of the urban regeneration program, in practice, some academics (Healey, 1991; Turok, 1992) contend that many problems exist in the local regeneration programs. For instance, in the urban policies in Britain in the 1980s, Healey argued that although the government claimed the broader economic and social ends of their urban regeneration policies, one might question what could be the "economic, social, physical, and environmental consequences of urban regeneration" policies, which focused on property development (Healey, 1991: 98). Turok (1992: 376, 377) sarcastically contended that as characterized by property development, urban regeneration policies in Britain were more like a placebo rather than panacea to solve its many urban problems.

Nevertheless, although the theory of urban

个全面的城市更新过程（图2-1）(Roberts, 2000：20)。

尽管上述理论讨论展现了城市更新项目的乐观情景，但在实际中，一些学者(Healey, 1991；Turok, 1992)认为地方城市更新项目中存在着很多问题。例如，在20世纪80年代，关于英国的城市政策，赫利(Healey)认为尽管英国政府宣称他们的城市更新政策符合更广泛的经济和社会目的，但是仍不免有人产生疑问，在以房地产主导的政策中，"城市更新的经济、社会、物质和环境效益"是怎样的？(Healey, 1991：98)。图若克(Turok, 1992：376, 377)曾讽刺地说，以房地产开发为特色的英国城市更新政策并不是解决众多城市问题的灵丹妙药，不过仅仅起到安慰剂的作用罢了。

不过，尽管这种城市更新的理论并不能轻易解决很多复杂的城市挑战，但它还是为研究者对现有的城市挑战和机遇的研究提供了一个有用的理论框架。另外，在特定的城市背景中，地方城市更新项目中所采取的更新规划以及发展动力也是不同的。简要回顾发达国家的城市更新政策，有助于了解对关键因素的相互作用以及参与城市更新项目中的动力因素。

2.1.1 西方国家

本书将简要回顾西方国家的城市更新政策的演变。在西方国家（主要是欧洲和北美），许多学者根据城市更新项目的特征和策略，倾向于把从二战后到现在这段历史时期大致分为四个阶段(Carmon, 1997；Ng, 1998；Ng & Tang, 2002a)。

regeneration by no means could easily solve the many complicated urban challenges, it still provides a useful theoretical framework to guide researchers on the existing urban challenges and opportunities. In addition, given specific urban contexts, there are various planning and development dynamics in the local urban regeneration programs. The brief historical overview of the evolution of urban regeneration policy in the context of developed countries will contribute to the reflections on the interactions of the key factors and dynamics involved in the urban regeneration programs.

2.1.1 Western world (from post World War II to present)

This book will briefly review the evolution of urban regeneration policies in Western countries. In the Western world (mainly Europe and North America), many academics tend to roughly divide the historical period of urban regeneration from the post-World War II to the present time into four stages based on the characteristics and strategies taken by the urban "regeneration" programs (Carmon, 1997; Ng, 1998; Ng & Tang, 2002a).

The Bulldozing era (immediate post World War II ~ end of the 1960s)

The immediate post-war time until the end of the 1960s was faced with derelict housing conditions. The slums experienced massive destructions as a result of World War II. Thus, urban "regeneration" or redevelopment policies in many western countries were characterized by massive clearance and the bulldozing orientation. In the U.K., beginning 1954, slum clearance and rehousing slum habitants were led by the central government. Although the private sectors' interests were considered, the government still took full responsibility in all slum-clearance decisions (Short, 1982). In the US, with the Housing Act passed by the congress in 1949, urban renewal practices were characterized by the demolition of old residential areas

"推土机"式的清除阶段（二战后～20世纪60年代末）

从二战结束到20世纪60年代，这段时间西方世界的居住条件都非常破败。贫民窟在二战中遭受了巨大的破坏。因此，在很多西方国家，城市更新和再发展政策的特征是进行大量的清理工作，倾向于大量清除。1954年初，在英国中央政府的领导下，英国上下进行了贫民窟清理和重新安置贫民窟居民。尽管充分考虑了私营部门的利益，政府仍然倾力进行清理贫民窟（Short，1982）。1949年，美国国会通过了住宅法，其城市改造的特色是拆除旧居民区，修建新道路、振兴商业以及建造住宅区来进行再开发，在这过程中，私人开发商得到了最大利益，但是那些被强制搬出的原住居民却被忽视了（Carmon，1997；Gans，1967）。不少学者对此持批评态度：尽管原住居民的住房条件可能会有所改善，但是清理贫民窟和让原住居民迁移到其他地方不可能减少贫困。让他们迁移只能极大地改变他们的生活水平（Carmon，1997；Gans，1962；Western et al.，1973；Young & Willmott，1957）。

邻里修复阶段（20世纪60年代末～20世纪70年代末）

20世纪70年代期间，推土机清除计划减少，和整体拆除完全相反的邻里修复的方法备受青睐。20世纪60年代末，随着"贫困的再发现"运动的展开，一些学者认为"犯罪、违法、酗酒、学业失败、失业、精神疾病和其他低收入家庭的问题并没有仅仅因为自然环境的改变而有所减少"（Western et al.，1973：589）。正如德里（DelliQuadri）所说，城市复兴政策在物质和社会两方面影响着城市生活。

and redevelopment for paving new roads, commercial purposes, and sometimes for residential houses. During this process, the interests of private developers were intensified, but indigenous residents who were forced out of their original living sites were seldom cared for (Carmon, 1997; Gans, 1967). Many academics criticized that although the housing conditions might be improved, clearing the slums and moving the indigenous occupants elsewhere could not reduce their poverty. Moving them could only change their living levels radically (Carmon, 1997; Gans, 1962; Western et al., 1973; Young & Willmott, 1957).

Neighborhood rehabilitation (end of the 1960s ~ end of the 1970s)

During the 1970s, the bulldozer clearance schemes decreased, and approaches like neighborhood rehabilitation as opposed to entire demolition were favored. With the "rediscovery of poverty" by the end of the 1960s, some academics contended that "crimes, delinquency, alcoholism, school failure, unemployment, mental illness, and other disabilities of low income families were not reduced simply by alterations in the physical environment" (Western et al., 1973: 589). As DelliQuadri eloquently argued, the policy of urban rehabilitation dealt with urban life both "physically and socially." It endeavored to bring about a system of "education to help people plan for themselves in their own home and their own neighborhood" (DelliQuadri, 1963: 59). Accordingly, multiple goals were proposed in urban redevelopment programs covering both the rebuilding of physical environments and the renewal of social life (Carmon, 1997; Holcomb & Beauregard, 1981). In carrying out these urban rehabilitation programs, the government began to integrate the private sectors and market resources to obtain the necessary funds. The role of the local government transferred from a "previous uni-sector of decision-making and implementation to a coordinator, policymaker and monitor" (Williams, 1985: 142).

这种政策努力形成一种体制，通过教育让人们学会在家庭和社区里为自己打算（DelIiQuadri，1963：59）。相应地，城市再开发项目中提出了多个目标，既包括物质环境的重建也包括社会生活的改造（Carmon，1997；Holcomb & Beauregard，1981）。在这些城市复兴项目的实施中，政府开始融合私人部门和市场资源来获得必要的资金。地方政府的角色实现了从"先前单一的决策和执行部门到协调者、决策者和监督者三位为一体"的转变。

经济复苏和公私合营阶段（20世纪80年代早期~20世纪90年代早期）

在20世纪70年代，许多西方国家的国民经济遭遇了"停滞"阶段。在美国，这一时期居高不下的失业率和严重的通货膨胀，影响了家庭和企业的财务状况。20世纪80年代初，美国政府政策提议联邦政府减少在城市事务的干预，并鼓励私营部门增加投资。当然，在自由市场体系下，私营部门是热衷于投资城市复兴和更新活动的（Carmon，1997：137）。关于政府和私营部门之间的关系，郝克姆（Holcomb）和布里格（Beauregard）认为，在任何情况下，"只有在政府和私人开发商同时支持的情况下"，城市再开发才会实现（Holcomb & Beauregard，1981：12）。对美国城市的情况，司铎克（Stoker）表示，在20世纪80年代，很多市区规划忽视了公平问题，使得上层阶级和下层阶级的差距没有减少反而扩大了（Stoker，1989：160）。

在整个20世纪80年代和20世纪90年代早期，英国城市政策的特点是通过公私合营进行以房地产为主导的更新规划，其城市发展实践的特点是城市绅士化以及为了促进社

Economic revitalization and public-private partnership (early of the 1980s ~ early of the 1990s)

In the 1970s, the national economy of many Western countries underwent a period of stagnation. In the U.S. at the time, there was a high unemployment rate and double-digit inflation affecting the financial situation of households and businesses. Beginning early 1980s, the U.S. government policies proposed the diminishing involvement of the federal government in urban affairs, and private sectors were encouraged to increase their investment. Of course, within free-market institutions, private sectors were keen to invest in urban revitalization and renewal activities (Carmon, 1997: 137). Regarding the relationships between public and private sectors, Holcomb and Beauregard contended that urban redevelopment at any given time "seldom occurs without the simultaneous support of government and private developers" (Holcomb & Beauregard, 1981: 12). In the context of the U.S. cities, Stoker indicated that many downtown plans in the 1980s neglected the equity issue, and the disparities between the upper and low strata were even increased rather than reduced (Stoker, 1989: 160).

In the U.K., throughout the 1980s and early 1990s, urban policies were characterized by property-led regeneration through public-private partnerships, with the features of gentrification and rising land values for socio-economic development in urban development practices (Lees, 2003: 66). With the policy of "public subsidies, tax breaks, and the reduction in planning and other regulatory controls," urban regeneration mechanisms created a context to promote the investment of development companies (Imrie & Raco, 2003: 3). The logic behind this was that the government would generate the private investment, and the local communities and the whole society would then benefit from it through a "trickle-down" system (DOE, 1985). Nevertheless, with the benefits of property-

会经济的发展而提高土地价值（Lees，2003：66）。随着"公共补贴、减税、减少规划开支和其他监管控制"的政策的实行，城市更新机制为吸引开发公司的投资创建了环境（Imrie & Raco，2003：3）。这背后的逻辑是，政府将促进私人投资的生成，而地方社区和整个社会都将受益于"渗入式"机制（DOE，1985）。然而，伴随着以房地产主导的城市更新带来的利益，即，为地方社区提供就业机会、合适的培训和教育机会以及不断增长的税基（Turok，1992：377），它也导致了城市中不公平和贫困现象的日益加剧（Healey，1991：98）。

通过多方合作、全面的城市更新阶段（始于20世纪90年代早期）

20世纪80年代末，随着可持续发展战略的提出（WCED，1987），进行城市更新需要更彻底更全面的方法。多维化城市更新的目标，就是通过社会各部门的合作，实现自然、社会和经济方面的提高。例如，英国新时代的城市政策已经不仅仅停留在物质环境目标，而是倾向"关注社会融合、财富创造、可持续发展、城市管治、健康和福利、预防犯罪、教育机会和自由运动、环境质量以及好的设计方案"（Lees，2003：66）。1997年，随着工党政府的上台，社区参与、多边合作关系被视为一种有效解决社会分化和经济不平等的方法（Imrie & Raco，2003：7）。"合作关系"包含多层含义，它是一个多机构参与的方法，这种方法是基于"自下而上"的概念，即从社区咨询到社区参与到最终成为城市更新的主人的过程（Hughes & Carmichael，1998：205-206）。此外，为了提高政府部门的办事效率，政府和其所服务的社区之间建立紧密的关系

led urban regeneration, that is, the provision of employment and possible training and education opportunities to local communities and the increasing tax base (Turok, 1992: 377), it resulted in the growing inequality and poverty in the cities (Healey, 1991: 98).

Comprehensive regeneration concerns through multi-dimensional partnerships (beginning early 1990s)

With the advocacy for sustainable development in the late 1980s (WCED, 1987), the capacity of urban regeneration required more radical and comprehensive approaches (Carley & Kirk, 1998). Multi-dimensional urban regeneration goals, that is, physical, social, and economic improvements were proposed through the partnerships of all sectors in the society. For instance, urban policies in the new era in the U.K. went beyond the physical environmental objectives towards "concerns for social inclusion, wealth creation, sustainable development, urban governance, health and welfare, crime prevention, educational opportunity and freedom of movement, as well as of environmental quality and good design" (Lees, 2003: 66). With the Labour government coming to power in 1997, community engagement and multi-level partnerships were seen as an effective way to solve the problems of social fragmentation and economic inequalities (Imrie & Raco, 2003: 7). The concept of "partnership" encompassed more meanings, and it was considered a "more inclusive multi-agency approach which is premised on the 'bottom-up' notion of community consultation, involvement, and ultimately owners of the regeneration process" (Hughes & Carmichael, 1998: 205-206). In addition, to increase the effectiveness of the public sector, the smooth relationships between the government and the communities they served became more important than the loose relationships. In the U.K., a greater engagement of multi-dimensional sectors was seen as an effective way to achieve the goal (Campbell & Marshall, 2000: 321).

比松散的关系更加重要。在英国，多方部门更好的管理被视为达到这一目标的有效方法。

总结

通过对西方文献中的城市更新政策的简要回顾，可以发现城市更新政策实践的重点已经发生了转变，从20世纪50年代的注重物质环境的改善到目前更加全面的考虑，也就是，不仅仅只是关注自然环境的改善，同时也关注在经济发展、政府政策及可持续发展倡导影响下的社会、经济、文化、政治的繁荣发展。因此，政府对城市更新的期望也已经从追求物质环境的改善和最大化的经济回报转为了追求更全面的目标。基于上述文献回顾，表2-1总结了在不同时期，西方国家的城市更新实践的特点。

Summary

The brief historical overview of urban regeneration policies in Western literature shows that the focus of urban regeneration practices has transformed mainly from the improvement of the physical environment in the 1950s to the currently more holistic concerns, that is, the enhancement of not only the physical environment but also social, cultural, political, and economic prosperities under the influence of economic development, government policies, and the advocacy of sustainable development. Accordingly, government expectations from urban regeneration have also shifted from the improvement of physical environments and the maximization of economic returns to the achievement of more comprehensive objectives. Based on the above literature review, a brief summary of the characteristics of urban regeneration practices in the Western world of different periodization is revealed in Table 2-1.

西方城市的城市更新演变过程（二战之后到当前） 表2-1
Evolution of urban regeneration in Western cities (from post World War II ~ present) Table 2-1

时期 Phases	城市更新模式 Modes of Urban Regeneration	关键人物和利益相关者 Key Actors and Stakeholders	更新重点 Regeneration Emphasis
二战结束后~20世纪60年代 Immediate post World War II~1960s	"推土机"式的清除 The Bulldozer Era	国家政府主导的再发展 到20世纪60年代逐渐发展为力求政府部门和私营部门之间的平衡 National government-led redevelopment Gradual movement towards a balance between public and private sectors by 1960s	通常是基于总体规划进行"推土机"清理以及大规模清除旧的破败的地区 郊区发展 Bulldozing clearance and massive demolition of old, declining areas, often based on master plans Suburban growth
20世纪60年代~20世纪70年代 1960s~1970s	邻里修复 Neighborhood Rehabilitation	私营部门的作用和参与日渐增加，地方政府权力的下降 Growing role and participation of private sectors and the decentralization in local government	主要是在城镇边缘进行街区修复计划以及逐步改造 Neighborhood rehabilitation schemes and gradual integrative redevelopment primarily on the periphery of towns
20世纪80年代~20世纪90年代 1980s~1990s	在房地产为主导的发展中，经济复苏和公私合营关系 Economic Revitalization and Public-private Partnership in Property-led Development	强调私营部门和公私合营关系的发展 Emphasis on private sectors and growth of public-private partnerships	许多大规模以及著名的再开发项目促进了经济复苏 城外项目 Many large-scale and prestigious redevelopment projects contributing to economic revitalization Out-town projects
自20世纪90年代~目前 Since 1990s~Present	多维方法（为了经济、自然和社会更新的持续性和同步性）、全面的城市更新 Multi-dimensional Approaches (for sustainable and simultaneous economic, physical and social regeneration)	多机构（政府、私营部门和社区之间）的合作是主要的方法 Multi-agency partnerships (between state, private sectors, and communities) is the dominant approach	转向更全面的政策和实践形式 更强调综合对待和全面的方法 Moving towards more comprehensive forms of policies and practices More emphasis on integrated treatments and holistic approach

资料来源：(Campbell & Marshall, 2000; Carmon, 1997; Healey, 2006; Holcomb & Beauregard, 1981; Hughes & Carmichael, 1998; Ng & Tang, 2002a; Short, 1982; Turok, 1992; Western, Weldon, & Haung, 1973)

Source: (Campbell & Marshall, 2000; Carmon, 1997; Healey, 2006; Holcomb & Beauregard, 1981; Hughes & Carmichael, 1998; Ng & Tang, 2002a; Short, 1982; Turok, 1992; Western, Weldon, & Haung, 1973)

上述文献综述表明，在可持续发展的主题下，城市更新强调社会可持续性以及多方利益相关者的参与，主要包括地方城市更新项目中各个部门的意见，例如，公共部门、私营部门以及社区。然而，很明显，这些结论主要是在西方国家采取的政策和实践中得出的，具有特定的政治、经济背景。因此，这些理论讨论只在一定的范围内有助于对中国历史城市中的城市更新进行分析。在中国历史城市的城市更新项目中，一个突出的问题是如何在快速城市化的进程中保护当地的历史遗产。在亚洲很多发展中国家中，这对当地历史环境的保护和更新来说是一个严峻和紧迫的挑战（Stovel，2002：107）。第二部分回顾了在两个亚洲城市（京都和吴哥遗址）的以保护为主导的城市更新政策和实践，体现了当地的城市保护政策，同时也暴露了一些问题。公众意见的参与，特别是地方社区的参与，在历史城市环境保护项目中有很大的作用。

2.1.2 亚洲城市

一些亚洲历史城市的城市保护和更新政策和策略体现了以社区为基础的城市更新政策。一般来说，如果城市更新过程主要关注经济的发展，那么许多城市会采取以再发展为导向的更新政策，而这种政策会对当地历史城市环境产生重大影响。但是，在这些历史城市的发展政策中，城市保护一直是一个很重要的问题。地方城市保护政策往往更注重保护当地的遗址古迹，而忽视了对原住居

The above literature review implies that, under the rubric of sustainable development, social sustainability is underlined, and the participation of multi-stakeholders in urban regeneration significantly includes the opinions of various sectors in local urban regeneration programs, for example, the public sector, private sector, and communities. Nevertheless, it is obvious that these conclusions are mainly obtained based the policies and practices in the context of Western countries, which have a particular political and economic background. Therefore, these theoretical discussions can contribute to the analysis of urban regeneration in the context of Chinese historic cities at a limited scale. With regard to the urban regeneration programs in Chinese historic cities, one salient issue is how to conserve local historical heritage during rapid urbanization. This is actually an acute and pressing challenge to the conservation and regeneration of local historic environments in many Asian developing countries (Stovel, 2002: 107). The next section reviews the conservation-led urban regeneration policies and practices in two Asian cities, namely, Kyoto and the Angkor heritage site, which revealed local urban conservation policies and some problems as well. The involvement of public opinions, especially of the local communities, is especially important to the regeneration or conservation programs of historic urban environments.

2.1.2 Asian cities

The urban conservation and regeneration policies and strategies in some Asian historic cities shed light on the community-based urban regeneration policies. Generally, when economic development is a big concern in the local urban regeneration process, the regeneration policies in many cities promote the redevelopment-led approach, which has great influence on the local historic urban environments. Nevertheless, urban conservation has been an important issue in the development policies of these historic cities. Local urban conservation policies often stress more on the

民生活的保护。这些城市更新项目不仅揭示了当地城市发展面临的挑战，而且也强调了当地社区对实现整体更新目标的作用。

本书所选的两个亚洲案例分别是日本的京都和柬埔寨的吴哥遗址。作为日本的古都之一，京都历史城区内部的更新反映了许多由快速城市化带来的城市挑战。同时，吴哥遗址的保护政策响应了在联合国教科文组织议程公布的"活态遗产"和综合城市保护的概念。因此，这两个亚洲历史城市的整体城市发展目标和实践，反映了在地方更新项目中原住居民参与的重要性。

2.1.2.1 日本：快速城市化和"本造町运动"

二战后日本的快速城市化

战后日本的城市化已经非常显著。1945年开始，日本迅速的城市重建和经济发展产生了快速城市化现象。一方面，这与农村人口的转移有关，从之前被疏散到的农村转移到城市（Kuroda, 2005：384）。快速城市化率从1945年的27.8%提升到1955年的56.1%。同样，1975年，城市化率增加到75.9%，2000年增加到78.7%，2005年增长到86.3%（Zheng, 2008：57）。随着这一趋势，日本城市也出现了许多城市问题和挑战，如城市扩张、自然灾害、城市衰落和其他类似的问题。面对这些问题，日本政府一般采取了支持框架立法的形式和通过层次结构的规划直接干预城市发展的形式。

20世纪90年代初，日本城市规划，采取自上而下的方法，偏向于经济增长。尽管自20世纪90年代以来，当地的发展已经开始考虑城市发展的其他方面如社会和文化方面，

conservation of local heritage monuments, while indigenous lives are excluded. These urban regeneration programs not only reveal local urban development challenges but also underline the contribution of local communities to the holistic regeneration objectives.

The two selected Asian cases are Kyoto city in Japan and the Angkor heritage site in Cambodia. As one of the ancient capital cities in Japan, the regeneration of the historic inner urban areas of Kyoto reflects many urban challenges with the rapid urbanization. At the same time, the conservation policies in the Angkor heritage site reverberate with the notion of "living heritage" and integrated urban conservation, as promulgated based on UNESCO's agenda. Thus, the holistic urban development objectives and practices of the two Asian historic cities imply the significance of the involvement of local indigenous residents in the local regeneration programs.

2.1.2.1 Japan: Rapid urbanization and "machizukuri" movement

Rapid urbanization in Japan beginning the post World War II

The post-war urbanization in Japan has been very remarkable. Beginning 1945, rapid urban reconstruction and economic development brought about rapid urbanization in Japan. On one hand, this is related to the shift of population from the rural, from where they had been evacuated, to the urban areas (Kuroda, 2005: 384). The rapid urbanization rate ascended from 27.8 percent in 1945 to 56.1 percent in 1955. The rate likewise increased to 75.9 percent in 1975, to 78.7 percent in 2000, and to 86.3 percent in 2005 (Zheng, 2008: 57). With this trend, many urban problems and challenges also arise in Japanese cities such as urban sprawl, natural hazards, urban decline, and other similar problems. In facing these problems, government policies in Japan have generally taken the form of supporting framework legislation and direct intervention to urban development through the hierarchy levels of planning (Zetter, 1994: 26).

By the early 1990s, urban planning in

经济增长仍然在当地发展目标中占据了最大的份额。在地方政府推行层次结构的规划期间，当地规划主要遵循《城市规划法》和《建设标准法案》，这些是用来规范日本城市土地利用的国家法律（Callies，1994：60）。值得注意的是，自20世纪90年代末以来，人们已经关注一个新的概念"本造町运动"，这是一个以社区为基础的城市发展战略，以实现整体城市发展目标。

"本造町运动"：日本市民参与地方环境管理和管治的运动

与日本之前自上而下的城市规划体系相比较，在20世纪90年代末推行的"本造町运动"，以市民参与地方管理为特色（Sorensen & Funck，2007b：2）。正如渡边（Watanabe，2007：44）认为的，自上而下的日本城市规划系统在二战后经济快速发展时期起了主导作用。但在20世纪60年代，人们开始意识到，由一小群市民和专家基于旧的城市规划体系所提出的规划，很难说服大多数市民和土地主；因此，很多居民组织和社会运动便应运而生了[1]（Watanabe，2007：44）。

20世纪80年代引进的区规划[2]，打算通过引进"小区域"原则或者"麻吉（社区）"来促进旧城市规划体系和本造町体系之间的互动（Watanabe，2007：51）。为了促进居民参与到"本造町运动"中，一些城市颁布了"本造町条例"，如1981年神户和1982年东京

---

[1] 神户丸山的居民运动，是居民通过社会运动以及和地方政府进行艰难的谈判成功实现他们的一些要求的一个很好的例子（在（Hirohaya，1989）可见更多资料）。

[2] 根据渡边（Watanabe）的定义，区规划是"一个系统，其中市政府在一个相对较小的区域内实行详细规划并进行土地利用控制"（Watanabe，2007：51）。

Japan featured the top-down approach and the bias for economic growth. Although other perspectives of urban development such as social and cultural aspects had begun to be considered in local development since the 1990s, economic growth still took the lion's share in local development aims. During the implementation of the hierarchy levels of planning at the local government levels, the local plans mainly followed the requirements of the Urban Planning Act and the Construction Standards Act, which are national laws to regulate the land use in Japanese cities (Callies, 1994: 60). Notably, since the late 1990s, a new concept of "machizukuri" has attracted the people's attention as a community-based urban development strategy towards the achievement of holistic urban development goals.

Machizukuri: Citizen engagement in the local environmental management and governance in Japan (Nunokawa, 2007: 172)

Compared with the previous top-down city planning system in Japan, machizukuri, implemented beginning the late 1990s, was characterized by citizen engagement in place of management (Sorensen & Funck, 2007b: 2). As Watanabe (2007: 44) argued, the top-down Japanese city planning system played a dominant role during the postwar rapid economic development period. In the 1960s, people began to realize that the planning produced by a small group of citizens and experts, based on the old city planning system, could hardly convince the majority of the citizens and landowners; thus, many residential organizations were formed and social movements were carried out (Watanabe, 2007: 44)[1].

The District Planning[2] introduced in 1980 intended to bridge the interaction

---

[1] The residential movements in Maruyama, Kobe city, were a good example for the residents to successfully obtain some of their demands through social movements and many difficult negotiations with local governments (see more on the social movements in Maruyama at (Hirohara, 1989)).

[2] As Watanabe indicates, District Planning is "a system where municipality carries out detailed planning and land use controls in a comparatively small area as a unit" (Watanabe, 2007: 51).

世田谷区颁布的"本造町条例"(Watanabe, 2007：52)。据小林(Kobayashi)称，在20世纪90年代后期，为了引起居民广泛的兴趣，补充城市规划法律，日本几乎所有的市政当局都颁布了"本造町法令"(Kobayashi, 1999：3)。伴随着"小区域"原则的引入，1992年城市修改法也提出了"参与"原则，该原则要求所有的市政当局要通过公众参与的方式制定总体规划(Murayama, 2007：206)。所有这些政策都极大地促进了以社区为基础的城市规划进程。

本书接下来将以一个日本典型的历史城市——京都来进一步讨论基于社区的城市更新实践。随着快速城市化的进程，这座历史城市面临的一个大的挑战是当地倾向经济发展的政策带来的影响。一些学者认为在应用日本之前的城市规划之前，应该对其做一些改变或一些限制(Mimura, Kanki, & Kobayashi, 1998：39)。因为无节制的城市再开发和空间重组几乎使京都成了一个"濒临灭绝的历史城市"(Ryoichi, 2003：367)。因此，必须平衡当地发展和历史城市的保护两者之间的关系，而以社区为基础的发展政策恰恰阐明了这一点。

以日本古都：京都为例

1) 城市建设与扩张

日本迁都到京都以后，京都(时称"平安")在794～1868年大约1000年的时间里一直是日本的都城。京都建于794年，其城市形态模仿了当年唐代中国都城长安(今西安)。平安城建立之初，其唯一的作用就是作为国家政策的仪式中心没有经济目的。在很长一段时间内，古都的转换与政治权力的变迁息息

between the old city planning system and machizukuri system by way of introducing the "small area" principle or "machi (community)" (Watanabe, 2007: 51). In order to facilitate the machizukuri activities of the residents, some cities enacted machizukuri ordinances such as the machizukuri ordinance in Kobe city in 1981 and Tokyo's Setagaya Ward in 1982 (Watanabe, 2007: 52). According to Kobayashi, in the late 1990s, nearly all municipalities in Japan enacted machizukuri ordinances to arouse wider interest from the residents and to supplement the City Planning Law (Kobayashi, 1999: 3). Accompanying the introduction of the small area principle, the participation principle was also proposed in the City Planning Amendment Law of 1992, which required all the municipalities to prepare their Master Plans through public participation (Murayama, 2007: 206). All these policies strongly facilitated community-based city planning processes.

In the next part, this book will further discuss the community-based urban regeneration practices in Kyoto, a typical Japanese historic city. With the rapid urbanization, one big challenge facing this historic city is the influence from local pro-growth development policies. Some academics argue that there should be some changes or "limitations to applying Japan's former concepts of town planning" (Mimura, Kanki, & Kobayashi, 1998: 39) because the unchecked urban redevelopment and spatial restructuring have almost brought Kyoto to "the verge of extinction as a historic city" (Ryoichi, 2003: 367). Therefore, it is imperative to achieve a balance between local development and conservation for this historic city, a balance where the community-based development policy shed light on.

The case of Kyoto: The ancient capital of Japan

1) A brief overview of urban construction and expansion

Kyoto (then named Heian-Kyo) had been the capital city of Japan for approximately 1,000 years from 794 A.D. to 1868 A.D., when

相关(Mimura, et al., 1998: 41)。明治时期始于1868年，随着工业产品、军事活动、高等教育及商业活动的发展，京都和日本其他城市快速成长。这座城市逐渐向外扩张到新的外围区。到20世纪90年代后期，通过京都的城市扩张，大型的农业地区转变为郊区。

2）城市保护与发展管理结构

二战结束后，在京都城市的发展中，就其历史城区环境的发展和保护，提出了多层次的关注。相应地，颁布了一些法律法规，通过保护和开发历史文化资源和旅游设施，将京都开发成文化旅游城市。这些法律法规包括1950年颁布的《京都国际文化和旅游城市建设法》，1966年颁布的《历史景观保护特别法》(Callies, 1994: 67, 1997: 150)，1971年颁布的《开发与保护政策》(Masafumi & Waley, 2003: 348)，1995年颁布的《京都城市景观规范制度》以及《自然景观保护条例》(Ryoichi, 2003: 373)。

3）挑战：效率低下的保护政策和以开发为主导的体制(Masafumi & Waley, 2003: 366; Ryoichi, 2003: 374)

虽然上述保护法律法规极大地保护了历史城市的环境，但就像很多亚洲城市遇到的问题一样，京都快速的城市发展给其历史城市环境带来了很多挑战和破坏，萨拉提(Salastie)表示这些挑战或者破坏主要来自两个方面："城市规划的负面影响以及缺失合理的城市保护政策。"(Salastie, 1999: 9)。但实际上，快速城市化带来的历史环境破坏是如此的惊人，以至政文(Masafumi)声称说，尽管这些历史城市在二战时幸免于难，但是它们却在二战之后的半个世纪里消失了

the capital was moved to Tokyo (Callies, 1997: 150). When Kyoto was built in 794 A.D., in the urban form, it followed that of Chang'an city (today's Xi'an), which was then the Chinese capital of the Tang dynasty. When Heian-Kyo was first built, its sole role was to serve as the ceremonial centre for national policy without economic intentions. For a long time, the transformation of the ancient capital had been closely related with the vicissitudes of the political powers (Mimura et al., 1998: 41). When the Meiji period began in 1868, Kyoto grew quickly as other Japanese cities did with the development of manufactured products, military campaigns, higher education, and commercial activities. The city gradually expanded outward to the new peripheral zones. By the late 1990s, the urban expansion of Kyoto transformed large farming areas into suburban districts (Mimura et al., 1998: 42).

2) Regulatory structure for urban conservation and development

In the immediate post-World War II, multi-levels of concern in the development of Kyoto city, as a historic capital, were proposed regarding both the development and conservation of its historic urban environments. Accordingly, some laws and regulations were enacted to develop Kyoto into a city of culture and tourism by conserving and developing its historical and cultural resources and tourist facilities. These laws and regulations included the Kyoto International Cultural and Tourist City Construction Law in 1950, the Special Law for the Preservation of Historical Landscapes in 1966 (Callies, 1994: 67, 1997: 150), the Development and Conservation Policy in 1971 (Masafumi & Waley, 2003: 348), the Kyoto System for the Regulation of the Urban Landscape, and the Natural Landscape Preservation Ordinance in 1995 (Ryoichi, 2003: 373).

3) Challenges: Ineffective conservation policies and development-oriented system (Masafumi & Waley, 2003: 366; Ryoichi, 2003: 374)

Although above-mentioned conservation laws and regulations contributed much to the preservation of its historic urban environments,

(Masafumi & Waley, 2003: 347)。良一（Ryoichi）坦率地说：" 京都能否保持其著名历史城市的地位，还有待商榷"（Ryoichi, 2003：367）。柯尼（Keane）直接指出了对京都町屋的历史木质建筑所造成的巨大破坏。

"调查显示，仅在1978年和1988年之间，就有5万个木质建筑消失了——10年之间平均每天就有13个木质建筑被摧毁（Keane, 2000：77）。"

虽然在历史遗址保护方面有很多法律法规，但是这些法规存在明显的缺陷，并且在保护历史建筑和风景地貌方面也有明显不足。例如，保护法规不适用于一些拥有传统建筑的地区，即町屋。另外，管理这些地区的法规内容非常模糊[1]（Ryoichi, 2003：374）。由于缺乏有效的保护系统，城市中心地区的历史建筑环境已经遭到了很多破坏。但是，一些学者（Callies, 1997；Masafumi & Waley, 2003；Ryoichi, 2003；Waley & Fiévé, 2003）争论说造成这些破坏更根本的原因是，京都的城市规划采取了通过建设大型建筑的措施来实现历史城区的现代化。在城市现代化的过程中，为了迎合当前建筑标准法的消防要求，无论是传统的木制建筑物还是现有的狭窄的街道在建设和保护方面都存在问题（Masafumi & Waley, 2003：364）。许多传统的当地乡土建筑没有被列为官方文化遗产，而且也没包

---

[1] 例如，一些法规只是要求"开发商高于20米的规划建筑……必须以官方文件请示市长办公室，并尽力避免使用那些和周边城镇景观不相称的花哨颜色或装饰"（Ryoichi, 2003：374）。

the rapid urban development of Kyoto impose many challenges and damages on its historic urban environments similar to the problems encountered by many Asian cities. Salastie argued that these challenges or damages mainly came from two aspects: "the threatening effects of town planning as well as the lack of appropriate urban conservation policies" (Salastie, 1999: 9). In fact, the damages to historical environments because of rapid urban development are so tremendous that Masafumi indicated that while the historical buildings escaped from World War II, they have been lost in the last half-century since then (Masafumi & Waley, 2003: 347). Ryoichi bluntly pointed out that "Kyoto's ability to maintain its status as a great historic city is in question" (Ryoichi, 2003: 367). The massive destruction to the historic wooden structures in Kyoto, *machiya*, was pointedly singled out by Keane,

A survey shows that 50,000 wooden structures were torn down between 1978 and 1988 alone — and average of 13 demolitions a day, every day, for ten full years (Keane, 2000: 77).

There are various legislations and regulations on the conservation of historical heritages, but the defects in these regulations and their insufficiencies in conserving historical buildings and landscapes are also obvious. For instance, the conservation regulations do not apply to some areas with many traditional buildings, that is, machiya. In addition, the regulations governing these areas are very vague[1] (Ryoichi, 2003: 374). The lack of effective conservation system has resulted in many damages to the historically built environments in the urban central areas. Nevertheless, some academics (Callies, 1997; Masafumi & Waley, 2003; Ryoichi, 2003;

---

[1] For example, some regulations only require that "developers planning buildings over twenty meters ... must file an official notice with the mayor's office and that they do their best to refrain from using garish colors or decorations that would not fit in with the surrounding townscape" (Ryoichi, 2003: 374).

括在1993年的联合国教科文组织的世界遗产名录中。因为这个原因，这些当地乡土建筑迅速消失（Waley & Fiévé, 2003: 32）。城市改造和调整已经显著改变了城市建筑的日益恶化和公共设施不足的现象[1]。当这些改造活动以追求新的经济增长和新建筑为目标时，以发展为导向的规划就对已经确立的保护目标带来了挑战（Callies, 1997: 136）。

4) 对策：以社区为基础的保护和发展政策

为了解决京都历史内城区快速城市化所带来的问题，采取了三个措施，旨在形成一个更合适的城市管治和开发控制机制。第一个是通过组织城市居民以社区参与的方式（本造町运动）参与到当地城市发展中。第二个是引入一个附加的详细分区制和地区规划系统，这个系统是通过地方居民和政府的合作产生的（Mimura et al., 1998: 50; Watanabe, 1994: 432）。第三个是修正政策和提高地方政府政策水平。例如，遗产税条例已经得到了修改和调整，以前它的作用就是阻止原居民住在他们原有的房子里，但是现在，这项条例在当地发展进程中提供了更多的遗址保护功能（Keane, 2000: 79, 213）。

这些措施最显著的特点就是采取地方居民组织的形式让社员参与到城市规划体系中。居民参与的组织有京都市政厅复兴组织、京

Waley & Fiévé, 2003) contend that more fundamental causes behind the damages lie in the tendency of city planning in Kyoto to modernize historic areas through the construction of large buildings. In the process of urban modernization, both the traditional wooden buildings and existing narrow streets are problematic in terms of construction and conservation to meet the fire fighting requirements of the current Building Standards Law (Masafumi & Waley, 2003: 364). Many traditional local vernacular architecture are not listed as official cultural heritages and are not included in UNESCO's 1993 world heritage list. For this reason, they are disappearing very rapidly (Waley & Fiévé, 2003: 32). Urban renewal and readjustment have been significant tools to revamp the urban areas with deteriorating buildings and insufficient public utilities[1]. When these renewal activities are aimed at new growth and new constructions, the development-oriented planning forms a big challenge to locally established conservation intentions (Callies, 1997: 136).

4) Countermeasures: Community-based conservation and development policies

In order to solve the problems accompanying rapid urbanization in the historic inner urban areas of Kyoto, three measures have been taken towards a more suitable urban governance and development control mechanism. The first is by means of community participation (Machizukuri Kensho) in local urban development policies through the organizations of the town residents. The second is the introduction of an additional

---

[1] 一般而言，内部城区中，关于社区文化特征以及现存的城市布局保护方面的城市问题包括以下四个方面（Mimura et al., 1998: 50）：（1）建筑群质量退化；（2）传统建筑被现代化建筑取代；（3）交通堵塞和社区解体引起的环境质量恶化；（4）人口数量减少。另外，正如研究者所说，由于之前表演这些节目的居民的离开，一些传统的每年表演的节目消失了。

[1] Generally, These urban problems of the inner urban area, regarding the preservation of the character of both the communities' culture and existing urban layout, include the following four aspects (Mimura et al., 1998: 50): (1) the deterioration of the quality of building stock; (2) the replacement of traditional buildings by modern structures; (3) the deterioration of environmental quality because of excess traffic and the dissolution of neighborhood community; (4) the decrease in the population. Besides, as Mimura argues, some traditional staging annual festivals have been disappearing, because the communities who previously conduct these festivals have left.

都模拟社团[1]、京都国际艺术文化交流委员会（简称KICACE）以及其他组织（Ryoichi，2003：382）。通过举办类似"音乐会、现代艺术展览和讲座"的活动（Ryoichi，2003：383），这些地方组织鼓励地方居民向城市保护和社区建立的目标发展。这样一来，地方居民之间的关系得到了巩固，曾经一度濒临灭绝的社区现在也开始复苏了。同时，在兴建新建筑的时候，要综合考虑地方居民的知识和经验以及传统住房的原则（Mimura et al.，1998：50）。地方居民参与的活动和对一些无人居住的历史建筑的适当重新使用，不仅复兴了地方文化和社区生活，同时也通过旅游业带来了经济效益。通常情况下，这些活动都得到了地方居民的大力支持。同时，京都市政府对其历史环境的保护和复兴的态度也有了很明显的变化（Ryoichi，2003：383）。为了促进历史城区的发展，国家和地方都颁布了一系列的法律法规，如，1994年颁布的《建筑标准法》（Callies，1997：139）、1995年颁布的《有关城市景观规定的京都体系》（Ryoichi，2003：373）以及1996年颁布的《有关城市景观规定的条例》（Masafumi & Waley，2003：351）。

总结日本城市的经验：以地方社会资本为基础，社区参与城市管治

以社区为基础的保护和发展策略以及实践，极大地促进了当地以保护为主的城市发展政策的制定。正如所罗森和唐克（Sorensen & Funck，2007：269）指出的，社区参与可以让城市，尤其是历史城区更宜于居住，更

---

[1] 良一认为，京都模拟社团之前也被称为拯救京都国际社团。（Ryoichi，2003：382）。

detailed zoning and a district planning system, which are produced through the partnership of local residents and authorities (Mimura et al., 1998: 50; Watanabe, 1994: 432). The third is the revision and improvement of local government policies. For example, the inheritance tax policy has been revisited and adjusted. It previously served as an obstacle for original residents to live in their original houses but now provides more heritage conservation possibilities in the context of local development (Keane, 2000: 79, 213).

In these measures, community participation in the city planning system through local residents' organizations is a salient characteristic. The many local residents-involved organizations include the Kyoto Townhouse Revitalization Society, the Kyoto Mitate[1], the Kyoto International Committee of Art and Cultural Exchange (KICACE), and other organizations (Ryoichi, 2003: 382). These local groups play the role of mobilizing local residents towards the goal of urban conservation and community build-up (Keane, 2000: 213) through such activities as "concerts, contemporary art exhibitions, and lectures" (Ryoichi, 2003: 383). This way, the relationships among local residents are intensified, and the communities, once on the verge of dying out, begin to be revived. At the same time, when there are new structures to be built, the knowledge and experiences of local residents and traditional housing principles are integrated (Mimura et al., 1998: 50). These local residents-engaged activities, as well as the adaptive reuse of some uninhabited historical buildings, not only revive local cultural and community lives but also bring economic returns through tourism industries. Normally, these activities are strongly supported by local residents. At the same time, the Kyoto city government's attitude also has many palpable changes towards the conservation and revitalization of its historical environments (Ryoichi, 2003: 383). Some laws and regulations

---

[1] According to Ryoichi, Kyoto Mitate was formally known as the International Society to Save Kyoto (Ryoichi, 2003: 382).

具有可持续性，因为一旦这些历史环境开始消失，地方居民会明显感觉到。在现实生活中，"很多日本人和自己的邻居们在同一个项目中工作时，表现出极大的意愿和才华"，因为他们意识到，"一个美好的城市对人们生活的质量有很大的影响"（Sorensen & Funck, 2007a：277）。总结地方居民对城市发展政策的认识能够极大促进对历史城市环境的保护。研究进一步表明，地方居民之间的关系以及地方政府和社区之间的关系对地方社区的参与有很大影响。这反映了社会资本在历史城市环境保护和发展中的影响。

石田（Ishida）认为在城市居住环境和建筑环境的改善中，两个方面起到了至关重要的作用：基于地方居民合作关系的社区参与和本市周密的规划和行政工作（Ishida, 2007：130）。这表明社区联系和基于地方社会资本的组织能够促进他们参与到当地的保护项目中。同时，周密的规划和管理机制是很重要的，它强调在保护和发展项目中地方社区之间真正的合作的重要性。因此，日本城市规划体系要进行实质性的转变，改善地方城市管治模式以及市民和地方政府之间的关系是至关重要的（Sorensen & Funck, 2007a：272）。但是，如果城市规划体系没有采纳地方社区或组织的意见，那么他们关于历史保护的意见和目标在预想的总体规划中就没有多大影响了。

2.1.2.2 吴哥遗址：对"活态遗址"的保护

通过对吴哥遗址案例的分析，进一步探讨地方社区参与在历史城区保护中的重要性。按照"活态遗址"的概念，将原住民的生活

at both the national and local levels have been enacted to instigate the harmonious development of the historic urban areas, such as the 1994 Building Standard Law (Callies, 1997: 139), the 1995 Kyoto System for the Regulations of the Urban Landscape (Ryoichi, 2003: 373), and the 1996 Ordinance for the Regulation of the Urban Landscape (Masafumi & Waley, 2003: 351).

Summary of experiences from Japanese cities: Community engagement in urban governance based on local social capital

The community-based conservation and development strategies and practices strongly promote local conservation-led urban development policies. As Sorensen and Funck (2007a: 269) indicated, community participation can contribute to more livable and sustainable cities, especially in the historic urban contexts because local residents will show an increasing awareness once the historic surroundings begin to disappear. In reality, "many Japanese people demonstrate an enormous willingness and talent for working together with their neighbors in common projects" because they realize that "good places can make a great difference in people's quality of life" (Sorensen & Funck, 2007a: 277). The integration of local indigenous knowledge in the urban development policies significantly contributes to the conservation of historic urban environments. Studies further indicate that the relationships among local residents and the relationships between local governments and communities have a substantive influence on the participation of local communities. This implies the impact of social capital on the conservation and development of historic urban environments.

Ishida eloquently argued that two perspectives play a critical role in the improvement of urban living conditions and built environment: community participation based on the cooperative relationships among local residents and careful planning and administrative efforts of the city (Ishida, 2007: 130). This indicates that community connections and organizations based on local social capital contribute to their engagement

融合到地方保护项目中，已经列入了联合国教科文组织整体城市保护的议程中，下面章节将会对此进行讨论。

吴哥遗址和联合国教科文组织的保护计划简介

吴哥位于柬埔寨暹粒省的北部，是古代高棉帝国的首都（9～15世纪）。自1989年以来，联合国教科文组织就开始参与这个遗产的保护了。如今，在这个5000多平方公里的地方，存在着宗教遗迹、防御工事、定居的山岗和其他由古高棉人建造的建筑（Wager，1995：515）。1992年12月，吴哥中心地区被列入世界遗产名录，该指定区域的占地面积约401平方公里，该区域有森林、纪念碑、80多个村庄，总人口8万余人（Miura，2005：3）。

一般来说，该区的大部分地方村民都非常贫穷，人口增长速度很快。他们以当地农业、渔业和林产品为生。因此，由于过度开采，当地历史环境遭到了破坏（Wager，1995：520）。为了实现经济的可持续发展，当地城市保护和更新规划应该考虑当地的经济结构。另外，地方旅游业的发展可以为农村人口提供很多就业机会。在这方面，一些地方政府的政策得到了地方居民的支持。例如，吴哥保护办公室招聘维护和修复工作人员时，就会优先考虑保护区内的当地居民，以便让他们赚取额外的收入。而且，这可以减少对农业资源和森林资源的集中使用（Wager，1995：522）。此外，让当地居民参与吴哥历史遗址的管理，能够在当地城市保护进程中融入原住民的努力。实际上，地方社区参与保护规划，被许多专家学者视为促进整体城

in local conservation projects. At the same time, the careful planning and management mechanism is important, which underlines the genuine incorporation of local communities in the conservation and development programs. Therefore, the improvement of local urban governance and the relations between citizens and local governments is critical for a real paradigm shift in the Japanese urban planning system (Sorensen & Funck, 2007a: 272). However, when the opinions of local communities or organizations are not considered in the planning system, their suggestions and intentions on historic conservation will have little influence on the preconceived master plan.

2.1.2.2 Angkor heritage site: The conservation of the "living heritage site"

The significance of integrating local communities in the conservation of historic urban areas is further explored through the case of the Angkor heritage site. The integration of indigenous lives into the local conservation programs in terms of the concept of "living heritage site" has been on UNESCO's agenda of integrated urban conservation, which will be discussed in the next section.

Introduction to the Angkor site and the UNESCO's conservation program

Situated in the Siem Reap province of northern Cambodia, Angkor was the capital of the ancient Khmer Empire (9th to 15th century). Since 1989, UNESCO began to be involved in the protection of this heritage site. Today, in an area of more than 5,000 km^2, the remains of religious monuments, fortifications, settlement mounds, and other structures built by the ancient Khmers can be found (Wager, 1995: 515). In December 1992, the central part of Angkor was part of the World Heritage List including the designated site covering an area of approximately 401 km^2 made up of forests, monuments, and more than 80 villages with a combined population of over 80,000 (Miura, 2005: 3).

Generally, most local rural residents are very poor, and the population is growing very

市保护政策的实质性方法（Fletcher, Johnson, Bruce, & Khun-Neay, 2007；Miura, 2005）。

吴哥遗址和原住居民之间的关系

吴哥遗址作为一个居住城市，直到1992年才被列入世界遗产名录，因此对该遗址理解的转变，在保护、经济发展和社会公平方面引发了很多挑战和冲突。弗莱彻认为（Fletcher, 2007：387），目前柬埔寨政府把旅游业看作是提高本国社会经济的重要手段，并且吴哥遗址的发展已经成为政府政策的重中之重。因此，在当地管理中，如何对待遗址中现存的农村生活显得很是棘手。通过分析吴哥遗址对地方村民的重要性以及地方村民是如何利用当地资源生活的，苗拉在两个方面探讨了吴哥遗址和地方村民之间的错综复杂的关系。首先，由于地方居民和古庙与寺院有着宗教和精神上的关系，他们与吴哥保持着亲密的联系。例如，许多寺庙的僧侣和寺院雕像的看护人都来自附近的村庄。第二，很多寺庙和寺院附近的森林被视为"村庄的公共遗产"。例如，不同地段的树木，都是由个人家庭来管理，生产树脂，然后将其出售来补充家用（Miura, 2005：7）。因此，无论是在精神上还是生活上，地方居民与历史遗址都有非常密切的关系。

当前城市保护政策面临的挑战：将原住民隔离在遗址之外

在历史上，自20世纪90年代以来，由于各种原因，当地居民被逐出吴哥遗址，搬迁到其他地方（Miura, 2005：9）。随着地方政府在2000年4月实施禁令，许多当地的活动都因保护和促进旅游业的发展的名义而被禁止（Miura, 2005：11）。苗拉强烈地认为当

rapidly. Their livelihood depends mainly on local agriculture, fisheries, and forest products. As a result, the local historic environments suffer from over-exploitation (Wager, 1995: 520). In order to achieve sustainable economic development, local urban conservation and regeneration plans should take local economic restructuring into consideration. In addition, by virtue of the local tourism industry, many employment opportunities should be created among the rural population. In this regard, several local government policies are supported by the local residents. For instance, when the Angkor Conservation Office employed people for maintenance and rehabilitation works, the local residents in the protected area were given the priority in employment to earn additional income. In addition, this may reduce the local intensive use of agricultural and forest resources (Wager, 1995: 522). Further, the participation of local residents in the management of the Angkor historic site is a significant way to integrate indigenous efforts in local urban conservation processes. In fact, the engagement of local communities in the management of the conservation planning is seen as a substantive way to promote the integrated urban conservation policy by many academics (Fletcher, Johnson, Bruce, & Khun-Neay, 2007; Miura, 2005).

Relationships between the Angkor heritage site and indigenous lives

Since the Angkor site as a living city was not included in the World Heritage list until 1992, the transforming interpretation of the heritage site poses many challenges and conflicts in the conservation, economic development, and social equity aspects. According to Fletcher (2007: 387), the current Cambodian government views the tourism industry as critically important to the improvement of the country's socio-economic situation, and the development of Angkor has been central to the government policy. Therefore, how to treat local existing rural lives in the heritage site also becomes a critical problem in local management. By demonstrating what Angkor means to the local villagers and how local villagers live their everyday lives by using local resources, Miura explored the intricate

地居民不应该被隔离在吴哥遗址外，对该历史遗址进行全面了解以及建立一个参与管理机制是很有必要的。正如苗拉所指出的：

"历史的价值可以被欣赏、维持和支持，而这只有通过那些已经意识到历史价值并且急切地把这些传授给后代的人才能实现，同时他们也能意识到持续性变化的必要性，以及能够适应可持续发展的可能性（Miura，2005：15）。"

这一观点获得了弗莱彻的支持（Fletcher，2007：386）。他表示，正因为柬埔寨居民知道有关吴哥的历史和故事，所以他们是活态遗址的宝贵的无形资源。有研究者认为"地方居民会远比那些总是坐在会场里面的人更了解自然资源的价值"，同时"即便一些当地居民由于非法伐木或钓鱼而破坏了自然环境"，适当的解决方法应该是提出替代选择，而不是简单地禁止这样的活动（Khouri-Dagher，1999：11）。事实上，据联合国教科文组织的经验，如果原住民社区消失了，地方保护工作会变得更加困难。因此，一个明确的目标应该是提高当地居民自己保护和管理遗址的能力（Khouri-Dagher，1999：11）。

与此同时，很显然，由于缺乏必要的保护和开发控制机制和法规，可能会发生不同程度的有害活动。例如，虽然对吴哥遗址的当地村民有很多严格的禁令，但正是当地遗产政策在多方面违反了这些禁令，比如村民贿赂遗址区的警察，以便进行放牧、伐木取火、照看佛像以及其他活动（Miura，2005：11）。实际上，主要是出于生存安全考虑，许多贫

relationships between the Angkor site and local villagers mainly in two ways. First, local residents have a close association with Angkor through their religious and spiritual relationships with the ancient temples and monasteries. For instance, many monks in the temple and caretakers of the statues of the monasteries are from nearby villages. Second, the forests adjacent to many temples and monasteries are seen as the "communal heritage of local villages." For instance, some trees in the different sections are managed by individual families to produce resin, which then will be sold to supplement the families' income (Miura, 2005: 7). Therefore, local residents actually have a very close relationship with the historic site through their everyday lives both spiritually and physically.

Challenges of existing urban conservation policies: Excluding indigenous residents from the heritage site

Historically, since the beginning of the 1900s, local inhabitants were excluded from and relocated out of the Angkor site to other locations for various reasons (Miura, 2005: 9). With the bans implemented by the local government in April 2000, many local activities have been prohibited in the site in the name of conservation and tourism promotion (Miura, 2005: 11). Miura maintained that local residents should not be excluded from the site, and a comprehensive understanding of the historic site and a participatory management mechanism are necessary. As Miura comments:

The values of the past can be appreciated, sustained and upheld only through those people living in the present who recognize and are eager to transmit them to future generations, while recognizing the need for, and accommodating the possibility for, sustainable change (Miura, 2005: 15).

This viewpoint was supported by Fletcher (2007: 386) when he articulated that the presence of the Cambodian residents, with their experiences and stories about Angkor, is an invaluable intangible resource of living

困地区的居民也愿意贿赂警察，让他们继续从事当地所禁止的活动。

总结

简而言之，吴哥城的案例说明了原住民的生活对历史城市环境的保护有重大意义，因此，在地方政府鼓励保护遗址和发展旅游业时，他们不应该简单地把原住民分离或隔离在文化遗址之外。除了遗址保护和旅游开发之间存在的冲突，案例进一步表明，由于缺乏一个有效的监控和管理机制，地方保护工作常常被迫妥协。

如今，在诠释世界遗产时，"活态遗产"被视为当地遗址不可或缺的一部分，从而突出了原住民生活的重要性。这个观点得到许多学者的支持（Egloff & Newby, 2005; Filippi, 2005; Takaki & Shimotruma, 2003），并且也在许多国际文件中有所强调（UNESCO, 2003, 2007）。在全面保护当地历史遗址或综合城市保护方面，讨论中强调了两个因素。首先，地方居民和历史环境之间的关系或联系对当地"活态遗址"的保护有一定的影响。这种联系在地方居民的日常生活中促进了地方居民之间的关系或联系。因此，对当地历史环境的保护也和社会资本有着密切的联系。第二，以社区为基础的保护政策在保护地方原住民的生活的进程中是必不可少的。吴哥城的案例说明，建立有效的保护和发展控制机制以及管理系统是非常重要的，这样可以保证地方居民真正参与到保护进程中，以及确保已有的保护政策顺利进行。

通过对西方文献中不同时期城市更新的演变过程以及两个亚洲城市的城市更新政策

heritage. Natarajan Ishwaran argued that "[l]ocal people know the value of nature more than people sitting in meetings" and that "[e]Even if some locals harm the environment with such practices as illegal tree-felling" or fishing, the appropriate solution to this should be the suggestion of alternatives rather than simply banning such activities (Khouri-Dagher, 1999: 11). In fact, according to UNESCO's experiences, when the indigenous community is gone, local conservation work often becomes even more difficult. Therefore, one conspicuous aim should be to boost the ability of the local people to conserve and manage the heritage site by themselves (Khouri-Dagher, 1999: 11).

At the same time, it is obvious that with the lack of the necessary conservation and development control mechanisms and regulations, many harmful activities may exist at various levels. For instance, while there are many strict bans imposed upon the local villagers in the Angkor site, it is the local heritage policies that often violate such bans in ways such as the bribery to the heritage polices by local villagers in order to graze cattle, collect firewood from fallen trees, look after statues, and other activities (Miura, 2005: 11). In fact, many impoverished villagers are willing to bribe the police for the continuation of locally banned activities mainly for the reason of subsistence security (Miura, 2005: 11).

Summary

In a nutshell, the case of Angkor city illustrates that indigenous lives have significant contributions to the conservation of historic urban environments, and they should not be simply separated or excluded from the heritage site when heritage conservation and tourism development are promoted by the local government. In addition to the conflicting demands of heritage conservation and tourism development, the case further reveals that local conservation efforts often become compromised due to the lack of an effective monitoring and management mechanism.

Today, in the interpretation of world heritage, living heritage is seen as an indispensable part of the composition of the local heritage site,

和实践的概述,可以清楚地看到,地方政府机关在不同的城市背景下,对历史城市的更新起到了非常重要而独特的作用。同时,在几乎所有的研究中,经济发展问题都是影响城市更新或保护目标的关键因素。在倾向经济增长的政策中,历史城市的保护面临着来自快速城市化或偏向旅游发展的挑战。虽然地方社区在可持续发展和全面城市更新的框架中有着非常重要和积极的作用,但是社区问题在不同的城市背景下,都被不同程度地忽视或隔离。具体到关于亚洲发展中国家对"活态遗址"的保护的讨论,地方原住民的生活对当地宝贵的无形遗产有实质性的贡献。为了促进对原住民生活的保护,让地方居民参与地方社区的保护政策和管理机制是很重要的。文献研究表明,地方居民之间的联系和关系以及社区社会资本的存量有助于推进地方居民的组织参与到地方保护和开发活动中。同时,地方居民和政府之间的关系也影响他们参与保护项目,这突显了建立一个参与式的保护和管理机制的重要性。文献研究在理论上表明,建立一个参与式的城市管理机制对实现全面城市保护和更新目标是必不可少的。

which highlights the significance of invaluable indigenous lives. This concept is supported by many academics (Egloff & Newby, 2005; Filippi, 2005; Takaki & Shimotruma, 2003) and underlined in many international documents (UNESCO, 2003, 2007). In terms of the comprehensive conservation of local historic site or integrated urban conservation, two factors are stressed in the discussion. First, the relationships or connections between local residents and the historic environment have an impact on the conservation of local living heritages. These connections between the local residents and the historic environment actually promote the relationships or connections among local residents through their daily activities. Therefore, the conservation of the local historic environment also has a close connection with the stock of social capital. Second, the community-based conservation policy is essential in conducting the conservation of local indigenous lives. The case of Angkor city illustrates that well established conservation and development control mechanisms and management systems are very important, which guarantee the genuine participation of local residents in the conservation process and the smooth implementation of established conservation policies.

In the overview of the evolution of urban regeneration of different periodization in Western literature and the urban conservation policies and practices in two Asian cities, it is clear that local authorities play a very important yet different role in the regeneration of historic cities of different urban contexts. At the same time, the issue of economic development is a crucial factor influencing the urban regeneration or conservation goals in almost all the study contexts. In the pro-growth urban policies, the conservation of historic cities has met many challenges from rapid urbanization or tourism-biased development. The issue of the local community has been neglected or excluded in different extents in various urban contexts, although it plays an important and active role in the framework of sustainable development and comprehensive urban regeneration.

Specific to the discussion on the conservation of living heritages in the context of Asian developing countries, local indigenous lives substantively contribute to the conservation of local invaluable intangible heritages. In order to promote the conservation of indigenous lives, the participation of the local community in the conservation policy and management mechanism is significant. Literature implies that the connections and relationships among local residents or the stock of community social capital promote the involvement of local residents' organizations in the local conservation or development activities. At the same time, the relationship between the local residents and the government also affect their engagement in the conservation programs, which underlines the significance of a participatory conservation and management mechanism. The literature theoretically suggests that the participatory urban governance mechanism is essential in order to achieve the comprehensive urban conservation and regeneration goals.

2.2 历史城市的更新

上述对西方国家的城市更新以及两个亚洲城市的城市保护实践的文献综述，表明地方社区居民在历史城市的更新过程中的重要意义，但是在实践中，在不同的政治经济背景中，地方居民经常被隔离在外。文献进一步强调了对历史城市更新有重大影响的两个方面：地方倾向经济发展的政策，以城市绅士化和居民迁移为特色；参与式的城市管治机制和保护政策。而社区社会资本的存量对地方居民在保护和更新过程的参与中起着至关重要的作用。基于这种理解，本章将进一步探讨这些关键变量的内容和它们之间的关系。

2.2 Urban Regeneration in Historic Cities

The above literature review on urban regeneration in Western countries and urban conservation practices in two Asian cities underlines the significance of local communities in the urban regeneration of historic cities, although in practice, local residents have often been marginalized in various political and economic contexts. The literature further highlights two aspects with a substantive influence on the regeneration of historic cities: local pro-growth urban development policy often characterized by urban gentrification and resident displacement, and participatory urban governance mechanism and conservation policies. In terms of the participation of local residents in the conservation and regeneration processes, the stock of community social capital plays a significant role. Based on this understanding, this chapter will further explore the contents of these key variables and the relationships among them.

2.2.1 城市"绅士化"与居民重置

城市"绅士化"指的是在伦敦，城市中产阶级以上阶层取代低收入阶级重新由郊区返回内城市中心区的现象，这个名词最初由20世纪60年代的英国社会学家格拉斯（Ruth Glass）所提出。城市"绅士化"对贫困地区以及破旧的城市地区的修复和改善起了很大的作用（Glass，1964：2）。

一般来说，对城市"绅士化"有各种各样的专门的解释，包括人口、生态、社会文化和政治经济方面的解释。随着中产阶层进入市中心，"绅士化"过程造成了空间和社会分化的过程。除了显而易见的空间变化，它还反映了一个深刻的社会变化趋势。正如祖秦（Zukin）所说，中产阶级及其相对富裕的生活方式的涌入意味着企业和政府部门中的高收入人员的扩张（Zukin，1987：139）。城市"绅士化"带来的自然环境的变化，作为对贫困地区的入侵，被地方政府看作是提高贫困地区的生存能力和蓝领阶层的机会（Zukin，1987：144）。而且，"绅士化"被看作有利于历史建筑的保护。一些学者（Hannigan，1995；Wright，1985；Zukin，1982，1987）认为随着资产阶级对老城区的投资，历史建筑或原建筑在审美品位方面会得到更好的修复。一般来说，搬进来的绅士更加富裕，他们能够"表现天生的好品味而创造一个新的生活方式"（Hannigan，1995：179）。在经济发展方面，"绅士化"现象通常为资产阶级和房地产开发商提供投资机会（Zukin，1987：137）。因此，地方政府非常看好"绅士化"。有时，颁布实质性的法律法规来促进"绅士化"运动（Hannigan，1995；

2.2.1 Gentrification and displacement

The use of the term "gentrification," which describes the movement of middle-class residents into the low-income urban areas in London, can be traced back to the British sociologist Ruth Glass in the early 1960s. Gentrification was responsible for the rehabilitation and improvement of the disadvantaged neighborhood and the dilapidated urban areas, respectively (Glass, 1964: 2).

Generally, there are various trends in ad hoc explanations on urban gentrification including demographic, ecological, socio-cultural, and political-economic interpretations (London, Lee, & Lipton, 1986: 371). With the movement of middle-class residents to the city-centre, the process of gentrification engenders the process of spatial and social differentiations. Aside from the visible spatial changes, it also represents a more profound trend of social changes. As Zukin indicated, the influx of the middle-class, as well as their relative affluent lifestyles implies the "expansion of high-income personnel in corporations and governments and producers' services" (Zukin, 1987: 139). The changes in the physical environment because of gentrification, as an invasion of the disadvantaged urban areas, are often seen by the local government as opportunities to improve the viability of disadvantaged urban areas and blue-collar neighborhoods (Zukin, 1987: 144). Moreover, gentrification is seen to be in favor of the preservation of historical buildings. Some academics (Hannigan, 1995; Wright, 1985; Zukin, 1982, 1987) argued that with the investment in the old areas by the gentrifiers, the restoration and rehabilitation of historical or original buildings could be conducted better in terms of aesthetic tastes. Generally, the in-moving gentrifiers are more affluent, and it is possible for them to "construct a new lifestyle around the expression of an innate sense of good taste" (Hannigan, 1995: 179). In terms of economic development, gentrification activities usually provide opportunities for speculation for both gentrifiers and real estate developers (Zukin, 1987: 137); therefore, the local government usually takes a positive attitude towards it.

虽然，城市"绅士化"有上述积极的作用，但是有文献研究表明居住隔离促进"绅士化"进程同时也导致了许多城市问题。通常，土地主或财主为了利用土地或财产获取更高的经济利益，他们会迫使当地贫穷居民搬出去（London et al.，1986：374）。香港的城市再开发研究表明居住隔离导致了原本适当的社会层面内容的失衡。研究结果证明通过隔离进行再开发的结果也不符合最初的设定目标（Lai，1990）。例如，在分析了香港城市更新项目中隔离产生的结果之后，苏尼克（Susnik）总结说，缺乏对"'空间'整合和'社会'整合的支持"，经常导致不良的社会整合（Susnik & Ganesan，1998：115）。下面的章节将会进一步讨论在城市更新过程中，居民迁移产生的社会代价。

此外，正如上面提到的，尽管之前学术界一致认为"绅士化"有助于市中心破败的历史建筑的修复和保护（Hannigan，1995：179），但是现在随着综合城市保护意识的提高，这个论点受到了极大的挑战，综合城市保护在城市保护策略中强调对物质和非物质文化遗产的保护。随着《保护非物质文化遗产公约》的颁布（UNESCO，2003），人们更加关注对宝贵的非物质文化遗产的保护，例如，地方原住民的生活和传统活动。在城市"绅士化"过程中，由于地方居民的搬迁或位移，像原住民的生活一样的非物质文化遗产经常遭到破坏。

2.2.2 综合的城市保护

对历史城市的城市更新而言，城市保护

Sometimes, substantive laws and regulations are enacted to promote gentrification activities (Hannigan, 1995; Zukin, 1991).

However, although gentrification has its positive aspects as mentioned above, literature shows that residential displacement, as measures and a result of gentrification, entails many urban problems. Forced residential displacement usually occurs when the land or property owners have the motivation and right to force local disadvantaged residents out in the pursuit of higher economic profits of the land or property (London et al., 1986: 374). Studies on urban redevelopment in Hong Kong show the failure of balancing a proper social perspective through residential displacement. Research findings prove that the social outcomes of the redevelopment through displacement do not meet the initial objectives (Lai, 1990). For instance, after analyzing the displacement outcomes in the urban renewal projects of Hong Kong, Susnik concluded that the "support of 'spatial' as well as 'social' integration was lacking" and that physical separation often led to poor social integration (Susnik & Ganesan, 1998: 115). The social cost of urban regeneration through residential displacement will be further discussed in the next section.

Moreover, as mentioned above, although academics agreed that gentrification facilitated the restoration and preservation of dilapidated historical buildings in the city-centre (Hannigan, 1995: 179), this contention is highly challenged nowadays with the rise of consciousness on integrated urban conservation, which stresses on both tangible and intangible heritage in urban conservation strategies. With the promulgation of the *Convention for the safeguarding of the intangible cultural heritage* (UNESCO, 2003), people pay more attention to the invaluable intangible heritage, for example, local indigenous lives and traditional activities. These intangible aspects in terms of indigenous lives are often undermined through the relocation or displacement of local residents in urban gentrification.

2.2.2 Integrated urban conservation

As regards urban regeneration in historic

是一个重要的方法。一般来说，国际文件中的保护重点已经从重点保护物质文化遗产扩展到对非物质文化遗产的保护。这突显了保护地方原住民的生活的重要性，并且已经列入联合国教科文组织议程中。

2.2.2.1 联合国教科文组织议程中的城市保护

20世纪90年代初以来，国际层面上关于遗产的内涵已经得到了延伸，既包括自然环境也包括物质和非物质文化遗产（Ahmad, 2006；UNESCO, 2005）。相应地，城市保护范围已经从重点保护物质文化遗产延伸为既保护物质也保护非物质文化遗产。

到目前为止，国际上以及各个国家已经颁布了40多个保护文件，而且大部分是由联合国教科文组织和国际古迹遗址理事会（ICOMOS）发起。其中，《国际古迹保护与修复宪章》，俗称1964威尼斯宪章，是第一个也是最重要的指导方针（Ahmad, 2006：293）。另外，《保护世界文化和自然遗产公约》，也称为1972公约，是众多起草的公约中最成功的一个。但是，这些公约更关注于对古迹和物质文化遗产的保护。直到1992年，世界遗产委员会提议对《操作指南》，也称为《实施世界遗产公约的操作指南》进行修订，在遗产保护中增加非物质文化遗产的部分，非物质文化遗产才受到关注[1]（Blake, 2002；UNESCO, 2005）。

2003年，联合国教科文组织颁布《保护

---

[1]《操作指南》的目的是促进《保护世界文化和自然遗产公约》的实施，并且《操作指南》会定期进行修订，以反映世界遗产委员会的决定（UNESCO, 2005）。

cities, urban conservation is an important approach. Generally, the conservation focus in international documents has expanded from tangible heritage-focused to intangible elements-included. This underlines the significance of the conservation of local indigenous lives, which has been outlined on the UNESCO's agenda.

2.2.2.1 Urban conservation on the UNESCO's agenda

Since the early 1990s, the concept of heritage at the international level has been broadened to include tangible and intangible heritage as well as natural environments (Ahmad, 2006; UNESCO, 2005). Accordingly, urban conservation has evolved from the conservation of tangible heritage-focused to the conservation of both tangible and intangible heritage.

To date, more than 40 conservation documents have been enacted at both the international and national levels, and most of them have been initiated by the UNESCO and the International Council on Monuments and Sites (ICOMOS). Among them, the *International Charter for the Conservation and Restoration of Monuments and Sites*, commonly known as the Venice Charter 1964, was the first and also the most significant guideline (Ahmad, 2006: 293). On the other hand, the *Convention Concerning the Protection of the World Cultural and Natural Heritage*, known as the 1972 Convention, has been proven to be the most successful among the many initiated conventions. Nonetheless, these conventions focus much on the conservation of monuments and tangible heritage. The inclusion of intangible heritage did not come into discussion until 1992, when the World Heritage Committee proposed to include intangible components in heritage conservation through the revision of Operational Guidelines, also known as *Operational Guidelines for the Implementation of the World Heritage Convention*[1] (Blake, 2002; UNESCO, 2005).

---

[1] The *Operational Guidelines* aim to facilitate the implementation of the *Convention concerning the Protection of the World Cultural and Natural Heritage*. And *Operational Guidelines* are periodically revised to reflect the decisions of the World Heritage Committee (UNESCO, 2005).

In 2003, the *Convention for the Safeguarding of the Intangible Cultural Heritage* was issued by the UNESCO (2003). This is the most inclusive convention regarding the conservation of intangible cultural heritage so far. In this convention, intangible cultural heritage is defined as follows:

> the practices, representations, expressions, knowledge, skills—as well as the instruments, objects, artifacts and cultural spaces associated therewith—that communities, groups and, in some cases, individuals recognize as part of their cultural heritage. This intangible cultural heritage, transmitted from generation to generation, is constantly recreated by communities and groups in response to their environment, their interaction with nature and their history, and provides them with a sense of identity and continuity, thus promoting respect for cultural diversity and human creativity (UNESCO, 2003: 1).

Compared with what was delineated in the *Operational Guidelines*, this concept widens the interpretation of intangible cultural heritage and underlines that intangible cultural heritage can be recognized as part of the local cultural heritage not necessarily by the whole society but sometimes by local communities and even individuals. This concept can better cater for the actual situations of many historic urban areas in China. Especially when Chinese historic cities have gone through the "Cultural Revolution" and economic-biased urban development periods in the 1990s, many historic cities actually possess very few tangible heritages. However, intangible heritage in terms of local traditional lives and custom is exuberant. This calls for the recognition of diversified cultural heritage in specific study contexts, especially the types of intangible heritage when physical remains scarcely exist due to various reasons (Blake, 2002: 72). More discussions on this will continue in the next chapter.

2.2.2.2 "Living heritage site" and "indigenous lives"

In the discussion on integrated urban

社会和文化因素，强调了地方社区居民和原住民在城市保护中的作用。1987年，华盛顿宪章提出，"对历史城镇和其他历史城区的保护应成为经济与社会发展政策的完整组成部分，并应当列入各级城市和地区规划中"（ICOMOS，1987：1）。这表明了城市保护的社会方面应该和经济方面同样重要（Imon，2003：1）。城市保护中的社会关注与"活态遗产"的概念相关。

2003年，由ICCROM组织的曼谷活态遗产项目战略会议中，首次提出"活态遗产"的概念，该会议广泛讨论了通过本地原住民的参与加强对原住民生活的保护，"同时承认遗产的社会动态以及人们和当地遗产之间的互动"（Miura，2005：5）。据菲利比（Filippi）称，"活态遗产"指的不仅仅是建筑环境，而且也包括活态环境，如地方居民的日常生活和活动。因此，对待"活态遗产"是一个不断提高意识的过程，在全面策略中以参与式的行动把地方发展和保护统一起来（Filippi，2005：4）。塔克和石谋赞成地方原住居民参与到对地方历史环境的保护中（Takaki & Shimotsuma，2003：1），他们说"活态遗产"是通过"人们和文化遗产之间的交流和互动来体现的"。他们认为，"活态遗产"中的"活态因素"是指"那些有利于人们和文化遗产之间创造和加强联系的因素，或者是那些激发人们为了他们共同的未来而努力配合的因素"。苗拉（Miura）较确切地指出，"活态遗产"突出"强调在保护活动中创造发展文化遗产以及当今社会之间的良好沟通"（Miura，2005：5-6）。基于对澳大利亚文化遗产管理过程的分析，艾格罗夫和纽比（Egloff &

conservation, social and cultural issues have been frequently referred to, where the role of local communities and indigenous residents in urban conservation has been underlined. As proposed in the 1987 Washington Charter, "the conservation of historic towns and other historic urban areas should be an integral part of coherent policies of economic and social development and of urban and regional planning at every level" (ICOMOS, 1987: 1). This indicates that the social perspectives of urban conservation should be weighed as equally important with the economic perspectives (Imon, 2003: 1). Social concerns in urban conservation are related to the concept of the "living heritage site."

The concept of the "living heritage site" was first proposed at the *Living Heritage Site Program Strategy Meeting* in Bangkok organized by ICCROM in 2003. In this meeting, the trend of paying more attention to the conservation of local indigenous residents through their participation was widely discussed, "while recognizing the social dynamics of heritage values and the interaction between people and their heritage [sites]" (Miura, 2005: 5). According to Filippi, living heritage concerns not only built-environments but also "living" environments in terms of the local residential daily lives and activities. Therefore, to deal with living heritage is a process of developing awareness-raising and participatory actions to integrate local development and conservation in comprehensive strategies (Filippi, 2005: 4). The participation of local indigenous residents in the conservation of local historic environments is endorsed by Takaki and Shimotsuma (2003: 1) by indicating that a "living heritage site" is expressed by "the communication or interaction between cultural properties and the populations." According to Takaki and Shimotsuma (2003: 1), the "living elements" in the "living heritage site" refer to "what brings opportunities to create or strengthen the relation between cultural properties and the populations, or what motivates the population to cooperate in achieving their common future vision." In short,

Newby)认为历史地区亚瑟港[1]是由不熟知该遗产的个人管理的。在当地文化遗产的保护和管理实践中,他们强调多方利益相关者或"群众监督"(Egloff & Newby, 2005:29)的重要性,目标是"在保留社区和遗产价值的同时促进可持续发展"(Egloff & Newby, 2005:28)。

上述讨论表明,地方社区和原住民在综合城市保护中起了重要的作用。一方面,在非物质文化遗产方面,地方原住民的生活是地方文化遗产不可分割的一部分。另一方面,保护地方社区或原住民的生活有助于实现地方综合城市发展的目标,或通过一个参与式的保护管理机制促进历史遗产的可持续发展。

2.2.2.3 城市保护和城市更新

自20世纪50年代以来,随着城市保护原则的发展演变,在历史城镇和城市的更新中,城市保护一直被认为是可贵的策略。研究表明,把更新和保护结合起来的方法有助于综合城市更新目标的实现,例如,实现经济、社会和文化方面的进步(Pearce, 1994:88)。

皮尔斯(Pearce)提到以保护为主导的更新方法有助于推进城市改造的过程。具体来说,它有助于以下几个方面(Pearce, 1994:90):

- 通过强化未被利用的建筑资源的功能,使其改善和再利用;

---

[1] 亚瑟港被认为是澳大利亚最重要的文化遗产遗址,是被考虑在世界遗产名录提名的众多遗址之一(Egloff & Newby, 2005:22, 28)。

as Miura nicely pointed out, "living heritage site" highlights the "focuses in the conservation activities to create and develop favorable communication between cultural heritage and the current society" (Miura, 2005: 5-6). Based on the analysis of the Australian cultural heritage management processes, Egloff and Newby argued that the historic area Port Arthur[1] is managed by individuals who are not familiar with the site. They highlighted the importance of involving multi-dimensional stakeholders or "public scrutiny" (Egloff & Newby, 2005: 29) in the practices of the conservation and management of local cultural heritages, with the aim of "facilitat[ing] sustainable operations as well as maintain[ing] community and heritage values" (Egloff & Newby, 2005: 28).

The above discussions show that local community and indigenous residents play a significant role under the rubric of integrated urban conservation. On one hand, indigenous lives are the indispensable components of local cultural heritage in terms of valuable intangible cultural heritage. On the other hand, the conservation of local community or indigenous lives contributes to the local comprehensive urban development goals or sustainable development of the historic heritage sites through a participatory conservation and management system.

2.2.2.3 Urban conservation and urban regeneration

With the evolution of the conservation principles since the 1950s, urban conservation has been regarded as a valuable strategy in the regeneration of historic towns and cities. Studies show that the approach of combining regeneration and conservation strategies contributes to the achievement of comprehensive urban regeneration goals, for example, the improvement of economic, social, and cultural perspectives (Pearce, 1994: 88).

---

[1] Port Arthur is regarded as one of Australia's foremost cultural heritage sites, and is one of a series of places being considered for nomination to the World Heritage List (Egloff & Newby, 2005: 22, 28).

●产生强烈的视觉影响，从而引起关注、展示信誉，提高城镇形象；

●提高商业信心，从而提高效率，在一定程度上增加就业机会；

●利用各种政府资助，可以引进大量私营部门对该区域的投资；

●可以改善住房条件；

●可以开发旅游潜力；

●可以加强城镇的文化认同感。

显然，历史城市背景下，城市保护策略远不止这些优点，当具体到特定的历史城市背景时，还有其他社会文化价值。除了上述几个方面，斯坦伯格（Steinberg）指出历史城区的保护策略也有助于维护历史城市建筑物，保留当地社区生活的本质以及其他社会方面（Steinberg，1996：467）。

相关文献表明，城市更新项目能否成功，地方城市管治模式和政府机构是关键，因为他们决定了如何促进城市更新政策、实施以及决策交互前进（Healey，2006：5）。为了促使城市保护策略和城市更新政策相结合，促进历史城市更新工作顺利进行，地方政府应该确定一个包含目标、目的和指标的明确策略（Pearce，1994：92）。这点强调了建立一个有力的城市更新机构的重要性，地方政府在城市更新项目中对综合城市保护策略的制定有着积极的作用,已经得到了其他人的支持。斯坦伯格说，促进保护政策在历史城市更新中的实施，需要一个灵活多变的政治环境，政府机构也应该做相应的调整（Steinberg，1996：473）。为了实现全面城市更新的目标，地方政府应该平衡各部门包括私营部门和政府部门之间的利

Pearce mentioned that by adopting the conservation-led regeneration approach helps to facilitate the processes of urban adaptation. To be specific, it contributes to the following aspects (Pearce, 1994: 90):

● Building resources, which might otherwise remain under-utilized, are improved and reused thereby enhancing their functional attention;

● A strong visual impact is achieved, increasing awareness, demonstrating commitment and enhancing the image of the town;

● Business confidence can be raised, leading to improved efficiency and, in some instances, additional employment opportunities;

● Packages of private sector investment can be levered into the area, using a variety of public grants;

● Housing conditions can be improved;

● Tourist potential can be realized;

● The cultural identity of the town can be reinforced.

Obviously, these merits are not limited within the context of historic cities. When it comes to concrete historic urban contexts, there are other added socio-cultural values. In addition to the above aspects, Steinberg indicated that the conservation policies in historic urban areas also contribute to the maintenance of the historic urban fabric, essential qualities of local community lives, and other social perspectives (Steinberg, 1996: 467).

Literature demonstrates that the mode of local urban governance and institutions is fundamentally important to the success of urban regeneration programs, because it determines how urban regeneration policies, implementation and decision-making can interactively advance (Healey, 2006: 5). In order to integrate urban conservation strategy with urban regeneration policy and to promote holistic urban regeneration initiatives, local authorities are expected to "establish a clearly documented strategy, containing aims, objectives and targets" (Pearce,

益。例如，在新加坡的案例研究中，学者指出，新加坡市区重建局，在早期带领新加坡通过再开发实现城市转型，到后期才意识到保护的问题过程中，显得颇为"强硬"。(Kong & Yeoh, 1994：248)。新加坡的另一个案例研究中，之所以能够成功地保护历史街区和保留原住民的生活，李（Lee）将其归功于当地实施的城市政策和参与式的城市管治模式（Lee, 1996：400）。

上述讨论表明以保护为主导的方法有助于实现历史城市的全面城市更新目标。然而，在城市更新过程中，有力的政府机构和参与式的城市管治模式是实施和管理以保护为主导的更新项目的基础。

2.2.3 城市更新与管治模式

自 20 世纪 90 年代末，随着一些发达国家，如英国和美国在城市发展中更加全面的考虑，人们更加广泛地关注和讨论以社区为基础的合作和蹉商。这种覆盖面广，多机构参与的方法已经对地方城市管治产生了重大影响，如人口问责方面（Hughes & Carmichael, 1998：206）。为了建立一个参与式的城市管治模式，关键是进行制度改革，其中地方政府起着非常重要的作用（Hamdi & Majale, 2005：35）。然而，研究发现，在很多情况下，制度改革最大的阻力也是来自于地方政府机构或政策（Burns, Hambleton, & Hoggett, 1994：13）。因此，为了对参与式的管治模式有一个更好的理解，需要进一步讨论地方管治模式和地方政府的作用。

1994: 92). This highlights the importance of an enabling urban regeneration institution. The active role of local authorities in integrating urban conservation strategy within the urban regeneration program has been supported by others. According to Steinberg, in facilitating the conservation policies in the regeneration of historic cities, "a changed political environment" is necessary and government institutions should be modified accordingly (Steinberg, 1996: 473). To achieve the comprehensive urban regeneration aim, it should be part of the local authorities' roles to balance the interest of various agencies including both private and public sectors. For instance, in the case studies of Singapore, academics revealed that the quite "heavy-handed" governmental Urban Redevelopment Authority (URA) led Singapore's urban transformation from an early concern with redevelopment to the later recognition of conservation issues (Kong & Yeoh, 1994: 248). In another case study of Singapore, Lee also ascribed the success of the preservation in historic districts and the retention of indigenous community lives to local facilitating urban policy and a participatory urban governance (Lee, 1996: 400).

These discussions show that a conservation-led approach contributes to the comprehensive urban regeneration aims of historic urban environments. Nonetheless, the facilitating government agencies and a participatory mode of governance in urban regeneration processes are fundamental to the implementation and management of local conservation-led regeneration programs.

2.2.3 Urban regeneration and mode of governance

With more holistic concerns in the urban development of some developed countries such as the U.K. and the U.S.A since the late 1990s, community-based partnership and consultation have gained wide attention and discussion. This kind of inclusive, multi-agency approach has posed significant impact on local urban governance in terms of demographic

2.2.3.1 地方参与式的城市管治模式
城市管治和以社区为基础的管治模式

马卡妮认为，"管治不同于政府，它是指公民社会和国家之间、统治者和被统治者之间、管理和被管理之间的关系（McCarney, 1995：95）。"这个概念清楚地表明了管治应该强调各个机构之间的关系而不是某一特定的机构。在其他学者的不同解释中可以发现此类基本的理解。在豪斯和海尼特（Haus & Heinelt）看来，城市管治强调公众和广泛参与者之间的互动（Haus & Heinelt, 2005；Jessop, 1998）。此外，在最常见的使用中，解索普（Jessop）将"管治"定义为"特定机构或组织的管理模式，其中包括多重利益相关者、公私合营关系的作用以及自治却又相互依存组织之间的战略联盟"（Jessop, 1998：30）。

在讨论多机构参与的管治模式时，学者认为，多方参与和赋予公民权利是形成一个可信的管理机构和解决众多城市问题的合理途径（Wu, 2002a：1072）。沃尔什（Walsh）认为，实施分权以及社区赋权的策略，目的就是建立一个基于社区的管治模式（Walsh, 1997：7）。在讨论地方管治时，学者相信通过国家对地方下放权力，地方政府有望根据自己的需要，进行决策、管理以及负责维护服务体系（Mutchler, Mays, & Pollard, 1993：14）。通过分配资源，不同级别的机构，包括"州、县或地方政府机构、私营营利或非营利机构"之间的决策，地方管治结构会成为一个有力的制度结构（Mutchler et al., 1993：14）。

在以下对制度化的地方管治结构的讨论

accountability (Hughes & Carmichael, 1998: 206). In order to achieve participatory urban governance, institutional changes are critical where local government plays an important role (Hamdi & Majale, 2005: 35). Nevertheless, studies find that in many situations, the biggest resistance in institutional changes also comes from the local governmental agencies or politics (Burns, Hambleton, & Hoggett, 1994: 13). Hence, in order to understand better the participatory urban governance, it is necessary to discuss more about local urban governance and the role of local government.

2.2.3.1 Towards local participatory urban governance
Urban governance and community-based governance

According to McCarney (1995: 95), "Governance, as distinct from government, refers to the relationship between civil society and the state, between rulers and the ruled, the government and the governed." This definition clearly indicates that governance should emphasize on the relationships between various agencies rather than any specific agency. This basic understanding can be found in various interpretations by other academics. In Haus and Heinelt's viewpoint, urban governance highlights the interactive arrangements among the public and the wide range of participants (Haus & Heinelt, 2005; Jessop, 1998). Furthermore, in the most common usage of this term, Jessop defined governance as "the mode of conduct of specific institutions or organizations with multiple stakeholders, the role of public-private partnerships, and other kinds of strategic alliances among autonomous but interdependent organizations" (Jessop, 1998: 30).

In discussing multi-agency governance, academics argue that multi-participation and empowerment to civil society are proper paths to arrive at credible governing institutions and solve many urban problems (Wu, 2002a: 1072). Walsh contended that, with the strategies of decentralization and community empowerment, the aim is community-based

中，司铎克（Stoker）宣称，一个"好"的地方管治模式应该具备三个特点（Stoker, 1996: 189）。

- 开放能力：这样人们就有权利和机会在当地公共场所表演。
- 商议能力：解决社区面临的问题，既代表市民领导也代表普通市民。
- 行动能力：能够让来自国家、私营部门和志愿部门的机构和参与者集中人力物力，来实现共同的目标

要达到具备上述特征的良好的地方管治模式关键是要赋予公民决策权和政策执行权，这样才能达到真正的社区管治。博克斯（Box, 1998: 22）指出，通过公民参与的方式，制度化的地方管治有助于实现基于社区的城市管治模式。博克斯还进一步补充了好的地方管治模式应有的另外两个原则：问责制和合理性。问责制强调政策执行者的代表性和支持性特点，他们应该"在社区生活中发挥重要作用"（Box, 1998: 21）。另一方面，合理性强调参与其中的公民或代表应该非常明确地表达他们的价值观、设想以及理由（Box, 1998: 23）。

上述分析强调了这样一个事实，制度化的地方管治是实现真正的基于社区的城市管治模式的关键。社区参与有助于推进基于社区的管治模式的进程。一些因素和社区参与紧密相关。在基于社区的管治模式中，这些因素包括各个机构之间的关系和联系，或社会中的社会资本水平。后面的章节将会对此做进一步讨论。

governance (Walsh, 1997: 7). In the discussion on local governance, academics believe that by empowering the localities from the state or national government, the localities are expected to determine, govern, and maintain accountability for service systems to meet their own needs (Mutchler, Mays, & Pollard, 1993: 14). By means of allocating resources and making decisions among various levels of agencies including "state, county, or local governmental agencies, and private for-profit and/or non-profit agencies," the local governance structure reveals an enabling institutional structure (Mutchler et al., 1993: 14).

Following the discussion on institutionalized local governance structure, Stoker asserted that "good" local governance should possess three features (Stoker, 1996: 189).

- Capacity for openness: so that people can have the right and opportunity to act in local public life;
- Capacity for deliberation: on the issues confronting the community both on the part of civic leaders and "ordinary" citizens;
- Capacity for action: to enable institutions and actors from public, private and voluntary sectors to blend their resources and skills to achieve common purposes.

The above features of "good" local governance imply that it is fundamental to empower the public with the rights of decision-making and policy implementation in order to achieve genuine community governance. Box (1998: 22) also pointed out that the institutionalized local governance contributes to community-based urban governance through public participation. Further, Box supplemented two other principles underlying "good" local governance: accountability and rationality. The term accountability emphasizes on the representative and supportive features of the policy practitioners, who should "have important roles to play in community life" (Box, 1998: 21). Rationality, on the other hand, stresses that involved citizens or representatives should

参与式社区管治面临的挑战

文献表明，关于社区参与的讨论起于20世纪60年代和20世纪70年代，那时美国政府在社会福利项目中推行社区参与。自那以后，许多学者和组织，例如阿恩斯坦（Arnstein，1969）、英国环境部（DOE.，1994b）以及哈姆迪（Hamdi，2005）都对社区参与进行了研究。实践中的一些因素会破坏社区参与的进程。博克斯（Box，1988）强调了两个主要因素。一个是"拥有强大政治或经济势力的人把公民参与看作是一种威胁，因而抵制公民参与"，另一个是缺乏"代议制民主的结构"（Box，1998：87）。除了这两个主要因素，就如博克斯指出的，其他不利于居民参与当地城市发展项目的因素包括以下几个（Box，1998：88）：

- 一些咨询机构在公共政策中有所作为的能力是有限的，或是因为他们被授予的责任范围有限或有很少实权，或两者兼有。
- 参与机会有可能受限，因为只有在一些非常有限的地区才有参与机会，而这些地区需要特定的或专业的知识。
- 咨询机构和社区参与结构，有可能被持有特定议程的少数人掌控。
- 公共对话的环境可能会让市民感觉自己在自治中不受欢迎。
- 社区参与结构可能为自治提供的机会有限，因为公民完全理解他们所监督的机构的能力是有限的，这会造成一个理性和建设性的影响。

express their values, assumptions, and reasons very explicitly (Box, 1998: 23).

The above analyses underline the fact that institutionalized local governance is vital to achieve genuine community-based urban governance. Community participation contributes to the process of community-based governance. Some factors are closely related to community participation. In community-based governance these factors include the relationships or connections among various agencies or the level of social capital in the society. These will be further discussed in a later section.

Challenges to the participatory community governance

Literature indicates that discussions on community-participation started from the 1960s and 1970s when the U.S. government implemented community-participation in social welfare programs. Since then, community participation has been studied by many academics and organizations such as Sherry Arnstein (Arnstein, 1969), the Department of Environment in Great Britain (DOE., 1994b), and Nabeel Hamdi (Hamdi, 2005). In practice, there are some factors that tend to undermine the processes of community participation. There are two major factors stressed by Box (1998). The first one is "the presence of politically or economically powerful people who resist citizen involvement as potential threat," and the other is the lack of "the structure of representative democracy" (Box, 1998: 87). In addition to these two major factors, as Box indicated, the other factors that tend to discourage community participation in local urban development programs include the following (Box, 1998: 88):

- some advisory bodies might have limited ability to make significant differences in public policy, either because they are granted a small area of responsibility or have very little real authority, or both.
- participatory opportunities might be limited because they only occur in some very limited areas, where specific or professional knowledge are needed.

其他研究者也已经注意到了影响社区参与的这些不利因素，他们以此来说明个人或社区不愿参与到当地城市发展的原因（Reid，2000）。为了建立一个真正的社区自治模式，已经提出了一些建议，如，以社区为核心来制定政策、管理社区项目和服务；创建开放热情的参与机制（Box，1998：87；Reid，2000：3）。毕竟，就如施密特（Schmitter）坚持认为的，影响社区参与的最重要的因素是有关机构的态度和行为，例如，要尊重公民，倾听他们的心声，向他们学习，还有其他行为（Schmitter，2002：52）。这些建议和讨论强调了有助于推动社区参与的两个重要方面：地方政府的作用和有关部门之间的关系或联系，以及社会资本的水平。

2.2.3.2 地方政府在城市管治中的作用

古德温与佩因特对地方政府的理解是"当选的地方政治体系中正式的机构，它们影响地方区域的生活模式和经济结构"（Goodwin & Painter，1996：636）。另外，在联合国教科文组织议程中第21条中已经强调了地方政府在城市可持续发展中的重要作用，"地方政府的参与和合作在促进可持续发展中起决定性作用"（UNESCO，1992）。在实践中，英国环境部已经列出了地方政府在地方发展中的重要作用：

"地方政府的创新能力、预见问题的能力、领导能力以及参与其他组织的处理能力，能够极大促进发展战略的可持续性，这是地方需要也是优先考虑的（DOE，1994a：200）。"

- advisory bodies, as with the community involvement structures, may be dominated by a few people with a particular agenda.
- the setting of public-dialogue may not make the citizen feel welcome to participate in self-governance.
- community participation structure probably provides limited opportunities for self-governance, because of the ability of citizens to fully understand the services they oversee and have a rational and constructive impact.

These disincentives to community participation have also been noticed by other researchers to account for why individuals or communities are reluctant to be involved in local urban development issues (Reid, 2000). To achieve genuine community self-governance, some suggestions have been put forward such as bringing the community to the heart of the process of setting policy and administering community programs and services, and creating settings for participation, which are open and welcoming (Box, 1998: 87; Reid, 2000: 3). After all, as Schmitter maintained, the most important element affecting community participation is the attitude and actions of the agency concerned, for example, treating the public with respect, listening and learning from them, and other actions (Schmitter, 2002: 52). These suggestions and discussions highlight the two significant perspectives contributing to community participation: the role of local government and the relationships or connections among the agencies concerned and the level of social capital.

2.2.3.2 The role of local government in urban governance

As defined by Goodwin and Painter, local government refers to "the formal agencies of elected local political institutions which exert influence over the pattern of life and economic make-up of local areas" (Goodwin & Painter, 1996: 636). In addition, the important role of local government in sustainable urban development has been underlined in Agenda 21, where "the participation and cooperation

这些对地方政府的表述表明它与地方管治有密切的关系，而它们共同对城市经济、社会和文化活动有实质性影响。社区发展和地方政府政策之间的亲密关系体现在两个方面。一方面，来自社会领域和地方社区的方案有可能会列入政府政策的制定中，有助于对政府实践创建假想。另一方面，政府的方案或政策会对社区活动有显著的影响（Healey，2006：91）。

一般来说，在可持续发展的框架中，地方政府的职能已经从"提供或领导的角色（建设房屋、集中资源、规范操作）转变为授权和促进的角色（管理资源、分配资源、促进多样性和灵活性）"（Hamdi & Majale，2005：35）。这个观点得到了许多人的支持，汉布尔顿（Hambleton，1991：53）表示，像私营部门或自愿组织不可能发挥这样的作用，因为只有民众选举的地方政府是制定共同社会目标唯一的合法机构。

联合国教科文组织议程的第20条指出，地方政府在实现其目标的过程中，参与和合作策略在地方政府机制中是必不可少的（UNESCO，1992）。这要求政府更加开放，更有责任感。关于地方政府实践和机制的典范转移，有许多讨论。例如，希利（Healey）认为，"管理"的方法或更"创业型"的管治形式可能会融入地方政府机制中。这些表述反映了政府职能的转变，由"供应型"到"政策推动者"的转变，有助于推动多元机构的活动和参与（Healey，2006）。

1）制定授权型政策和制度框架

"授权型"机构或政府最初指的是"当选的地方政府的主要职责只是简单地向其他机

of local authorities will be a determining factor" to promote sustainable development (UNESCO, 1992). In practice, the significant role of local government in local development has been expressed by the Department of the Environment, Great Britain as the following:

> Local government's ability to innovate, to anticipate problems, to provide local leadership and processes for involving other groups, represents an important contribution towards the development of strategies for sustainability which reflect local needs and priorities (DOE, 1994a: 200).

These statements on the local government indicate that it has close relations with local governance, and together they have substantive impacts on urban economic, social, and cultural activities. The intimate relationships between community development and local government policies are reflected in two aspects. On the one hand, the initiatives from social arenas and local communities may be drawn into the development of government policies and help to create the assumptions on government practices. On the other hand, government initiatives or policies have an obvious impact on community activities (Healey, 2006: 91).

Generally, in the framework of sustainable development, the role of local government has shifted "from a providing or lead role (building houses, centralizing resources, standardizing operations) to enabling and facilitating role (managing resources, decentralizing resources, promoting variety and flexibility)" (Hamdi & Majale, 2005: 35). This viewpoint is supported by others. Hambleton (1991: 53) revealed that sectors like the private or voluntary ones could not fulfill this role because the popularly elected local government is the only legitimized agency to deliver common social goals.

As indicated in Agenda 21, the participation and cooperation strategies are imperative in the mechanisms of local government in fulfilling its objectives (UNESCO, 1992). It requires the government to be more open and accountable.

构发布文件,然后让他们处理"(Cochrane, 1993:69)。这个观点是非常重要也是非常必要的,因为由于这种特征,地方政府可以作为一名仲裁员,在实力强大的利益相关者之间进行斡旋,检查特定群体的统治,解除约束,制定授权型政策(Hamdi & Majale, 2005:35)。学者认为,在某种程度上,和其他部门相比,分权型地方政府更依赖于实力强大的部门(Cochrane, 1993:70)。然而,根据上述对良好的地方管治的讨论,有效的合作需要政府部门和非政府部门两方面的资源(Stoker, 1996:195)。这个特点要求地方政府"采取一个比以往更加多元化的方式,除了各种各样的政府机构外,还有私营和志愿机构"(Ridley, 1988:8)。事实上,地方政府的一项任务就是激励更多的机构参与到地方管治体系中。

2)促进各个机构的参与

一个良好的地方管治模式中,地方城市发展项目中应该有各个部门的参与。有学者声称,地方政府要认识到自己的局限性,实行权力分享,改变他们对地方社区能力的偏见(Hamdi & Majale, 2005:35)。司铎克表示,在政府再造方面,地方政府应该起催化剂的作用。而且,地方政府在团结、忠诚、信任和互惠的基础上通过建立关系发挥重要的引导作用(Stoker, 1996:204)。希利进一步表示说,地方政府如果没有抑制和引导,社区可能会按照房地产开发商的投资利益发展。离开政府的促进功能,起初约束市场活动的监管机制,有可能成为他们管理的产业的"牺牲品"(Healey, 2006:200)。

以上讨论清楚地表明了在地方城市发展

There are many discussions regarding the paradigm shift of the local government's practices and mechanisms. For instance, as Healey (2006: 131) argued, a "managerial" approach or a more "entrepreneurial" style of governance may be integrated into the local government mechanism. These statements reflect the shift of the government's role from "provider state" to a "strategic enabler," contributing to the activities and the inclusion of multi-agencies (Healey, 2006).

1) Enabling policies and institutional framework

The notion of "enabling" authority or government initially means "the main role of an elected local authority might simply be to issue contracts to other agencies and let them (enable them) get on with it" (Cochrane, 1993: 69). This idea is important and necessary because with this quality, the local government can work as an arbitrator, mediate between powerful stakeholders, provide checks on the dominance of specific groups, remove constraints, and establish enabling policies (Hamdi & Majale, 2005: 35). Academics argue that, to some extent, the enabling local authority relies on the comparatively more powerful position than other sectors (Cochrane, 1993: 70). Nevertheless, according to the above discussion on "good" local governance, effective partnerships require the blending of resources from both government and non-government actors (Stoker, 1996: 195). This feature needs the local government to "operate in a more pluralist way than in the past, alongside a wide variety of public, private and voluntary agencies" (Ridley, 1988: 8). In fact, one of the local governments' tasks is to stimulate more agencies to be involved in local governance institutions.

2) Facilitating the engagements of various agencies

In the context of "good" local governance, various actors should be involved and considered in the local urban development programs. Academics assert that local governments need to recognize their own limitations, conduct power-sharing, and modify their prejudice on the local community's competence (Hamdi &

的过程中，地方政府起着很重要的作用。为了促进多方参与的进程，良好的地方管治要求地方政府在当地城市发展策略中起到支持和促进的作用。

2.2.4 小结

总之，在前面的章节中，已经围绕历史城市更新回顾了与之相关的主要议题以及它们之间的关系，即城市更新概念、城市保护项目和地方城市管治之间的关系。一般而言，社区参与当地以保护为主导的更新项目有助于实现综合城市保护的目标。文献进一步强调了有助于社区参与的两个方面：地方参与式的城市管治，地方政府发挥重要作用；各个机构之间的关系或联系，或社会资本的水平。关于这些，社区团结和联系或社区的社会资本对社区参与地方发展项目有很大的影响。在下一节中，将进一步讨论在地方城市更新进程中社会资本的问题。

Majale, 2005: 35). Stoker said that the local government should play a catalytic role in terms of government reinvention. Moreover, the local government has an important facilitator role to play through the establishment of relationships based on solidarity, loyalty, trust, and reciprocity (Stoker, 1996: 204). Healey further indicated that if unchecked or unfacilitated by the local government, community development might follow the path in favor of the vested interests of property developers following the market rules. Without government facilitating functions, the regulatory mechanisms, which are initiated to constrain market activities, may be "captured" by the industries they regulate (Healey, 2006: 200).

The above discussions clearly indicate that the local government plays a significant role in local urban development processes. To promote the process of multi-participation, the mode of "good" local governance requires that the local government needs to provide the enabling and facilitating role in the local urban development strategies.

2.2.4 Summary

In sum, the previous sections have reviewed the related key issues of centering urban regeneration in historic cities and their relationships, that is, the concepts of urban regeneration, urban conservation programs, and local urban governance. Generally, community participation in local conservation-led regeneration programs contributes to the integrated urban conservation aims. Literature further underlines two perspectives conducive to community participation: local participatory urban governance where the local government plays a significant role and the relationships or connections among the agencies or the level of social capital. In relation to these, such issues as community ties and connections or community social capital have substantial influence on community involvement in the local development programs. In the next section, this book will discuss more about the issue of social capital in the local urban generation processes.

2.3 社会资本和城市更新

为了探索社会资本和历史城市更新之间的关系,本节首先对社会资本进行界定,其次,简要回顾了社会资本测量的过程,并探讨了一些社区层面上的社会资本常见指标。最后,讨论了地方城市更新中影响社会资本的因素。

2.3.1 社会资本及其分类

一般来说,关于社会资本的研究可以追溯到1916年,哈尼芬(Hanifan)把社会资本定义为"在人们生活中,使有形物质在人类的日常生活中产生最大价值的事物,即那些构成社会单位的群体之间和家庭之间的善意、友谊、互相同情和社会交往……"(Hanifan,1916:130)。在这个定义中,哈尼强调社区作为一个整体,应该从其成员的合作中受益,"一个人在其团体中将会发现乐于助人、同情他人以及邻里之间的友谊的好处"(Hanifan,1916:131)。显然,哈尼芬认为,通过参加社区合作和社团的活动,个人和社区获得的一种社会幸福感(或社会资本的收益)。20世纪90年代以来,西方国家已经有许多关于社会资本的多维性的讨论。各种对社会资本的界定和讨论都基于哈尼芬对社会资本的基本理解。这些界定和讨论各不相同,主要是讨论的范围不同。一些人在个人层面上讨论而有些人则在国家层面上讨论。

皮埃尔·布迪厄(Pierre Bourdieu)

许多研究认为皮埃尔·布迪厄(Pierre Bourdieu)是系统分析社会资本的第一人,他对社会资本的界定是"社会资本是现实或潜在的资源集合体,这些资源与拥有或多或

2.3 Social Capital and Urban Regeneration

In exploring the relationships between social capital and the regeneration of historic cities, this section first defines social capital. Second, the process of measuring social capital is briefly reviewed, and some common indicators on social capital at the community level are explored. Finally, the factors influencing social capital in local urban regeneration are discussed.

2.3.1 Social capital and its forms

Generally, the study on social capital can be traced back to Hanifan in 1916, when he defined social capital as "that in life which tends to make these tangible substances count for most in the daily lives of a people, namely, goodwill, fellowship, mutual sympathy and social intercourse among a group of individuals and families who make up a social unit ... " (Hanifan, 1916: 130). In this definition, Hanifan stressed that the community as a whole should benefit from the cooperation of all its parts, "while the individual will find in his associations the advantage of the help, the sympathy, and the fellowship of his neighbors" (Hanifan, 1916: 131). Apparently, Hanifan underlined a kind of social well-being (or the accrual of social capital), both for the communities and individuals through the activities of community cooperation and association. Since the 1990s, there have been pervasive discussions on the multi-dimensions of social capital in the Western world. Various definitions and discussions on social capital followed the general understanding as depicted by Hanifan. They were all different from each other, mainly at the different spheres of discussions. While some talked about it at the individual level, others discussed it at the national level.

Pierre Bourdieu

Many studies considered Pierre Bourdieu as the first person to have systematically analyzed social capital, which he defined as "*the aggregate of the actual or potential resources, which are linked to possession of a*

少制度化的共同熟识和认可的关系网络有关……它从集体拥有的角度为每个成员提供支持"（Bourdieu, 1986: 248）。布迪厄（Bourdieu）声称，通过参与团体和社会环境的建设，可以积累社会资本，个人利益也会相应地增加。显然，布迪厄（Bourdieu）在他的定义中强调了社会资本的两个方面：在社会关系中，个人可以要求使用其同伴拥有的资源，以及这些资源的数量和质量（Portes, 1998: 3-4）。

詹姆斯·科尔曼（James Coleman）

在科尔曼对社会资本的界定中，也体现了上述两个方面。不同之处就是科尔曼强调通过参与者的行动来生产公共资源。根据科尔曼，社会资本是指"有两个相同点的各种实体：在这个结构中，他们都包含社会结构的某些方面，他们都能激励一些人的特定行动——无论是个人还是群体"（Coleman, 1988: 598）。波茨在他的界定中也强调了这两个方面，即在这种网络或关系中所获得的福利或利益；在这种社会网络或其他社会结构中的会员资格（Portes, 1998: 5）。

罗伯特·普特南（Robert D.Putnam）

最著名的社会资本的界定是由普特南提出的。他将社会资本定义为"社会组织的特征，如网络、规范和社会信任，促进协调合作，实现互利共赢"（Putnam, 1993: 35）[1]。根据这个定义，社会资本的根本特点是促进协会成员的协调与合作（Serageldin & Grootaert, 2000: 40）。考虑到社会资本有助于社会协

---

[1] 与该表述类似，普特南在他的著作《独自打保龄球：美国社区的衰落与复兴》中阐述了社会资本的概念："物质资本是指物质事物，人力资本是指个体的属性，社会资本是指人与人之间的联系——社会网络和互惠规范以及从中产生的信任（Putnam, 2000: 19）"。

durable network of more or less institutionalized relationships of mutual acquaintance or recognition … which provides each of its members with the backing of collectively-owned capital" (Bourdieu, 1986: 248). Bourdieu purported that by participating in groups and intentional construction of the social environment, social capital can accrue, and accordingly the profits to the individuals can accrue. Clearly, Bourdieu underlined two perspectives of social capital in his definition: social relationship where individuals can claim access to resources possessed by their associates, and the amount and quality of those resources (Portes, 1998: 3-4).

James Coleman

These two perspectives are echoed somehow in Coleman's definition of social capital. The difference however, is that Coleman stressed on the process of the production of common resources through the actions of participants. According to Coleman, social capital refers to "*a variety of entities with two elements in common: They all consist of some aspect of social structures, and they facilitate certain action of actors—whether persons or corporate actors—within the structure*" (Coleman, 1988: 598). Portes also highlighted the two perspectives in his definitions, that is, the amount of benefits or interests within the networks or relationships, and the memberships in such social networks or other social structures (Portes, 1998: 5).

Robert D. Putnam

The most well-known definition of social capital was proposed by Putnam. He defined social capital as "*features of social organization, such as networks, norms and social trust that facilitate coordination and cooperation for mutual benefit*" (Putnam, 1993: 35)[1]. According to this definition, the

---

[1] Similar to this expression, Putnam elaborates the concept of social capital in his seminal book "*Bowling alone: the collapse and revival of American community*" as the following: "Whereas physical capital refers to physical objects and human capital refers to properties of individuals, social capital refers to connections among individuals–social networks and the norms of reciprocity and trustworthiness that arise from them. (Putnam, 2000: 19)"

调与合作的特点，普特南认为社会资本和政治参与有着密切联系，就是"我们和政治制度的关系"。因此，普特南提出了"公民参与"这个术语，表明了人们与社区生活以及政治的联系（Putnam，1995b：665）。普特南（Putnam）强调社会信任和公民参与之间的密切关系。他认为相互联系的人越多，他们彼此就会越信任；加入或参与组织活动的人越多，他们就会更加信任彼此，反之亦然（Putnam，1995b：665–666）。

此外，就像其他形式的资本一样，普特南概述了社会资本的经济属性。他说拥有大量社会资本的社区，其经济和政治发展也会较好（Putnam，1993：5）。自普特南的研究以来，社会资本在政治和经济发展中被视为一个决定因素，在城市政策的制定中，决策者和规划者已经关注社会资本。

林南（Nan Lin）

林（Lin）通过社会网络，在社区投资和预期成果方面探索了社会资本。在林（Lin）的社会资本定义中，社会网络是一个显著的特点。基于这种理解，林（Lin）对社会资本的假设是显而易见的："社会关系的投资，及其带来的预期回报"（Lin，1999a：30）。因此，社会资本可以被定义为"嵌入于一种社会结构中的可以在有目的的行动中涉取或动员的资源"（Lin，1999a：35）。林（Lin）明确地将社会资本表达为"个人在社会关系的投资，这样他们可以使用嵌入的资源来提高实际行动或表达带来的预期回报"，他指出其中的三个过程："（1）社会资本的投资；（2）获取和调动社会资本；（3）社会资本的回报"（Lin，1999a：39）。

key feature of social capital is to facilitate the coordination and cooperation of the members of the association (Serageldin & Grootaert, 2000: 40). Considering the features of contributing to social coordination and cooperation, Putnam argued that social capital has a close relationship with political participation, which means "our relations with political institution." Hence, Putnam introduced the term "civic engagement" through which he indicated people's connections with the life of communities as well as politics (Putnam, 1995b: 665). Putnam stressed the intimate relationships between social trust and civic engagement. He claimed that the more people connect with each other, the more they tend to trust; the more people join or participate in groups' activities, and the more they trust, and vice versa (Putnam, 1995b: 665-666).

In addition, just like other forms of capital, Putnam outlined the economic prosperity of social capital. He stated that communities with larger stocks of social capital also show better economic and political development outcomes (Putnam, 1993: 5). Since Putnam's study, social capital has been seen as a determinant role in political as well as economic development, and it has attracted the attention of decision-makers and planners in urban policy-making.

Nan Lin

Lin explored social capital in terms of community investment and expected outcomes through social networks. The social network is one outstanding feature of Lin's definition of social capital. Based on this understanding, Lin's premise on social capital is obvious: "investment in social relations with expected returns" (Lin, 1999a: 30). As such, social capital is defined as *"resources embedded in a social structure which are accessed and/or mobilized in purposive actions"* (Lin, 1999a: 35). When Lin sharpened the expression of social capital as *"investment in social relations by individuals through which they gain access to embedded resources to enhance expected returns of instrumental or expressive actions."* He indicated the three processes involved: "(1) investment in social capital,

一般来说，社会资本有很多分类方法。例如，根据"利益是为群体还是个人积累的"（Lin, 1999a: 31），社会资本可以分为两类：第一类侧重于由个人使用社会资本，即个体如何访问和使用嵌入在社会网络的资源来获得回报。这一类的代表人包括林南（Lin & Bian, 1991），波特（Burt, 1992; 1997a），波茨（Portes, 1998）等学者。第二类关注社会资本在群体层面的使用，即，群体是如何作为一个集体资产，开发和维护社会资本，以提高群体成员的生活水平的。这一类的代表包括布迪厄（Bourdieu, 1986），科尔曼（Coleman, 1988; 1990），普特南（Putnam, 1995a; Putnam, Leonardi, & Nanetti, 1993），亚特（Bryant & Norris, 2002），以及其他学者。

另一个讨论社会资本类型的非常重要的方式是区分联合型（或包容性）社会资本和聚合型（或专属性）社会资本。正常情况下，聚合型社会资本倾向于进一步"加强专属身份和同源群体"的独有特征，例如民族团体和教会组织成员间的关系，而联合型社会资本往往包括不同社会群体的人们之间的关系，如民权运动和社会服务团体（Putnam, 2000: 22）。相比较而言，聚合型社会资本有利于支持特定的互惠和促进团结，而联合型社会资本倾向于链接外部资产，并利于信息扩散。因此，聚合型社会资本往往有助于小范围内的社区福祉，而联合型社会资本"可以产生更广泛的身份认同和互惠范围"（Putnam, 2000: 23）。

尽管普特南指出，这两种社会资本在巩固社会参与和使人们团结一致方面，有强大的积极社会影响，他也明确表示，聚合型社会资本，

(2) access to and mobilization of social capital, and (3) returns of social capital" (Lin, 1999a: 39).

Generally, there are various ways to categorize social capital. For instance, according to "whether the profit is accrued for the group or for the individuals" (Lin, 1999a: 31), social capital can be divided into two categories. The first category focuses on the use of social capital by individuals, that is, how individuals access and use resources embedded in social networks to gain returns. Representatives of this category include Lin (Lin & Bian, 1991), Burt (1992; 1997a), Portes (1998), and other scholars. The second category focuses on the use of social capital at the group level, that is, how groups develop and maintain social capital as a collective asset in order to enhance group members' lives. Representatives of this category include Bourdieu (1986), Coleman (1988; 1990), Putnam (1995a; Putnam, Leonardi, & Nanetti, 1993), Bryant and Norris (2002), and others.

Another very important way to discuss the dimensions of social capital is to distinguish between bridging (or inclusive) social capital and bonding (or exclusive) social capital. Normally, bonding social capital tends to "reinforce exclusive identities and homogenous groups" such as ethnic organizations and church-based groups. In addition, bridging social capital tends to include people across diverse social groups such as civil rights movement and social service groups (Putnam, 2000: 22). Comparatively, bonding social capital is good for supporting specific reciprocity and mobilizing solidarity, while bridging social capital tends to link external assets and does well for information diffusion. Therefore, bonding social capital tends to contribute to community well-being in a smaller circle, while bridging social capital "can generate broader identities and reciprocity" (Putnam, 2000: 23).

Even though Putnam suggested that both kinds of social capital have powerful positive social effects in terms of undergirding social participation and bringing people together, he also explicitly indicated that bonding social capital, "by creating strong in-group loyalty"

"通过在群体内形成强大的忠诚"能够"形成对抗外团体强大力量"(Putnam, 2000：23)。这个观点已经得到了其他学者的支持。在少数民族的背景下,有学者认为,在少数民族社区内部紧密团结的情况下,联合型社会资本有利于形成社会凝聚力。鉴于其群体内种族分离的影响和融入外面更广的社会中可能产生的代价(Cheong, Edwards, Goulbourne, & Solomos, 2007：29),一些人(Uslaner & Conley, 2003)对聚合型社会资本持消极态度。因此,在某种程度上,以聚合型群体为组织形式的文化和宗教领地被看作是整个国家的团结和社会凝聚力的威胁(Cheong et al., 2007：25)。然而,除了这些外部的负面影响,普特南认为聚合型和联合型社会资本不是社会网络划分的"非此即彼","而是'或大或小'的范围,这样我们可以对不同形式的社会资本进行比较"(Putnam, 2000：23)。

2.3.2 社会资本在城市发展中的作用

学术界普遍认为,社会资本在社会经济、政治、制度、和经济发展方面对城市发展有巨大的贡献(Coleman, 1988；Forrest & Kearns, 2001；Fukuyama, 2001；Grootaert, 1998；Kleinhans, Priemus, & Engbersen, 2007；Middleton, Murie, & Groves, 2005)。在其经典著作《社会资本在创造人力资本中的作用》一书中,科尔曼(Coleman, 1988)探索了各种形式的社会资本,他在多个方面,展示了社会资本在社会经济发展方面的价值。社会资本的价值在于借助其功能,能够识别特定的社会组织,这些社会组织对不同人员的价值在于他们可以在其中获益。科尔

tends to "create strong out-group antagonism" (Putnam, 2000: 23). This argument has been backed by others. In the context of the minority ethnic groups, Cheong et al. argued that bridging social capital is conducive to building social cohesion when the minority ethnic groups tie closely within their communities. Given its group segregation effects and the cost of the possible integration into the wider outside society (Cheong, Edwards, Goulbourne, & Solomos, 2007: 29), some academics (Uslaner & Conley, 2003) even cast negative tones to bonding social capital. Accordingly to some extent, the cultural and religious enclaves in forms of bonding groups are seen as threats to the solidarity and social cohesion of a bigger picture (Cheong et al., 2007: 25). Nevertheless, in spite of these negative external effects, Putnam maintained that bonding and bridging social capitals are not "either-or" categories into which social networks can be divided "but 'more or less' dimensions along which we can compare different forms of social capital" (Putnam, 2000: 23).

2.3.2 Role of social capital in urban development

Academics generally agree that social capital has a significant contribution to urban development in the aspect of socio-economic, political, institutional, and economic development (Coleman, 1988; Forrest & Kearns, 2001; Fukuyama, 2001; Grootaert, 1998; Kleinhans, Priemus, & Engbersen, 2007; Middleton, Murie, & Groves, 2005). In his seminal work *Social Capital in the Creation of Human Capital*, which explores various forms of social capital, Coleman (1988) demonstrated the value of social capital in terms of socio-economic development in a number of ways. The value of social capital lies in the fact that it identifies certain social structures by their functions, and the value of these social structures to the actors lies in the fact that they can use these structures for their interests. Coleman opined that social capital provides useful resources for individuals or organizations. Social capital contributes to the

曼（Coleman）认为，社会资本为个人或组织提供了有用的资源。社会资本有助于促进社会组织形成责任感，期望值和可信度。这也是通过社会互动获得信息的一个途径。此外，社会资本有利于形成社会规范，"人应该摒弃自我利益并为集体利益而努力"（Coleman, 1988：104）。因此，一个有丰富社会资本的社区，不仅能促进某些行为规范的形成，同时也对其他一些行为起了限制作用。

符库（Fukuyama, 2001）强调了社会资本在城市发展中对经济和政治发展的贡献。在经济发展方面，社会资本有助于减少与正式的协调机制相关的交易成本例如、合同、层次结构、官僚统治和其他成本。在政治发展方面，社会资本有助于公民参与和协会，这是保障建立有限政府和现代民主体制的必要条件（Fukuyama, 2001：7）。符库（Fukuyama）表示，一个丰富的社会资本存量能够形成一个团结的公民社会，它是实现现代自由民主的必要条件。此外，通过社会信任和参与，社会资本是实现正式的公共机构正常运转的关键条件（Fukuyama, 2001：12）。其他学者也已经强调了社会资本对社会经济、政治和经济发展的贡献。格鲁特（Grootaert, 1998：11）认为，通过当地的协会组织和网络，社会资本对地方居民的福祉有积极的影响。多方面显示社会资本有助于促进社会公平、消除贫困、提高教育质量、保护环境（Grootaert, 1998：12）。但是，学者也指出一个可能破坏社会资本，形成社会和经济衰退的恶性循环的不良发展趋势（Grootaert, 1998：18）。符库（Fukuyama）指出，国家对社会资本造成的负面影响可以这样解释：当国家经常干涉

obligations, expectations, and trustworthiness of the social structure. It is also a way to gain information through social interactions. In addition, social capital is conducive to forming social norms, where "one should forgo self-interest and act in the interests of collectivity" (Coleman, 1988: 104). Hence, a community with rich social capital cannot only facilitate certain actions, but also constrain others.

Fukuyama (2001) emphasized the contribution of social capital to the economic and political development in urban development. In the dimension of economic development, social capital helps to reduce the transaction costs normally associated with formal coordination mechanism for example, contracts, hierarchies, bureaucratic rule, and other costs. In the dimension of political development, social capital contributes to civic engagement and association, which are necessary for the success of a limited government and modern democracy (Fukuyama, 2001: 7). According to Fukuyama, an abundant stock of social capital presumably produces a dense civil society, which has been seen as a necessary condition for modern liberal democracy. In addition, by way of social trust and engagement, social capital is essential to the proper functioning of formal public institutions (Fukuyama, 2001: 12). The contribution of social capital to the socio-economic, political, and economic development has also been undergirded by others. Grootaert (1998: 11) opined that social capital, through local associations and networks, has a positive impact on the well-being of local households. It has been shown in various ways to affect the growth of social equity, poverty alleviation, improvement of education quality, and environmental management (Grootaert, 1998: 12). Nevertheless, academics also point out that an inappropriate path of development can destroy social capital and set forth a vicious circle of social and economic decline (Grootaert, 1998: 18). Fukuyama claimed that the negative impact of states on social capital is illustrated when the states often undertake

"私营部门或公民社会"的活动（Fukuyama, 2001：18）。

学者表示，除了能够促进社会经济和政治的进步，社会资本还有利于规范经济行为和商业活动。格拉诺维特（Granovetter）认为商业活动受到社会关系的严格约束，因此"把它们解释为独立的个体是严重的误解"（Granovetter, 1985：482。基于这一点，许多研究试图探索企业内部关系和相互合作的社会结构，以及在这些关系中社会资本的形成。商人做的决定不仅仅是"基于价格和质量因素，而且也基于其他属性,如信任和社会价值观"（Boschma, Lambooy, & Schutjens, 2002：19）。建立在个人关系和联系之上的诚信以及商业合作伙伴之间稳定的关系有助于规范经济行为并促进经济发展（Hakansson & Johanson, 1993：35）。实际上，学者相信社会关系中嵌入或嵌入式经济价值观不仅与基于信任的网络有密切的关系，例如，道德标准、亲属关系、友谊或同情，而且还与其他社会文化和制度因素相关（Storper, 1997：291）。司徒珀（Storper）甚至认为，如果企业之间没有相互依赖关系以及经济实体之间缺乏信任，那么企业在更广泛的发展空间里很难持续下去（Storper, 1997：292）。在不同的城市环境中，社会关系对企业的重要性已经得到了广泛认可。边燕杰与丘海雄（Bian & Qiu, 2000：88, 89）认为企业和生意的成功和他们拥有的网络或社会关系息息相关。下一章将会以中国城市的社会资本为研究案例继续讨论这一点。

2.3.3 社会资本的测量

正如上述所讨论的，既然社会资本的界定

activities that should be left to "the private sector or to civil society" (Fukuyama, 2001: 18).

In addition to the socio-economic and political progress, academics also claim that social capital is conducive to economic behavior and business activities. Granovetter argued that business activities are so constrained by social relations that "to construe them as independent is a grievous misunderstanding" (Granovetter, 1985: 482). Based on this, many studies try to explore the social construction of inter-firm relationships and collaborative interactions, as well as the creation of social capital among them. It is believed that businessmen make their decisions not only based "on information about prices and quantities but also on other attributes like trust and social values" (Boschma, Lambooy, & Schutjens, 2002: 19). Trustworthiness based on individual ties and connections and stable relations between business partners contribute to economic behavior and development (Hakansson & Johanson, 1993: 35). In fact, academics believe that embeddedness or embedded economic value in social relations has close relations with not only trust-based networks, for example, ethics, kinship, friendship or empathy, but also other socio-cultural and institutional factors (Storper, 1997: 291). Storper even contended that enterprises could hardly be sustained without the interdependencies of firms and trust among economic actors in the wider development structures (Storper, 1997: 292). The significance of social relations to enterprises has been acknowledged in various urban contexts. Bian and Qiu (2000: 88, 89) assumed that the success of enterprises or businesses closely relates to the networks or social relationships they have. The discussion on this will be continued in the study of social capital of urban China in the next chapter.

2.3.3 Social capital measurement

When there are different forms of social capital and various ways to define it as discussed

有不同的形式和方法，那么其测量方法也有不同的标准（Fukuyama, 2001：12）。尽管如此，许多学者在不同的层面，例如，微观层面、宏观层面中观（或中级）层面对社会资本测量的一些共同特征进行了概括。通常，微观或个人层面的社会资本指标与财富、社会地位、教育程度、和其他参数有关。中观层面，如社区团体和协会的社会资本指标包括会员数量、会议频率、种族亲属以及其他会员的范围。宏观层面的社会指标与国家机构的设计、法律和司法系统以及政府运作相关（Grootaert, 1998：15；Krishna, 2002：445）。

在普特南等人（Putnam et al., 1993）的著作中，通过评估信任和公民参与程度的调查数据，详细阐述了对中观层面社会资本的研究（Fukuyama, 2001：12）。通过计算对各种类型的会员的统计，如，在体育俱乐部、保龄球联盟、文学社团和其他组织，普特南认为"在过去的25年里，协会的会员的平均数量已经下降了1/4"，这也是美国社会的社会资本下降的一个迹象（Putnam, 1995a：71）。林（Lin）赞成用网络来测量社会资本。他认为，既然社会资本来源于社会网络和社会互相交流，那么它应该"根据其来源来测量"（Lin, 1999a：35）。与参与特定群体相比较，社会网络受地理和位置的约束较少。一般来说，有三个工具测量社会资本网络：提名生成法、位置生成法和资源生成法。前两种方法倾向于通过网络和互相交流评估社会资本，第三种方法试图在网络中，通过评估嵌入式资源测量社会资本。

above, it is also natural that there should be no consensus on how to measure it (Fukuyama, 2001: 12). Nonetheless, some common features of social capital measurement have been generalized by academics at different levels, for example, the micro, macro, and meso (or intermediate) levels. Usually, the indicators of social capital at the micro or individual level relate to wealth, social status, education level, and other parameters. The indicators of social capital at the intermediate level such as community groups and associations include the number of members, frequency of meetings, and dimensions of membership along ethnic, kinship, or other lines. The indicators of social capital at the macro level relate to the design of state institutions, legal and judicial systems, and government functioning (e.g., the ability to enforce contracts) (Grootaert, 1998: 15; Krishna, 2002: 445).

The studies of social capital at the intermediate level are well demonstrated in Putnam, Leonardi, and Nanetti's (1993) work by evaluating the survey data on grades of trust and civic engagement (Fukuyama, 2001: 12). By counting all types of group memberships, for example, in sports clubs, bowling leagues, literary societies, and other organizations, Putnam concluded that "the average number of associational memberships has fallen by about a fourth over the last quarter-century," which is an indication of the decline of social capital in American society (Putnam, 1995a: 71).

Lin supported the social capital measurement through networks, as he contended that considering social capital being rooted in social networks and social interactions, it should be "measured relative to its root" (Lin, 1999a: 35). Compared with participation in specific groups, social networks have less geographical and locational constraints. Generally, there are three tools of measuring social capital networks: name generator, position generator, and resource generator. The first two methods tend to assess social capital through networks and interactions, and the third one attempts to measure social capital by assessing the embedded resources within networks.

2.3.3.1 三种测量社会资本网络的方法
提名生成法/解释法

最全面也是最常用的社会资本测量方法是提名生成法/解释法（McCallister & Fischer, 1978：133）。这种方法把以自我为中心的社会网络当作之后社会资源库存的一个起点（Van Der Gaag & Snijders, 2003：6），它的目的是调查受访者和其所在网络的成员之间的关系（PRI, 2005a：22）。为了确定网络中的成员，这种方法通常提出下面一个典型的问题：

"你会和谁谈论个人问题？（Van Der Gaag & Snijders, 2003：6）"

通过分析受访者的答案，可以理解受访者和他们所指出的成员之间的关系，同时也可以了解这些成员的性格特点（Laumann, 1973：27）。尽管如此，一些学者表明，这种测量社会资本的不足之处是其具有模糊性，因为这种分布在很大程度上"受目录列表的内容或角色和数量的影响"（Lin, 1999b：476）。因此，提名生成法能更好地反映出封闭的地理区域中的更强的团结、联系或关系（Campbell & Lee, 1991：203）。

位置生成法

位置生成法是林南与杜敏（Lin & Dumin, 1986）最早发明的测量社会资本的方法，他们选定了很多行业，包括威望高的，也包括没有名气的，来询问受访者是否认识每个行业中的人。因此，这种方法不是询问与具体某个人的关系，而是调查他们与特定社会职位的社会联系，这里通常存在嵌入式的资源（Lin & Erickson, 2008：9）。这种方法使探索

2.3.3.1 Three tools of measuring Social capital networks
Name generator/interpreter

The most comprehensive, as well as the most commonly used social capital measurement method is the name generator/interpreter (McCallister & Fischer, 1978: 133). This method takes the ego-centered social network as a starting point for subsequent social resource inventory (Van Der Gaag & Snijders, 2003: 6), and it aims at investigating the relationship between the respondents and the members of their networks (PRI, 2005a: 22). To identify the members of the networks, this tool usually proposes a typical question such as the following:

With whom do you talk about personal matters? (Van Der Gaag & Snijders, 2003: 6)

By analyzing the respondents' answers, relationships between the respondents and their indicated members as well as the characteristics of these group members can be understood (Laumann, 1973: 27). Nonetheless, some academics indicate that the deficiency of this method as a social capital measurement is obscure because the distributions are very much "affected by the content or role and number of names" (Lin, 1999b: 476). Therefore, name generator better reflects the stronger ties, stronger relations, or relationships of close geographic limits (Campbell & Lee, 1991: 203).

Position generator

Lin and Dumin (1986) were the first to develop the method of social capital measurement known as position generator, wherein they selected a number of occupations ranging from high to low in prestige and asked respondents whether they knew anyone in each of the occupations. Hence, rather than asking about the relationships with particular people, this method investigates the social links to specific social positions in which various resources are usually embedded (Lin & Erickson, 2008: 9). This technique makes it possible to explore the potential resources

与微弱关系或次要关系相关的潜在资源成为可能。与提名生成法方法中常见的问题相比，位置生成法中通常会提到以下主要问题：

"你认识某行业中的某人吗？（PRI，2005a：24）"

根据加拿大政策研究机构（PRI）的研究，当人们成为更富裕、更多元化网络的一份子时，他们通常也会拥有一种丰富的、多样化的社会资本（PRI，2005a：24）。因此，这种方法经常被用于调查应聘工作或职务和社会关系之间的联系（Bian & Zhang, 2001；Flap & Volker, 2008；Lin & Dumin, 1986）。边燕杰和张文宏认为，社会网络和人际关系有助于得到工作，社交网络的形式以两种方式存在，一是主要方式，即提供个人友谊（人情）；第二种是次要方式，即传递信息（Bian & Zhang, 2001：77）。

资源生成法

基于以上两种方法，一些荷兰学者（Van Der Gaag & Snijders, 2004）已经研究了第三种测量社会资本的方法：资源生成法，其中问题直接涉及个人在他们扩展的网络里访问各种资源。学者认为，资源生成法通过一个清单来测量社会资本，当"在一个面试中下，一个人是否被录用依赖于一个有用的、具体的社会资源的清单"。其背后的逻辑是，通过社交网络进行交换资源（Van Der Gaag, Snijders, & Flap, 2008：29）。资源生成法的主要问题是：

"你认识可以让你使用x类型资源的人吗？（PRI，2005a：24）"

associated with weak ties or peripheral relationships (Burt, 1997b). Compared with the common question asked in the name generator technique, the main question of the position generator is usually posed as follows:

Do you know someone in the X profession? (PRI, 2005a: 24)

According to the Canadian Policy Research Initiative (PRI), when people belong to the wealthier and more diversified networks, they also usually possess a wealthy and diversified social capital (PRI, 2005a: 24). Therefore, this technique has been frequently used to investigate the relationships between job attainment or employment positions and social relationships (Bian & Zhang, 2001; Flap & Volker, 2008; Lin & Dumin, 1986). Bian and Zhang argued that social networks and human relationships contribute to job attainment, and the forms of social networks exist mainly in two ways, providing individual friendship (*Renqing*) as the predominant way and delivering information as the subordinate way (Bian & Zhang, 2001: 77).

Resource generator

Based on the previous two methods, some Dutch scholars (Van Der Gaag & Snijders, 2004) have developed a third one: resource generator, in which the questions directly concern the types of resources that individuals can access within their extended networks. According to Van Der Gaag, Snijders, and Flap, resource generator measures social capital by means of a checklist, when "in an interview situation, access is checked against a list of useful and concrete social resources." The logic behind this is that resource exchanges take place through social networks (Van Der Gaag, Snijders, & Flap, 2008: 29). The main resource generator question is usually as follows:

Do you know someone who can potentially give you access to type X resources? (PRI, 2005a: 24)

许多研究中都已经使用了资源生成法的方法。格鲁特（Grootaert）进行了测量社会资本的综合调查（简称SC-IQ），通过网络使用权和各种个人和家庭层面的参与，深入探究了宝贵的资源（Grootaert et al., 2004）。此外，英国国家统计局（简称ONS）所进行的社会资本调查，实际上通过网络和在国家层面上，"嫁接"社会资本或关系，调查了嵌入式资源（Bryant & Norris, 2002）。这种方法也被世界银行所采用，在发展中国家的社区层面上探讨嵌入式资源，特别是在社区主导型发展（简称CDD）的嵌入式资源[1]。

以上对三种社会资本测量方法的简要回顾表明，每种方法都有优缺点，只有根据我们自己特有的需要和目标来制定，才是一个理想的测量方案。在每个社会资本测量方法中，为了取得一些具体指标，通常遵循两个路线，即演绎和归纳的方法（VanDerGaag & Snijders, 2003：5）。本文将在下一部分对其进一步讨论。

2.3.3.2 社区层面的社会资本常见指标

根据演绎方法，首先要区分重要的生活领域或方面。因此，一个特定的社会资本要在每个领域分别进行测量，这就需要制定覆盖多重领域的问卷调查项目（Van Der Gaag & Snijders, 2003：5）。但是在现实中，很难

The resource generator method has been used in a number of studies. Grootaert explored the Integrated Questionnaire for the Measurement of Social Capital (SC-IQ), which delves into the valued resources through network access and various forms of participation at the individual and household level (Grootaert et al., 2004). Furthermore, the Social Capital Surveys developed by the Office of National Statistics (ONS) in the U.K. actually look into the embedded resources through networks and "bridge" social capital or ties at the country level (Bryant & Norris, 2002). This method has also been adopted by the Word Bank in examining the embedded resources at the community level in the developing countries, particularly in the projects of the Community Driven Development (CDD)[1].

The above brief overview of the three social capital measurement methods indicates that each method has its own merits and disadvantages, and an ideal measurement scheme can be constructed only according to our specific needs and goals. In each social capital measurement method, in order to obtain some specific indicators, two routes can be usually followed, namely, deductive and inductive approaches (Van Der Gaag & Snijders, 2003: 5). This book will discuss more on this in the next section.

2.3.3.2 Common indicators of social capital at the community level

In the deductive way, the important life domains or perspectives need to be distinguished first. Consequently, a specific social capital will be measured separately in each domain, which will lead to the multiple domain-covering questionnaire items (Van Der Gaag & Snijders, 2003: 5). In reality, however, specific clues on the social capital

---

[1] 关于社区主导型发展（简称CCD）更多的细节讨论以及测量工具——社会资本测量工具（SOCAY）以及社会资本综合调查问卷（SC-IQ）：是由世界银开发采用的，用来研究发展中国家社区层面的社会资本，参见格鲁特等人（Grootaert et al., 2004）的研究。

[1] More detailed discussions on Community Driven Development (CDD) and the measuring tools—the Social Capital Assessment Tool (SOCAT) and the Social Capital Integrated Questionnaire (SC-IQ)—developed and adopted by World Bank to examine the social capital at community levels in developing countries, see the study by Grootaert (Grootaert et al., 2004).

分清各种生活领域的社会资本具体线索（Flap，2002：29）。上面提到的一些实证研究就采用了这种方法，如2002年，国家统计局进行的社会资本调查（Ruston，2002）。哈比特（Habbitt）的工作对在社区层面上社会资本指标的产生，提供了一个很好的例子。在探讨社区内部的社会资本中，建立和探讨了社区参与的一些观点，互相信任和期望，代表性的横向和纵向的联系，社会资本形成的协同效应（Hibbitt et al., 2001：159）。

相比之下，归纳方法通常是建立或收集多重社会资本项目。然后通过调查这些社会资本项目之间的相关性，确定单独的领域（Van Der Gaag & Snijders, 2003：5）。通过在洛杉矶微微联盟区域的一个案例研究，哈金森（Hutchinson, 2004）探讨了社会资本积累和社区建设之间的关系。在她的研究中，首先收集本地信息收集，包括社区参与、依附感、居住期限、相识、互惠、信任和其他信息。用许多当前的和常用的指标，来比较这些诱导因素，例如，"对社区的依附感、地理区域内的交际圈、邻里之间信任程度、邻里之间互惠的水平，参与地方社区发展的积极性"（Hutchinson, 2004：170），从而探索了或测量了一个特定社区的社会资本存量，也有助于理解社会资本存量和一个特定社区的建立之间的关系。

基于上述两种方法，通过对社区层面的社会资本指标的探索，本文概括了一些常见指标，见表2-2。然而，应该指出的是，这些指标只适用于社区层面上。特定的社会资本指标只能通过一个特定的研究背景生成。有时，即使指标是相同的，在不同的背景下可能会产生不同的结果。

of various life domains are not easy to distinguish (Flap, 2002: 29). Some of the empirical studies mentioned above belong to this approach, such as the social capital surveys developed by the ONS in 2002 (Ruston, 2002). In generating social capital indicators at the community level, Hibbitt's work offers a good example. Several perspectives on community involvement, trust and expectations of reciprocity, cross-sectional horizontal and vertical linkages, and the synergy of social capital formation are set up and explored in examining social capital within neighborhoods (Hibbitt et al., 2001: 159).

Comparatively, in the inductive approach, multiple sets of social capital items are usually created or collected first. Separate domains are then established by investigating the correlation among these social capital items (Van Der Gaag & Snijders, 2003: 5). Through a case study in the Pico Union area of Los Angeles, Hutchinson (2004) explored the relationship between the social capital accrual and community building. In her study, local information was collected first, covering community involvement, attachment, length of residence, acquaintances, reciprocity, trust, and other information. By comparing these induced items with many current and commonly used indicators, for example, "attachment to the community, range of acquaintances in the geographical area, levels of trust in neighbors, levels of reciprocity among neighbors, and involvement in local initiatives for neighborhood improvement" (Hutchinson, 2004: 170), it examines or measures the stock of social capital in a particular community. This also helps to understand the relationships between the stock of social capital and the building of a specific community.

By exploring the studies on social capital indicators at the community level based on the above two ways, this book generalizes some commonly referred to indicators shown in Table 2-2. However, it should be noted that these indicators only provide a general picture of what usually are considered at the

社会资本的常见指标　表2-2
Common indicators on social capital　Table 2-2

社会资本的分类 Categorization of Social Capital	内容 Contents	预期影响或结果 Expected Implication or Outcomes
网络和群体成员 Network and group membership	社区组织和网络 Community organizations and networks	社会凝聚力和包容性 Social cohesion and inclusion
社区观念 Community perceptions	信任、规范和团结 Trust, norms and solidarity	权力和公民参与 利益互惠 Empowerment and civic engagement; Reciprocity
个人或社区互动 Individual or community interactions	社区互动和合作 Community interactions and cooperation 社区参与 Community involvement 信息共享 Information sharing	

资料来源：作者汇总自（Bryant & Norris, 2002; Grootaert, Narayan, Jones, & Woolcock, 2004; Hibbitt, Jones, & Meegan, 2001; Hutchinson, 2004; Ng, 2005a; Putnam, 1993, 1995b; Putnam, 2000; Ruston, 2002）

Source: Author synthesized from (Bryant & Norris, 2002; Grootaert, Narayan, Jones, & Woolcock, 2004; Hibbitt, Jones, & Meegan, 2001; Hutchinson, 2004; Ng, 2005a; Putnam, 1993, 1995b; Putnam, 2000; Ruston, 2002)

2.3.4 影响社会资本的因素

虽然不同城市背景下，影响社会资本生成的因素是多种多样的，但本节主要讨论两个因素，即，城市管治模式和民族、宗教问题，这些对城市更新政策有显著的相关性。文献表明，城市管治模式影响联合型社会资本的水平，而民族和宗教问题影响聚合型社会资本水平。其他一些影响特定社区中的社会资本的因素，将在下面具体的案例分析中进行讨论。

城市管治模式

一般来说，社区和地方政府之间的互动和关系常常反映城市管治模式。因此，城市管治模式往往会影响社区与地方政府之间联合型社会资本的水平。斯图特（Stewart，2005：160）认为，可以通过改善社区和政府之间的关系来累积社会资本。这种关系在格拉诺维特

community level. The specific social capital indicators can only be generated through a specific study context. Sometimes, even if the indicators are the same, the outcomes may be interpreted very differently in different contexts.

2.3.4 Factors affecting social capital

Although there are various factors influencing the generation of social capital in different urban contexts, this chapter mainly focuses on two factors, namely, mode of urban governance and ethnic and religious issues, which have significant relevance to the urban regeneration policy. Literature shows that the mode of urban governance exerts an impact on the level of bridging social capital, and ethnic and religious issues have an impact on the level of bonding social capital. Some other factors affecting social capital within particular communities will be discussed in later specific case studies.

Mode of urban governance

Generally, urban governance is often revealed through the interactions and the relationships between the communities and the local government. Therefore, the mode of urban governance tends to exert an impact on the level of bridging social capital between the communities and the local government. Stewart (2005: 160) contended that social capital could be accrued by improving the relationships between the community and the government. This relationship belongs to the weak tie in Granovetter's theory, and he indicated that these "weak ties provide people with access to information and resources beyond those available in their own social circle" (Granovetter, 1983: 209). By studying the relationships between social capital and the local government, Lowndes and Wilson pointed out that the institutions of local governance shape the conditions wherein social networks and individual relationships are generated and prospected (Lowndes & Wilson, 2001: 629).

At the same time, higher level of bridging social capital facilitates the community-based urban governance through community

(Granovetter)的理论中属于微弱关系，并且他认为这种"微弱的关系可以让人们使用在他们自己社交圈之外的信息和资源"(Granovetter, 1983：209)。通过研究社会资本和地方政府之间的关系，朗底斯(Lowndes)和维尔信(Wilson)指出，地方管治体系形成一种情况，在这种情况下，可以生成和展望社会网络和人际关系(Lowndes & Wilson, 2001：629)。

同时，通过社区参与城市发展和管理问题，高水平的联合型社会资本促进以社区为基础的城市管治模式的建立(Putnam, 1998；Saegert & Winkel, 1998)。在任何一种城市管治模式的实施中，交易成本是不可避免的，因为在市场规则下，在谈判和交易的过程中都会产生费用。不过，斯图特声称，如果社区成员认识或相信对方，就可以减少地方管治的成本(Stewart, 2005：149)。普特南认为，社会资本"对公民之间以及公民和政府之间的合作行动起到润滑剂的作用"。此外，"如果不能提供充足的社会资本——就是没有公民参与、健康社区机构、互利互惠的规范和信任——那么社会制度就会摇摆不定"(Putnam, 1998：v)。意识到联合型社会资本和地方城市管治模式之间的关系，能够促进社区管治的进程，进而促进地方社区的发展。

民族和宗教问题

大量的研究已经证实了民族问题对社会资本的生成有很大的影响，特别是对特定少数民族成员之间的聚合型社会资本(Anthias, 2007；Goulbourne & Solomos, 2003；Granovetter, 1985；Zhou, 2005c)。一些学者通过引进"种族划分作为社会资本"的概念，探讨了社会资本和民族社区之间的关系(Goulbourne & Solomos,

participation in urban development and management issues (Putnam, 1998; Saegert & Winkel, 1998). In the implementation of any kind of governance, transaction costs are inevitable because under market rules there are costs on such processes as negotiation and exchange. However, the costs in local governance, according to Stewart, could be lessened if the community members knew or trusted each other (Stewart, 2005: 149). Putnam contended that social capital "lubricates co-operative action among both citizens and their institutions." In addition, "without adequate supplies of social capital—that is, without civic engagement, healthy community institutions, norms of mutual reciprocity, and trust—social institutions falter" (Putnam, 1998: v). The growing realization on the relationships between the bridging social capital and the local mode of urban governance promotes the progress of institutionalized community governance and therefore local community development.

Ethnic and religious issues

A number of studies have identified ethnic issues as a significant factor affecting the generation of social capital, especially of bonding social capital among the members of particular ethnic groups (Anthias, 2007; Goulbourne & Solomos, 2003; Granovetter, 1985; Zhou, 2005c). Some academics have explored the relationship between social capital and ethnic communities through the concept of "ethnicity as social capital" (Goulbourne & Solomos, 2003; Zhou, 2005c). Shah defined the "ethnicity of social capital" as *a set of resources, norms, obligations and expectations, information channels and cultural endowments that inheres in the structure of social relations within an ethnic community*" (Shah, 2007: 29). Anthias claimed that the ethnic resources could be seen as bonding social capital in the form of social ties and networks when they can be effectively mobilized. It implies that ethnic resources could be transferred to social resources, and they potentially play a part in building the relationships among the ethnic

2003；Zhou, 2005c）。莎（Shah）将"社会资本的种族划分"定义为是"少数民族中社会关系结构中固有的一组资源、规范、义务和期望、信息渠道和文化禀赋"（Shah, 2007：29）。安迪亚斯（Anthias）声称当民族资源可以被有效利用，可以视其为社会关系和网络形式的聚合型社会资本。它意味着民族资源可能转换为社会资源，他们对建立少数民族成员之间的关系可能发挥作用（Anthias, 2007：789）。

其他学者已经强调了宗教在社区建设中的意义（Durkheim, 1969；Hutchinson, 2004；Misztal & Shupe, 1998）。就像敦克海姆（Durkheim）所言，"宗教是社会的突出现象"。这是因为宗教的表现形式通常是以集体活动的形式，如宗教仪式，"在这些群体中，激励、保持或重新创建特定的心理状态"（Durkheim, 1969：22）。集体活动会给宗教成员带来许多社会互动机会。此外，宗教教义往往形成许多共享的生活原则和规则。遵循这些共同的生活规范，相应地，宗教成员在他们的特定的团体中，就会建立更多的信任。所有这些极大促进了聚合型社会资本在宗教团体成员中的积累。例如，在哈金森（Hutchinson）的研究中，关于宗教对洛杉矶微微联盟地区社区的建设的贡献，哈金森（Hutchinson）指出参加教堂活动的社区成员之间更容易建立信任，同时他们也比那些不参与教堂活动的人更加积极地参与社区活动中（Hutchinson, 2004：171）。

2.3.5 社会资本和城市更新政策

一般来说，社会资本和城市更新实践之间的关系是相互的。一方面，基于社区存量的社区参与助于促进城市更新过程；另一方

group members (Anthias, 2007: 789).

The significance of religion in community building has been underlined by others (Durkheim, 1969; Hutchinson, 2004; Misztal & Shupe, 1998). As Durkheim asserted, "religion is something eminently social." This is because religious representations are often collective representations, such as religious rites, "to excite, maintain or recreate certain mental states in these groups" (Durkheim, 1969: 22). The collective activities bring many social interaction opportunities for the religious members. In addition, religious doctrines tend to form many shared living principles and rules. Following these common living norms, the religious members accordingly tend to build more trust among their particular groups. All these substantially contribute to the accrual of bonding social capital among the religious group members. For example, Hutchinson's study, on the contribution of religion to community building in the Pico Union area of Los Angeles, indicated that community members attending churches are more likely to build trust among them and also show more participation in community activities than non-church attendees (Hutchinson, 2004: 171).

2.3.5 Social capital and urban regeneration policies

Generally, mutual relationships between social capital and urban regeneration practices exist. On one hand, community involvement based on the stock of social capital contributes to urban regeneration processes; on the other hand, social capital existing in community groups and organizations is often undermined during pro-growth urban regeneration processes.

2.3.5.1 Social cost and residential relocation strategies

The social impact of urban development policies has been discussed by academics for a long time. In Young and Willmott's (1957) study on the impact of the post-war redevelopment in London's East End, it

面，存在于社区团体和组织的社会资本在倾向经济发展的城市更新进程中常常被破坏。

2.3.5.1 社会成本和住宅重置策略

长期以来，许多学者已经对城市更新政策的社会影响进行了讨论。在对二战后伦敦东区的再开发的研究中，杨等学者（Young & Willmott, 1957）指出，再开发项目对工人阶层社区中历史悠久的家族和血缘造成了严重的破坏。但是，直到最近才开始从社会资本的角度重视这个问题（Forrest & Kearns, 2001；Ginsburg, 1999；Kearns, 2003；Kleinhans et al., 2007；Middleton et al., 2005；Saegert & Winkel, 1998）。这在很大程度上，与最近在关于城市更新的政治辩论中把社会资本作为一个主题有关（Kleinhans et al., 2007：1069）。

文献表明，城市工人阶级地区固有的稳定性[1]往往会使居民之间会产生更多的血缘关系和长期友好关系。在许多更新项目中，住宅重置策略的实施对本地已有的社区关系有很大的影响（Gans, 1962；Young & Willmott, 1957）。虽然城市更新的预期结果是实现"经济、物质、社会和环境条件的改善"（Roberts, 2000：17），但是 Ginsburg 指出，实际上，城市更新经常采取以房地产为主导或倾向经济发展的策略，而忽略了对社会方面的考虑（Ginsburg, 1999：56）。金斯博（Ginsburg）总结了在不发达地区城市更新项目中的社会经济成本，其中包括破坏了当地支持和信任的社会网络、没有人愿意住在新建的社区、忽略大部分被隔离在外的人的需求（Ginsburg, 1999：56）。

---

[1] 这里固有的稳定性指的是居民没有空间的、社会的或职业的流动性。

points out that the redevelopment program severely destroys many long-established family and kinship networks of working-class communities. Nevertheless, it is rather recent to address this issue from the perspective of social capital (Forrest & Kearns, 2001; Ginsburg, 1999; Kearns, 2003; Kleinhans et al., 2007; Middleton et al., 2005; Saegert & Winkel, 1998). This has much to do with the recent introduction of social capital as a topic in the political debate on urban regeneration (Kleinhans et al., 2007: 1069).

Literature indicates that the inherent stability[1] within city working class areas tends to generate more kinship ties and long-term friendly relationships among communities. With the implementation of residential relocation strategies in many regeneration programs, it often has substantial impacts on locally established community relationships (Gans, 1962; Young & Willmott, 1957). Although the output of urban regeneration is expected to be the "improvements in the economic, physical, social, and environmental condition" (Roberts, 2000: 17), as Ginsburg put it, the reality is that urban regeneration frequently turns out to be property-led and economic growth-biased development without the consideration of the social aspect (Ginsburg, 1999: 56). Ginsburg summarized the socio-economic costs of urban regeneration programs in underdeveloped areas in terms of damaging local networks of support and trust, building new neighborhoods that no one tends to live in, and ignoring the needs of the most excluded (Ginsburg, 1999: 56).

The essential characteristics of urban renewal, the one strategy supported in many cities, are expressed by Couch as bringing about the "changes in the use or occupancy of urban land and buildings and therefore results[resulting] in changes in where, how and under what conditions people live" (Couch, 1990: 79). In the context of Hong Kong, Susnik

---

[1] Here, inherent stability means the lack of spatial, social or occupational mobility of residents.

许多城市都支持城市改造战略，它的本质特征，就像考齐（Couch）认为的那样，能够带来"城市土地和建筑使用或占用方式的变化，从而改变了人们的居住环境、生活方式，以及生活条件"（Couch，1990：79）。在香港的城市改造中，苏尼克（Susnik）表示，尽管城市更新有其积极的一面，但是居民重置项目的实施也产生了巨大的社会成本，尤其是对那些低收入的居民而言（Susnik，1997：11）。这个观点已经得到其他学者的支持。吴（Wu）认为，"一般情况下，家庭可以在搬迁中获益"，但是，"利益的分布倾向于那些有较高的教育水平，他们在住房制度中能够得到更好的安置"（Wu，2004：466）。

2.3.5.2 促进城市更新进程的社会资本

已经有许多学者肯定了社会资本对城市更新进程的促进作用（Saegert & Winkel, 1998; Temkin & Rohe, 1998）。通过对纽约贫困城市更新中社会资本的研究，学者们总结道，社会资本能够积极促进破败建筑的改善，社会资本能够使在贫困居住区的政府投资增值。除此之外，他们的研究还证明了社会资本对贫困家庭的生计、提高他们的经济水平以及社区参与不良环境的改善，都有重大影响（Saegert & Winkel, 1998）。

在城市更新实践中，在社会网络和社会联系方面的社会资本，既关系到地方社区和政府机构之间的纵向关系，也关系到社区居民之间内部的横向关系（PRI, 2005b, 2005c; Temkin & Rohe, 1998）。研究表明，社会资本通过集中各个层面的意见来影响决策过程。例如，社会资本问题已经渗透到了世界银行

argued that although urban renewal has its positive side, it exacts a great social cost due to the residential relocation programs, especially to those residents with low incomes (Susnik, 1997: 11). This proposition has been supported by many others. Wu asserted that "households in general gain a benefit through relocation" but "the distribution of the benefits is skewed towards those who have higher education levels and are better positioned in the housing system" (Wu, 2004: 466).

2.3.5.2 Social capital facilitating the urban regeneration process

The contribution of social capital to urban regeneration processes has been supported by many scholars (Saegert & Winkel, 1998; Temkin & Rohe, 1998). By examining the social capital in the regeneration of the distressed inner cities of New York, Saegert and Winkel concluded that social capital can be an effective element to improve distressed buildings and that social capital adds more value to the government investment in disadvantaged living areas. Moreover, their study proves that social capital does have implications in the survival and economic improvement of poor households and community participation in distressed environments (Saegert & Winkel, 1998).

During urban regeneration practices, social capital in terms of social networks and connections concerns both vertical ties between the local community and government agencies and horizontal ties among the communities themselves (PRI, 2005b, 2005c; Temkin & Rohe, 1998). Studies show that social capital has an influence on policy-making through the integration of opinions at various levels. For example, social capital is involved with the policy discourse of the World Bank, especially during the urban regeneration practices in underdeveloped countries (WB, 1998: 6).

The urban regeneration studies in Merseyside, U.K., indicate that community participation in the regeneration program

的政策讨论环节中，尤其是针对欠发达国家或地区的城市更新实践当中（WB，1998：6）。

英国默西赛德郡（Merseyside）的城市更新研究表明，在更新项目中，基于社会关系和联系的公民参与，有助于"打破居民和决策者之间的壁垒"（Hibbitt et al.，2001：160）。米斯塔（Misztal）也支持这个观点，说应该关注"人与人之间以及人民与决策者之间的关系"（Misztal，1996：269）。因此，社会资本在城市更新中的作用"需要根据潜在联系和网络来判断，这种联系和网络可以通过社区居民和更广泛的权力结构以及机构之间的社会资本的"扩大"的元素来培养（Hibbitt et al.，2001：159）。这些讨论突出了社会资本对社区参与的促进作用，这对城市更新的进程有极其重大的意义。

2.4 总结

通过回顾西方文献中城市更新的演变过程，本文发现对于城市更新的研究，经历了从以改变自然环境为目标到以综合发展目标的转变。然而，研究发现，在实践中，城市更新远离理想的综合目标，往往倾向经济发展，而忽略或损害了其他方面，尤其是社会发展方面。在回顾了亚洲城市以保护为主导的更新实践之后，本章指出，社区参与有助于实现历史城市全面更新的目标。本文认为，在历史城市的城市更新实践中，不同阶层的参与是非常重要的，并且，制度化的参与城市管治是必不可少的。

在实践中，破旧城区采取的以再开发为主导的更新策略，通常伴随着居民重新安置和搬迁，从而在许多地方产生了城市"绅士化"现象。这对社会公平和社区凝聚力带来

based on social relationships and connections contributes to "breaking down barriers between residents and policy-makers" (Hibbitt et al., 2001: 160). Barbara Misztal also supported the idea that attention should be given to "the relationships among people and between people and decision makers" (Misztal, 1996: 269). Accordingly, the role of social capital in urban regeneration "needs to be judged in terms of the potential linkages and networks that it can nurture through the 'scaling-up' of elements of social capital between communities and wider power structures and institutions" (Hibbitt et al., 2001: 159). These discussions highlight the contribution of social capital to community-based participation, which is essentially significant to the urban regeneration process.

2.4 Conclusion

By reviewing the evolution of urban regeneration in Western literature, this book was able to discover studies on urban regeneration that have experienced the transition from the physical environment-centered to holistic development goals. However, it found that despite the presumed comprehensive goals, urban regeneration in reality is often economic-biased and some other aspects, social development in particular, have been neglected or undermined. After the overview of the conservation-led regeneration practices in Asian cities, this chapter pointed out that community participation contributes to the comprehensive regeneration aims of historic cities. This book contends that the involvement of various actors in urban regeneration practices of historic cities is important, and institutionalized participatory urban governance is necessary.

In practice, the redevelopment-led regeneration strategies in dilapidated inner-

了许多挑战。因此，在城市更新规划中需要通过社区居民的参与来对地方社区进行维护。同时，在历史城市更新中，城市保护是一个很重要的问题。在可持续发展的框架中，城市保护不仅是对物质遗产的保护也是对非物质文化遗产的保护，其中已经对"活态遗产"和"本地生活"两个概念进行了讨论。这种趋势进一步说明，地方原住民在历史居民区的保护中的重要性。

为了促进社区参与到城市更新进程中，以社区为基础的管治极其重要，而且地方政府应该发挥关键作用。除了参与式的城市管治模式之外，文献也表明社会资本有助于促进社区参与到城市更新实践中。通过回顾社会资本测量的各种方法，已经确定了社区层面上的一些常见社会资本指标，包括社区网络、社区信任和规范以及社区参与。此外，本文还探讨了在城市更新中，影响社会资本性能的两个关键因素，即，地方城市管治模式以及民族和宗教问题。通过探讨社会资本和城市更新政策之间的相互关系，本文认为社会资本有利于推动参与式城市更新的过程。

urban areas often undertake residential resettlement and relocation, which result in urban gentrification in many places. This poses many challenges to social equity and to the establishment of cohesive communities. Therefore, it calls for the maintenance of local communities through their involvement in the regeneration plans. At the same time, in the regeneration of historic cities, urban conservation is an important issue. In the framework of sustainable development, the concept of urban conservation concerns not only the tangible but also intangible heritage, where the notions of "living heritage site" and "indigenous lives" have been discussed. This trend further pinpoints the significance of involving local indigenous residents in the conservation of historic residential areas.

To promote community participation in the regeneration process, community-based governance is significant, and the local government should play a key role. In addition to participatory urban governance, the literature shows that social capital contributes to community participation in the urban regeneration practices. By reviewing the various methods of social capital measurement, some common social capital indicators at the community level have been identified including community networks, community trust and norms, and community involvement. In addition, two key factors affecting the performance of social capital in urban regeneration have been explored, that is, the mode of local urban governance and ethnic and religious issues. By exploring the interactions between social capital and urban regeneration policies, this book contends that social capital contributes to participatory urban regeneration processes.

第三章 转型期中国历史城市的更新与社会资本：研究架构的建立

CHAPTER 3 Urban Regeneration and Social Capital in Historic Cities of Transitional China: Towards an Analytical Framework

对城市更新、城市保护、社会资本、西方文献中的相关因素以及它们之间的关系的回顾有助于对历史城市更新中社会资本作用的正确理解。接下来，本章将探讨中国历史城市背景下的有关问题。

首先，本章简要回顾了中国城市的发展和更新演变过程，旨在对中国城市的改造和更新的城市政策演变过程提供一种历史和语境下的解读。其次，本章探讨了两个重要的城市更新方法：以再开发为主导的方法和城市保护，这都和中国历史城市中的更新政策和策略相关。同时，本章也回顾和分析了中国城市的保护政策和法规的演变。再次，本章还探讨了中国城市更新管治模式的转变，这种模式和以再开发为主导、倾向经济发展的更新政策紧密相关。此外，本章还探讨了当前企业的管治模式。最后，为了促进社区居民参与到中国历史城市的保护和更新过程中，本章探讨了社会资本问题，及其对更新过程中社区参与的促进作用。基于之前和现在对城市更新政策的演变、城市管治模式、城市保护法律法规、社会资本状况以及它们之间错综复杂的关系的讨论，本章建立了一个分析框架，以便更好地理解中国历史城市的社会资本和城市更新问题，本章最后展示了该分析框架。

The review on the issues of urban regeneration, urban conservation, social capital, and related factors in Western literature as well as the relationships among them contributes to the general understanding of social capital in the regeneration of historic urban areas. Subsequently, this chapter explores the corresponding issues in the context of urban China.

First, this chapter briefly reviews urban development and the regeneration evolution in urban China, which provides a historical and contextual understanding of Chinese cities' evolving urban policies on urban renewal and regeneration. Second, this chapter probes into two outstanding urban approaches: redevelopment-led approach and urban conservation, which border on urban regeneration policies and strategies in Chinese historic cities. In relation to this, the evolving urban policies and regulations on urban conservation in China are also reviewed and analyzed. Third, the transforming mode of urban governance in Chinese cities, which is closely related to the redevelopment-led and pro-growth urban regeneration policy, is explored. In addition, this chapter examines the current entrepreneurial mode of governance. Finally, in order to promote community participation in the conservation and regeneration processes of Chinese historic cities, this chapter investigates the issue of social capital and its contribution to community participation in the urban regeneration process. Based on the discussions in the preceding and current chapters on the transformation of urban regeneration policies, mode of urban governance, laws and regulations on urban conservation, and the situation of social capital, as well as the intricate relationships among them, an analytical framework towards a better understanding of social capital and urban regeneration in Chinese historic cities is developed, and it is presented at the end of this chapter.

3.1 自1949年以来中国历史城市的发展与更新

自1949年中华人民共和国成立以来，全国各地城市发生了很大的发展与改变。总体而言，20世纪70年代以前，我国的城市发展以中央集中规划和对共产主义价值观倾向的发展政策为主导（Xie & Costa, 1993：111）。随着1978年改革开放政策的实施，更多关于城市发展目标的话题开始出现。中央政府逐渐把发展经济作为国家发展的重要目标。由于城市经济发展受到之前计划经济体制的束缚，所以进行了大量的制度改革。随着中国政治和经济体制从过去计划经济体制转变为目前的市场经济体制，财政权力的分散和再分配为地方市政发展提供了很大的空间和机遇（Wu, 1997：641）。但是，随着转型期经济体制的建立，地方政府开始关注当地发展的需求，其中一个就是需要增加地方财政收入，而这些之前在计划经济体制下是由中央政府按计划进行划拨的（Ng & Wu, 1995：291）。1986年的土地改革和1988年的住房改革，为地方政府通过房地产开发来增加收入提供了契机（Xie & Ingram, 2008：9）。20世纪90年代是中国经济快速发展的重要时期，其特点是全国范围内开展了倾向于经济发展的再开发活动。

3.1 Urban Development and Regeneration in China from 1949 to Present

Since the establishment of the People's Republic of China (P.R. China) in 1949, massive urban development and changes have taken place in the whole country. Generally, before the 1980s, urban development in China was dominated by the centralized planning and ideology-biased national policies towards socialist values (Xie & Costa, 1993: 111). With the open-door policy introduced in 1978, more concerns on urban development objectives began to show up. Since then, the central government put economic growth as the first priority of national development. When urban economic development was shackled by the previous planned-economic institutions, massive institutional changes took place. With the transition of Chinese political and economic institutions from the previous socialist to the current more market-oriented ones, the decentralization and redistribution of fiscal power provided much space and opportunities for local municipal development (Wu, 1997: 641). Nonetheless, with this transitional economic institution, local governments also began to face their needs, one of which was the need to boost their own revenues, which for a time had been allocated by the central government in the previous planned-economic institution (Ng & Wu, 1995: 291). The following land reform in 1986 and housing reform in 1988 offered many incentives to local governments in terms of economic returns from real estate development activities (Xie & Ingram, 2008: 9). The 1990s witnessed a massive economic growth period in China,

这也给很多历史城市带来不可逆转的改变。到21世纪20年代为止，之前以快速城市化和经济增长为特点的实践发展给许多城市带来了社会和环境方面的挑战。相应地，国家和地方的城市发展政策也开始呈现出很多变化，并且城市发展目标中更加"全面的"思路[1]开始显现（Hu，2001；NDRC，2007；Wang，2000；Zhou，2002，2004a）。所有这些对中国历史城市的更新政策与实践演变都产生了很大影响。

要全面理解中国的城市更新不仅需要对其城市转型历史的回顾，而且也需要对一些城市发展的问题，诸如城市规划、地方政府、政治和经济体制及其他关键因素进行重新解释。尽管1949年以来就有许多关于中国城市的研究，像城市政策和体制研究（Ma，1979，2002；Oi，1995；Sit，2009；Walder，1995b）、城市规划机制和发展动态（Ng & Tang，2004a，2004b）以及一些对特定城市的改建或更新的讨论（Bao，1993；Chen，1993；Ng，Cook，& Chui，2001；Ng & Liang，2007；Ng & Tang，2002a，2002b；Qu，1993），但是很少有针对转型期中国背景下的城市改造或更新的研究。

根据1949年以来的社会经济、政治发展和转型期城市更新政策的特点，对中国城市更新动态及影响因素的探讨可以划分为以下五个阶段：(1)按照总规划进行的以工业发展为主导和破旧城区的修复（1949—1965）；(2)无规划下的零星的城市建设及旧城区的

---

[1] 尽管经济增长仍在城市更新政策中占主导地位，但是自21世纪初以来，很多地方政府开始在其更新政策中进行更加全面的考量。

which was characterized by economic-biased urban redevelopment throughout the country and brought many irreversible changes to many Chinese historic cities. By the early 2000s, previous pragmatic developments characterized by rapid urbanization and economic growth imposed many social and environmental challenges to urban China. Accordingly, many changes began to take place in the national and local urban development policies, and more "holistic" concerns[1] in urban development objectives began to show up (Hu, 2001; NDRC, 2007; Wang, 2000; Zhou, 2002, 2004a). All these had a substantive influence on the evolving regeneration policies and practices in Chinese historic cities.

A good understanding of urban regeneration in China requires not only the review of her urban transformation history but also a new reinterpretation of some urban development issues, such as the role of urban planning, local government, political-economic institutions, and other key factors. While there have been many studies on Chinese cities since 1949, such as studies related to urban politics and institutions (Ma, 1979, 2002; Oi, 1995; Sit, 2009; Walder, 1995b), urban planning mechanisms and development dynamics (Ng & Tang, 2004a, 2004b), and discussions on urban renewal or regeneration of some specific cities (Bao, 1993; Ng, Cook, & Chui, 2001; Ng & Liang, 2007; Ng & Tang, 2002a, 2002b; Qu, 1993), there are very few studies on urban renewal or regeneration in the context of a transitional China.

According to the characteristics of the socio-economic and political development and transformative urban regeneration policies since 1949, five stages have been identified in exploring the dynamics and factors of urban regeneration in China: (1) industrial development-centered and rehabilitation of dilapidated urban areas within general plans (1949-1965); (2) fragmented urban development

---

[1] Even though economic growth still plays a significant role, many local governments began to take more comprehensive concerns in its urban policies since the early 2000s.

充分利用（1966—1976）；（3）根据地方总体规划对破败地区的复兴和重建，以及对历史城市的保护。(1978—20 世纪 80 年代末）；(4) 倾向经济发展的以房地产主导的改造和重建，这往往违反地方总体规划（20 世纪 90 年代的 10 年）；(5) 考虑更加全面的倾向经济发展的政策（始于 21 世纪初）。

3.1.1 基于总体规划的、以工业发展为中心和对破败城区的修复（1949—1965）

中国在 1949 年新中国成立初期，由于国家饱经战争的破坏和创伤，全国许多拥有 80 或上百年历史的城市都受到了严重破坏（Yang & Wu, 1999：21）。城市中的生活环境遭到了巨大的破坏。居住区的问题也是显而易见：低收入群体居住在密度极高的旧居民楼中、落后的城市基础设施以及残破的居住环境。因此，那时的当务之急就是改善居民的居住环境和城市状况（Yang & Wu, 1999：21）。另一方面，整个中国的社会和政治经济情况也严重衰退。这从当时的城市化比例和国民生产总值可以反映出来。根据 1949 年的官方数据显示，当时中国城市化比例为 10.6%，然而世界平均比例是 28%。至于国民生产总值，与 1936 年最高工业生产高峰期相比，1949 重工业生产总值下降了 70%，轻工业生产总值下降了 30% (Dai, 1992：378)。

在国家无法为城市居民提供足够的经济支援，以改善当地生活环境的情况下，不少城市中来自各行各业的居民开始自发地改善他们的居住环境。结果，这些居民也得到了他们应有的回报而不是等待政府的直接救济（以工代赈）。城市居民也被动员起来改

and full utilization of old urban areas without planning (1966-1976); (3) revitalization and redevelopment of derelict urban areas following local general plans and conservation of historic cities (1978-late 1980s); (4) economic-biased and property-led renewal and redevelopment that often violated local general plans (decade of the 1990s); and (5) pro-growth urban regeneration policies with more comprehensive concerns (beginning early 2000s).

3.1.1 Industrial development-centered and rehabilitation of dilapidated urban areas within general plans (1949-1965)

When the P.R. China was established in 1949, in the midst of damage and destruction from the Wars, most cities in China with a history of 80 or hundreds of years were dilapidated (Yang & Wu, 1999: 21). The living environments in the urban areas were dramatically disadvantaged. The problems of the residential houses were evident: high densities of old residential buildings populated by low income residents, poor urban infrastructures, and derelict residential environments. Thus, one of the urgent issues at the time was to improve the living environments and urban conditions (Yang & Wu, 1999: 21). On the other hand, the entire social and political economy in China was seriously recessed. This was reflected in the urbanization ratio and GDP at that time. According to the official statistics in 1949, the average urbanization ratio in China was 10.6 percent, while the average figure in the world was 28 percent at the time. Regarding the GDP, compared with the highest industrial production period in 1936, the heavy industrial production in 1949 decreased by 70 percent, and the light industrial production decreased by 30 percent (Dai, 1992: 378).

When the nation was unable to supply enough economic assistance to the urban residents to ameliorate local living environments, many cities involved residents from all walks of life to improve their living environments by themselves. Consequently, these residents

造和修复当地最残破的区域（Dai，1992；EBUCCC，1990；Yang & Wu，1999）。与此同时，为了国民经济的尽快复苏，中央政府也鼓励大力发展工业，并提出将消费型城市向生产型城市的转变（Buck，1975：25）。因此，当时城市发展最显著的一个特点就是生产能力满足工业增长的需要。面对大规模的破败的环境以及资金不足的状况，中央政府的政策主张对旧城区进行充分利用（EBUCCC，1990：38）。在这种背景下，中国城市更新实践的特点就是建设一些新住宅和城市基础设施。因此，旧城区的城市更新政策可以总结为"充分利用、逐步改造、加强维护"（EBUCCC，1990：200）。

经过3年的经济恢复期，1953年中国提出了第一个五年计划，这在很多人看来是中国城市发展的一个黄金时期（Dong，1955；Zhang，2002a；Zhou，2005a）。从那时起，经济发展就被列为国家发展计划的重要目标，并以大规模的工业发展为特色。至于当地城市发展，中央政府的政策要求当地的城市建设要为工业发展服务（Thompson，1975：599，600）。不过，漫无目的的城市建设常常无视总体规划的要求，也为日后的工厂、道路和交通系统的合理布局造成很多障碍。这些活动也为以后的城市发展带来了不少问题。例如，很多"新城镇"随着工厂的肆意扩张也被建造起来（Dong，1955：3）。今天，人们在不少居住区仍然可以看到当时兴建的工厂或工业区。

1958年5月，中国实行第二个五年计划（1958—1962）实施初期，颁布了"又快又好地建设社会主义，力争上游"的发展方针，

received their corresponding rewards instead of waiting for the direct distribution of relief by the nation (*yi gong dai zhen*). Urban residents were mobilized to renew and revitalize their most dilapidated and rundown urban areas (Dai, 1992; EBUCCC, 1990; Yang & Wu, 1999). At the same time, in order to revitalize the national economy quickly, the central government advocated the industrial development forcefully and proposed to transform the consumer cities into producer cities (Buck, 1975: 25). Accordingly, one outstanding characteristic of urban development at that time was the capacity to meet the industrial growth necessities. Confronting large scales of dilapidated environments and insufficient funding, the central governmental policies advocated the full utilization of old urban bases (EBUCCC, 1990: 38). In this background, urban regeneration practices in China were characterized by the building of some new residential buildings and basic urban infrastructure. Thus, urban regeneration policies in old urban areas can be summarized as "maximum utilization, gradual renewal, and enhancement of repair and maintenance" (*chongfen liyong, zhubu gaizao, jiaqiang weixiu*) (EBUCCC, 1990: 200).

Following the three years of national economic recovery, China proposed its First Five-year Plan (henceforth the FFP) in 1953, which was considered by many as a golden period of urban development in many Chinese cities (Dong, 1955; Zhang, 2002a; Zhou, 2005a). Since then, economic development was prioritized in national development schemes, characterized by massive industrial development. As regards local urban development, the central government policy required local urban constructions to serve the industrial growth scheme (Thompson, 1975: 599, 600). However, the aimless urban constructions often ignored the requirements of comprehensive plans and set many barriers to the rational layouts of industrial factories, roads, and transportation system. These activities posed many problems for the later development of these cites. For instance, many "new towns" were built mindlessly along with the random

这标志着以高指标、盲目指挥、浮夸风、意识形态倾向等为特点的"大跃进"时期的开始。在这种背景下，国家建工部提出了"将工业发展大跃进与城市建设相结合"的方针。一个显著特征就是全国的城市数量从1957底的177座突增到1961年的208座。这也是新中国成立以后的12年中设市城市最多的一个时期（EBUCCC，1990：71-72）。1960年建工部还提出了城市改造的指导方针："在未来十到十五年的时间里，旧城区应当重新建成社会主义现代化城市"（Wang，1990：76）。根据这些要求，为了尽快改变城市外观，很多城市开始盲目地扩张面积并声称已经建立了大城市或街道。自1960年以来的独立自主和自力更生的趋势使这种情况不断恶化，那时许多城市甚至是在国家没有充分协调或合作的基础下各自发展各自的工业体系（Buck，1975）。

3.1.2 缺少总体规划的碎片式城市建设和对历史城区的充分利用（1966—1976）

在1966年"文化大革命"期间，在新旧城区肆意建设、占用公共空间成了司空见惯的现象（Xie & Costa，1993：104）。1967年随着在旧城区见缝插针地建设新建筑政策的出台，旧城区受到了更大的破坏。该政策的提出旨在节省土地空间，并使空间的破坏能够降到最低（EBUCCC，1990：92）。但由于没有详细的规划，居住区内建立的很多工厂存在火灾隐患，并且容易产生噪声污染。结果是这些活动不仅占用了公共绿化带和庭院，同时也扰乱了城市交通和布局系统（Hua，2006：92，123）。在"文化大革命"

expansions of industrial factories (Dong, 1955: 3). Today, in many residential areas, there are still some factories or industrial sites standing.

In May 1958, at the beginning of China's Second Five-year Plan (1958-1962), the Chinese Communist Party promulgated the development guideline of "strengthening to pursue upper levels in constructing socialism quickly and frugally," which indicated the beginning of the Great Leap Forward in China, characterized by high indicators, mindless guidance, over-exaggeration, communist tendency, and other indicators. In this background, the Ministry of Civil Engineering (henceforth MCE) put forth the policy of "uniting the Great Leap Forward in industrial growth with that in urban development." One conspicuous phenomenon was the increasing city numbers in the whole country, from 177 at the end of 1957 to 208 in 1961. This was the quickest time of setting up cities since the establishment of China (EBUCCC, 1990: 71-72). Urban renewal guidelines were proposed by the MCE in 1960: "The old cities should be reconstructed into socialist modern cities within 10-15 years" (Wang, 1990: 76). According to these requirements, many cities had expanded their urban areas irrationally and claimed to build the big cities or streets in order to change the urban appearances abruptly (Hua, 2006: 92). These practices were exacerbated by the tendency of self-reliance and independence since 1960, when many cities and even counties were constructing their own complete industrial systems without the full coordination or cooperation among them (Buck, 1975).

3.1.2 Fragmented urban development and full utilization of old urban areas without planning (1966-1976)

During the Cultural Revolution in 1966, the random constructions in old or new urban areas, and the occupation of public space could be seen everywhere (Xie & Costa, 1993: 104). The destruction in old urban areas was intensified in 1967, with the policy of erecting new buildings in

期间对历史悠久区域的破坏主要表现为"破四旧"运动，即旧思想、旧文化、旧风俗和旧习惯。许多城市公园、纪念碑和历史遗产都遭到了空前的破坏（EBUCCC，1990：93）。1964 年基于当时国际形势的考虑，中央提出了三线建设（1964—1971）的发展方针。该方针要求重要的沿海企业和国防项目转移到内陆，即"靠山、分散、隐蔽"的地区（Sit，2009：9）。在国家方针的指导下，许多城市都建立起了自己的"小三线"，即当地工业、工厂和学校都远离城区转移到山区。这些运动导致了全国范围内零星的和碎片式的城市发展过程（Dai，1992；Zhuang & Zhang，2002）。

3.1.3 基于总体规划的历史城市保护，和破败城区的复兴与再开发（1978—20 世纪 80 年代末）

1978 年 3 月，国家提出改革国家经济体制和改革开放的政策。1980 年，在总结历史经验教训时，国家基础设施建设委员会在北京举行了关于国家城市规划的会议，期间详细规定了市政府是在城市规划、建设和管理中的责任。到 1986 年底，96% 的城市和 85% 的乡镇完成了总体规划。但这次与以往不同，城市规划开始在地方城市发展过程中发挥重要作用（Xie & Costa，1993：104）。这对许多城市中的城市更新过程有着显著影响。根据当地总体规划的要求，一些严重影响旧城区居民生活环境的工厂及工业搬迁到了新区域。为了改善居民的生活条件，中国在第六个五年计划期间建立了许多居住小区（1981—

some old urban areas as densely as possible (*jian feng cha zhen*). The policy was proposed by the National Infrastructure Construction Committee (henceforth NICC) in order to save land space and minimize the areas' destruction. Without conceivable plans, many factories with fire hazard and noise were built in the residential areas. As a result, these activities not only occupied the public green space and courtyards but also disturbed the urban transportation and layout systems in many cities (Hua, 2006: 92, 123). Destruction in historic areas during the Cultural Revolution was outstandingly reflected in the movement of "destroying the four olds" (*po si jiu*), namely, the old ideas, old culture, old tradition, and old habits. Many urban gardens, monuments, and historic heritage went through unprecedented destruction (EBUCCC, 1990: 93). In 1964, considering the international situations at the time, Mao Zedong initiated the Third Front Construction (*san xian jian she*) (1964-1971). This guideline required the important coastal enterprises and national defensive projects to move towards the hinterland zones, that is, "in hills, of dispersed mode and within caves" (Sit, 2009: 9). Following this national guideline, many cities also constructed their local "minor third front" (*xiao san xian*), which means that local industries, factories, and schools were relocated from urban areas to remote mountainous sites. These activities highly promoted the fragmented urban development in the entire country (Dai, 1992; Zhuang & Zhang, 2002).

3.1.3 Revitalization and redevelopment of derelict urban areas following local general plans and conservation of historic cities (1978-late 1980s)

In March 1978, the Chinese Communist Party put forth the reform on national economic system, and the open door policy was adopted. In 1980, in the summary of historical empirical lessons, the NICC held the conference on national urban planning in Beijing, which specified the responsibility of the municipal government in urban planning, construction, and management.

1985)。随着单位[1]制度的改革所产生的工作—住房分离趋势，在中国逐渐加强（Wang & Chai, 2009）。此外，随着城市居住区的重建，相应的城市基础设施社会福利制度也建立起来（Wang, 2006b；Zhou & Jiang, 2008）。

总的来说，在该时期内中国的城市更新具有以下特点：

1）按照地方总体规划的逐步复兴与重建

随着许多城市完善其总体规划，旧城区的城市更新开始按照地方总体规划开始实施。具体的更新方法包括拆除废旧城区，对该区域内多层建筑或者高层建筑进行重建（Chen, 1993：122；EBUCCC, 1990：108）。

2）为多样化的更新方法制定相应的法律法规

有些城市就居民重置和城市管理制定了法规，这使得城市更新有法可依（Zhuang & Zhang, 2002：237）。根据法规，这些城市提出了地方多样化的更新策略，例如，旧城区的拆除、重建和修复（Chen, 1993：122）。

3）多渠道筹集必要的资金

许多城市为地方更新计划筹集必要资金采取了不同的方式。例如，基于互惠互利的原则，一些城市的房产管理机构联合当地企业一起开展更新项目，如烟台、沈阳和大连。在其他城市，城市改造项目也通过多种方法得以实现，例如，通过企业、地方居民或商品房建设（EBUCCC, 1990：201）。

4）历史文化名城的保护

随着经济的快速发展，大规模的城市建

---

[1] 布雷认为，单位是"代表中国社会主义时期工作场所的通用术语，以及它所代表的特定的实践内容"（Bray, 2005：3）。

By the end of 1986, 96 percent of the cities and 85 percent of the towns finished their general plans. Therefore, unlike the previous time, urban planning began to play an important role in local urban development (Xie & Costa, 1993: 104). This had significant impacts on the urban regeneration processes in many cities. Some factories and industries, which seriously affected the residential living environments of old urban areas, were moved out to new zones according to local general plans. In order to improve the residential living conditions, many residential estates (*ju zhu xiao qu*) were first built during the period of the Sixth Five-year Plan of China (1981-1985). These activities were strengthened with the trend of separated jobs-housing relationships in line with the reform of the Danwei[1] system in China (Wang & Chai, 2009). Moreover, with the restructuring of urban residential areas, corresponding urban infrastructure was built and social welfare systems began to be established in the society (Wang, 2006b; Zhou & Jiang, 2008).

Generally, urban regeneration in many Chinese cities in this period had the following characteristics:

1) Gradual revitalization and redevelopment following the local general plans

With many cities finishing their general plans, urban regeneration in the old urban areas began to take place according to the local general plans. The specific urban regeneration approaches included the demolition of derelict areas and the redevelopment of multiple-storey or high-rise buildings in the areas (EBUCCC, 1990: 108).

2) Laws and regulations enacted for diversified regeneration approaches

Some cities enacted the regulations on residential relocation and urban management, which made it possible for urban regeneration to follow (Zhuang & Zhang, 2002: 237). Under the regulations, these cities put forth local diversified regeneration strategies, for example,

---

[1] According to Bray, Danwei is "a generic term denoting the Chinese socialist workplace and the specific range of practices that it embodies" (Bray, 2005: 3).

设对很多城市的历史环境造成了相当大的"建设性"破坏。鉴于此,自1982年以来国务院已经公布了99个国家历史文化名城。这意味着中国要平衡经济发展和城市保护的进程。此后,很多省级和市级城市也制定了地方性保护规划和法规,并且这些政策对这些城市的多维更新策略和目标产生了巨大影响[1](Yang, 2003:35)。

3.1.4 以经济发展为导向的房地产再开发政策常突破地方的既定规划(20世纪90年代)

20世纪80年代以来,中国的城市发展以快速城市化和经济发展为特色。图3-1显示了中国城市居民所占全国人口的比例[2],体现了中国城市化的现状(Johnson, 1990:4)。一方面,城市改造或再开发对历史建成环境造成了很大影响(Li, 2003b; Shan, 2006; Zhang, 2006a)。随着快速城市化进程,中国很多旧城区都进行了重建以满足当地的需要。学者已经讨论了20世纪90年代旧城区的重建和改造带来的城市问题,如,地方城市特色的丧失、对地方原住民的影响。(Li, 2003a; Ma, 2007c; Pearce, 1994; Shan,

---

[1] 例如,苏州是一个拥有两千五百多年历史的城市。20世纪80年代初,面临城市中大量的破败地区,国务院要求地方规划中必须包括对传统文化和历史环境的保护。1986年,国务院在苏州的总体规划中强调"尽管城市基础设施需要更新,新城区需要建设,但是不能忽视对历史环境和优秀文化遗产的保护"。这些地方性保护政策极大地促进了历史城市多维更新目标的实现。

[2] 约翰森认为,"城市化"指的是城市影响力不断扩大的过程,城市化率的测量方式是人口普查确定的城区居民占全国总人口的比例(Johnson, 1990:4)。

demolition, redevelopment, and rehabilitation of old urban areas (Chen, 1993: 122).

3) Multiple channels to raise necessary funds

In order to raise the necessary funds for local regeneration programs, many cities adopted various ways. For instance, based on mutual reciprocity, the urban real estate management agencies in some cities united local enterprises, for example, in Yantai, Shenyang, and Dalian. In other cities, urban renewal programs were implemented in various ways, for example, by enterprises, by local residents, or by way of building commodity houses (EBUCCC, 1990: 201).

4) Conservation of Historic and Cultural Cities (Historic Cities) (*li shi wen hua ming cheng*);

With the rapid economic development, massive urban constructions led to considerable "constructive" destruction in many historic environments. Seeing this, the State Council has declared 99 National Historic and Cultural Cities since 1982. This signifies the balancing process between economic development and urban conservation in China. Following this, many provincial and municipal cities also enacted local conservation planning or regulations, and these policies had substantive influences on the multi-dimensional regeneration strategies and objectives of these cities[1] (Yang, 2003: 35).

3.1.4 Economic-biased and property-led renewal and redevelopment that often violated local general plans (1990s)

Since the late 1980s, rapid urbanization and economic development have characterized the urban development in China. The urbanization status in China in terms of the proportion of urban

---

[1] For instance, Suzhou is a city with a history of more than 2,500 years. Facing the immense derelict areas in the city at the beginning of the 1980s, the State Council required that the conservation of traditional culture and historic environments to be an imperative part in local plans. In 1986, in the master plan of Suzhou, the State Council highlighted that "while the urban infrastructures need to be renewed and new urban areas to be built, the historic environments and excellent cultural heritage should be conserved." These local conservation policies forcefully contributed to the multi-dimensional regeneration objectives of this historic city (Ruan & Xiang, 1997).

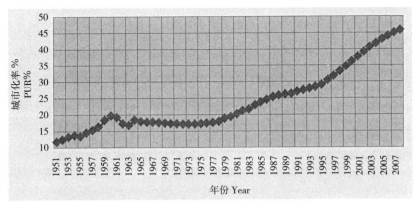

图 3-1：城市人口占全国总人口的比例（1951—2008）
Figure 3-1: Proportion of urban residents to the total national population (1951-2008)
注：该数据不包括香港、澳门和台湾的人口
资料来源：（CSSA，2009；DED，2009）
Note: Data exclude the population of Hong Kong SAR, Macao SAR and Taiwan Province
Source: Author synthesized from (CSSA, 2009; DED, 2009)

2006；Wang，2007b；Yang & Wu，1999；Zhang，2006a；Zhao，2007）。这些影响反映在各个方面，如追求商业目的的再开发、追求政治业绩的短期活动（政绩工程）。基于这些目的，历史城区中的城市再开发项目通常造成千城一面的景观或仿造的古城现象，如"历史一条街"（Shan，2009：24）。另一方面，学者认为，为了给当地大规模的重建工程让步，许多原住居民被迫搬迁或背井离乡，这引发了很多社会问题，像社会不公平、分散的街区以及社区不团结（Wu，2004）。

对于这些历史环境中的城市再开发活动，学者指出了两个主要原因（Ruan & Sun，2001；Yang & Wu，1999；Zhao，Lee，& Sun，2006）：

1）破旧的城区不能满足当地社会经济的需要。此外，鉴于旧城区持续增加的人口，城市改造和再开发活动为提供更多的居住空间，并满足了当地需要。

2）在郊区进行了几年的大规模建设

residents to the entire national population (PUR)[1] is presented in Figure 3-1 (Johnson, 1990: 4). With rapid urbanization, many old urban areas in Chinese cities were redeveloped to serve local needs. Urban problems accompanying the redevelopment or renewal of old urban areas in the 1990s have been discussed by many academics, such as losing local urban features and the impacts to the local original neighborhoods (Li, 2003a; Ma, 2007c; Pearce, 1994; Shan, 2006; Wang, 2007b; Yang & Wu, 1999; Zhang, 2006a; Zhao, 2007). On the one hand, urban renewal or redevelopment has great impacts on the historic built environments (Li, 2003b; Shan, 2006; Zhang, 2006a). These impacts are reflected in various aspects such as in the total redevelopment with commercial goals and short-term activities for the sake of political progress (*zheng ji gong cheng*). With these intentions, urban redevelopment of historic urban areas often results in similar urban scenes (*qian cheng yi mian*) or fake antiques like "one historic street" (*li shi yi tiao jie*) (Shan, 2009: 24). On the

---

[1] According to Johnson, the term 'urbanization' refers to the process of expanding urban influence, and the measure of urbanization is taken as the proportion of the total population in a country that lives in urban areas as defined in the census (Johnson, 1990: 4).

和城市扩张之后，很多城市中用于未来发展的土地减少了；因此，许多城市开始将他们的注意力转移到老城区或市中心区（SCXPC, 1991：5）。

此外，为了从地方再开发工程中追求最大的经济回报，也驱使着这样的再开发活动向旧城区转移。这和20世纪80年代末城市土地改革和住房改革制度的变化有很大关系。总的来说，20世纪90年代的中国城市发展和更新具有以下特点：

1）从地方城市改造与再开发工程追求最大的经济回报

随着政治体制的变化，地方政府逐渐开始介入当地的再开发项目并开始从中取得经济效益。在当前的城市发展制度下，将土地使用权出租给私人开发商也是为地方城市改造活动筹集必要资金的主要方式（Yang & Wu, 1999：116）。不过，由于私人开发商的参与，追求最大的经济回报往往也成为其唯一的目的。

2）多方、多渠道为地方城市的改造工程筹集必要资金

在市场经济体制下，地方城市改造工程的必要资金通常来源于各方参与者，例如，政府、企业、金融公司以及其他渠道。这种不断变化的情况也使地方政府由之前唯一的项目参与者变为后期的其他私营部门的合作者。琼·爱（Jean Oi）尖锐地指出中国的地方政府不再是"行政服务的提供者"，而成为"完全成熟的经济参与者"（Oi, 1995：1137）。在这种情形下，许多城市扩张和再开发活动也很容易就获得地方政府的正式批准而开展。

other hand, in order to give way to local massive redevelopment projects, many original residents are relocated or displaced, and this causes many social problems for example, social inequity, dismantled neighborhoods, and less community solidarity, as argued by academics (Wu, 2004).

With regard to these urban redevelopment activities in historic environments, some academics point to two main reasons (Ruan & Sun, 2001; Yang & Wu, 1999; Zhao, Lee, & Sun, 2006):

1) Derelict urban areas could not meet the requirements of local socio-economic development. Moreover, considering the increasing population in old urban areas, urban renewal and redevelopment offered an effective way to provide more building space and satisfy local needs.

2) After years of extensive construction and urban sprawl in suburban areas, the land for the potential future development in many cities decreased; hence, many cities turned their sights to the old urban areas or city center areas (SCXPC, 1991: 5).

Furthermore, the pursuit of maximum economic returns from local redevelopment projects has also forcefully driven such redevelopment activities into the old urban areas. This has much to do with the changing institutions of urban land reform and housing reform since the late 1980s. Generally, urban development and regeneration of urban China in the 1990s have the following characteristics:

1) The pursuit of maximum economic returns from local urban renewal and redevelopment projects

With the political institutional changes, local municipal governments tended to intervene with the local redevelopment projects and were provided with great incentives to explore economic profits from these projects. In the current urban development institution, leasing land-use rights to private developers is the main means of raising the necessary funds for local urban renewal activities (Yang & Wu, 1999: 116). At the same time, with the involvement of private developers, the sole objective becomes

3.1.5 促进地方发展的城市更新政策与整合的城市发展理念（始于21世纪初）

20世纪90年代，快速的城市改造与再开发活动使中国很多城区都发生了巨大变化。不过，随着物质环境的改善，很多城市问题也开始出现，如：

1) 过于简化的城市改造实践和社会环境面临的挑战

20世纪90年代，中国很多城市改造的特点是对破败城区进行简单重建（Lin, 2007：32）。然而，城市基础设施的建设，如供水和排污系统、道路维护和其他服务等投资大而经济利益小的项目，则很难吸引私营开发商的兴趣（Yang & Wu, 1999：28）。同时，由于没有整体的改造计划，许多地方政府也未想改善当地破败的基础设施。持续增加的人口使这种情况趋于恶化，也给原本就已破败的城市基础设施增添了更多的负担。

2) 历史环境特色的逐渐消失

房地产再开发主导的改造工程为历史城市环境的保护带来很多挑战（Ma, 2007c；Pan, 2004；Wu, 1998；Zhang, 2002b；Zhao, 2007）。尽管当地的保护法规试图为历史城市环境的保护提供法律依据，但在实践中则常常被突破或违反。私人发展商有时会违背历史城区中关于容积率、限高和建筑密度的要求。在这种情况下，为了给再开发活动让步，一些历史建筑物甚至也遭到拆除（Ma, 2007a, 2007b, 2007c；Wang, 2007a, 2007b；Zhao, 2007）。

3) 旧城区原社区邻里关系的解体

学者认为，在城市建设过程中几乎所有 the quest for maximum economic returns.

2) Various actors and means of raising the necessary funds for local urban renewal projects

Within the market-oriented economic institution, the necessary funds for local urban renewal projects often come from various actors such as the state, enterprises, financial companies, and other sources. This changing situation has also changed the status of local governments from being the previously sole actor to the latter cooperator of other private sectors. Jean Oi commented that local governments in China are no more of the previous "administrative-service provider" but "fully fledged economic actors" (Oi, 1995: 1137). In this background, many urban sprawl and redevelopment activities can be conducted with local authoritarian approval.

3.1.5 Pro-growth urban regeneration policies with more comprehensive concerns (beginning early 2000s)

After the rapid urban renewal and redevelopment activities in the 1990s, many urban areas in China underwent great changes. With the improvement of the physical environments, many urban problems also began to show up, such as the following:

1) Over-simplified urban renewal practices and social and environmental challenges

Urban renewal activities in many cities in China in the 1990s were characterized by the simplified process of replacing the dilapidated urban areas with new redevelopment (Lin, 2007: 32). Nevertheless, the development of urban infrastructures, for example, water and sewage systems, pavements, and other services, which need more investment and generate less economic interest, could hardly attract the interest of private developers (Yang & Wu, 1999: 28). At the same time, many local governments did not want to improve local dilapidated infrastructural systems without the overall renewal projects. This situation often worsened with the increasing local population, which put more burdens to the already dilapidated urban infrastructures.

的矛盾都来源于土地使用权属的变化（Xie & Ingram, 2008：164）。在用土地使用权置换经济利益的过程中，很多地方政府制定的城市再开发规划倾向于改变原来土地的使用功能，如用办公或商业区来替代原来的居住区的功能。根据20世纪90年代的中国城市调查，世界银行提供了一份再开发前后重点城市中心带典型的土地使用变化数据表。（表3-1）罗列出了这些数据。需要注意的是，通过再开发活动可以获得更多的容积率。

通常情况下，再开发过程需要对大量的原住民进行重置，这往往会造成原街区的解体。我国目前的城市改造工作机制已显现出这种现

城市中典型土地利用方式的变化
表 3–1
Typical land use conversion in city centers
Table 3–1

	重建前 Before–redevelopment			重建后 After–redevelopment		
	土地面积(%) Land-Area (%)	占地面积(%) Floor-Area (%)	容积率(%) Floor-Area Ratio (%)	土地面积(%) Land-Area (%)	居住面积(%) Floor-Area (%)	容积率(%) Floor-Area Ratio (%)
道路 Street	8.0			18.0		
住宅 Residential	55.0	49.7	0.6	30.0	24.3	2.5
办公 Office	20.0	24.1	0.8	25.0	36.4	4.5
商业 Commercial	12.0	21.7	1.2	27.0	39.3	4.5
工业 Industrial	5.0	4.5	0.6	0.0	0.0	0.0
总共 Total	100.0	100.0	0.7	100.0	100.0	3.1

资料来源：世界银行（WB, 1993：108）
Source: World Bank (WB, 1993: 108)

2) Disappearing features of historic environments

Redevelopment-led urban renewal projects often posed challenges to the conservation of historic urban environments (Ma, 2007c; Pan, 2004; Wu, 1998; Zhang, 2002b; Zhao, 2007). Although local conservation regulations endeavored to provide legal documents for the conservation of historic urban environments, they were often overlooked or violated by local policy-makers in practice. Some requirements on the floor-area-ratio, height control, and building density in historic areas were breached by private developers. In this background, some historic buildings were removed to give way to redevelopment activities (Ma, 2007a, 2007b, 2007c; Wang, 2007a, 2007b; Zhao, 2007).

3) Dismantled original community neighborhoods in old urban areas

Some academics claimed that nearly all the contradictions in urban renewal processes come from the changes in land-use rights (Xie & Ingram, 2008: 164). In the exchanges of land-use rights for economic returns, many urban redevelopment plans by local governments tend to change the functions of the original land use, for example, replacing the original residential areas with office or commercial areas. Based on the surveys of Chinese cities in the early 1990s, the World Bank (1993) came out with a list of the typical land use changes in the city centers before and after redevelopment programs. The list is presented in Table 3-1. It should be noted that usually through redevelopment could more floor-area-ratio be realized.

Normally, the redevelopment process requires the relocation of large numbers of original residents, often leading to the dismantlement of original neighborhoods. This is revealed in the current urban renewal procedure in China as illustrated in Figure 3-2. The procedure includes the improvement of urban infrastructure, public facilities, and urban living environment by intensifying the land-use of old urban areas and by involving private developers.

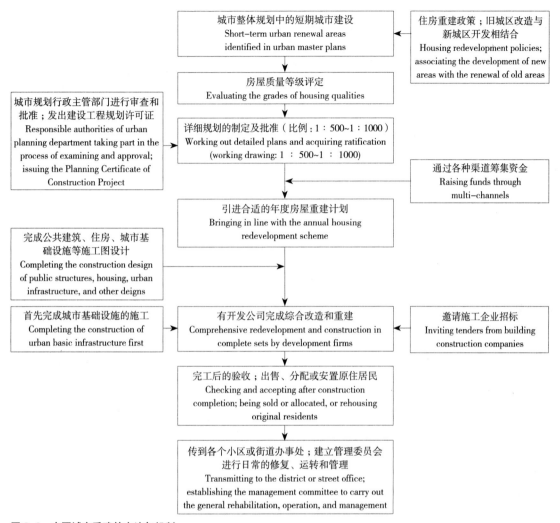

图 3-2：中国城市重建的方法与机制
Figure 3-2: An illustration of the methods and procedure of urban renewal in China
资料来源：(Wang, 2005：319)
Source: (Wang, 2005: 319)

象，如（图 3-2）所示。这个过程包括，通过加强旧城区的土地使用和私人开发商的参与，来改善城市基础设施、公共设施和城区生活环境。不过，在整个城市建设过程中，原住居民的参与未被包括在内（Wang, 2005：319）。

学者认为上述历史城市环境中的城市改造问题，与当前我国尚不完善的城市改造法律和缺乏公众参与机制相关（Yang & Wu,

In the whole urban renewal process, the original residents are not included (Wang, 2005: 319).

Academics argue that the abovementioned urban renewal problems in historic urban environments have much to do with the current imperfect laws on urban renewal and the lack of public participation mechanism (Yang & Wu, 1999; Zhao et al., 2006):

Currently in China, there are still no specific laws and regulations on urban renewal or regeneration, although there are related regulations

1999；Zhao et al.，2006）。

目前我国虽然已有诸如《城市规划法》、《城乡规划法》、《城市房地产管理法》和《协议出让国有土地使用权规定》等相关法律法规，但仍缺乏关于城市改造和更新的具体法律规范。而且，在有关法律法规的施行过程中，经常出现对具体的再开发项目模棱两可的条例，这也严重影响了地方更新项目的实施和管理。同时，由于公众参与机制仍处于初级阶段，所以行政部门、私人开发商以及社区居民也普遍缺乏对公众参与的价值和意义的正确认识和理解。由于在城市改建中缺乏公众参与，地方社区居民没有一个正式途径参与到地方城市改造过程中（Zhao et al.，2006）。这也不利于对当地历史环境的保护，尤其是在综合原住民的智慧与知识方面。

全面的城市发展问题

有学者（Wang & Fan，2005：383）指出，在20世纪90年代中国政府逐渐开始认识到城市发展过程一味追求经济增长所带来的不少问题，即"牺牲环境、占用和污染耕地、破坏生态环境、造成的严重污染和垃圾处理的问题"[1]（Ng，2004：i）。自那时起，

---

[1] 中国关于全面的城市发展的学术性讨论主要围绕以下几个方面：追求社会公平和公众利益（Zheng&Yang，2005）；建立城市特征和身份（Zhou&Cheng，2005；Zhou&He，2005）；促进可持续发展（Dai，Wang，&Fu，2006；Wang，Sun，&Wu，2006；Wang，He，&Bao，2006；Xia，2006；Zhao et al.，2006；Zhou，2002）；强调旧城区的社会经济发展（Cheng，2006b；Dong，2005；Li，2004；Wang，2006c；Zhao，2006）；地方发展中政治和制度问题（Zhang，2000）。此外，还强调了新城区的城市规划和规划制定者在促进社会公平和公众利益中的重要作用，而这往往在地方发展规划中被边缘化（Zhou，2001，2002b）。

like the City Planning Act (cheng shi gui hua fa), the Urban and Rural Planning Law (cheng xiang gui hua fa), the Law on Urban Real Estate Administration (cheng shi fang di chan guan li fa), and the Regulations on the Compromising Relinquishment of National Land-use Rights (xie yi chu rang guo you tu di shi yong quan gui ding). In the implementation of these regulations, ambiguous articles on specific redevelopment projects are often present, and this seriously affects the implementation and management of local renewal projects. At the same time, the mechanism of public participation is still at the rudimentary period, and therefore lacks the understanding of the value and significance of public participation of the different levels from the administration sector and private developers to the communities. Due to the lack of a participation mechanism in urban renewal, local communities do not have the official way to become involved in local urban renewal processes (Zhao et al., 2006). This is detrimental to the conservation of local historic environments especially in terms of integrating local indigenous knowledge.

Comprehensive urban development concerns in China

Some academics (Wang & Fan, 2005: 383) pointed out that in the early 1990s, the Chinese government actually began to realize the problems accompanying the sole aim of pursuing economic growth in urban development, which were "at the expense of the natural environment, encroaching and contaminating agricultural land, destroying ecosystem, and causing serious pollution and waste disposal problems"[1] (Ng, 2004:

---

[1] The academic discussions on the comprehensive urban development in China mainly centre around the following aspects: pursuing social equity and mutual interests (Zheng & Yang, 2005), establishing urban characteristics and identities (Zhou & Cheng, 2005; Zhou & He, 2005), promoting sustainable development (Dai, Wang, & Fu, 2006; Wang, Sun, & Wu, 2006; Wang, He, & Bao, 2006; Xia, 2006; Zhao et al., 2006; Zhou, 2002), underling socio-economic development of old urban areas (Cheng, 2006b; Dong, 2005; Li, 2004; Wang, 2006c; Zhao, 2006) and political and institutional concerns in local development (Zhang, 2000). In addition, the role of urban planning and planners in the new era have also been emphasized as one critically factor in pursuing social equity and public interests which are often marginalized in local development schemes (Zhou, 2001, 2002b).

我国政府就开始积极参与探讨有关可持续发展的国际研讨会。例如，1994年3月发布的中国21世纪议程明确提出，中国要"综合考虑人口、经济、社会、自然资源和环境等综合因素来制定发展方案"的决心（SCC，1994：1）[1]。

我国城市发展政策之所以进行调整的原因主要是吸取了之前资源消耗型发展政策带来的教训。在第十个五年计划中（2001～2005年），中央政府明确指出，过去的资源消耗型的发展政策不再符合中国当前的发展现状。因此，更全面的国家和社会发展方案提上了国家议程（Hu，2001：3）。在第十一个五年计划中（2006～2010年），这种全面的发展观念进一步融入了科学发展观、和谐社会的政治意识形态（Lin，2006；NDRC，2007）。在2004年中国共产党全国人民代表大会期间，根据2003年提出的"科学发展观"和强调社会发展和经济、政治、文化发展同样重要的政策，中央政府提出了构建和谐社会的目标[2]（Cheng，2006b：1）。

为了实现综合发展的目标，地方城市改造和更新政策开始有所改变。例如，就历史城市的更新（Xia，2006），南京、苏州和西安的地方市政府提出了各自的城市保护和发展规划以及当地特有的环境保护法（Zhou & Tong，2005）。除此之外，很多历史城市都制

---

[1] 更多关于中国城市规划和可持续发展议程之间关系的描述可见（Ng，2004）。
[2] 国家十一五规划提出了该原则，内容为"用科学的方法谋发展，我们应努力独立创新、改善制度和机制、促进和谐社会、增强中国的综合国力"（NDRC，2007：3）。

i). Since then, the Chinese government has actively participated in international discussions on sustainable development. For instance, in the national Agenda 21 adopted in March 1994, China explicitly expressed to "find a path for development, wherein considerations of population, economy, society, natural resources and environment are coordinated as a whole" (SCC, 1994: 1)[1].

This kind of changing mindset of China's urban development policy mainly came from the lessons of previous resource-consumption development policies. In its 10th Five-year plan (2001-2005), the central government explicitly indicated that previous resource-consumption-led development policy could not fit China's current development situation any longer. Consequently, more comprehensive concerns of national and societal development came to the national agenda (Hu, 2001: 3). In the 11th Five-year plan (2006-2010), this kind of comprehensive development concept has been further integrated with political ideology in terms of scientific development and harmonious society (Lin, 2006; NDRC, 2007). During the National Congress of the Communist Party of China in 2004, the central government put forth the goal of establishing a harmonious society, following the 2003 policy of pursuing "scientific concept of development" and strengthening the social development to be as important as the economic, political, and cultural developments[2] (Cheng, 2006: 1).

In view of comprehensive development concerns, some changes have begun to show up in the local urban renewal and regeneration policies. For example, regarding the regeneration of historic

---

[1] More elaborations on the relationships between urban planning and sustainable development agenda in China may be found at (Ng, 2004).
[2] This principle is listed in the National Eleventh Five-Year Plan, where it states that "based on scientific approaches to development, we should focus on independent innovation, improve institutions and mechanisms, promote social harmony, and enhance China's overall national strength,…" (NDRC, 2007: 3).

定了关于保护历史城区环境的法规（SCFPC, 1997；SCGPC, 1999；SCXPC, 2002）。2008年城乡规划法中第31条明确传达了历史城市的整体发展意图：

"旧城区的改建，应当保护历史文化遗产和传统风貌，合理确定拆迁和建设规模，有计划地对危房集中、基础设施落后等地段进行改建。历史文化名城、名镇、名村的保护以及受保护建筑物的维护和使用，应当遵守有关法律、行政法规和国务院的规定（SCNPC, 2008：第31条）。"

随着城市发展目标全方位的考虑，在地方城市更新规划中，很多历史城市的更新政策把经济增长和文化遗产的保护放在同等重要的位置。此外，这些因素都对地方城市规划的结果有直接的影响。

中国的综合更新计划与西方国家经验之间的比较

总体来说，学术界认为在一个全面的城市发展框架中，中国城市的更新项目应突出以下几方面内容（Zhai & Ng, 2009；Zhang, 2004）：

- 强调城市规划和设计的重要性；
- 保护城市环境和历史城市特色；
- 促进社会经济全面发展；
- 旧城区的改造和新城区的再开发相结合；
- 由地方政府领导，制定具体和详细的政策，并鼓励私人开发商参与其中；

cities (Xia, 2006), local municipal governments like Nanjing, Suzhou, and Xi'an have put forth their own urban conservation and development plans as local special sectional plans (Zhou & Tong, 2005). In addition, some historic cities have enacted local regulations on the conservation of historic urban environments (SCFPC, 1997; SCGPC, 1999; SCXPC, 2002). These comprehensive development concerns in historic cities have been explicitly conveyed in Article 31 of the 2008 Urban and Rural Planning Law:

As for the reconstruction of old urban areas, it is necessary to protect historical and cultural heritage and traditional style, reasonably determine the demolition and construction scale, and reconstruct the places where there are many dilapidated houses and the infrastructure is relatively backward. The protection of famous historical and cultural cities as well as the preservation and use of protected structures shall be conducted in accordance with the related laws, administrative regulations and the provisions of the State Council (SCNPC, 2008: Article 31).

With these multi-faceted concerns in urban development aims, many urban regeneration policies in historic cities consider economic growth and the conservation of cultural heritages as equally important in the local urban regeneration plans. In addition, they directly influence the outcomes of local regeneration plans.

Comparison between China's comprehensive regeneration programs and the experiences of Western countries

Generally, academics contend that in the framework of a comprehensive urban development, the urban regeneration programs of Chinese cities stress the following aspects (Zhai & Ng, 2009; Zhang, 2004):

- intensifying the importance of urban planning and design
- protecting the urban environment and

- 多渠道为城市更新集资。

西方国家和中国的更新政策实践存在着明显的差异。通过对城市更新项目的理论分析，罗伯茨（Roberts）总结了一些体现西方城市更新特征的原则，反映了城市变化及其结果所带来的挑战。罗伯茨认为，城市更新应该根据如下原则进行（Roberts, 2000：18）：

- 应该基于对城区情况的详细分析；
- 目标应该是城区的物质结构、社会结构、经济基础和环境条件同时得到改善；
- 通过更新和一个全面的综合的策略努力实现同时改善的目标，平衡、有序、积极地解决问题；
- 确保策略和相应的实施项目符合可持续发展的目标；
- 根据可持续发展确保政策的落实和实施；
- 制定清晰的可行目标，如果可能应该量化；
- 充分利用自然、经济、人力及其他资源，包括土地和现存的城市环境特色；
- 通过在城区的更新过程中具有合法权利的利益相关者之间的合作和参与，努力达成共识；这可以通过合作或其他工作形式实现；
- 意识到通过具体目标的完成情况衡量策略进程的重要性，并对影响城区的内外因素的性质和影响的变化进

conserving the historic urban features
- promoting overall socio-economic development
- integrating the renewal of old urban areas with the redevelopment of new areas
- led by the local governments, formulating specific policies and encouraging the involvement of private developers
- raising the funds for urban regeneration in various ways

It is easy to see the differences in the regeneration policies between the experiences of Western countries and those of China. Based on the theoretical analysis of urban regeneration programs, Roberts generalized some principles as the hallmarks of urban regeneration, which reflect the challenges of urban change and their outcomes. According to Roberts, urban regeneration should be as follows (Roberts, 2000: 18):

- be based upon a detailed analysis of the condition of an urban area;
- be aimed at the simultaneous adaptation of the physical fabric, social structures, economic base and environmental condition of an urban area;
- attempt to achieve this task of simultaneous adaptation through the generation and implementation of a comprehensive and integrated strategy that deals with the resolution of problems in a balanced, ordered and positive manner;
- ensure that a strategy and the resulting programs of implementation are developed in accord with the aims of sustainable development;
- set clear operational objectives which should, whether possible, be quantified;
- make the best possible use of natural, economic, human and other resources, including land and existing features of the built environment;
- seek to ensure consensus through the fullest possible participation and co-operation of all stakeholders with a legitimate interest in the regeneration of an urban area; this may be achieved through partnership or other modes of working;

行检查；

● 由于这些变化的发生，实施的初步规划可能需要修改；

● 意识到策略的不同因素可能会以不同的速度取得进展；这可能需要资源重置或提供额外的资源，以确保城市更新规划制定的目标和允许所有战略目标的实现之间保持平衡。

上述西方国家制定的城市更新原则表明，从各种主题的输出到更新结果的测量，这样一个综合化的更新机制，其中重点强调了全面因素的综合。图3-3显示了这些城市更新的原则，体现了相互影响的城市更新过程。

与西方国家城市更新实践相比较，尽管我国各级政府已经意识到城市全面发展目标的重要性，并在地方城市规划和更新过程中体现出来，但是我国仍未建立制度化的更新体制。这也会从一开始就削弱不同领域参与者对更新过程的影响。尽管中国很多城市在城市规划和更新过程中也颁布了公众参与的政策，但是仍没有具体的法律法规保护参与者的合法权益。当地方城市的发展计划要突出强调经济发展的重要性时，类似原住居民的参与和对原住居民社会文化的保护往往会被置于次要地位。结果，很多城市的更新实践往往成为追求政治进步或"政治表现"的一种手段，而并没有像原始规划中提到的那样来提升公众利益。这些措施也阻碍了城市更新全面目标的实现。因此，一方面，地方政府应意识到全面更新目标的重要性。另一方面，还必须意识到如果不对城市更新体制

● recognize the importance of measuring the progress of strategy towards the achievement of specified objectives and monitoring the changing nature and influence of the internal and external forces which act upon urban areas;

● accept the likelihood that initial programs of implementation will need to be revised in-line with such changes as occur;

● recognize the reality that the various elements of a strategy are likely to make progress at different speeds; this may require the redirection of resources or the provision of additional resources in order to maintain a broad balance between the aims encompassed in a scheme of urban regeneration and to allow for the achievement of all of the strategic objectives.

The above principles of urban regeneration in the Western context show the integrated regeneration mechanism from the inputs of various themes to the assessment of the regeneration outputs afterwards, during which the integration of comprehensive factors are outstandingly emphasized. These principles of urban regeneration are illustrated in Figure 3-3, which indicates the interacted urban regeneration process.

Comparing the urban regeneration programs in Chinese cities with the experiences of Western countries, it is clear that although the importance of the comprehensive development goals in China has been realized and promulgated in local urban planning and the regeneration programs by different levels of administrative governments, there is no institutionalized regeneration system in China yet. This severely undermines the inputs of various integrated actors at the very beginning. Although public participation has been promulgated in urban planning and regeneration processes in many Chinese cities, there are no specific laws or regulations to guarantee the legitimate involvement of various stakeholders. When economic development is outstandingly stressed in local urban development, other aspects such as the involvement of local residents and

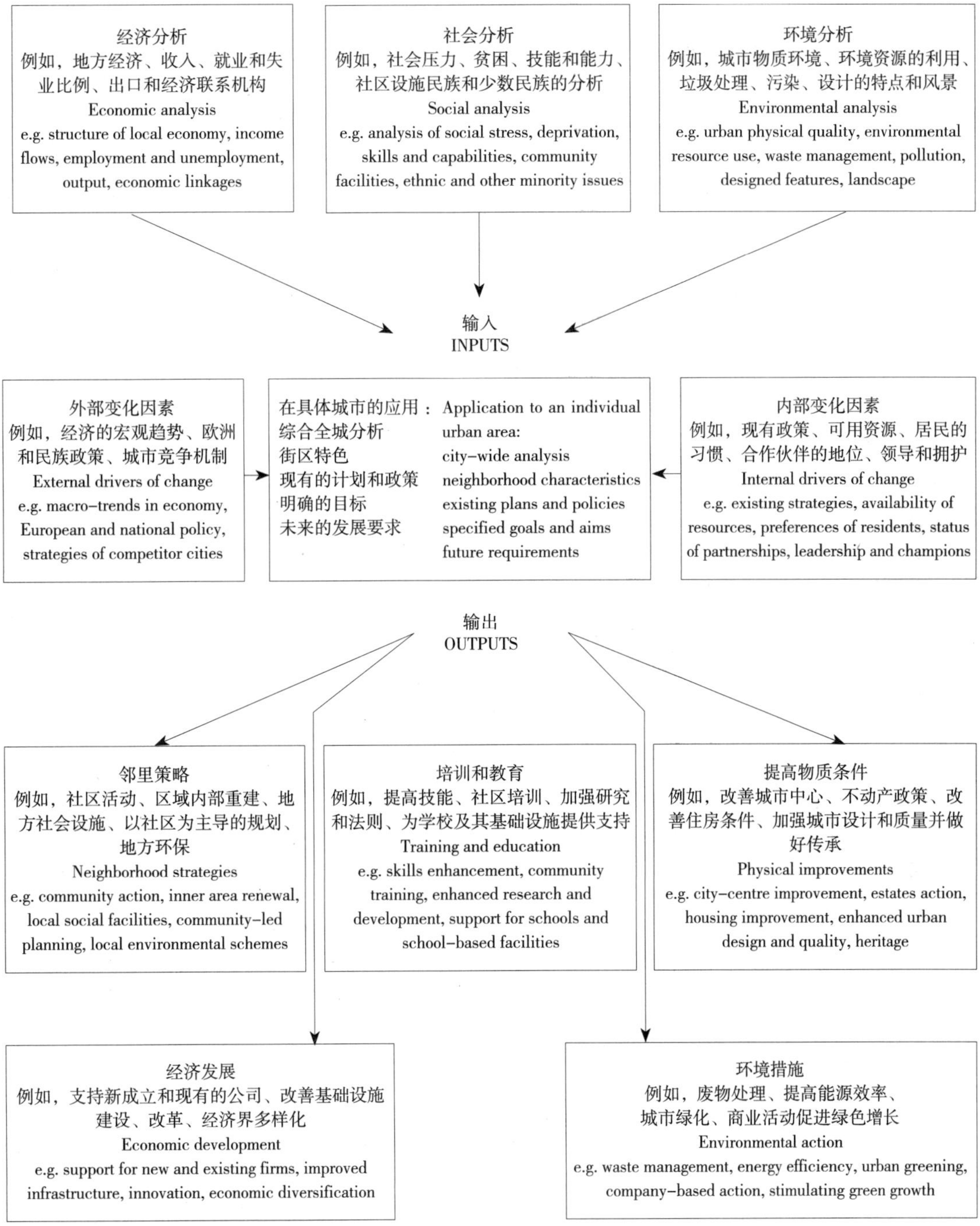

经济分析	社会分析	环境分析
例如，地方经济、收入、就业和失业比例、出口和经济联系机构	例如，社会压力、贫困、技能和能力、社区设施民族和少数民族的分析	例如，城市物质环境、环境资源的利用、垃圾处理、污染、设计的特点和风景
Economic analysis e.g. structure of local economy, income flows, employment and unemployment, output, economic linkages	Social analysis e.g. analysis of social stress, deprivation, skills and capabilities, community facilities, ethnic and other minority issues	Environmental analysis e.g. urban physical quality, environmental resource use, waste management, pollution, designed features, landscape

输入
INPUTS

外部变化因素	在具体城市的应用：综合全城分析 街区特色 现有的计划和政策 明确的目标 未来的发展要求	内部变化因素
例如，经济的宏观趋势、欧洲和民族政策、城市竞争机制	Application to an individual urban area: city-wide analysis neighborhood characteristics existing plans and policies specified goals and aims future requirements	例如，现有政策、可用资源、居民的习惯、合作伙伴的地位、领导和拥护
External drivers of change e.g. macro-trends in economy, European and national policy, strategies of competitor cities		Internal drivers of change e.g. existing strategies, availability of resources, preferences of residents, status of partnerships, leadership and champions

输出
OUTPUTS

邻里策略	培训和教育	提高物质条件
例如，社区活动、区域内部重建、地方社会设施、以社区为主导的规划、地方环保	例如，提高技能、社区培训、加强研究和法则、为学校及其基础设施提供支持	例如，改善城市中心、不动产政策、改善住房条件、加强城市设计和质量并做好传承
Neighborhood strategies e.g. community action, inner area renewal, local social facilities, community-led planning, local environmental schemes	Training and education e.g. skills enhancement, community training, enhanced research and development, support for schools and school-based facilities	Physical improvements e.g. city-centre improvement, estates action, housing improvement, enhanced urban design and quality, heritage

经济发展		环境措施
例如，支持新成立和现有的公司、改善基础设施建设、改革、经济界多样化		例如，废物处理、提高能源效率、城市绿化、商业活动促进绿色增长
Economic development e.g. support for new and existing firms, improved infrastructure, innovation, economic diversification		Environmental action e.g. waste management, energy efficiency, urban greening, company-based action, stimulating green growth

图 3-3：西方文献中的城市更新过程
Figure 3-3: The urban regeneration process in Western literature
资料来源：（Roberts, 2000: 20）
Source: (Roberts, 2000: 20)

和工作机制做出实质性的改进，例如，则很难实现真正的全面更新目标。

3.1.6 小结
计划经济体制下的城市更新

在1978年改革开放政策实施之前，中国的城市发展速度缓慢且不均衡。当时，城市发展最突出的特点就是以中央政府的方针和政治意识形态为主导。20世纪50年代，国家建设发展规划以工业发展为中心，地方城市建设要围绕工业建设开展。在动荡的"文化大革命"期间，城市发展与建设都受到了严重影响（Sit，2009：9）。1978年之前我国城市发展的另一个特点，就是集中经济下的碎片式城市发展。1949年中华人民共和国成立初期，地方自力更生和自给自足的思想导致出现了碎片式城市区域发展，而缺乏各地间的良好协调与合作。这使得不同地区出现了基础设施和机构的重复建设，阻碍了资源的有效利用，不利于城市建设。

1949年以后，我国破败城区的城市更新政策主要集中于对破旧城区的修复和必要基础设施的建设。此外，"大跃进"和"文革"期间很多遗留的误导性的城市政策，严重阻碍了地方城市功能系统的进行，例如公共绿化带被肆意占用，城市交通系统瘫痪（Hua，2006）。根据三线建设的指导方针，城市建设呈现碎片化，这进一步恶化了很多城市的城市布局。随着地方大量的零星发展活动，这都给1978年之后市场经济体制下的城市发展和开发带来很多挑战。在文化遗产和历史城市的保护方面，很多城市在中央计划经济时

the conservation of indigenous socio-cultural elements are often pushed aside and overlooked. As a result, many urban regeneration practices turn out not to be towards the improvement of public interest as planned but only as means of political progress or "political performance" (Zhang, 2002c: 498). These severely undermine the urban regeneration comprehensive goals. Therefore, it is necessary for the Chinese government to realize the importance of comprehensive regeneration goals. However, it has to realize as well that without the substantive improvement of the urban regeneration system, for example, the necessary political and institutional improvements, it is difficult to achieve the genuine comprehensive regeneration aims.

3.1.6 Summary
Urban regeneration in a centrally-planned economy

Before the open-door policy was adopted in 1978, urban development in China was slow and sporadic. During this period, one outstanding feature of urban development was the central government guidelines and political ideology playing the dominant role in urban development. When industrial development was central in the national development schemes of the 1950s, local urban constructions also needed to center around industrial development. During the Cultural Revolution, which was characterized by political turmoil, urban development and construction was severely affected (Sit, 2009: 9). Another feature of urban development before 1978 was the fragmented urban development in a cellular economy. In the Maoist era beginning 1949, the ideology of local self-reliance and self-sufficiency led to the regional fragmented development, where coordination and cooperation were not promoted. This resulted in the duplicated facilities and institutions constructed in different localities, which brought inefficiency to the use of resources and became detrimental to urban structures.

In this setting, the urban policies on the

期，尤其是在政治动荡的年代，都遭受了前所未有的重创。但是，这并不意味着当时没有关于遗产和城市保护的计划。本文后面将进一步讨论此问题。

市场经济转型期的城市更新

在1978年之后，中国城市更新具有鲜明的特点，因为很多政治与机构的改革为地方城市的改造和更新实践都提供了巨大的鼓励措施。20世纪80年代以来，地方政府获得了"前所未有的规划土地使用权，这得益于中央政府经济规划权力的下放、金融和银行制度的改革以及对外贸易管理责任的转移"[1]（Xu & Ng，1998：37）。由于这些变化，很多因素的作用也发生了变化。例如，城市规划起着至关重要的作用，而且很多城市更新活动要在地方总体规划的指导下进行。同时，各个利益相关者的作用也发生了转变。

随着20世纪80年代末城市土地改革和住房改革的进行，整个更新过程开始受经济驱动，私营部门开始积极参与到地方以项目开发为主导的重建过程中。为了提高财政收入，地方政府倾向于与私人开发商展开合作。在地方再开发过程中，地方政府和私人开发商的合作关系可概括为：地方政府提供土地使用权，私人开发商进行再开发活动（He，2002；Xie & Ingram，2008），或者两者作为公私合营的开发商共同进行再开发活动（Li，2002）。无论哪种方式，其目的都是追求最大的经济回报。在当前中国城市改造和更新实践中，地方政府既是规划方案的提出者、决策者和实施主体，同时还是工程

[1] 后面将会对城市发展的动力作进一步阐述。

regeneration of derelict inner urban areas after 1949 focused on the rehabilitation of dilapidated urban areas and the building of necessary infrastructure. Further, the existence of many misleading urban policies during the Great Leap Forward and the Cultural Revolution caused many obstacles in local urban functional systems, for example, public greenery areas being occupied and urban transportation systems being disrupted (Hua, 2006). With fragmented urban restructuring according to the guideline of the Third Front Construction, urban layout in many cities were further exacerbated. Accompanied by many local fragmented development activities, all these posed many challenges to urban growth and development in the market-oriented economy after 1978. With regard to the conservation of cultural heritage and historic cities, many historic urban areas experienced unprecedented damages during the centrally planned economy, especially in the period of political turmoil. However, it does not mean that there were no initiatives on heritage and urban conservation at the time. This book will address this issue in a latter section.

Urban regeneration in a transitional market-oriented economy

Urban regeneration in Chinese cities after 1978 produced conspicuous features, and this closely relates to the many political and institutional changes that provide plenty of incentives for local urban renewal and regeneration practices. Since the 1980s, local municipal governments have obtained "unprecedented power to plan the use of their land as a result of the devolution of economic planning power from the central government, the reform of financial and banking systems and the devolution of management responsibilities for foreign trade"[1] (Xu & Ng, 1998: 37). With these changes, many factors played different roles as before. For instance, urban planning played an important role, and many urban regeneration activities were conducted according to local master

[1] More elaborations on the dynamics of urban development will continue in a later section.

项目管理者和国有资源拥有者（Ng & Wu，1995；Xu & Ng，1998）。地方政府过于集中的职能身份，对有效的城市发展控制机制来说构成了挑战，并会影响全面城市更新目标的实现。

2000年初以来，很多历史城市根据中央政府颁布的政策在地方城市发展中提出了全面的更新目标。这些更新规划除了提到要促进经济发展之外，也着重提出对地方历史城市特色的保护。不过，由于缺乏一个制度化的参与机制，社区居民较少能够直接参与到城市更新过程中，而他们往往成为地方城市更新发展的利益受损者。然而，一个重要却通常被忽略的问题是，地方原住居民有助于保护和发展历史环境中的有形和无形文化遗产。

表3-2显示出转型期我国城市更新的关键内容与主要特点

plans. At the same time, the roles of various stakeholders in urban regeneration evolved.

With the urban land reform and housing reform in the late 1980s, the entire regeneration process appeared to be economics-driven, and the private sectors became very actively involved in local project-oriented redevelopment activities. In order to generate local revenues, local governments frequently develop a cooperative relationship with private developers. The relationship between local governments and private developers in local redevelopment can be summarized as the local governments providing the land-use rights and private developers conducting redevelopment (He, 2002; Xie & Ingram, 2008), or the local governments and private developers conducting redevelopment together as public-private developers (Li, 2002). No matter which way, the aim is often to pursue the maximum economic returns. It is obvious that in the current urban renewal and regeneration practices of Chinese cities, local governments play the role of decision-maker in planning, implementation body, manager of local projects, and owners of state-owned resources (Ng & Wu, 1995; Xu & Ng, 1998). These roles challenge the effective urban development control and compromise the comprehensive urban regeneration objectives.

Since the early 2000, many historic cities have proposed comprehensive regeneration objectives in local urban development as promulgated by the central government. These regeneration plans propose the conservation of local historic urban features in addition to economic development. Nevertheless, it is also clear that in the lack of an institutionalized participation mechanism, local communities generally play very little active roles, and they often become victims in local pro-growth developments. However, a significant but often overlooked issue is the contribution of local indigenous communities to the conservation and development of both tangible and intangible qualities of historic environments.

The key factors and characteristics of urban regeneration in transitional China are presented in Table 3-2.

表 3-2
Table 3-2

中国的城市更新特点（1949 ~ 当前）
Characteristics of urban regeneration in the history of P.R.China (1949 ~ present)

阶段 Phases	城市更新特点 Characteristics of Urban Regeneration	主要参与者和特色 Key Actors and Features — 政府：（中央，市以及区政府） State: (central, municipal, and local district governments)	主要参与者和特色 Key Actors and Features — 地方社区 Local community	主要参与者和特色 Key Actors and Features — 开发行业 Development industry	更新重点 Regeneration Focuses
1949~1965	以工业建设为中心，在总体规划范围内的对破旧城区进行修复 Industrial development-centered rehabilitation of dilapidated urban areas within the general plans	中央政府主导的资源配置和产业发展；按照地方整体规划，修复破败的城区并建设城市基础设施 Central government—led resource allocation and industrial development; rehabilitation of dilapidated urban areas and construction of urban infrastructures following the local general plans	地方政府动员地方社区对破旧城区进行更新和修复；鼓励自力更生和独立自主 Mobilized by local governments to renew and revitalize the most dilapidated and rundown areas; promoted self-reliance and independence	从中央政府到地级地方社区建立了国有房地产管理机构 State-owned real estate management institutes set up hierarchically from central government to local municipalities	"最大化利用，逐渐更新，增强修复和维护" "Maximum utilization, gradual renewal, and enhancement of repair and maintenance"
1966~1976	分散的城市发展和无规划的旧城滥用 Fragmented urban development and full utilization of old urban areas without planning	政治动乱和无政府主义；单一经济中分散独立自主和自力更生 Political turmoil and anarchism; fragmented urban development in a cellular economy towards self-reliance and self-sufficiency	受政府意识形态的鼓动，倾向"破四旧"运动 Mobilized by governmental ideologies to destroy the "Four Olds"	废除了国有房地产管理机构和相应的政策和法规 State-owned real estate management institutes, and corresponding policies and regulations were dismantled	对旧城区和文化遗址造成了巨大破坏 Massive destruction in old urban areas and cultural heritage
1978~20世纪80年代末 (1978-late 1980s)	根据地方总体规划对破旧城区的修复和重建以反对历史遗址的保护 Revitalization and redevelopment of derelict urban areas following local general plans and conservation of historic cities	中央和地方政府之间经济利益的再分配和地方政府财政权力的下放；地方政府负责制定地方全面规划，建设和管理 Decentralization and redistribution of economic benefits and fiscal power between central and local governments; local governments are responsible for generating local comprehensive plans, construction, and management	作为必要资金的来源，参与到当地城市改造和复兴项目中 Integrated in local urban renewal and revitalization projects as a source of necessary funds	中国私营房地产公司的蓬勃发展并积极参与商业城市开发；私营开发商兴起 Private real estate companies boomed in China and became actively involved in local urban development; private developers co-existed with public developers	根据地方总体规划进行改造和重建，通过各种途径筹集必要资金 Revitalization and redevelopment according to local general plans and various ways to raise the necessary funds
20世纪90年代 (1990s)	经常违背当地的再开发实行以地方政府主导的，以房地产主导的改造和重建 Economic-biased and property-led renewal and redevelopment that often violate local general plans	通过房地产主导的再开发实行以地方政府主导的城市改造政策和经济发展 Local government-led pro-growth urban renewal policies and economic development through property-led redevelopment; cooperated with private developers to pursue the maximum economic returns.	被隔离于地方开发项目之外，居民重置，社区解体 Excluded from local urban renewal and redevelopment; residential relocation and dismantling of community neighborhoods	私人发展商的蓬勃发展并积极参与地方城市发展，与地方政府合作追求最大的经济回报 Private developers boomed in China and became actively involved in local urban development; cooperated with local governments to pursue the maximum economic returns	在城市改造和重建的过程中追求最大经济回报，搁置城市总体规划 Pursuance of maximum economic returns from urban renewal and redevelopment; pushed aside local general plans
始于21世纪初 (Beginning early 2000s)	全面考虑的，倾向经济发展的城市更新政策；促进经济发展并保护历史城市特色 Pro-growth urban regeneration policies with more comprehensive concerns	全面考虑的，倾向经济发展的更新政策；促进经济发展并保护历史城市特色 Local government-led pro-growth urban regeneration policies with more comprehensive concerns; promoted economic development and conservation of historic urban features	被隔离于地方开发项目之外，居民重置，社区解体 Excluded from local urban renewal and redevelopment; residential relocation and dismantling of community neighborhoods	积极参与地方城市改造与重建；保留地方历史城市特征的同时追求最大的经济回报 Became actively involved in local urban renewal and redevelopment activities; pursued the maximum economic returns while "conserving" local historic urban features	通过大规模开发和保护有限的历史建筑对"传统"城市的景观进行重塑 Re-fabrication of "traditional" urban scenes through massive redevelopment and conservation of limited historical buildings

资料来源：作者从不同的文献资料进行汇总 (Ma, 1979, 2002; Ng & Tang, 2004b; Oi, 1995; Tang & Liu, 2005; Walder, 1995a, 1995b; Wu, 2002; Wu, 2004; Yang, 2000; Yang & Wu, 1999; Zhao, Lee, & Sun, 2006)

Source: Author-synthesized from (Ma, 1979, 2002; Ng & Tang, 2004b; Oi, 1995; Tang & Liu, 2005; Walder, 1995a, 1995b; Wu, 2002; Wu, 2004; Yang, 2000; Yang & Wu, 1999; Zhao, Lee, & Sun, 2006)

3.2 影响中国历史城市更新的关键因素

上述讨论表明，在中国历史城市中，城市保护问题对城市更新政策的制定至关重要的。目前中国城市的更新机制下，地方政府在更新计划的制定、实施和管理方面起着主导作用，并拥有很多自由的裁量权。很多时候，为了促进经济发展，当地综合的城市更新目标往往做出让步。在地方城市的更新动力下，研究认为社区邻里联系或社会资本有助于促进地方居民参与到更新过程中，有助于促进对历史环境和地方原住民生活的保护，并从而也有助于历史城市综合更新目标的实现。因此，本章接下来首先探讨了中国的保护法律法规问题；其次，研究了中国城市的企业管理模式；最后，本章分析了中国城市的社区邻里关系和社会资本问题。

3.2.1 文化遗产与城市保护[1]

一般来说，从对可移动遗产，例如古董的保护方面来讲，中国的文化遗产保护[2]拥有悠久的历史。在古代，通过鉴赏和收藏这些

---

[1] 本章节中关于中国遗产和城市保护的官方文件主要是从（CAUPD&DTCPMC, 2008: 14, 15; Ruan&Sun, 2001; Wang, Ruan, &Wang, 1999: 9–12）等文件中汇总得来。

[2] 保存与保护的概念来自《巴拉宪章》（ICOMOS, 1999）文件的规定。根据《巴拉宪章》，保护意指为"保留一个地点的文化意义而采取的所有照管行动"。在这里，文化意义指的是，"为过去、现在或将来的人们保留其美学、历史、科学、社会或精神价值"。而保存是指"保持一个地点现有状态下的构建，延缓其恶化。并且大家公认，随着时间的流逝，所有地点及其组成部分都会以不同的速度发生变化"（ICOMOS, 1999）。

3.2 Key Factors in Urban Regeneration in Chinese Historic Cities

The above discussions show that the issue of urban conservation is critical to the urban regeneration policy in Chinese historic cities. In addition, in current Chinese urban regeneration mechanism, local governments play a dominant role in terms of regeneration plan-making, implementation, and management. This situation gives local governments much discretionary power, and local comprehensive urban regeneration objectives are often compromised in promoting economic development. Nevertheless, in the context of local urban regeneration dynamics, this book contends that community connections or social capital contributes to the involvement of indigenous residents in the regeneration processes. This facilitates the conservation of historic environments and local indigenous lives, which in turn contributes to the comprehensive urban regeneration goals in Chinese historic cities. Therefore, this section explores the issue of conservation laws and regulations in China. Next, the mode of entrepreneurial governance in Chinese cities is examined. Finally, this chapter probes into the issue of community relationships or social capital in urban China.

3.2.1 Cultural heritage and urban conservation in China[1]

Generally, heritage conservation[2] in China has a long history in terms of the protection of some movable heritage such as ancient antiques. In ancient times, appreciating and cherishing these antiques was a way to acquire the knowledge

---

[1] The official documents on heritage and urban conservation in China in this section are mainly summarized from (CAUPD & DTCPMC, 2008: 14, 15; Ruan & Sun, 2001; Wang, Ruan, & Wang, 1999: 9–12).

[2] The concepts of conservation and preservation follow the *Burra Charter* (ICOMOS, 1999). According to the *Burra Charter*, Conservation means "all the processes of looking after a place so as to retain its cultural significance", which means "aesthetic, historic, scientific, social or spiritual value for past, present or future generations". And preservation means "maintaining the fabric of a place in its existing state and retarding deterioration. It is recognized that all places and their components change over time at varying rates" (ICOMOS, 1999).

古董，可以了解历史，进行纪念。而且，在中国这些是关于遗产保护最早的活动（CAUPD & DTCPMC, 2008：14）。但是，在保护宫殿遗址、建筑物和其他不可动遗产方面，在中国历史上做得不能十分令人满意，这主要有两方面原因：主观或人为原因和客观原因。

1）主观或人为原因（例如，由于政治或宗教的原因）

学术界认为，在中国古代，两个人为的不利传统／习俗给古代建筑物造成很大的破坏。例如，早在公元前3世纪的中国，就有人企图用新建筑来代替前朝的宫殿建筑，因为他们相信这种以新代旧的做法可以彻底粉碎前朝政权复辟的可能性（Fu, 2002：28）。同时，即便要继续使用前朝的宫殿建筑，他们也会对其进行大肆修改，以表明这些建筑已经完全被革新和转变（Dong, 2004：46-47）。另一方面，关于古代寺庙建筑，很多朝圣者认为对古寺庙建筑和佛像进行重建是值得称赞和颂扬的（Fu, 2002：28）。结果，很多代表各个朝代建筑水平的寺庙，历经多次改建，没能对原布局和宝贵建筑进行保护。

2）客观因素（例如，战争和自然灾害）

由于很多古代木质建筑没有防火能力，他们在战争或自然灾害中很容易被烧毁。例如，紫禁城中的三座大殿是在明代永乐年间（1404～1424年）建造的，而就在建成的第二年就遭电击，被烧毁了。而且，这些建筑重建后，在几年后再一次被摧毁（Pan, 2004：2）。因此，保护原建筑变得很难。此外，随着1840年西方帝国主义的侵略，很多古建筑和遗产都被破坏殆尽（CAUPD & DTCPMC, 2008：14）。

of history and commemoration. Moreover, these were the earliest activities of heritage conservation in China (CAUPD & DTCPMC, 2008: 14). Nevertheless, the conservation of palace buildings, constructions, and other immovable heritage was not done well in Chinese history mainly because of two reasons: subjective or man-made reasons and objective ones.

1) Subjective or man-made factors (e.g., political or religious)

According to academics, two man-made unfavorable traditions in ancient China account for the destruction of ancient constructions. As early as the 3rd century B.C., there was a scheme to replace the palace buildings of preceding dynasties with new constructions because it was believed that the recovery possibilities of the preceding dynasty regimes could be annihilated completely with replacement (Fu, 2002: 28). At the same time, even if the preceding palace buildings continued to be used, massive changes were implemented to show they were already reformed and transformed completely (Dong, 2004: 46-47). On the other hand, with regard to the ancient temple structures, many pilgrims assumed it to be meritorious and virtuous to reconstruct ancient temple structures and Buddha figures (Fu, 2002: 28). As a result, many temples representing the architectural levels of various dynasties were reconstructed many times, which resulted in the failure of protecting the original layouts and valuable constructions (Luo, 1987: 59).

2) Objective factors (e.g., warfare or natural disasters)

Considering the vulnerability of many ancient wooden structures to fire, they were easily burned during wars or natural disasters. For instance, the three main halls within the Forbidden City were struck by lightning and burned down only in the second year after they were constructed in the Yongle period (A.D. 1403~1424) of the Ming dynasty. In addition, these buildings were reconstructed and burned down again in later years (Pan, 2004: 2). Thus, this made it very difficult to protect the original buildings. Further, with the aggression of Western imperialism in 1840, many ancient buildings and heritage were massively destroyed (CAUPD & DTCPMC, 2008: 14).

3.2.1.1 城市保护体系的历史回顾

我国现代意义上的文化遗产保护始于20世纪初，那时清政府设立了国家事务部，并于1906年颁布了《保存古物推广办法》。这是中国关于遗产保护的第一个官方文件。后来，在1949年中华人民共和国成立之前也颁布了几部遗产保护的法律法规[1]（CAUPD & DTCPMC, 2008: 14）。这些章程、法规和法律是中国关于文化遗产保护最早的官方文件。最早成立于1928年的文物保护中央委员会，是最早的遗产保护和管理的国家级特别组织。然而，由于那个年代混乱的政治斗争，没有一个持久稳定的管理体系，而且地方政府也没有建立相应的遗产管理机构。由于这些文件实施不力，全国许多文化遗产都缺乏适当的管理（CAUPD & DTCPMC, 2008: 14）。

总体而言，自中华人民共和国1949年成立以来，我国的文化遗产保护形成了四个方面：保护文化遗址、保护历史街区、保护历史城市（Ruan & Sun, 2001: 25）以及保护非物质文化遗产（CEFLAD, 2007）。

文化遗址的保护（始于1950年）

1950年开始，除了中央和地方管理机构颁布法令和法规，中央政府也推出了一系列法令和法规。1962年，国务院颁布文物保护管理暂行条例，在180个国家级文化遗址中试行，象征了重大历史古迹机构的成立（Wang, Ruan, & Wang, 1999: 10）。不过，随后于

---

[1] 1908年：《城镇乡地方自治章程》；1916年：《保存古物暂行办法》；1928年：《名胜古迹古物保存条例》；1930年：《古物保存法》；1931年：《古物保存法细则》；1932年：《中央古物保管委员会组织条例》（CAUPD&DTCPMC，2008：14）。

3.2.1.1 Historical overview of the urban conservation system

The modern sense of heritage conservation in China commenced in the early 20th century, when the Qing government set up the Home Affair Department and launched the *Promulgated Method on Heritages Preservation (Baocun Guwu Tuiguang Banfa)* in 1906. This was the first official document on heritage conservation in China. Later, several more laws and regulations on heritage conservation were enacted before the establishment of the P.R. China in 1949[1] (CAUPD & DTCPMC, 2008: 14). These charters, regulations, and laws were the earliest official documents on heritage conservation in China, and the Central Committee of Antiquities Conservation, which was first set up in 1928, was the earliest national special organization for heritage conservation and management. Nevertheless, due to the chaotic political struggle in those days, there was no lasting and steady management system, and the local government did not provide corresponding heritage management institutions as well. With the inefficient implementation of these documents, many cultural heritages in the whole country lacked proper management (CAUPD & DTCPMC, 2008: 14).

By and large, since the establishment of the People's Republic of China in 1949, cultural heritage conservation in China has incorporated four facets: the preservation of Cultural Sites, the conservation of Historic Districts, the conservation of Historic Cities (Ruan & Sun, 2001: 25), and the conservation of Intangible

---

[1] In 1908: Charter of Autonomous Management for Counties, Towns and Villages (*Chengzhenxiang Difang Zizhi Zhangcheng*); In 1916: Temporary Methods on Heritage Preservation (*Baocun Guwu Zanxing Banfa*); In 1928: Preservation Regulations of Historic Sites and Monuments (*Mingshenggujiguwu Baocun Tiaoli*); In 1930: Antiquities Preservation Law (*Guwu Baocunfa*); In 1931: Detailed Regulations on Antiquities Preservation Law (*Guwu Baocunfa Xize*); In 1932: Organization Regulations on the Central Committee of Antiquities Conservation (*Zhongyang Guwu Baoguan Weiyuanhui Zuzhi Tiaoli*) (CAUPD & DTCPMC, 2008: 14).

1966年爆发的"文化大革命"几乎摧毁了整个新成立的遗产保护机构。受到"文革"期间"破四旧"的破坏，中国几乎所有的历史古迹和遗址都遭受到了史无前例的毁坏。更严重的问题是，"文化大革命"之后很久仍然残留的破旧立新的观念（EBUCCC，1990；Luo，1987）。直到20世纪70年代中期，才开始恢复遗产保护工作。随着1982年《中华人民共和国文物保护法》的颁布，遗产保护的法律体系在国家层面上奠定了基础。该法案也象征着以保护单个文物古迹和遗址的历史文化遗产保护体系的形成（Wang et al.，1999：11）。

历史城市的保护（20世纪80年代早期）

随着20世纪80年代经济的快速发展，很多城市开始了大量的城市更新和再开发活动，使得历史环境的保护面临巨大挑战和威胁。为了保护历史遗迹免受大规模的破坏，保护问题由对文化遗址的保护转为对整个历史城市的保护。1982年，随着国务院颁布的《关于保护我国历史文化名城的请示的通知》，24座城市列为第一批国家历史城市（Wang et al.，1999：16）。随着以后指定的历史名城的增加，2007年，历史名城的总数达到了108座（Jia，2007：3）。

随着城市的快速发展，20世纪90年代中期又出现了很多问题（Chan，1999；Ruan，2007；Sun，1992）。由于历史城区通常位于城市的中心地区，会吸引很多商业和旅游业的目光。这种情形下，往往需要对历史城区的原有环境进行很大程度的改变。但是，由于历史城市中地方保护法规的限制，几乎不允许对城市中的物质环境进行改变，而且很

Cultural Heritages (CEFLAD, 2007).

The conservation of Cultural Sites (beginning 1950)

Beginning 1950, the central government launched a series of ordinances and regulations in addition to those of the central and local management institutions. In 1961, the State Council launched the Temporary Legislations on Heritage Conservation and Management (*wen wu bao hu guan li zan xing tiao li*), and 180 Cultural Sites at the national level were enacted, indicating the setting up of the institution of significant historic monuments (Wang, Ruan, & Wang, 1999: 10). Nevertheless, the subsequent Cultural Revolution in 1966 nearly wiped out the entire newly established heritage conservation institutions. Compelled by "destroying the four olds" (*po si jiu*), historic monuments and sites in China underwent unprecedented and overwhelming destruction. However, the more severe issue was that the tendency of destructing the old and erecting the new (*po jiu li xin*) remained in the society long after the Cultural Revolution (EBUCCC, 1990; Luo, 1987). The heritage conservation work was not reinstated until the mid-1970s. With the enactment of the Cultural Relics Conservation Act in 1982, the legal system for heritage conservation at the national level was provided its basis. The Act also symbolized the construction of the historic cultural heritage conservation system centering on the conservation of individual historic monuments and sites (Wang et al., 1999: 11).

The conservation of Historic Cities (early 1980s)

With the rapid economic development in the 1980s, massive urban renewal and redevelopment activities took place in many cities, posing great challenges and threats to the conservation of historic environments. In order to protect historic sites from massive destruction, the conservation issue turned from the conservation of Cultural Sites to the conservation of whole historic urban settings. In 1982, with the State Council enacting the *Notice of Requesting the Instructions of Conserving the Historic Cities in China* (*guan yu bao hu wo guo li shi wen hua ming cheng de qing shi de*

多地方当局觉得这样会降低现有土地的价值。这种情况就造成城市保护和经济发展之间的矛盾（Ruan，2007：14）。阮仪三总结了城市保护和经济发展之间的矛盾，通常表现在三个方面：第一，私人开发商不愿意进行基于保护的再开发项目，因为受保护政策的限制，他们得到的经济回报是有限的；第二，仅仅依靠地方政府对整个保护项目进行资金支持是很困难的；第三，大多数地方居民在当地自我建设中，希望取代传统建筑，建设现代化的建筑（Ruan，2007：14）。

面对这些矛盾，以及由于缺乏必要的城市保护法律和管理系统，不少历史城区已经受到了严重破坏（Jin & Xu，2000：69；Shan，2006）。另外，对历史城市中心进行大量的再开发活动，通常会使这些地区面临巨大压力，来提供多种功能需求，包括解决人口过剩、人口密度大以及其他类似功能（Yang，2003：38）。本研究认为，与必要的法律法规相比，有效的保护和开发管理体系显得更加重要，因为这样可以通过多种途径来处理违反现有保护法规和无视地方保护法规的行为（Gao，2002：25）。

历史街区的保护（20世纪90年代中期）

自20世纪90年代以来，由于大量地方城市更新和发展对历史城区造成了相当大的破坏，所以历史街区已经引起了广泛地关注（Ruan & Sun，2001；Wang，2002a）。1996年在黄山市屯溪区举行了历史街区保护研讨会。在研讨会上，对历史街区的保护被认为是中国遗产保护体系中一个重要组成部分。这象征着中国遗产保护的第三个阶段的开始（Wang et al.，1999：12）。

tong zhi), 24 cities were selected as the first batch of the national Historic Cities (Wang et al., 1999: 16). With the addition of the Historic Cities later designated, the total number of the Cities reached 108 in 2007 (Jia, 2007: 3).

Nevertheless, with the rapid urban development, many problems emerged by the mid-1990s (Chan, 1999; Ruan, 2007; Sun, 1992). Often located in the central urban areas, historic urban sites attract much commercial and tourism attention. In this situation, many changes and improvements on the original environments are needed. However, considering the restrictions of local conservation regulations in historic cities, very few changes are normally allowed to take place in the physical environments, and this is believed by many local authorities to lower down the existing land values. This situation forms the basic contradiction between urban conservation and economic development (Ruan, 2007: 14). Ruan summarized that usually the contradictions between urban conservation and economic development rest on three aspects: first, private developers are not willing to carry out conservation-based redevelopment projects considering the limited economic returns from the constraints of conservation policies; second, it is difficult for local governments to support financially the entire conservation projects alone; and third, most local residents would like to replace traditional buildings and forms with modern ones in local self-constructions by themselves (Ruan, 2007: 14).

In the face of these contradictions and with the lack of the necessary urban conservation laws and management system, many historic urban areas have been destroyed (Jin & Xu, 2000: 69; Shan, 2006). In addition, the massive redevelopment in historic urban centers often puts these areas under pressure with too many functions, including those that concern overpopulation, high density, and similar functions (Yang, 2003: 38). This book argues that compared with the necessary laws and regulations, effective conservation and development management system are more important, which in many ways can account for the violation of existing conservation regulations and the neglect of local

根据 1986 年国务院文件[1]，中国的历史街区指的是"拥有集中体现文化遗迹和历史古迹，或者是能够完全体现一个特定历史时期的传统、民族和地方色彩的街道或街区"（Wang, 2002a：1）。由于意识到历史街区对历史城市的意义和贡献，许多城市在当地城市保护规划中都强调了对历史街区的保护（Wang, 2002a：1）。不过，在实践中，当地决策机构往往把这些保护建议放到一边或进行"重新解释"，例如，有些城市倾向于把整个历史地区推倒，然后重建仿古建筑（Ruan, 2000：46）。阮仪三和孙萌（Ruan & Sun, 2001）认为，20 世纪 90 年代的国家政策在一定程度上吸引了地方政府和私人开发商参与到历史街区的再开发项目中。20 世纪 90 年代中期，中央政府提出了几项规定（SCC, 1992, 1997；SCXPC, 2004a），目的是为了抑制对耕地的随意占领，在新开发区中的大规模建设，以及自 20 世纪 80 年代以来出现的城市肆意扩张现象（Ruan & Sun, 2001：27）。受这些政策的限制，城市规划管理部门和私营开发商已经开始转向在旧城区寻求再开发的机会，城市改造和再开发实践已对当地的历史环境产生了很大影响。概括来讲，这些影响主要反映在如下几方面：

1) 相似的城市景观（如"千城一面"现象）

在地方城市的再开发中，很多城市丧失了当地城市的特色，变得千城一面，如，摩天大楼、欧式建筑、大型广场及其他特色（Shan, 2006）。

urban conservation plans (Gao, 2002: 25).

The conservation of Historic Districts (mid-1990s)

Facing the considerable destruction in historic urban areas in massive local urban renewal and development since the mid-1990s, much attention has been given to Historic Districts (Ruan & Sun, 2001; Wang, 2002a). In 1996, the Symposium on the Preservation of Historic Districts (*Lishijiequ Baohu Yantaohui*) was held in Tunxi district, Huangshan city. In the symposium, the conservation of Historic Districts was regarded as an important component in the heritage conservation system in China. This indicated the beginning of the third stage of heritage conservation in China (Wang et al., 1999: 12).

According to the State Council document[1] in 1986, Historic Districts in China refer to "the streets or blocks of buildings with concentrations of cultural relics and historic monuments or complete embodiments of traditional, ethnic and local features of a certain historic period" (Wang, 2002a: 1). With the recognition of the significance and contribution of Historic Districts to historic cities, many cities have outlined the conservation of Historic Districts in local urban conservation plans (Wang, 2002a: 1). In practice however, these conservation proposals are often pushed aside or "reinterpreted" by local decision-makers, and many cities tend to bulldoze whole historic areas and reconstruct fake historic buildings (Ruan, 2000: 46). Ruan and Sun (2001) revealed that national policies in the 1990s partially played the role of attracting local governments and private developers' attention to redevelop Historic Districts. In the mid 1990s, the central government put forth several regulations (SCC, 1992, 1997; SCXPC, 2004a) in order to curb the random occupation of farmland and the massive construction of new development zones and urban sprawl that had occurred since the 1980s (Ruan & Sun, 2001: 27). Due to the constraints

---

[1] 1986 年《批准建设部、文化部关于请公布第二批国家历史文化名城名单的报告的通知》（Ruan & Sun, 2001）。

[1] The Ratifying the Document on Historic Cities of the Second Batch by the Ministry of Construction and the Ministry of Culture of P.R.China (*Pizhun Jianshebu Wenhuabu Guanyu Qing Gongbu Di'erpi Guojialishiwenhuamingcheng Mingdan De Baogao De Tongzhi*) in 1986 (Ruan & Sun, 2001).

2）为了迎合地方旅游业的需求，拆毁历史建筑，建造"假古董"。（如仿古一条街）

地方政府和私人开发商，通常把利用历史城市作为发展旅游业的资源，看作是一种盈利方式。在这个意义上说，城市保护仅仅被当做促进旅游业的一种手段（Ruan & Sun, 2001：28）。此外，许多原有历史建筑被拆毁，取而代之的是"假古董"。这些重建活动也完全违背了文物保护政策中"真实性"的原则。如《奈良真实性文件》（1994年）所述，许多方面都必须遵循真实性的规则，包括"形式和设计、材料和物质、使用和功能、传统和技术、位置和设置，精神和感觉，以及其他内部和外部因素"（ICOMOS, 1994：2）。本研究认为，从非物质文化遗产保护方面来看，如果在当地城市再开发项目中，原住民被排斥或隔离于项目之外，也是违反"真实性"原则的。

3）居民重置和社区解体

在许多历史城市区的再开发进程中，经常伴随居民迁移和重置现象。阮仪三和孙萌认为，没有原住民的历史街区缺乏真实生活，而且会失去其历史特性，而这些不能被仿建的遗迹或仿造的性能所替代（Ruan & Sun, 2001：28）。随着非物质文化遗产保护的普及，不少学者认为（Cheng, 2006a；Fang, 2006；Li, 2007a；Wang, 2006c；Zhang, 2007a），应该把生活在历史街区的原住民作为当地非物质文化遗产的载体来保护。后面的章节中将会进一步来讨论这一点。另一个随着居民迁移和重置产生的弊端就是它影响社会的凝聚力。吴缚龙认为，虽然居民在搬迁的过程中会得到一些好处，但是在迁移的过程中，利益的分配"倾向于那些有较高教育水平的人，他们会得到更好的安置"

caused by these policies, local urban planning management departments and private developers have tended to conduct redevelopment in old urban areas since then, and these urban renewal and redevelopment practices have considerably affected local historic environments. Generally, these impacts are reflected in many perspectives such as the following:

1) Similar cityscapes (e.g., "Thousand cities one face" [*qian cheng yi mian*])

In local urban redevelopment, many cities lose local urban features and feature similar cityscapes (*qian cheng yi mian*) such as skyscrapers, European-style buildings, big squares, and other features (Shan, 2006).

2) Demolishing historic buildings and constructing fake antiques catering to the needs of local tourism (e.g., "One Historic Street" [*li shi yi tiao jie*])

Local authorities and private developers often see it as a profitable way to utilize historic urban areas as resources to develop tourism industries, and in this sense, conservation is taken only as a means to facilitate tourism industries (Ruan & Sun, 2001: 28). Moreover, many original historic buildings are demolished and replaced by fake antiques. These redevelopment activities are totally against the conservation policies in terms of authenticity. As indicated in the *Nara Document on Authenticity* (*1994*), many perspectives are required to follow the rule of authenticity, including "form and design, materials and substance, use and function, traditions and techniques, location and setting, and spirit and feeling, and other internal and external factors" (ICOMOS, 1994: 2). This book argues that it is also against the authenticity principle in terms of the conservation of intangible cultural heritage when original residents are removed or relocated from local urban redevelopment.

3) Residential relocation and community disruption

In the redevelopment of many historic urban areas, residential displacement and relocation frequently occur. Ruan and Sun maintained that historic districts without indigenous residents lack authentic lives and lose their historic quality, which could never be replaced by fake antiques

(Wu, 2004：466)。这突显了一个事实，就是大多人的利益，尤其是"草根"阶层的利益，通常受到侵犯。这种情况已经得到了证实，因为在住房分配的过程中，有更多的寻求公平和正义的诉求已经被提交了。就像周岚与何流所表述的，在目前的城市发展进程中，公众对住房分配政策的抱怨和不满也越来越多（Zhou & He，2005）。根据对2006年中国城市的社会冲

2006年对中国个人经历的社会矛盾和冲突的调查
表 3-3
Investigation of self-experienced social contradictions and conflicts of China in 2006
Table 3-3

矛盾和冲突 Contradictions and conflicts	比例（%） Ratio (%)
学校不合理的收费问题 Unreasonable charges from schools	19.00
影响居住环境的污染问题 Pollution affecting living environments	18.40
政府部门不合理的收费问题 Unreasonable charges from government departments	16.70
解雇后受忽视 Overlooked after being laid off	11.00
拖欠工资/加班 Be behind in payment / overtime work	10.40
征地拆迁、房屋搬迁、拆迁和不合理的赔偿 Land appropriation, housing removal, displacement and unreasonable compensation	7.80
司法不公/暴力执法 Jurisdiction injustice / violent law-enforcement	5.30
恶劣的工作环境/雇主的恶劣态度 Adverse working environment / rude attitudes of employers	4.20
医患纠纷 Doctor-patient dispute	3.40
贪污受贿/侵犯集体财产 Bribery and corruption / encroaching collective-owned property	3.20
社会保障争议 Social security dispute	2.80
巨额消费，如购买住房的争议 Disputes in big bulk of consumption such as buying a flat	1.60

资料来源：（Wang, Yang, & Chen, 2006：72）
Source: (Wang, Yang, & Chen, 2006: 72)

or artificial performances (Ruan & Sun, 2001: 28). With the promulgation of the conservation of intangible cultural heritage, many academics (Cheng, 2006; Fang, 2006; Li, 2007a; Wang, 2006c; Zhang, 2007a) believe that indigenous community lives in historic districts should be conserved as the carriers of local intangible cultural heritage. Further discussion on this will continue in a latter section. The other drawback accompanying residential displacement and relocation in historic districts concerns social cohesion. Wu contended that although there are some benefits for residents during neighborhood relocation, the distribution of the benefits is "skewed towards those who have higher education levels and are better positioned" in the relocation processes (Wu, 2004: 466). This emphasizes the fact that the interest of most people, especially at the grassroots' level, is infringed upon. This situation is proven by the fact that more residential appeals to seek justice and equity in housing relocation processes have been lodged. As stated by Zhou and He, there are more and more public grudges against and dissatisfaction with residential displacement strategies in the current urban development processes (Zhou & He, 2005). According to the investigations on social conflicts and resolutions in Chinese cities in 2006, "land appropriation, housing removal, displacement, and unreasonable compensation" are some of the most serious social conflicts. Nearly 7.8 percent of the respondents had personal experiences of these conflicts. At the same time, about 6.08 percent of the respondents chose to go to court to solve these conflicts; 2.59 percent chose to settle the problems through violence; and 1-3 percent chose to go on strikes or sit-down protests (Wang et al., 2006: 72, 74). The investigation of the most serious social contradictions and conflicts in China in 2006 is displayed in Table 3-3.

The conservation of Intangible Cultural Heritage (since 2004)

In 2004, the Standing Committee of the National People's Congress (thereafter, SCNPC) ratified the decision on the UNESCO document, that is, the *Convention for the*

突及其解决方案的调查,"征地拆迁、住房搬迁、拆迁和不合理的补偿"是最严重的社会冲突里的一部分。近7.8%的受访者表示,个人曾经经历过这些冲突。同时,约有6.08%的受访者选择走司法程序来解决这些冲突;2.59%的人选择通过暴力来解决问题;1%到3%的人选择继续罢工或静坐抗议(Wang et al., 2006:72, 74)。表3-3显示出2006年我国比较严重的社会矛盾和冲突的调查结果。

非物质文化遗产的保护(2004年以来)

2004年,全国人民代表大会常务委员会(下文简称为SCNPC),批准了联合国教科文组织文件中的决定,即,《保护非物质文化遗产公约》,(下文称为《公约》),它象征着中国遗产保护体系的第四个阶段的开始(SCXPC, 2004b)。2005年3月,基于中国的宪法和有关法律、法规,国务院公布关于《加强我国非物质文化遗产保护工作的意见》(之后简称为《意见》),概述了中国非物质文化遗产的声明和评估标准(SCC, 2005a)。然而,在实践中,学者普遍认为,中国的非物质文化遗产的保护面临着三个挑战。

1)联合国教科文组织对非物质文化遗产的定义与中国城市的现实;

上述《意见》指出,中国非物质文化遗产的概念[1]和联合国教科文组织公约的界定

---

[1] 根据《保护非物质文化遗产公约》,非物质文化遗产指的是"被各社区、群体,有时是个人,视为其文化遗产组成部分的各种社会实践、观念表述、表现形式、知识、技能以及相关的工具、实物、手工艺品和文化场所。这种非物质文化遗产世代相传,在各社区和群体适应周围环境以及与自然和历史的互动中,被不断地再创造,为这些社区和群体提供认同感和持续感,从而增强对文化多样性和人类创造力的尊重"(UNESCO, 2003)。

Safeguarding of the Intangible Cultural Heritage (thereafter the **Convention**), which symbolizes the beginning of the fourth facet of the heritage conservation system in China (SCXPC, 2004b). In March 2005, the State Council announced the *Opinions on Intensifying the Conservation of Intangible Cultural Heritage in China* (thereafter the *Opinions*), which, based on the Constitution of China and related laws and regulations, outlines the declaration and assessment criteria for intangible cultural heritage in China (SCC, 2005a). Nevertheless, in practice, academics agree that generally there are three challenges facing the conservation of intangible cultural heritage in China.

1) Discrepancies between the definition of intangible cultural heritage adopted by the UNESCO and its realities in urban China

As laid out in the *Opinions*, the definition of intangible cultural heritage[1] in China largely follows that in the UNESCO *Convention*. This definition stresses the culture created and inherited among folk communities, which Duan believed as the reflection of the concept of culture in a broad sense, which includes not only the literati culture (or high culture) but also the folk culture created collectively by ordinary people (or low culture) (Duan, 2007: 1). Nevertheless, in China today, the understanding of intangible cultural heritage focuses much on the literati or elite culture in a narrow sense (Li, 2007a: 45). In many people's eyes, folk culture can hardly be seen as a kind of culture. Therefore, in order to protect intangible cultural heritage in China, it is necessary to recognize

---

[1] As laid out in the *Convention for the Safeguarding of the Intangible Cultural Heritage*, intangible cultural heritage refers to "the practices, representations, expressions, knowledge, skills—as well as the instruments, objects, artifacts and cultural spaces associated therewith—that communities, groups and, in some cases, individuals recognize as part of their cultural heritage. This intangible cultural heritage, transmitted from generation to generation, is constantly recreated by communities and groups in response to their environment, their interaction with nature and their history, and provides them with a sense of identity and continuity, thus promoting respect for cultural diversity and human creativity" (UNESCO, 2003).

大致相同。这个定义强调了文化是在民间社区创造和继承的。段宝林认为，从广义上说，这是文化概念的反映，其中不仅包括文人文化（或高雅文化）也包括普通人集体创造的民间文化（或通俗文化）（Duan，2007：1）。然而，在今天的中国，对非物质文化遗产的理解,更多关注狭义上的人文或精英文化（Li，2007a：45）。在很多人看来，民间文化几乎不能被视为一种文化。因此，在中国，为了保护非物质文化遗产，意识到社区中的地方文化的意义是非常必要的（Wu，2007：33）。

2）历史街区中原住民生活内容的忽视；

虽然本文讨论了地方社区文化在快速城市化发展进程中的重要意义，但同时也指出，许多作为非物质文化遗产载体的原住民的生活，往往被忽视。随着原住民生活的消失，地方非物质文化遗产也很有可能随之消失。方李莉认为生态博物馆理论[1]可以为人们提供许多思路，也突出强调了文化遗产与其环境之间的关联（Fang，2006：186）。根据这一理论，从保护当地居民的"灵感、智慧以及个人特征"角度而言，保护原住居民也是文化遗产保护当中不可或缺的一部分内容（Rivière，1985：182）。学者断言，如果地方居民成为遗产保护的核心，当地非物质文化遗产的保护就会随着原住居民生活的改善而不断加强（Liu et al., 2005）。

3）缺乏有关非物质文化遗产保护的法律法规，同时缺乏参与式的发展机制；

---

[1] 生态博物馆的概念最早于1971年由Georges-Henri Riviere 和 HugodeVarine 提出（Liu，Liu，& Wall，2005）。正如瑞费（Rivière）所定义的，生态博物馆是"由政府机关和地方居民共同构想、使用并掌握的一种工具"（Rivière，1985：182）。

the significance of local culture among the communities (Wu, 2007: 33).

2) Neglect of indigenous lives in historic districts

While discussing the significance of local community culture amidst the rapid urban development, it is noted that many indigenous lives, the carriers of intangible cultural heritages, are often neglected. With the disappearance of these indigenous lives, it is very likely that local intangible heritage will also vanish with them. As a countermeasure, Fang (2006) suggested the theory of eco-museum[1] that can provide many clues, wherein the links between heritage and its environment are underlined (Fang, 2006: 186). According to this theory, indigenous residents are assumed indispensable in heritage conservation in terms of their "aspirations, knowledge, and individual approaches" (Rivière, 1985: 182). Academics assert that when local residents come to the center of heritage conservation, local intangible heritage will go hand-in-hand with the improvement of indigenous lives (Liu et al., 2005).

3) Lack of necessary laws and regulations on the conservation of intangible cultural heritage, as well as a participatory development mechanism;

In China, legislations on the conservation of intangible cultural heritage generally go through a process starting from the local to the national level (Li, 2006; Zhou, 2005b). In the late 1990s, Ningxia and Jiangsu provinces announced their conservation legislations on folk fine arts and folk art works, and this was the beginning of conservation legislations on intangible cultural heritage at the local level (Li, 2006: 237). The promulgation of conservation legislation at the national level was issued in 2005, and so far it is still under ardent discussions by the legislation group consisting of members from the SCNPC, Ministry of Publicity, Ministry of Culture, and other organizations (Li, 2006: 238; Zhou, 2005b).

---

[1] The concept of eco-museum was first put forth by Georges-Henri Riviere and Hugo de Varine in 1971 (Liu, Liu, & Wall, 2005). As defined by Rivière, eco-museum is "an instrument conceived, fashioned and operated jointly by a public authority and a local population" (Rivière, 1985: 182).

在中国，关于非物质文化遗产保护的立法一般经历了从地方到国家的过程（Li, 2006；Zhou, 2005b）。20世纪90年代，宁夏和江苏两省宣布成立对民间美术和民间艺术作品的保护立法，这标志着在地方层面，关于非物质文化遗产保护立法的开端（Li, 2006：237）。国家层面的保护立法始于2005年，而且迄今为止，由人大常委会、外交部、文化部、宣传部以及其他部门所组成的立法团体仍在研讨有关立法（Li, 2006：238；Zhou, 2005b）。除了强调立法的必要性之外，本文指出更重要的是建立一个参与式的管理和发展控制机制。目前中国的立法体系，逐渐授权于各个政府机构，让他们制定相应的法律，履行各自职责（Li, 2006：239），而且公民社会和地方社区的权力都受到了监督（Zhang, 2007a；Zhang, 2007b）。由于缺少一个参与式的管理和决策机制，保护规划完全是由地方政府执行和管理的。考虑到地方政府在城市发展中的既得利益，很多之前的保护规划或规定往往被搁置或突破。

3.2.1.2 中国当前关于城市保护的法律法规

目前为止，中国关于文化遗产保护的国家级法律体系包括《中华人民共和国文物保护法》[1]（SCNPC, 2007）和《中华人民共和国城乡规划法》[2]（SCNPC, 2008）。遵照这两部国家法律，《中国文物古迹保护准则》（下文简称《准则》）是唯一一套国家级别的遗产保护

---

[1] 《中华人民共和国文物保护法》是由人大常委会在1982年颁布的，并于2007年做了最新修订。
[2] 随着全国人大常委会对1990年颁布的《中华人民共和国城市规划法》的废除，由全国人大常委会批准的《中华人民共和国城乡规划法》在2008年开始施行。

In addition to the necessary legislations, this book asserts that a participatory management and a development control mechanism are more important. The legislation system of China today tends to authorize respective government bodies to implement corresponding laws and their duties (Li, 2006: 239), and the force of the civil society and local communities are overlooked (Zhang, 2007a; Zhang, 2007b). With the lack of a participatory management and decision-making mechanism, it is totally up to the local authorities to implement and manage the conservation plans. Frequently, due to the vested interest of local authorities in urban development, many previous conservation plans or regulations have been pushed aside or violated.

3.2.1.2 Current laws and regulations on urban conservation in China

So far, the legal system concerning cultural heritage conservation in China at the national level consists of the Law of the P.R.China on the Protection of Cultural Relics (*wen wu bao hu fa*)[1] (SCNPC, 2007) and the Urban and Rural Planning Law of the P.R.China (*cheng xiang gui hua fa*)[2] (SCNPC, 2008). In compliance with these two national laws, the *Principles for the Conservation of Heritage Sites in China* (hereinafter referred to as the *Principles*) is the only set of guidelines for heritage conservation and management at the national level. Promulgated in the late 2000 by China ICOMOS with the approval of the State Administration of Cultural Heritage[3], the *Principles* provides an integrated and methodological approach to the conservation and management of heritage sites (ICOMOS, 2000). Based on the *Principles* and the above two laws, one code

---

[1] Law of the P.R.China on Protection of Cultural Relics was first enacted by the SCNPC in 1982, and the latest revision took place in 2007.
[2] With the abolition of the City Planning Act of the P.R.China promulgated by the SCNPC in 1990, Urban and Rural Planning Law of the P.R.China endorsed by the SCNPC came into force in 2008.
[3] The *Principles* was initiated by the State Administration of Cultural Heritage in cooperation with the Getty Conservation Institute and the Australian Heritage Commission (ICOMOS, 2000).

和管理方针。经国家文物局批准，中国国际古迹遗址理事会在2000年末颁布了《准则》[1]，《准则》为遗产保护和管理提出了一个综合的方法论的方法（ICOMOS，2000）。基于《准则》和上述两部法律，一部国家级的准则和保护规则已经颁布了，即，2005年颁布的《历史文化名城保护规划规范》（下文简称《规范》）（MC & GAQSIQ，2005）和2008年7月颁布的《历史文化名城名镇名村保护条例》（下文简称《条例》）（SCC，2008a）。

《规范》

《规范》在基础建设方面为历史城市和地区提供了必要的指导方针，地方保护规划应该在建筑高度控制、保护边界和交通控制方面来遵循这些基础建设（MC & GAQSIQ，2005）。然而,《规范》具有明显的缺陷。首先，它只是一个参考，在实践中不具有法律效力的。因此，许多城市可能在发展计划中不遵循《规范》。事实上，在制定地方保护规划时，《规范》只听取了本地专家的意见，所以，许多当地决策者可能从未听说过《规范》。第二，由于《规范》是在国家层面上产生的，其中有许多条款也显得过于笼统或模糊，而削弱了它的有效性[2]（MC & GAQSIQ，2005：8）。

《条例》

2008年，由国务院颁布的《条例》主要关注个别城市应该遵守的要求和程序，以便

---

[1] 《准则》是由国家文物局和盖蒂保护研究所以及澳大利亚遗产委员会联合发起的（ICOMOS，2000）。
[2] 例如，关于交通管制,《规范》表示对富有特色的街巷，应保持原有的空间尺度（MC & GAQSIQ，2005：8）。但是，关于有多少特色能被认为是"富有特色"并值得保护，是很难界定的。可能完全需要凭借地方当局的判断。

and one conservation regulation at the national level have been enacted, namely, the *Code of Conservation Planning for Historic Cities* (hereinafter referred to as the *Code*) in 2005 (MC & GAQSIQ, 2005) and the *Conservation Regulation for Historic Cities, Towns and Villages* (hereinafter referred to as the *Regulation*) in July 2008 (SCC, 2008a).

The Code

The *Code* provides the necessary guidelines to the planning and development of Historic Cities and Districts in terms of the basic structure to which local conservation plans should conform such as the building height control, conservation boundaries, and traffic control (MC & GAQSIQ, 2005). The *Code*, however, has obvious defects. First, it is only a suggestion and does not have the legal force in practice. Therefore, many cities may not follow it in their development plans. In fact, it is only consulted by local professionals in making local conservation plans, and many local decision-makers have never heard of it. Second, as the *Code* is produced at the national level, many inside articles are generalized or vague, severely undermining its efficiency[1] (MC & GAQSIQ, 2005: 8).

The Regulation

Enacted by the State Council in 2008, the *Regulation* mainly focuses on the requirements and procedures that individual cities should follow in obtaining the title of Historic Cities, producing conservation plans, and applying for related conservation funds. According to the *Regulation*, once the city is given the title of a Historic City, a corresponding conservation plan should be produced within one year. What should be concluded in the conservation plan is also underlined. The *Regulation* states that the conservation plan is legally binding during the validity of the local master plan, and throughout this period, if the conservation

---

[1] For instance, regarding traffic control, it indicates that the scale of original street in historic districts should be conserved if it is rich of characteristics (MC & GAQSIQ, 2005: 8). However, how many characteristics can be considered as "rich of characteristics" and deserve to be conserved, is difficult to define. It is absolutely up to local authorities' judgments.

得到历史城市称号、制定保护规划以及申请相关保护基金。根据《条例》，一旦城市被授予历史城市的称号，一年之内必须制定相应的保护规划。同时还强调了保护规划中应该包含什么内容。《条例》声明，在地方总规划有效期间，保护规划具有法律约束力，在这期间，如果保护计划需要修正，要得到原检查机构的批准。

总之，通过回顾中国目前的保护法律法规，本文发现这些法律法规更注重对物质遗产或物质环境的保护。这从《规范》(ICOMOS, 2000) 中对文物古迹[1]的定义中以及上述历史街区的定义中，就能明显看出来。然而，由于历史原因，在这快速城市再开发过程中，由于不当的保护策略，很多历史街区的物质元素已经破坏殆尽或完全消失了。显然，尽管在这些历史街区中，有相当多的非物质文化遗产，比如，传统的居民生活、宗教活动、地方习俗以及其他形式，但是尚未有一部专门的法律法规适用于这些历史街区。

3.2.2 城市更新机制与企业化管治模式

除了保护法律法规之外，制度化的城市更新管治模式对地方城市规划的程序和结果产生巨大影响。这是由于制度变革通常会给城市发展或更新的各个阶层都带来很多激励与限制 (Walder, 1995a: 978)。在目前以市场为导向的城市发展中，地方政府在规

---

[1] 文物古迹是指人类在历史上创造或人类活动遗留的具有价值的不可移动的实物遗存，包括地面与地下的古文化遗址、古墓葬、古建筑、石窟寺、石刻、近现代史迹及纪念建筑、由国家公布应予保护的历史文化街区（村镇），以及其中原有的附属文物。（ICOMOS, 2000: 4）。

plan needs to be revised, it should obtain the approval from its original inspection institution.

To summarize, by reviewing current conservation laws and regulations in China, this book finds that they focus more on the conservation of tangible or physical environments. This is evident from the definition of heritage sites[1] in the *Principles* (ICOMOS, 2000) and in the definition of the abovementioned Historic District. However, due to historical reasons, many tangible elements in historic districts have been severely disrupted or completely disappeared during the previous rapid urban redevelopment through improper conservation strategies. Obviously, none of these laws and regulations applies to such historic districts, although there are plenty of intangible cultural heritage in these districts such as traditional community lives, religious activities, local customs, and other forms.

3.2.2 Urban regeneration mechanism and the mode of entrepreneurial governance in urban China

In addition to the conservation laws and regulations, institutionalized urban regeneration governance substantively affects the procedure and outcome of local urban regeneration plans as well. This is because institutional changes often provide the fundamental incentives and constraints facing all actors in urban development or regeneration (Walder, 1995a: 978). In the current market-oriented urban development, the local government plays a significant role in planning, decision-making, and implementation. Academics (Shen, 2005; Wu, Xu, & Yeh, 2007; Zhang, 2002c) argue that with the economic reform since 1978, the entrepreneurial mode of governance is a salient feature of the urban regeneration policies in China.

---

[1] Heritage sites are the immovable physical remains that were created during the history of humankind and that have significance; they include archaeological sites and ruins, tombs, traditional architecture, cave temples, stone carvings, sculpture, inscriptions, stele, and petroglyphs, as well as modern and contemporary places and commemorative buildings, and those historic precincts (villages or towns), together with their original heritage components, that are officially declared protected sites (ICOMOS, 2000: 4).

划、决策和实施方面发挥着重要作用。学者认为（Shen，2005；Wu，Xu，& Yeh，2007；Zhang，2002c），自1978年经济改革以来，我国城市更新政策的一个显著特点是实行企业化的管治模式。

3.2.2.1 中国的房地产开发企业

在中国，土地行政管理通常是一个至关重要的政府职能，它是通过多层政府，如中央政府、省级政府、市政府以及区政府来实现。在改革开放政策实施之前，中国的城市房地产问题主要集中城市土地管理上，体现在对征地拆迁和由中央政府分配的城市土地的"免费供应"方面。1978年以后，中国城市的房地产开发行业和管理首次在经济特区和沿海开放城市出现。之后迅速在整个国家扩展开来（Wang，1990：83）。1984年之后，中国房地产行业进入蓬勃发展的时期。1980年第一批房地产行业建立，到1986年，房地产行业数量达到2200，并且这仅仅代表那些在中国人民建设银行已经开户的房地产行业（EBUCCC，1990：189）。这段时期的快速发展在很大程度上得益于1984年实行的国家政策，那时经济改革的焦点已经从农村转移到城市地区。许多城市利用房地产行业，将城市发展模式从先前的维护和管理转变为再开发和房地产活动（Dai，1992：395）。随着中国（1980—1984）第一个财政政策的实施，地方政府被鼓励通过刺激经济发展来增加税收，而在地方发展中，房地产公司的税收是地方收入的主要来源是减少中央政府财政补贴的重要途径（Ma，2004：242；Shen，2005：43）。随着有偿土地使用机制的引进，1986年

3.2.2.1 Rise of the development industries in China

In China, land administration management is generally a crucial government function performed through the multi-tiers of governments, that is, central government, provincial government, municipal government and district government. Before the open door policy, urban real estate issues in China mainly focused on urban land management in terms of land appropriation and "free supply" of urban land allocated by the central government (Wu et al., 2007: 134). After 1978, real estate development industries and management in urban China first emerged in the areas of Special Economic Zones (*jing ji te qu*) and Coastal Cities of Open Economy (*yan hai kai fang cheng shi*). These were quickly expanded over the entire country (Wang, 1990: 83). After 1984, the real estate industries in China came to a booming period. With the first real estate industry established in 1980, the number grew to 2,200 by 1986, and this only represents those who had opened their accounts in the People's Construction Bank of China (EBUCCC, 1990: 189). The rapid development at the time had much to do with the national policy in 1984, when the focus of economic reform had shifted from rural to urban areas. Many cities took advantage of the real estate industries and transformed the urban development mode from the previous thrusts of maintenance and management to redevelopment and real estate activities (Dai, 1992: 395). With the first fiscal reform in China (1980-1984), local governments were encouraged to stimulate economic development to increase their revenue, and the tax from real estate companies in the local development was a significant means to generate local revenues and alleviate the financial subsidy from the central government (Ma, 2004: 242; Shen, 2005: 43). The means of the local government to generate its revenues were transformed from property tax-based to land leasing-oriented with the promulgation of the Land Administration Law in 1986, when the mechanism of a paid land-use right was introduced in China (EBUCCC, 1990: 188). The urban land reform in 1986 symbolized

颁布了《土地管理法》，从此，地方政府获取收入的方式从依靠房地产税收转变为出租土地使用权（EBUCCC，1990：188）。1986年进行的城市土地改革象征着中国城市土地市场的形成。土地改革支持国家对国有土地享有垄断控制权，并通过支付土地保障金使土地租赁合法化（Tang & Liu，2005：202）。连同1994年的财政体制改革，不少地方政府在当地房地产中拥有很大股份。由于分税制的实施，他们"为了避免损失地方财富而不愿意在自己的财产上征税"。因此，地方税收主要来源于土地租赁方面，而不是房地产税（Zhang，2002c：483）。

3.2.2.2 地方政府与企业化管治模式

由于中国财政改革和土地改革之间的相互作用，中央政府颁布了分税的政策以及与地方市政府分享发展决策权。这背后的逻辑是为了提高政府的效率，并促进地方经济增长。因此，与1978年之前，以中央政府控制为特色的城市发展体制相比，20世纪80年代以来的中国城市发展政策更多地受到地方政府自己规划的影响（Zhang，2002c：480）。然而，随着行政和财政权力的分散，地方政府面临巨大的压力，他们要靠自己来增加税收收入，尤其是在他们在中央政府得到的预算资金越来越少的情景下（Ng & Wu，1995）。根据1994年财政改革的规定，地方政府需要将他们税收的55.7%上交给中央政府，而在1993年的比例是22%（Shen，2005：44）。为了促进他们所在的辖区内的经济增长，地方政府开始采取各种措施。

首先，政府制定了许多优惠政策来吸引

the establishment of the urban land market in China. The land reform supported the national monopolistic control of all urban land supply and legitimized land leasing through the payment of land premium (Tang & Liu, 2005: 202). Together with the fiscal system reform in 1994, many municipal governments, as the owners of a large share of local property, were "reluctant to impose tax on their own property to avoid the 'loss' of local wealth to the central government" due to the tax sharing mechanism. Thus, the municipal tax revenue was mainly derived from land-leasing income rather than from the property tax (Zhang, 2002c: 483).

3.2.2.2 Rise of the local governments and the mode of entrepreneurial governance

With the interactions between the fiscal system reform and land reform in China, the central government promulgated the policy of tax-sharing and development decision power sharing with the local municipal governments. The rationale behind this was to promote government efficiency and local economic growth. Therefore, compared with the urban development system characterized by the domination of the central government before 1978, urban development policies in China since the 1980s have been shaped more by initiatives from the local governments (Zhang, 2002c: 480). However, with the decentralization of administrative and fiscal power, local governments also face great pressure to generate revenues by themselves, especially in this setup where they are confronted with the decreasing budgetary funding from the central government (Ng & Wu, 1995). According to the 1994 fiscal reform, 55.7 percent of local governments' revenues need to be attributed to the central government, which have increased from the 22 percent in 1993 (Shen, 2005: 44). In order to promote the economic growth within their jurisdictions, local governments have taken various measures.

First, local governments propose many privileged policies to attract investment, either foreign or national. These policies are significant to the local development, especially with the intensive urban and regional competitiveness (Wu & Zhang, 2007a: 67). Second, local

国内的投资。这些政策对地方发展极其重要，尤其是在巨大的城区竞争力下（Wu & Zhang, 2007a：67）。其次，地方政府干预地方企业的商业活动（Wu, 2002b；Yang & Chang, 2007）。在此期间，地方政府需要对地方公共企业进行保护和补贴（Zhu, 1999：539）。在这个过程中，地方政府如同一个地方性堡垒政府，把地方企业看做"大公司的组成一部分"，琼·爱（Oi）称之为"地方性的国家企业主义"（Oi, 1995：1132）。由于政府和地方企业之间的亲密关系，许多城市实施的城市更新政策，如地方政府主导的房地产再开发政策，都开始变为以追求最大的经济回报为特点。这些经济回报在耕地转变为工业、商业或居住用地之后，来源于巨大的土地价值差异（Shen, 2005：47）。在现实中，这已成为驱使城市空间重组的主要动力，因为这样，地方政府能够在土地市场获得更大的土地税收（Ng & Wu, 1995；Yang & Wu, 1999）。

不过，在追求土地税收最大化的过程中，地方政府在政策规划、实施和开发控制上也面临两难境地。正如有学者所指出的，为了实施地方城市规划，地方政府需要出租土地来增加税收；但是，在出租土地的过程中，经常需要使现有城市规划作出让步、改变、甚至推到一边的情况（Ng & Wu, 1995：291）。这种情况在地方城市的更新过程中也普遍存在，而这与随意的地方发展控制和管理体系密切相关，这种体系给了地方政府太大的空间来调整政策以及操控发展结果（Wu et al., 2007：115）。

如今，在中国，作为城市更新过程中重要一员的公民社会或社区力量，仍然显得比

governments tend to intervene in the business activities of local enterprises (Wu, 2002b; Yang & Chang, 2007), during which local governments need to protect and subsidize local public enterprises (Zhu, 1999: 539). In this process, local governments behave like a local state and treat local enterprises "as one component of a larger corporate whole," which Oi termed as "local state corporatism" (Oi, 1995: 1132). With this close relationship between the government and local enterprises, some urban regeneration policies such as the local government-led property-redevelopment can be achieved in many cities characterized by the pursuit of maximum economic returns. These returns come from the large differential land values when farm lands are converted into industrial, business, or residential use (Shen, 2005: 47). In reality, it has already become a major force to drive the urban space restructuring in China through the acquisition of significant land venues from the land market by local governments (Ng & Wu, 1995; Yang & Wu, 1999).

During the process of pursuing maximum land revenues, the municipal governments encounter the policy dilemma on the planning, implementation, and development control. As Ng and Wu pointed out, land needs to be sold to generate revenues to implement local urban plans; however, land sales often compromise, change, or even push aside existing urban plans (Ng & Wu, 1995: 291). This situation also exists during local urban regeneration processes, and it is closely related to the discretionary local development control and management system that provides much room for local governments to maneuver their policies and manipulate development outcomes (Wu et al., 2007: 115).

In China today, the civil society or community power, as one important actor in urban regeneration, has remained very weak[1]. With

---

[1] Zhang identifies four parties engaged in the urban development of China: the central and local governments, the marketplace or private sectors, and the community power. Among the four players, community power remains the weakest voice in local pro-growth redevelopment (Zhang, 2002d: 490, 491).

较微弱[1]。缺乏参与式的城市更新机制（Ng & Wu, 1995：292），甚至影响到了综合式城市更新目标的实现。关于目前中国城市中城市发展中的公众参与，我国官方渠道包括信访、举报制度以及其他方法（Diao, 1996：1）。这些体现公众参与的官方渠道，在很大程度上，有助于在地方城市更新过程中整合社区居民的意见。文献表明，在诸多影响社区组织和社区参与的因素中，社会资本作为社区互动和联系的嵌入资源，能够对促进社区参与到地方政府主导的更新过程，起到实质性的影响。同时，社会资本有助于在社区居民和地方政府之间建立更好的关系，促进地方城市更新规划的实施（Kearns, 2003：59）。

3.2.3 中国城市中的社会资本

3.2.3.1 社会资本及其在中国城市发展中的作用

自20世纪80年代末以来，随着社会资本在西方国家被广泛介绍和讨论，很多中国学者很快意识到了这个概念的重要性，它有助于解释20世纪90年代末中国城市更新中出现的不少问题（Bian, 2006；Li, 1999, 2000；Zhou, 2004b）。2001年，在国家层面上，中央政府提出了在2010年以前实现第十个五年规划，以及可持续发展的战略目标和政策建议。其中，关于社会资本对中国城市发展

---

[1] 张庭伟认为在中国参与城市发展的部门包括：中央政府；地方政府；市场或私营部门；社区的力量。在地方倾向经济发展的再开发活动中，社区力量的声音显得微弱（Zhang, 2002d：490，491）。

the lack of a participatory urban regeneration mechanism (Ng & Wu, 1995: 292), this situation undermines the comprehensive urban regeneration aims. As regards public involvement in the urban development processes of the current Chinese cities, the official methods include petition letters or appeals system (*xin fang*), offence reporting system (*ju bao zhi du*), and others (Diao, 1996: 1). These official methods of public participation significantly integrate communities' opinions in the local urban regeneration processes. Literature shows that among the many factors affecting community organizations and participation, social capital as resources embedded in community interactions and connections substantively promotes community involvement in local government-led regeneration processes. At the same time, social capital contributes to a better relationship between community members and local governments, facilitating the implementation of local urban regeneration plans (Kearns, 2003: 59).

3.2.3 Social capital in urban China

3.2.3.1 Social capital studies and its role in urban development of urban China

When social capital has been introduced and discussed pervasively in the Western world since the late 1980s, many Chinese academics quickly sensed its implications in explaining many urban issues in the urban regeneration of China in the late 1990s (Bian, 2006; Li, 1999, 2000; Zhou, 2004b). At the national level in 2001, the central government put forth the Objectives and Political Suggestions for Tenth Five-Year-Plan and Sustainable Development Strategies before 2010, within which the significance of social capital to China's urban development was underlined as follows:

> Previous urban development centers on the improvement of the physical dimension, featured by GDP increase, capital accumulation, and investment. However, empirical practices reveal that investment in human capital, social capital, and intangible capital can

的意义表述如下：

"之前的城市发展关注物质方面的提高，其特点是国民生产总值的增加、资本积累和投资。但是，实践表明，和在自然资源开发、物质资本和有形资本的投资相比，在人力资本、社会资本以及无形资本的投资能得到更多的回报。中国在21世纪的发展战略应该以人为本，促进可持续发展（Hu，2001：9）。"

边燕杰表示，在中国，对社会资本的基本理解和解释都是围绕社会网络和人际关系。但是，在中国不同的历史时期，对社会资本的理解不尽相同（Bian，2006：39）。这种观点也得到了其他学者的支持（Zhang，2006b；Zhou，2004b）。在中国城市的背景下，这种对社会资本的理解和人际关系在人们社交生活中的重要作用有很大关系（Wang，1989：49）。除了把社会资本理解为社会网络和社交关系，其他学者（Bian，2008；Bian & Qiu，2000）也把社会资本理解为一种嵌入在并且可以在社会网络或人际关系中获得的资源（或者人们可以使用这种资源的能力）（Bian，2008：82）。

一些学者（Zhang，2006b：73）认为，中国关于人际关系的研究，可以追溯到20世纪30年代。当时胡适认为，在以儒家思想为主的社会，如中国，良好行为的本质体现在如何处理和其他人的社会关系（Hu，1936：2）。梁漱溟也在他的一部重要著作《中国文化要义》中，认为在中国社会，本位主义既不是个人关系也不是社会关系，而是人和社会之间的相互

offer more rewards than what investment in natural resources exploration, physical capital, and tangible capital can realize. China's development strategies in the 21st century should be people-centered and promote sustainable development (Hu, 2001: 9).

Bian stated that the fundamental interpretation and understanding of social capital in China revolve around social networks and human relationships. However, the specific definition of social capital in China varies in different historical periods (Bian, 2006: 39). This statement is shared by other academics (Zhang, 2006b; Zhou, 2004b). This kind of understanding of social capital in the context of urban China has much to do with the reality that human relationships (or *ren ji guan xi*) play a significant part in people's social lives in China (Wang, 1989: 49). In addition to the definition of social capital in relation to networks and relationships, other scholars (Bian, 2008; Bian & Qiu, 2000) understand social capital as the resource (or the abilities with which people can obtain the resource) embedded in and can be obtained from the networks or social relationships among human actors (Bian, 2008: 82).

Nevertheless, in terms of the studies on human relationships in China, some academics (Zhang, 2006b: 73) argue that it actually dated back to the 1930s, when Hu eloquently stated that the essence of good conduct in a Confucian society such as China is to deal with the social relationships with others (Hu, 1936: 2). Liang also contended in his seminal book *Essentials on Chinese Culture* that in the Chinese society, the selfish departmentalism (*ben wei zhu yi*) is neither individual nor societal but of mutual relationships (Liang, 1963: 80). In the in-depth study and exploration of human relationships in China, many academics (Bu, 2005; Zhang, 2006b) refer to Fei's theory of "Diversity-orderly structure" (*cha xu ge ju*) to describe the relationships in a native soil society in China (*xiang tu she hui*) (Fei, 1998: 27). According to Fei's theory, the social network in a Chinese society is enlarged

关系（Liang, 1963：80）。在对中国人际关系的深入研究和探索中，很多学者（Bo, 2005；Zhang, 2006b）参考费孝通的"差序格局"理论来描述在乡土社会的中国的人际关系。根据费孝通的理论，中国社会的社会网络是通过个人联系的增加而扩大，同时社交范围是由个人联系形成的网络组成的。他将"差序格局"比喻为一粒石子投入水面所形成的同心扩散圈。每个人都处在圈的中心，以差序方式来构建与他人的社会关系。通常离中心越近，道德性和工具性的责任感也越强；相反，离中心距离越远，关系就越弱、越疏远。费孝通还进一步解释了根据一个人的社会需要，在亲人的指导下，一个人可以扩展和其他人的关系。这种关系不仅为个人提供了关系渠道，而且也是他们生存和发展的重要社会资源（Fei, 1998：27）。尽管费孝通的理论是在中国乡土社会的背景下发展起来的，但是它也和中国城市的社会关系和网络或社会资本的理解高度相关（Zhang, 2006b：74）。直到20世纪90年代，中国学者才开始通过社交关系或社会资本深入研究城市发展政策。例如，李培林强调了家庭和社会网络在中国城市发展过程中对特殊资源的分配所起的作用。与正式的立法和政治制度相比，社会关系和网络形成了一种非正式的社会结构制度（Li, 1994：11）。

学术界认为，企业和商业的表现如何以及是否成功和他们拥有的社会资本密切相关。在中国，学者主要通过社会网络来探讨企业当中的社会资本问题（Bian & Qiu, 2000；Liu, 2006b）。社会网络的一些特点有助于社会资本的生成，例如"网络关系、网络结构和网络资源"（Bian, 2008：83）。基于对人

through incremental individual connections, and the social scope is derived from the network composed of individual connections. People's relationships are likened to the ripples formed when a stone is dropped into the water. Every individual is located at the center of the ripple, and he/she establishes social relationships with others through the "diversity-orderly structure." Usually, the closer the individual is to the center, the stronger his/her ethnic and instrumental responsibility will be in relation to the one at the center. On the other hand, the farther away one is from the center, the weaker and more estranged the relationships will be. Fei further explained that one's relationship with others could be expanded in the direction of kinship according to one's social necessities. These relationships not only provide connection channels for individuals and families but also become the substantial resources for their living and development (Fei, 1998: 27). Although Fei's theory was developed in the context of the Chinese native soil society, it is highly related to the understanding of social relationship and network or social capital within urban China (Zhang, 2006b: 74). It is not until the 1990s that Chinese academics began to study pervasively urban development policies through the networks of social relationship or social capital. For instance, Li highlighted the role of family and social networks in the distribution of special resources in China's urban evolvement. Compared with the formal institutions of legislations and politics, social relationships and networks form a kind of informal institution of social structure (Li, 1994: 11).

Academics believe that the performances and successes of enterprises and businesses are closely related to the social capital they possess. In China, academics address the issue of social capital in enterprises mainly from the perspective of social networks (Bian & Qiu, 2000; Liu, 2006b). Some features of the networks contribute to the generation of social capital such as "network ties, network structures, and network resources" (Bian, 2008: 83). Based on the discussions and

际关系的讨论和理解，边燕杰认为企业和职业生活的社会关系和网络是社会资本形成的来源，因为企业或工作可能为开发社会关系网络提供机会（Bian, 2008：82）。边燕杰进一步指出，为了通过人际关系来产生社会资本，社会关系网络应该具备四个特点：网络规模、网络多样性、网络上限和网络构成（Bian, 2008：84-85）。基于对上述产生社会资本的网络的基本特点的理解，边燕杰和丘海雄认为，企业或商业的成功和他们拥有的网络或社交关系密切相关。他们总结在产生社会本的中国企业拥有的关系应该涵盖三方面的关系：和上级政府和机构之间的纵向关系；和其他企业之间的横向关系；企业中员工或雇主之间的社会关系（Bian & Qiu, 2000：88-89）。虽然这种通过网络来研究企业中的社会资本的方法，已经得到了其他学者的支持（Wei, 2008），不过刘林平指出，如果网络的规模太大，仍很难通过社会网络评估企业中的社会资本。因此，他提出了一种评估企业社会资本的方法，就是通过测量企业或商业为建立相互关系的网络所使用的成本（Liu, 2006b：213）。在接下来的部分，会进一步讨论不同背景下社会资本的测量。

3.2.3.2 中国城市中的社会关系和社区参与机制的转变

为了更好地理解中国城市背景下社会资本的概念以及它的测量方法，下文将简要回顾中国城市中社会关系和社区参与体制的转变。

转型期中国的社会关系

随着中国的政治经济改革以及资源占有

understanding of interpersonal relationships, Bian argued that social relations and networks of enterprises and occupational lives are sources of variations in the formation of social capital because enterprises or jobs potentially provide opportunities for developing networks of social relations (Bian, 2008: 82). Bian further pointed out that in order to generate social capital through interpersonal relations, the networks of social relations should comprise four features: network size, network diversity, network ceiling, and network composition (Bian, 2008: 84-85). Rooted on the understanding of the above four basic features of networks in generating social capital, Bian and Qiu contended that the success of enterprises or businesses is closely related to the networks or social relations they have. They generalized that the relationships Chinese enterprises have in generating social capital should cover three relationships: vertical relations with the high levels of government and authority, horizontal relations with other enterprises, and social relations of employees or employers of the enterprises (Bian & Qiu, 2000: 88-89). Although this kind of study on the social capital in enterprises through networks has been supported by others (Wei, 2008), Liu pointed out that it is not easy to assess social capital in enterprises through networks when the size of the networks is very large. Accordingly, he suggested a way to assess social capital in enterprises by measuring the cost that enterprises or businesses use to establish the networks of interrelationships (Liu, 2006b: 213). Further discussion on social capital measurement in various settings will continue in the next part.

3.2.3.2 Transforming social relationships and participation institutions in urban China

In order to understand better the concept of social capital as well as its assessment in the context of urban China, the following will briefly review the transforming social relationships and community participation system in urban China.

Social relationships in transitional China

With the transforming political economy and the modes of resource possession

和分配方式的转变，中国的整个社会环境和社会关系都发生了巨大变化。很多学者倾向于把中国的社会和体制改革划分为三个阶段（Bo，2005；He，2004；Sun，1996；Wang，2006c；Wu，1999）：1949年之前传统的中国社会；1949年至1978年期间，计划经济体制下的社会发展；1978年以来，逐步从计划经济向市场经济的转变期。

1949年以前中国社会的人际关系

自秦汉以来，在中国的传统社会中，农业是基础产业，社会组织的主要形式是以家庭组织和血缘关系为主（King，1999）。每个人和他们的家人关系亲密，家族或血缘关系和村民/村妇人际关系的建立中起着重大作用。因此，许多重要的社会资源根据这种关系来分配，尤其是在家族关系中。权力、社会地位和财产都是通过家庭/家族关系来继承。结果，这种人际关系和联系构成了家族权威地位（Smith，1894：242）。所以说，中国传统社会的社会资本的显著特点是社会关系是建立在家族和国家关系的基础上的。

计划经济体制下的人际关系

1949年中华人民共和国成立以来，随着政治经济体制和分配制度改革，社会上的人际关系也发生了许多根本性变化。从1953年第一个五年计划的实施到1978年改革开放之前，在中央计划经济体制下，中国的社会资源分配和管理是由中央政府以"单位"的形式进行，而不再是通过之前的家族与血缘方式。作为一种特殊的社会组织和国家组织的衍生物，"单位"的功能、实现方式和范围都被看作是国家分配制度的结果（Lu，1989）。当时的单位存在多种类型，包括政府、国有

and distribution, social environments and interrelationships in China have undergone massive changes. Many academics tend to divide the period of social and institutional transformation in China into three stages (Bu, 2005; He, 2004; Sun, 1996; Wang, 2006c; Wu, 1999): the traditional Chinese society before 1949, the period of 1949 to 1978 when a centrally-planned economy dominated societal development in China, and from 1978 onwards, with the gradual transition from a planned institution to a more market-oriented one.

Human relationships in a traditional Chinese society before 1949

In a traditional Chinese society since the Qin and Han dynasties, agriculture was the basic industry, and the family organization and kinships were the main modes of social organization (King, 1999). Individuals had close relationships with their families, and the kindred or blood relationship (*xue yuan*) and countryman/countrywoman (*di yuan*) played a significant role in building individual relationships. Thus, many important social resources were distributed according to this relationship, especially the kindred connection. Power, social status, and properties were inherited through the relationships of families or kindred. Hence, these interrelationships and connections formed the authoritative status of the kindred and country-fellow relationship (Smith, 1894: 242). Therefore, social capital in a traditional Chinese society has the conspicuous feature of having social relationships based on kindred and country-fellow relationship (Bu, 2005; Sun, 1996).

Human relationships in a centrally planned economy

Since the establishment of the P.R. China in 1949, human relationships have experienced substantial changes with the revolutionary institutions of political economy and resource distribution. During the centrally-planned economic institution from the First Five-year Plan in 1953 to the open-door policy in 1978, social resources distribution and management in China were carried out through the allocation from the central government in the way of Danwei rather

企业、集体企业和人民公社。国家采用不同的单位形式，同时也以不同的方式进行资源分配（Zhou & Yang，1999：38）。在这样一个体制下，个人不可能使用很多重要资源。计划经济的本质是将整个社会组织变为一个大型工厂，而资源是由中央政府通过管理方式进行分配。另外，它的特点是计划都是提前制定好的（Wu，1999：415）。1978年实行改革开放政策之前，国家资源几乎全部由国家掌控，由国家行政机构来决定个人身份以及他们相应的社会地位。在此期间，人们的思想与其概念价值观一致，具有政治倾向性（Bo，2005：411）。因此，随着社会资源的分配体制采用单位的形式，计划经济塑造了人际关系的典型特征。

随着国家政权的胜利，实现国家的奋斗目标逐步成为重中之重，任何个人关系都不应干扰国家目标的实现。这种情况下，人们担心会由于泄露任何信息，可能会对自己或朋友不利，所以，人们更倾向于和朋友之间保持一种适度关系。这种情况存在于早期运动中："农村土地改革、知识分子思想改革以及三反和五反运动"时期（Vogel，1965：50）。因此，人际关系由友谊关系逐步过渡为同志关系。随着同志关系得到普遍接受，在这种新型关系中，人们认为彼此之间是平等的，而且在这种新情况下，不允许其他任何亲密关系来干预这种平等（Wang，1989：188）。因为人们之间的关系成为同志关系，难免人们之间的友谊会被削弱，尽管这并不意味着同志关系中没有友谊的成分。不少学者指出，在社会主义中国时期，人们之间的同志关系比其他任何关系，包括家庭和血缘

than the previous kindred and blood relationship method. As a special kind of social organization and extended production of state organizations, the functions, implementation ways, and ranges of Danwei were all seen as the consequences of national distribution (Lu, 1989). In China, there were several kinds of Danwei at that time, including the political governments, state-owned enterprises, collectively-owned enterprises, and people's commune. When the state took various forms of Danwei, social resources distribution was conducted in this way as well (Zhou & Yang, 1999: 38). In such a system, it was impossible for individuals to avail of many important resources. The essence of a planned economy was to turn the entire social organization into a big factory, and the resources would be distributed by the central government through administrative ways. In addition, it was characterized by a set of previously agreed upon plans (Wu, 1999: 415). Before the open-door policy in 1978, the state owned almost the entire national resources, and the national administrative institutions decided on individual identities and their corresponding social statuses. During this time, people's thoughts were quite politically biased, which harmonized with their conceptual values (Bu, 2005: 411). Therefore, with the social resources distribution institutions by way of Danwei, the typical characteristics of human interrelationships were formed in a planned economy.

With the success of the regime, national goals and purposes were considered above everything, and any other personal relationships were not supposed to interfere with the national objectives. In this situation, for fear of supplying any information that might bring trouble to themselves or other friends, people tended to maintain a moderate relationship with their friends. These situations existed during the period of the early campaigns: "rural land reform, thought reform of the intellectuals, and 3-anti and 5-anti against cadres and businessmen" (Vogel, 1965: 50). Hence, there was a transition of personal relationships from friendship to comradeship. With the more universal

关系都显得更为"亲密"(Sun, 1996; Wang, 1999)。卜·长莉认为,人们之间的这种同志式关系也可归结为两大类:上级和下属之间相互庇护的纵向关系,以及人们之间没有领导或被领导关系的横向互惠关系(Bo, 2005: 419)。

市场经济转型期的人际关系

1978年实行改革开放政策以来,中国的政治经济体制逐渐由中央计划经济体制转变为市场化的经济体制,中国社会由以前物资匮乏的社会变成了物资充足的、更加发达的社会(Bing, 1997: 39)。随着资源分配方式越来越多,中国社会的人际关系也发生了根本性的变化。随着在资源的分配与管理中,政治权利的分散化,国家已经失去了作为单独的单位,通过单位的方式向社会成员提供资源和机会的作用。相反,市场在资源分配中发挥出越来越重要的作用。另外,1978年以后,随着政治层次社会体系的解体,人们拥有更多方法来获得不同的社会地位和身份。自20世纪80年代以来,孙立平(Sun)描述的"自由漂流资源"和"自由行动空间"的现象,就很好地解释了这种情况。它允许部分资源,在不受国家控制的情况下,以"自由漂流资源"的形式进入社会或市场。同时,"自由行动空间"的出现,是由于中国体制改革和政府政策调整的结果。它是指人们可以使用"自由漂流资源"的一定的空间范围。根据孙立平所述,"自由空间空间"是在被政策允许的空间内,得到必要的资源(Sun, 1993: 66)。因此,社会上形成了不同的价值观,这些价值观反过来也对人们之间的相互关系和人们之间社会资本的生成产生了很

concept of comradeship accepted, this new relationship assumed people to be equal under the state, and any other closeness in the new situation should not interfere with this equality (Wang, 1989: 188). Within such a universal comrade relationships, sometimes even friendship weakened, although it did not mean that comradeships did not exist among friends anymore. Many academics reason out that under the rubric of socialism, the comradeships were "closer" than any other relationship including family and blood relationships (Sun, 1996; Wang, 1999). In China's socialist planned economic institution, Bu (2005: 419) opined that this kind of comrade relationship among the public maybe categorized into two relationships: the vertical sheltering relationship between superiors and subordinates, and the horizontal reciprocal relationship among the members who do not lead or who are not led by each other.

Human relationships in a transitional market-oriented economy

Since the open-door policy in 1978 and with the gradual transition of China's political economy from a centrally planned to a more market-oriented one, the Chinese society has undergone a transition from a resource-scarce society to a more developed one (Bing, 1997: 39). With the soaring of the various means of resources distribution, social interrelationships in China also experienced fundamental changes. With the decentralization of political powers in resources distribution and management, the state lost the role as the unique unit that provides resources and opportunities to social members through Danwei. Instead, the market played a more and more important role in resource distribution. In addition, with the cancellation of the politically hierarchical social system after 1978, people found more methods in obtaining different social statuses and identities. Since the 1980s, this situation has been well illustrated in the phenomena of "free drifting resources" and "free action space" as described by Sun (1993: 64). One consequence of the Chinese institutional reforms is the dwindling state controlling scope

大影响。

许多学者认为，自20世纪80年代以来，中国的人际关系或社会资本的特点是趋向于工具化和商品化，这取代了之前人们之间的友谊关系和同志关系（Bo，2005；Gold，1985；Yue et al.，2002）。谷德认为（Gold，1985：659），工具化指的是人们通过与他人的联系或者某种特殊关系来解决事情的能力。有学者（Gold，1985；Luo，2008；Wang，1999；Zhai，2008）将这种基于互惠互利或者回报的联系称为"关系"。回报的概念是指一个人给另一个人帮忙，并期望之后能够得到某些好处。Luo认为，关系的本质是人与人之间的联系，同时意味着一个"持续的利益交换"（Luo，2000：2）。就像谷德所提的定义一样，关系是"一个利用非正式、非官方的关系来办事，既包括简单的事情，也包括重大的人生抉择"（Gold，1985：661）。有些学者觉得较难区分关系和社会资本这两个概念，因为关系本身有很多中国社会的特点（Luo，2008；Zhai，2008）。翟学伟（Zhai）表示，社会资本关注的是关系的实施或操作，而关系仅限于社交网络的结构（Zhai，2008：306）。在中国的市场经济体制下，人际关系的另一个显著特征是商品化（Bo，2005；Gold，1985；Yue et al.，2002）。因为商品和服务正逐渐由集体企业所掌控和管理，个人企业也会根据市场信号而做出反应。但是，有学者也担心，这可能使人们之间的关系建立在金钱的基础上，或者换句话说，一切都会向钱看（Gold，1985：662）。

在中国转型期的政治经济体制下，中国的人际关系除了上述两个特点，卜·长莉（Bo，

and power on resources. This allows partial resources to be beyond state control and enter the society or market in the form of "free drifting resources." At the same time, "free action space" emerged as a result of the institutional changes and government politics adjustments. It refers to the specific space where people can utilize the "free drifting resources." According to Sun, the "free action space" is defined as the space allowed by policies to obtain necessary resources (*zheng ce run xu fan wei nei*) (Sun, 1993: 66). Accordingly, it forms various values in the society, which in turn have substantive influences on the interrelationships and the generation of social capital among people.

According to many academics, human interrelationship or social capital in China since the 1980s has been characterized by instrumentalism and commoditization, which supplanted the previous friendship and comradeship (Bu, 2005; Gold, 1985; Yue et al., 2002). According to Gold (1985: 659), the tendency of instrumentalism has the characteristic of accomplishing tasks through human relationships or particularistic ties. Some academics (Gold, 1985; Luo, 2008; Wang, 1999; Zhai, 2008) define this kind of relationship as *Guanxi*, based on the concept of reciprocity or *Bao*. The concept of Bao implies that one does another a favor with the expectation of receiving benefits in return at a later time. Luo contended that the essence of Guanxi is the interpersonal linkages with the implication and expectation of "continued exchange of favors" (Luo, 2000: 2). As Gold defined it, Guanxi is "an informal, unofficial relationship utilized to get things done, from simple tasks to major life choices" (Gold, 1985: 661). At the same time, some academics try very hard to distinguish between the two concepts of Guanxi and social capital, as Guanxi inherently has many characteristics of the Chinese society (Luo, 2008; Zhai, 2008). According to Zhai, social capital focuses on the implementation or operation of Guanxi, while Guanxi is limited to the structures of social networks (Zhai, 2008: 306). Another salient feature of human interrelationships in

2005）指出，在一定程度上，中国社会仍存在着纵向的相互庇护关系或领导和客户之间的裙带关系。然而，由于不断变化的政治体制影响了控制资源和利益的专制权力和资源分配的原则，领导者和客户之间仍存在着合作关系。因此，卜·长莉认为，目前领导和客户之间的关系可以概括为一种多元的合作关系（Bo，2005：427）。

转型期中国的公众参与

城市管理中的政治参与

在对中国城市社会资本的讨论中，社区参与对城市发展也有着重要影响。一般来说，根据不同时期内，不同的思想指导方针和社会经济形式，有关中国城市发展中公众参与的研究可划分为两个阶段（Deng，2004；Diao，1996；Tao & Chen，1998；Wang，2006a）。

自1949年中华人民共和国成立以来，中国根据社会主义原则进行的政治改革和机构重组，为公众参与提供了更大的空间和更多机会。根据社会主义国家的"特点"，政治参与扩展到了整个国家，并且建立了一系列的参与式政治体系（Tao & Chen，1998：178）。不过，直至1978年以前，在以中央政府为主导、具有政治意识形态倾向的体制下，公众参与城市发展和管理还有不少限制。概括来讲，1978年之前，计划经济时期公众参与的特点是带有革命性的、动员式的以及"宏大"的公众参与（Wang，2006a：6）。陶东明（Tao）和陈明明（Chen）认为，"宏大"的公众参与是指席卷全国的意识形态思潮。这在20世纪60年代时尤为明显，当时国家陷入政治动荡期，一时间，人人都以国家主人公的身份，

China's market-oriented economic institutions is commoditization (Bu, 2005; Gold, 1985; Yue et al., 2002), which is related to the situation when goods and services become under the control and management of the collective, and individual enterprises respond to market signals. Nevertheless, academics argue that this tends to build human relationships based on money or, in other words, to cast all eyes on the money (*xiang qian kan*) (Gold, 1985: 662).

In addition to the above two features of human interrelationships in China's transitional political economic institutions, Bu (2005) claimed that the vertical sheltering or party-clientelism relationships between leaders and clients still exist in the society to some extent. Nevertheless, considering the changing political economic institutions, which affect the authoritarian powers on resource and interest controls and the principles on resource distributions, the cooperative relationships between leaders and clients still exist. Thus, Bu assumed that current relationships between leaders and clients could be called pluralistic cooperation relationships (Bu, 2005: 427).

Public participation in transitional China

Political participation in urban management

In the discussion of social capital in urban China, community participation in urban development is a significant issue. Generally, studies on public participation in urban development of China tend to categorize the period into two according to the different ideological guidelines and socio-economic situations of the specific time (Deng, 2004; Diao, 1996; Tao & Chen, 1998; Wang, 2006a).

With the fundamental political reform and institutional restructuring according to the socialist principles in China since 1949, much room and opportunities for public participation have been provided. According to the "characteristics" of a people's democratic dictatorship in a socialist country, political participation was extended to the entire nation, and a series of participatory political institutions were created (Tao & Chen, 1998: 178). Nevertheless, in the central government-led and political ideology-biased institution before 1978, public participation in

开始在国家层面上关心和讨论时事（Tao & Chen, 1998: 196）。

随着1978年的改革开放政策和社会主义市场经济的引入，社会上逐渐产生了个人自我意识，传统的自上而下的城市发展理念也开始与人们自我价值的实现相结合。这引起了公民参与新模式的出现。这是一种具有建设性的、自愿的以及更加理性的公众参与。这也表明公众参与的模式已经从之前的主要关注政治问题的模式，转变为目前的以社会和公众问题为核心的模式（Wang, 2006a: 5）。自20世纪90年代以来，中国的社区建设一直受不同层级的民政部门的监督和管理。国务院强调了民政部门在社区建设中的领导作用（Tang, 2000: 249-250）。学者认为，公众参与是中国社区建设的一个重要特征和必要条件（Deng, 2004；Tang, 2000；Wu & Zhang, 2007b）。关于社区参与机制，1995年，国务院颁布了"两层政府和三层管理"的方法，表明了市政府和区政府分别在市、区、街道的层面上，发挥着重要重要（Wu & Zhang, 2007b: 51）。同时，各种制度化的参与渠道也促进了社区建设和城市管理，例如信访制度（Cai, 2004）、举报制度、舆论参与、监督制度、专家咨询以及其他途径（Diao, 1996；Tao & Chen, 1998）。

城市规划中的社区参与

与中国社区建设中的政治参与相比，直至20世纪90年代中期，学者们才开始关注社区参与在中国城市规划和更新中的作用（Marmon, 1995）。不过即便是在今天，社区参与在不少城市制定当地的规划或城市更新过程中也未引起人们的足够重视。通常情

urban development and management had many constraints. Generally, public participation in a centrally planned economic institution before 1978 was characterized by revolutionary, mobilized, and "grand" democratic involvements (Wang, 2006a: 6). According to Tao and Chen, "grand" democracy refers to the ideological trend of overwhelming national democracy. This was apparent especially during the 1960s when the entire nation was thrown into political turmoil, and everyone suddenly took the responsibility to be the master of the country and began to care and discuss issues at the country level (Tao & Chen, 1998: 196).

With the open-door policy and socialist market-oriented economy introduced in 1978, individual self-consciousness emerged gradually and traditional urban development ideology of bottom-up began to unite with the trend of self-value realization. This aroused the emergence of new modes of citizen participation characterized by constructive, voluntary, and more rational public involvement. It also showed that the mode of public participation had evolved from previously being political issue-focused to currently being more social and public issue-centered (Wang, 2006a: 5). Community construction in China has been under the supervision and management of the different levels of the civil administration departments (*min zheng bu men*) since the early 1990s. The State Council reinforced the leading role of civil administration departments in community construction in 1998 (Tang, 2000: 249-250). Academics argue that public participation is the essential characteristic and necessary condition in Chinese community construction (Deng, 2004; Tang, 2000; Wu & Zhang, 2007b). With regard to the community participation mechanism, the State Council promulgated the method of "two tiers of government and three tiers management" in 1995. This specifies that the municipal and district governments play important roles in urban management at the municipal, district, and street levels (Wu & Zhang, 2007b: 51). At the same time, various and institutionalized participation accesses emerged to promote

况下，只有在规划方案公示和举行公开听证会期间，公众参与才进入人们的视野。（表3-4）显示了目前中国自上而下的城市发展机制下，在城市规划和城市更新进程中的公众参与。从该表可以看出，整个过程留给公众参与的机会比较有限。自20世纪90年代末以来，不少学者积极研究和探索可行的方法，以便使社区参与能够结合到城市规划过程中（Chen, 2000；Qian, Li, & Mao, 2004；Zhang, 2000；Zheng & Yang, 2005；Zhou & Lin, 2000）。在实践中，一些城市已经尝试把公众参与包括在当地的城市规划机制中。例如，1998年的深圳城市规划图则中，就已包含了公众参与的工作机制（Zhou & Lin, 2000：67）。

不少学者认为，当前我国城市规划中的公众参与，主要存在以下缺陷（Qian et al.,

community and urban management, for example, petition letters or collective appeals system (*xin fang*) (Cai, 2004), offence reporting system (*ju bao zhi du*), supervision by public opinion (*yu lun can yu*), monitoring/supervision or ombudsmanship system (*jian du zhi du*), professional consultation (*zhuan jia zi xun*), and other accesses (Diao, 1996; Tao & Chen, 1998).

Community participation in urban planning

Compared with political participation in Chinese community constructions, the significance of community participation in urban planning and urban regeneration in Chinese cities did not draw academics' attentions until the mid-1990s (Marmon, 1995). To this day, community participation is still a rare virtue in plan-making or urban regeneration processes in many cities. Generally, public participation in urban planning processes of many Chinese cities is reflected only during the promulgation and public hearing on urban planning proposals. In current China's top-down urban development mechanism, public participation in the urban planning and urban regeneration processes has been presented in the Table 3-4, which

中国城市规划和城市更新过程中的公众参与 表3-4
Public participation in urban planning and urban regeneration processes in China Table 3-4

城市规划和城市更新进程 Urban Planning and Urban Regeneration Processes	内容 Contents	特点 Features
阶段1 Stage 1	提出城市更新/复兴的要求，地方政府寻求开发公司，更新/复兴规划的公示 Proposing the needs of urban regeneration/renewal and seeking development companies by local government, followed by the propagation of the regeneration/renewal plans	公布和公众听证 Promulgation and public hearing
阶段2 Stage 2	根据地方总体规划评估城市更新/复兴规划 Evaluating the urban regeneration/renewal plans according to local master plan	决策 Decision-making
阶段3 Stage 3	由地方政府和开发公司制定城市更新/复兴规划 Deciding on the urban regeneration/renewal plans by local government and development companies	决策 Decision-making
阶段4 Stage 4	开发公司策划与准备项目草案 Project planned and prepared by development companies	投标或委托项目 Bidding or assigning projects
阶段5 Stage 5	根据城乡规划法制定的正式程序，地方政府审批项目 Project sanctioned by local government following the official procedures, laid out in the Urban and Rural Planning Law	决策 Decision-making

资料来源：（Lu, 2005：23）
Source: Author revised from (Lu, 2005: 23)

2004；Xu & Chow，2006；Zhang，2002c；Zheng & Yang，2005）：

  a）较少有城市真正采用公众参与的方式开展城市规划，而且不少城市的规划体系中未将公众参与放于足够重要的位置（Zheng & Yang，2005：64）。

  b）很多居民没有意识到城市规划或更新过程中公众参与的必要性，结果，只有部分人真正参与。这也与目前我国自上而下的规划和决策过程有关。由于目前城市规划立法机构中缺乏必要的参与机制，所以，在地方城市发展中，社区居民也很少了解城市规划的信息。结果，公众参与往往只在规划公示阶段才有所体现（Zhang，2002c：497）。

由于无法获得居民对自下而上的决策方式的支持，城市规划过程中的公民参与也会较难推广。有学者认为，基于社区的更新项目应给予社区在整个项目或社区面临的问题上，一定程度的控制和指导权利（Xu & Chow，2006：201）。学者认为，城市规划和更新过程中的社区参与有助于实现多元化的城市更新目标的实现。同时，社区参与还有助于促进地方政府和社区之间关系的改善或促进社会资本的积累（Kearns，2003；Putnam，1995；Zheng & Yang，2005）。

3.2.3.3 中国城市中社会资本的测量

  在前面的章节中提到，西方文献中的社会资本主要在三个层面上进行理解：微观层面、中观层面和宏观层面。同样，它的测量

`highlights the little community participation in this process. Since the late 1990s, many scholars have actively researched and explored feasible ways to integrate community participation in urban planning procedures (Chen, 2000; Qian, Li, & Mao, 2004; Zhang, 2000; Zheng & Yang, 2005; Zhou & Lin, 2000). In practices, some cities have attempted to include public participation in their local urban planning mechanism. For example, public participation mechanism was included in the *Shenzhen Urban Planning Regulation* in 1998 (Zhou & Lin, 2000: 67).

On the whole, academics affirm that some deficiencies in public participation in the urban planning of Chinese cities mainly include the following (Qian et al., 2004; Xu & Chow, 2006; Zhang, 2002c; Zheng & Yang, 2005):

  a) Public participation has been adopted only by a few cities in China, and urban planning systems in most cities do not involve the public at all (Zheng & Yang, 2005: 64).

  b) Many local residents do not see the necessity for public participation in urban planning or development processes, and as a result, only very few people are actually involved. This relates to the current China's top-down planning and decision-making process. With the lack of the necessary participatory institutions in current urban planning legislatures, the community has very little access to urban planning information during local development, and public participation in many cities only remains at the level of planning promulgation during its implementation (Zhang, 2002c: 497).

In the absence of the local people's serious commitment to the bottom-up decision-making, public participation in China's urban planning process can hardly be a promising path. Therefore, academics claim that community-based programs should provide the community with a degree of control and direction over the projects or the problems facing their communities (Xu & Chow, 2006: 201). Academics contend that community participation in urban planning

方式也有三种：提名生成法、位置生成法和资源生成法。文献表明，目前对于中国社会资本的测量也遵循同样的方式。通常情况下，不同的研究背景，就会有不同的社会资本测量指标。当在宏观层面和中观层面上进行研究社会资本时，研究者有时会面临大量的指标（Qiao & Mai，2003），相比较而言，个人或微观层面上的社会资本指标更加明确，也更容易产生（Bian，2008）。

宏观与中观层面的社会资本测量

为了使宏观层面上巨大的指标更加适用，一些学者引进了因素分析[1]的方法，来选取和分析最具有活力的社会资本指标。这有助于将多种变量缩减至几个关键变量，在测量或评估这种情况中，服务于同一个目的。至于选择关键因素的具体步骤，首先对因素分析的变量进行界定（信任、规范、信息共享渠道以及其他变量）。其次，对一系列相关变量进行计算，然后选定关键因素。最后，通过对这些因素进行数据分析，选定研究代表整个社会资本状况的指标（Xu & Yuan，2004：68）。边燕杰和丘海雄（Bian & Qiu，2000）对中国企业中的社会资本进行的探讨[2]，很好

---

[1] 通过对不同变量之间相互依存的关系的研究，探讨了这些变量的内在结构。几个有限的变量被假定代表基本的数据结构。这些假定变量能够体现很多之前的观察变量呈现的大量信息，并且解释这些观察变量之间的相互依存关系（Qiao&Mai，2003：90）。

[2] 基于中国企业的特点分析，边燕杰和丘海雄提出了测量当中社会资本的三个指标：第一个是看企业法人是否曾受雇于更资深的权威公司；第二是看企业法人是否曾受雇于其他跨境企业或担任过管理或执行领导；第三是看企业法人是否有广泛的社会网络和社交关系。用因素分析的方法将这三个指标的影响相结合，归纳总结的那个指标就是企业公司中的社会资本因素（Bian & Qiu，2000：89）。

and regeneration processes contributes to the multi-dimensional regeneration objectives. In addition, it facilitates the smooth interrelationships between local governments and the community or the accrual of the local social capital (Kearns, 2003; Kleinhans, 2006; Putnam, 1995; Zheng & Yang, 2005).

3.2.3.3 Social capital measurement in the context of urban China

As reviewed in the preceding chapter, social capital in Western literature is mainly understood at three different levels: micro level, intermediate level, and macro level. Likewise, it is measured in three ways: name generator, position generator, and resource generator. Literature shows that current studies on social capital measurement in the context of China largely follow the same way. Normally, according to different study contexts, various indicators on social capital measurement are generated. When studying social capital at the macro and intermediate levels, researchers sometimes need to face a great number of indicators (Qiao & Mai, 2003). Comparatively, social capital indicators at the individual or micro-level are more explicit and easier to generate (Bian, 2008).

Social capital measurement at the macro and intermediate levels

In order to make the tremendous indicators at the macro level more applicable, some academics introduce such ways as factor analysis[1] to select and analyze the most dynamic social capital indicators. This helps to reduce the various variables into several key ones, serving the same purpose in measuring or evaluating the situation. As for the specific procedures of selecting the critical factors, the variables for factor analysis are defined (trust,

---

[1] Through studying the dependent relationships between different variables, it explores the intrinsic structure of these variables. Several limited variables are assumed to represent the basic data structure. These assumed variables bear the capacity to express the bulk of information embodied in previous many observation variables, and can explain the dependent relationships among these observation variables (Qiao & Mai, 2003: 90).

地体现了这种方法。尽管这种方法有助于简化大量指标，甚至有时能够权衡各种指标，但是这种假定的指标能否代表实际情况仍有待论证。在这种情况下，一些学者提出用其他方法来评估社会资本的数量。例如，与上述边和邱所采用的因素分析法不同，刘林平（Liu）提出通过计算企业在社会关系中资金的投入和产出来测量其社会资本。刘林平认为，用同一种单位，更容易比较社会资本、物质资本和人力资本（Liu，2006b：209-210）。

个人或微观层面的社会资本测量

与宏观和中观层面的社会资本测量相比，个人或微观层面的社会资本测量更加直观、明确。目前对于个人或微观层面的社会资本的研究，很多学者采用网络的方法。边燕杰（Bian）表示，中国城市的社会资本容量可以通过测量人际关系中四个相关特征来进行评估：网络规模（产生更多社会资本的大网络）、网差（产生更多社会资本的高密度网络）、网顶（产生更多社会资本的较高社会地位或更高上限的网络）、网络构成（产生更多资本的包含各种阶级地位的网络）（Bian，2008：84-85）。社会资本形成中阐明了这四个特点，如图3-4所示。

与西方文献中提到的"提名生成法"、"位置生成法"和"资源生成法"相比，边燕杰强调这些方法不能很好地反映植根于关系网络的中国特殊的文化网络。边燕杰认为，"中国关系网络是利益交换的网络，对人们生活有重大影响"，因此"这些事件可能是网络生成设备的不可缺少的一部分，可以用来测量中国关系网络"（Bian，2008：91）。在此分析的基础上，他提出了"基于事件位置生成法"

norms, information sharing channels, and other variables). Second, related array of all the variables is calculated, and third, the critical factors are selected. Finally, through statistical analysis on these factors, the indicators representing the overall social capital status are selected and examined (Xu & Yuan, 2004: 68). A good example of the implication of this method is the study by Bian and Qiu (2000), where they explored social capital in Chinese enterprises[1]. Although this method helps to streamline the large number of indicators, which sometimes even play tradeoffs among themselves, it is still arguable whether the presumed indicators can entirely represent the actual situations. In this situation, some academics propose to assess the quantity of social capital in bulk by other means. For instance, unlike the factor analysis method adopted by Bian and Qiu, Liu proposed to calculate the monetary inputs and outputs in social relations to measure social capital in enterprises. According to Liu, it is easier to compare social capital, physical capital, and human capital when they are measured using the same unit (Liu, 2006b: 209-210).

Social capital measurement at the individual or micro-level

Compared with social capital measurement at the macro and intermediate levels, the measurement of social capital at the individual or micro level is straightforward and explicit. In current studies on social capital at the individual or micro level, many scholars take the approach of networks. According to Bian, the capacity of social capital among Chinese urbanites can be assessed by measuring the four relational

---

[1] Based on the characteristics of Chinese businesses, Bian and Qiu propose three indicators to measure the entailed social capital: the first is whether the Business's Initial Agent for Service of Process (BIASP) (*qi ye fa ren dai biao*) was ever employed in more senior authority firms; the second is whether the BIASP was ever employed in other trans-boundary firms or served as the leader in managing or executive statuses; and the third is whether the BIASP has a wide social network and relationships. Through combining the effects of these three indicators by means of factor analysis, the inductive one is seen as the social capital factor in the business corporation (Bian & Qiu, 2000: 89).

图 3-4：中国城市中社会资本的形成
Figure 3-4: The formation of social capital among Chinese urbanites
资料来源：（Bian，2008：85）
Source: (Bian, 2008: 85)

features of personal networks: network size (a large network generating more social capital), network diversity (a high-diversity network generating more social capital), network ceiling (a network with higher social positions or higher ceilings generating more social capital), and network composition (a network with a full combination of class positions generating more social capital) (Bian, 2008: 84-85). These four features are illustrated through the formation of social capital as shown in Figure 3-4.

Compared with the measurement of social capital in Western literature through "name generator," "position generator," and "resource generator," Bian emphasized that these methods could not reflect well the particular Chinese culture-based networks rooted from the Guanxi networks. Bian contended, "Chinese Guanxi networks are favor-exchange networks significant and sensitive to life events," and therefore, the "events may be a necessary part of any network-generating device for measuring Chinese Guanxi networks" (Bian, 2008: 91). Based on this analysis, he introduced the "event-based position generator" approach to measure personal networks in the Chinese context by asking respondents how many persons (or families) they visited during the celebration of the last Spring Festival (Bian, 2008: 91). This kind of network-based study on individual social capital can be found in other studies. For instance, in exploring the social capital in Beijing urbanites from the egocentric-networks view, Zhang classified related social capital indicators according to size and density, heterogeneity and homogeneity of personal networks, number and composition of various roles' relationships, and other parameters (Zhang, 2006b: 122).

The above discussion shows that social capital is measured in various methods according to how it is understood at the macro or micro level. Nevertheless, each method underlines the reality that social relationships or networks contribute to the understanding and exploration of social capital in different contexts. From this view, any factor affecting community relationships or networks may have an impact

的方法，通过询问受访者去年春节期间拜访了多少人（或亲人）来研究中国的人际网络（Bian，2008：91）。在其他研究中也可以发现基于网络对个人社会资本的研究，例如，张文宏（Zhang）则从个人网络的角度探讨北京市的社会资本，他根据网络规模和密度、个人网络的同质性和异质性、不同人际关系的数量和构成，以及其他参数，对相关社会资本指标进行了分类（Zhang，2006c：122）。

上述讨论表明，根据在宏观或微观层面对社会资本不同的理解，社会资本测量的方式多种多样。但是，每种方法都强调了社交关系或社会网络有助于对不同背景下社会资本的理解和探索。从这点来看，任何影响社区关系或网络的因素都会对宏观或微观层面上社会资本的生成产生影响。当然，那些已经被普遍认可的问题，如信任、规范、社区联系和参与也对中国城市的社会资本有重大影响 (Bryant & Norris, 2002；Grootaert, Narayan, Jones, & Woolcock, 2004)。

3.3 中国历史城市更新中的社会资本、城市保护与城市管治

本章通过对我国历史城市中城市更新的回顾，表明中国城市的更新政策自1949年以来已经发生了很大的变化。到了20世纪80年代早期，中国的城市更新政策当中已经包含了城市保护的问题。但是在20世纪90年代，在倾向于经济发展的城市再开发政策驱动下，城市保护常常被迫作出让步。到了2000年以后的新纪元，随着更加全面的城市更新目标的提出，城市保护策略的重点已经从之前保护有形的城市物质环境，逐渐演变为保护无形的社会文化环境。这与我国城市中政治权力的分化和体制改革也密切相关，在这种情况下，地方城市发展的管理机制形成了企业化的城市管治模式。此外，对当地居民和原住社区的保护也逐渐成为人们讨论的焦点。在城市更新进程中，基于社区网络的社会资本和参与有助于对历史城市环境的保护，包括对物质历史环境的保护和地方原住民生活的保护，从而实现更全面的城市更新目标。

在1949年到1978年的社会主义中央计划经济时期，中国的城市更新政策以政治意识形态倾向和突出工业发展为特点，这期间，不少历史城区也受到了很大的破坏。这种情况清楚地体现在"文革"时期的"破四旧"

on the generation of social capital at either the macro or micro level. Certainly, the common issues identified universally such as trust, norms, community interactions, and participation have a substantial influence on the social capital in urban China as well (Bryant & Norris, 2002; Grootaert, Narayan, Jones, & Woolcock, 2004).

3.3 Social Capital, Urban Conservation, and Urban Governance in the Regeneration of Chinese Historic Cities

Through the overview of urban regeneration in Chinese historic cities, this chapter shows that urban regeneration policies in urban China have transformed substantially since 1949. Generally, the issue of urban conservation has been integrated in the urban regeneration plans in historic cities since the early 1980s, but the urban conservation goals have often been compromised in an economic growth-biased urban development policy especially in the 1990s. This has much to do with the decentralization of political powers and institutional changes in urban China, under which an entrepreneurial mode of urban governance has formed. In the new era since 2000, together with the proposal of more comprehensive urban regeneration goals, urban conservation strategy has shifted its focus from tangible and physical environments to the intangible and socio-cultural environments. In addition, the conservation of local residents and indigenous community comes to the core of the discussion. Social capital based on community networks and participation in urban regeneration processes contributes to the conservation of historic urban environments in terms of the conservation of both the physical historic environments and local indigenous lives, hence to the more comprehensive urban regeneration goals.

During China's socialist centrally planned economy from 1949 to 1978, the urban regeneration policy in Chinese cities was generally characterized as political ideology biased and industrial development centered, during which many historic urban areas underwent massive

运动中。随着1978年之后城市规划体系的恢复，很多中国城市中的城市更新，都是在城市总规划的指导下进行城市更新和修复。20世纪80年代末期，政治体制改革涉及了财政改革、土地改革、住房改革等方面，这为地方政府在实施倾向于经济发展的城市更新政策，提供了很多激励机制与机会。在这种情况下，很多地方政府在当地破败的历史城区开展了以房地产为主导的城市再开发。20世纪90年代，中国城市发展的特点可概括为以地方政府为主导，主要进行私人项目的再开发过程。事实上，这个过程往往是地方政府通过出租土地使用权，来增加地方财政收入的主要方式。随着1988年中国土地市场的成立，中国城市形成了一种企业化的管治模式（Oi，1995：1132）。

不过，随着巨额的经济利益，城市发展当中也出现不少的问题。例如，当保护规划被修改或做出让步时，很多历史城市环境都遭到了破坏，并且为了发展当地的旅游业转而兴建仿古建筑。同时，由于城市物质环境的变化，当地居民也常常被重新安置，历史城区中的传统生活习俗也消失了。在中国历史城市的更新过程中，城市保护是一个重要的问题。通过对中国城市保护机制的回顾，我们可以看出城市保护的范围自20世纪50年代对文物古迹的保护已经扩展到，20世纪80年代对历史文化名城的保护，20世纪90年代对历史街区的保护，以及2004年以来的非物质文化遗产的保护。然而，中国当前关于城市保护的法律法规仍然比较关注物质环境的保护。相比之下，较少有法规来强调对非物质文化遗产的保护。而有关社会文化内

destruction. This situation was explicitly revealed in the movement of "destroying the four olds" during the Cultural Revolution. With the restoration of the urban planning system after 1978, urban regeneration in many Chinese cities, in the way of urban renewal and rehabilitation, was conducted under the direction of urban master plans. Political institutional reforms in terms of fiscal reforms, land reform, and housing reform in the late 1980s provided local governments many incentives and opportunities to implement economic growth-biased urban regeneration policies. In this situation, many local governments promoted property-led urban redevelopment in dilapidated historic urban areas. In the 1990s, urban China was characterized by local government-led and private project-oriented processes. During this process, local governments encouraged the involvement of private developers in the redevelopment projects. It was also a substantive way for the local governments to generate local revenues by leasing the land-use rights. With the creation of the land market in China in 1988, an entrepreneurial mode of governance formed in the regeneration of many Chinese cities (Oi, 1995: 1132).

Nevertheless, with the enormous economic returns came many problems. For example, many historic urban environments were destroyed and fake antique buildings were constructed to serve the needs of the local tourism industry, when the local conservation plan was revised or pushed aside. Furthermore, with the disappearing physical environments, local residents were often relocated, and traditional lives vanished. In the regeneration of Chinese historic cities, urban conservation is an important issue. The overview of urban conservation mechanism in China shows that the urban conservation scope has evolved to cover the cultural heritage site since the 1950s, historic cities since the 1980s, historic districts since the 1990s, and intangible cultural heritage since 2004. Nevertheless, current laws and regulations on urban conservation in China still focus much on the physical environments. Comparatively, very few regulations address the intangible perspectives. Many socio-

容的保护，如地方的传统生活内容还未引起人们的足够重视。

相关文献表明，在历史城市环境中，社区关系和社区参与方面的社会资本，有助于促进社区参与到城市更新的过程中。影响人际关系和社区参与的因素很多，例如城市更新机制、管治模式以及民族和宗教问题。由于在保护原住民生活方面缺少必要的法规，以及缺乏参与式管治模式，城市更新过程中的社区参与更多地表现为居民自发的、基于社区关系的、自下而上的方式。在中国城市的背景下，历史城区的社区关系也受多方面因素的影响，包括城市更新策略、社会经济因素以及民族和宗教问题。这些因素影响着社区成员之间的信任和规范，从而影响社区邻里关系以及他们在城市更新项目中的参与积极性。中国社会的人际关系自新中国成立以来也已经发生了巨大的变化，主要是由中央计划经济体制下的同志式关系，逐渐转变为目前市场经济体制下的更加商品化和工具化的人际关系（Bo，2005：427）。社区关系或社会资本的不同情况，直接影响社区居民在城市更新规划中的参与，从而对历史城市的城市更新结果产生影响。

通过前面两章，本研究对城市更新政策、城市管治模式、城市保护法律法规、社会资本以及他们之间错综复杂的关系进行了讨论。基于上述讨论，研究建立了一个分析理论架构，以对中国历史城市的社会资本和城市更新问题有较好的理解。图3-5显示了此理论架构。

cultural perspectives, in terms of local traditional lives, have been overlooked.

Literature reveals that social capital in terms of community relationships and participation facilitates the community involvement in the urban regeneration process in historic urban environments. There are various factors affecting human relationships and community involvement, for example, urban regeneration mechanism and mode of governance, and ethnic and religious issues. In the lack of necessary regulations on the conservation of indigenous lives and the participatory mode of governance, community involvement in urban regeneration processes has been presented more as an autonomous and in a bottom-up way based on community relationships. In the context of urban China, community relationships in historic urban areas are affected by many factors including urban regeneration strategies, socio-economic factors, and ethic and religious issues. These factors have an impact on the level of trust and norms among community members, hence affecting community relationships and their involvement in local urban regeneration programs. In China today, human relationships have generally transformed largely from the previous moderate comradeships in the centrally planned economy to the current relationships of commoditization and instrumentalism in China's market-oriented economy (Bu, 2005: 427). Different situations of community relationships or social capital directly affect community involvement in local urban regeneration plans, hence tending towards the readjustment of the urban regeneration outcomes in historic urban contexts.

Based on the discussions in the preceding two chapters on the transformation of urban regeneration policies, mode of urban governance, laws and regulations on urban conservation, and the situation of the social capital, as well as the intricate relationships among them, an analytical framework that will guide the better understanding of social capital and urban regeneration in Chinese historic cities has been developed. This framework is presented in Figure 3-5.

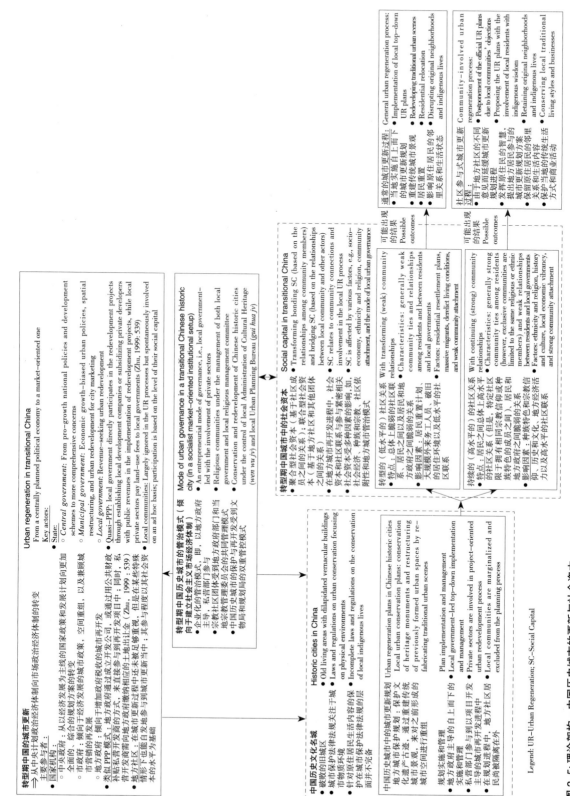

图 3-5: 理论架构：中国历史城市的更新与社会资本

Figure 3-5: Theoretical framework for the better understanding of social capital and urban regeneration in Chinese historic cities

第四章　西安的社会资本与城市更新
CHAPTER 4　Social Capital and Urban Regeneration in Xi'an

通过文献阅读与相关理论研究，前几章主要分析了关键影响因素及其在转型期中国历史城市更新、社会资本、城市保护、城市发展机制和管治模式中的相互关系。基于此，为了更好地理解中国历史城市中的社会资本与城市更新，本书将进一步探讨有关研究架构。

本章以西安为例开展案例研究。首先，简要介绍了西安的历史与地理状况。其次，回顾了西安自1949年以来的城市发展和更新政策的演变。正如前面章节所提到的，西安不同历史时期的城市更新政策，很大程度上体现了转型期中国历史城市更新的特点。本章还分析了西安城市历史核心区更新过程中的显著特色与更新动力。在概述了西安不断演变的城市更新政策之后，本书分析了当地的城市更新与保护制度以及规划发展控制机制，这对城市更新的规划和实施具有重要影响。随后，考虑到西安历史核心区更新中更加综合的更新政策，本研究从社会资本的角度分析了城市更新过程中的社区关系和社区参与问题。同时，还分析了影响当地社会资本生成的关键因素。最后，本章总结了社会资本在西安历史城市核心区的更新实践中所发挥的作用。

Through an extensive literature review and documentary studies, the previous chapters have explored the key variables in the book and their relationships in investigating urban regeneration, social capital, urban conservation, and urban development institutions and mode of governance in the context of transitional Chinese historic cities. Based on these, an analytical framework has been developed towards the better understanding of social capital and urban regeneration in Chinese historic cities.

Founded on this analytical framework, this chapter explores these issues in relation to a specific Chinese historic city, Xi'an. First, this book gives a brief historical and geographical introduction on Xi'an City. Second, this chapter reviews the transforming urban development and regeneration policies in Xi'an since 1949. Urban regeneration policies in the various periods of the history of Xi'an largely represent the characteristics of urban regeneration in transitional China as discussed in the previous chapters. Many outstanding local features and dynamics in the regeneration of the historic urban core of Xi'an are examined. Following the overview of the transforming urban regeneration policies, this book investigates urban regeneration and conservation institution, and the development control mechanism in Xi'an, substantively influencing the making of the local urban regeneration plans and implementation processes. Third, in light of the more comprehensive concerns in the regeneration of Xi'an's historic urban core, this book explores the issue of social capital in terms of community relationships and involvement in urban regeneration processes. At the same time, some key factors affecting the generation of

4.1 西安：地理介绍

历史简述

西安位于我国关中平原（渭河流域）的中部，其发展得益于大片肥沃的土壤（Wu, 1979：10）。它曾经是中国的六朝古都，西安的名字始于明朝，尽管"在明代之前该市已经发展得很繁荣了"（Ma, 1985：149）。中国历史上曾繁荣的都市，如西周时的丰京、镐京两京，秦朝时的咸阳，西汉、隋、唐时的长安都是西安的前身。虽然这些城市的地理位置不同，但都在西安直辖市范围内。这些城市的兴衰都反映了西安的发展变迁（Ma, 1985：149）。

在中国历史上，从西周到唐朝 1133 年期间有 16 个王朝在此建都。这意味着自公元前 11 世纪以来的 3000 多年中国历史中有超过 1/3 的时间都是以西安为政治、经济和文化中心（Zhu & Wu, 2003：3）。在汉和唐朝，著名的丝绸之路起始点就在西安。西安作为陆上的交通要道，在连接和促进亚洲、欧洲和非洲国家的贸易交流起到了关键作用（Wu, 1979：120）。更多关于唐朝之前西安的文献可以参看马正林（Ma, 1978），武伯纶（Wu, 1979），朱士光和吴宏岐（Zhu & Wu, 2003）等学者的著作。

在唐朝之后，西安不再是中国的首都，而是发展为一个重要的地方性都会。与"都城时代"相比，学者把唐朝之后的西安称为

social capital are studied. Finally, this chapter is concluded by underlining the role of social capital in the urban regeneration program of Xi'an's historic urban core.

4.1 The City of Xi'an: A Geographical Introduction

Xi'an in Chinese history

Xi'an is located at the center of the Guanzhong Plain (Wei River Valley) in China and owes its development to this large area of fertile soil (Wu, 1979: 10). As one of China's six greatest historic capitals, the name of Xi'an came into use during the Ming Dynasty, although "the city had grown and prospered well before that period" (Ma, 1985: 149). The prosperous cities in China's history, such as Fengjing and Gaojing of the Western Zhou Dynasty, Xianyang of the Qin Dynasty, and Chang'an of the Western Han, Sui, and Tang Dynasties, were all forerunners of Xi'an. Although all these cities occupied different sites, they are within the present Xi'an's municipality scope. The rise and fall of each of these cities reflect the vicissitudes in the development of Xi'an (Ma, 1985: 149).

In China's history, Xi'an was the capital city during the period of the Western Zhou to the Tang Dynasty, altogether comprising about 16 dynasties or 1,133 years. This means that Xi'an was the political, commercial, and cultural center for more than one-third of China's 3,000 years of history since 11th century B.C. (Zhu & Wu, 2003: 3). Xi'an was the starting venue of the famous Silk Road in China during the Han and Tang dynasties. It served as the transportation route on land and played an important role in connecting and facilitating the trades among Asian, European, and African countries (Wu, 1979: 120). More literature on Xi'an before the Tang Dynasty can be found in Ma (1978), Wu (1979), and Zhu and Wu (2003).

After the Tang Dynasty, Xi'an was no longer the capital of China, developing instead to an important regional city. Compared with the "Capital Period" (*du cheng shi dai*), academics refer to the time of Xi'an after the Tang Dynasty

"重镇时代"或"后都城时代"(Shi & Wu, 2007：1)。学者断言，研究西安的发展，尤其是唐朝之后10世纪到20世纪期间的发展具有重大意义，因为它连接了那些辉煌的历史和现在的条件，并促进了西安现在的城市和经济发展。(Shi & Wu, 2007：2)。更多关于西安"后都城时代"的讨论可以参看史红帅(Shi, 2008)，史红帅和吴宏岐(Shi & Wu, 2007)，吴宏岐(Wu, 2006)等学者的著作。

今天的西安

作为陕西省的省会城市，今日的西安还是中国西北地区最大的工业基地、经济中心、政治和教育中心。西安的支柱产业有纺织、电力机械和军工企业等。2011年西安市的行政区划（包括行政区、县）面积共约10108平方公里，东西跨度204公里，南北长116公里。2011年西安市常住总人口约851万人。其中主要市区（包括九个区）占地面积3582km^2，常住人口数为654万人(EBXSY, 2012)。

西安的历史城市格局主要指的是其著名的明城区或旧城区格局，其可以追溯到隋朝(581～618年)。但是西安的城市格局和面积在唐朝(618～907年)和明朝(1368～1644年)得到了扩张。在明朝洪武年间十一年(1378年)，在唐朝城墙区的基础上对西安进行了大规模的城市建设，如城墙、钟楼和鼓楼都是在当时建立起来的，为现在西安城市构建和历史范围打下了基础(He, 2006：42)。之后大量的城市建设和发展都以明皇城为主。图4-1展示了现在的西安城区。红色部分代表明皇城，也被称为历史城市核心区

as the "Important City Period" (*zhong zhen shi dai*) or the "Post-Capital Period" (*hou du cheng shi dai*) (Shi & Wu, 2007: 1). Academics assert that it is significant to study Xi'an's development especially from the 10th to the 20th century after the Tang Dynasty because it bridges those magnificent historic times with the present conditions and contributes to today's urban and economic development of Xi'an (Shi & Wu, 2007: 2). More discussions on the "Post-Capital Period" of Xi'an can be found in Shi (2008), Shi and Wu (2007), and Wu (2006).

The Xi'an City of today

Today, Xi'an is the capital of Shaanxi province, the biggest industrial base, commercial center, and political and educational core of Northwest China. Its local backbone industries include textile, electrical machinery, and military enterprises. In 2011, the municipality of Xi'an (including administrative districts and counties) has an area of 10,108 km^2, ranging 204 km from east to west and 116 km from south to north, with a permanent residents of about 8.51 million people in 2011. Within this administrative area, the central urban area including nine administrative districts is about 3,582 km^2, with a permanent residents of about 6.54 million people (EBXSY, 2012).

Xi'an's urban pattern, mainly referring to its urban street pattern known as Ming City (*ming cheng*), can be traced back to the Sui Dynasty (A.D. 581-618). Nonetheless, the Tang (A.D. 618-907) and Ming (A.D. 1368-1644) Dynasties contributed to and expanded most of its urban patterns and areas (XCUCRC, 2000: 35). During the 11th year of the Hongwu period (A.D. 1378), based on the Tang City wall area, massive urban constructions were carried out, for example, the City Wall, Bell-Tower, and Drum-Tower, which formed the basis of the current Xi'an's urban fabric and historical precincts (He, 2006: 42). Subsequent extensive urban constructions and development took place mainly based on Ming City. Figure 4-1 presents Xi'an's urban area today. The red part indicates the Ming City area, which is also known as the historic urban core or the City Wall Area (hereinafter, CWA). As the historic urban core, the CWA is about 12 km^2

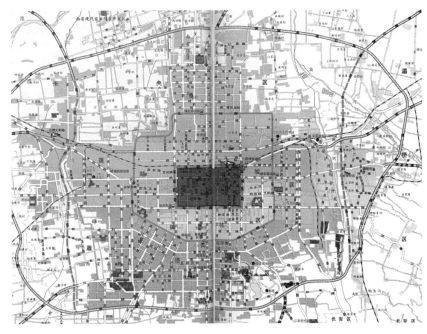

图 4-1：西安主要城区
Figure 4-1: Xi'an central urban area

或城墙区。作为历史城市核心区，2005 年，城墙区占地面积约 12 平方公里，拥有人口大约 4.2 万（XMPG，2005：104）。西安城区内仍保留了不少历史古迹和遗址，除了钟楼、鼓楼和大清真寺等历史古迹之外，还有三学街、德福巷和北院门（鼓楼回民区）之类的历史街道（SCXPC，2002：第 28 条）。

4.2 西安的城市发展与更新策略

如前面章节所述，自 1949 年中华人民共和国成立以来，西安的城市发展与建设的主要特点是开展城市再开发和更新活动。1949 年到 1957 年期间，西安经历了城市经济的复苏阶段。根据其第一次总体规划，西安奠定了其后来城市建设的格局和基础设施内容。1958 年至 1965 年期间，受到工业主导的城市发展政策和规划的影响，西安城市发展总体

and has a population of about 420,000 in 2005 (XMPG, 2005: 104). Many historical monuments and sites still remain within the Xi'an urban area. In addition to the historical monuments like the Bell-Tower, Drum-Tower, and the Great Mosque, there are some historical districts as well such as the Sanxuejie, Defuxiang, and Beiyuanmen historical districts (also known as the Drum Tower Muslim District) (SCXPC, 2002: Article 28).

4.2 Urban Development and the Implications on Regeneration in Xi'an

Since the establishment of P.R. China in 1949, the periodization of urban development and construction in Xi'an has taken in the characteristics of the urban development and regeneration practices in China discussed in the preceding chapters. From 1949 to 1957, Xi'an experienced the period of urban economic recovery, and based on its first master plan, Xi'an City set up its foundations for socialist urban patterns and infrastructure. From 1958 to 1965, Xi'an's urban development faltered because of the industrial-biased urban development policy

上停滞不前。1966年到1978年期间，受到政治斗争的影响，西安的城市发展和建设也遇到了前所未有的困难。随着1978年的政治体制改革，西安的城市发展和建设才又根据其既定总体规划的要求进行。在此背景下，当地政府于1980年提出了当地的第二次总体规划，其中尤其加强了关于文物保护的内容。20世纪90年代的10年期间，西安开展了以房地产主导的大规模城市改造活动，城市更新实践的主要特点是用土地使用权来置换经济效益。2000年，面对以前倾向于经济发展政策带来的问题以及新时期的机遇，如可持续发展的提出和1999年提出的中国西部大开发项目，西安市政府提出了最新的总体规划，突出强调了城市保护和再开发政策。这些总体规划对不同时期地方城市更新项目和实践产生了重要影响（He，2004；Wu，2004；Zhang，2006b）。

自1950年以来，西安市政府先后提出了1953～1972，1980～2000，1995～2010和2008～2020四个阶段性的总体规划。2008年5月，国务院批准了目前正在实施的第四次总体规划。除了这些不断演变的总体规划和当地的社会经济发展之外，西安城市更新政策的重点与观念也发生了许多变化。在第一次总体规划中，西安的发展目标是要发展成以精密仪器制造和纺织工业为主的轻工业城市。基于该总体规划并考虑当时的社会经济条件，西安的城市更新包括开发利用旧城区来满足工业发展的需要。从第一次总体规划中汲取到经验教训，第二次总体规划更加关注城市保护的问题。众所周知，对西安而言，城市保护与再开发同等重要。在20世纪

and deviation from the master plan. From 1966 to 1978, Xi'an's urban development and construction progressed with difficultly due to the effects of the political struggles. With the political institutional changes since 1978, urban development and construction have been revived under the direction of the master plan. In this background, Xi'an proposed its second master plan in 1980, and in it raised the issue of heritage conservation. The decade of the 1990s witnessed a massive property-led urban redevelopment, and urban renewal programs were characterized by the exchange between economic profits and land use rights. In 2000, facing many urban problems from previous pro-growth urban development policies and many opportunities in the new era, for example, the advocacy of sustainable development and the Western China Development Program of 1999, the Xi'an municipal government put forth the latest master plan highlighting the urban conservation and redevelopment policies. These master plans and policies have significant influences on the local urban regeneration programs and practices of the different periods (He, 2004; Wu, 2004; Zhang, 2006b).

Since 1950, the Xi'an municipal government has put forth four master plans in the following periods: 1953-1972, 1980-2000, 1995-2010, and 2008-2020. Approved by the State Council in May 2008, the fourth master plan is currently being implemented. Aside from these transformation master plans and local socio-economic development, the emphasis and ideologies in Xi'an's urban regeneration policies have also changed significantly. In the first master plan, the aim of Xi'an's urban development was to develop Xi'an into a light industrial city characterized by precision machinery manufacturing and textile industries. Based on the this master plan and considering the local socio-economic conditions at the time, the urban regeneration in Xi'an included the utilization of old urban areas to serve the needs of the industrial development. Through the experiences gained from the period of the first master plan, the second master plan dealt with the issue of urban conservation. It was

90年代中期，第三次总体规划将城市保护的范围从个别文物古迹进一步扩大到整个历史城区。然而，受到当时以经济发展为主导的城市政策和不够完善的发展控制机制的影响，城市保护规划常常被搁置，历史城区遭到了城市发展项目的破坏（Yang，2006：93）。因此，面对众多的城市问题，第四次总体体规划强调了城市更新过程中的城市保护问题，希望通过城市空间重组来解决目前历史区的压力，并为其他行业的发展提供空间（XCPB，2005b；XMPG，2005）。

4.2.1 西安城市发展与更新政策的历史回顾

4.2.1.1 基于第一次总体规划、以历史城区为基础、促进工业发展的阶段（1949～1957年）

整个明清时期，尤其是在国民党统治时期的内战期间，西安的明皇城发生了巨大的改变。到1949年，西安总的城市建筑面积约为13.2平方公里，其中主要是在明皇城内（SNU，1988：442）。当时的人口总数约为在77万人左右（XSB，1993：87）。由于历史原因，作为一个消费型城市，西安几乎没有现代工业，城市基础设和环境条件差（XMG & XSB，1999：111）。因此，1949年之后的几年里，西安的城市发展政策主要是改善城市基础设施和破旧的居民区，比如建设解放路、新城街区和其他地区的下水道系统（XMG & XSB，1999：115）。1950年，西安市政府开始着手设计总体规划（当时称为城市发展规划）。同年，成立了城市建设与规划委员会（Zhang，

acknowledged that urban conservation and redevelopment were equally important urban development policies for Xi'an. In the mid-1990s, the third master plan further expanded the scope of urban conservation from individual heritage buildings to the entire historic core area. Nevertheless, because of the local economic development-led urban policies and the loose development control mechanism, the urban conservation plan was pushed aside, and the historic urban core was disrupted by many urban redevelopment projects (Yang, 2006: 93). Thus, faced with many urban problems, the fourth master plan has underlined the issue of urban conservation in the local urban regeneration program by restructuring the urban space to lessen the pressures on the current historic urban core area and to provide more space for various industries (XCPB, 2005b; XMPG, 2005).

4.2.1 A historical overview of Xi'an's urban development and regeneration policies

4.2.1.1 Industrial development-oriented and historic core area-based development anchored in the first master plan (1949-1957)

Throughout the Ming and Qing Dynasties and especially during the reign of the Kuomintang during the civil war, Xi'an's historic Ming City area underwent tremendous changes. By 1949, Xi'an's total built urban areas were about 13.2 km^2, which was largely limited within the Ming City area (SNU, 1988: 442). The population at the time was about 770,988 (XSB, 1993: 87). As a consumption-led city due to historical reasons, Xi'an possessed very few modern industries, with poor urban infrastructures and derelict environments (XMG & XSB, 1999: 111). Hence, during the years immediately after 1949, Xi'an's urban development policies focused on the improvement of urban infrastructures and dilapidated residential areas, for example, the sewage systems of Jiefang road, Xincheng district, and other areas (XMG & XSB, 1999: 115). In 1950, the Xi'an municipal government set out to devise its master plan (then called Urban

1988：27）。根据20世纪50年代的国家经济发展计划，西安的城市发展与建设推动了城市功能区的重组和工业发展（Wang & Hague，1992：5）。到1954年末，西安市政府提出了第一次总体规划，包括对东西部郊区的详细规划和近期发展计划（图4-2显示了第一次土地使用总体规划）。1954年，国家建设委员会代表中央政府批准通过了这些规划（Zhang，1988：34）。在该规划中，西安城市发展的目标就是发展成"轻型精密机械制造和纺织为主的工业城市"（Wang & Hague，1992：5）。

第一次总体规划强调了为促进城市工业发展所需的土地利用规划，以及所需的基础设施建设（Wang & Hague，1992：5）。规划兴建的城市面积为131平方公里，人均城市面积108平方米（XCUCRC，2000：106）。根据中国第一个五年计划，作为重要的内陆发展中心，国家在西安规划并建立了17个

Development Plan). In the same year, Xi'an's urban construction and planning committee was established (Zhang, 1988: 27). Following the national economic development scheme of the 1950s, Xi'an's urban development and constructions promoted the restructuring of urban functional areas and urban-based industrial development (Wang & Hague, 1992: 5). By the end of 1954, Xi'an's municipal government proposed the first master plan (1953-1972) containing the detailed plan for its eastern and western suburban areas and recent development plan (see Figure 4-2 for the first master land-use plan). These plans obtained the approval of the State Construction Commission on behalf of the central government in 1954 (Zhang, 1988: 34). In this plan, the aim of Xi'an's urban development was to develop the city into "a precision machinery and instrument manufacturing and textile industry city" (Wang & Hague, 1992: 5).

The first master plan was, above all, a technical document addressing the engineering infrastructure and land use requirements for the major promotion of urban industrial development (Wang & Hague, 1992: 5). The proposed build-

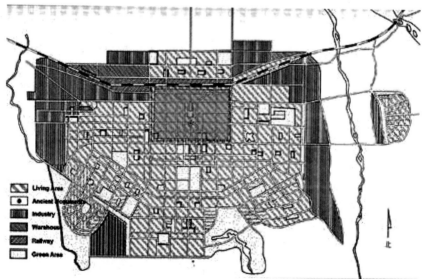

图4-2：西安第一次总体规划（1953—1972）
Figure 4-2: Xi'an master plan (1953-1972)
资料来源：(XCUCRC，2000)
Source: (XCUCRC, 2000)

大型工业项目，并且这些项目属于当时由苏联支持的 156 个项目之内（Zhang，2002a：12）。但是，在中央计划经济体制下，这些任务并没有分配给为西安地方政府工作的规划团队，而是给了在苏联技术援助之下的由中央政府组织的规划团队（Wang & Hague，1992：5）。

关于历史核心区的城市更新政策，该规划强调其历史核心区（或旧城区）应"充分利用、基本不做改变"，以保护历史都城的特色（Zhang，1988：36）。结果，该规划以城墙区为中心，向东、西、南部扩张。在距城墙区 4 公里处的东部和西部发展了新工业，而东南部为以后的工业发展区域（Wang & Hague，1992：5）。居民区位于城墙区和新工业发展区之间（SNU，1988：442）。南部的郊区计划建设成为居住、教育、研究和文化用途的"文教区"。为了保护北郊汉朝长安城遗迹和唐朝明宫遗迹，该区域内没有进行大规模的建设（Wang & Hague，1992：6）。不过，在当时城市发展和建设由中央政府进行统一领导，并以工业发展为中心的情况下，历史文化遗产保护在第一次总体规划中并未发挥十分突出的作用（XCUCRC，2000；XMG & XSB，1999）。

4.2.1.2 对历史城市核心区的充分利用与碎片化的城市发展阶段（1958～1977 年）

1958 年，在"大跃进，大都市"和"建设独立和完整的工业体系"的口号下，西安的城市发展在城市特色、人口密度、发展范围和规划布局方面发生了很多变化。其中很多内容也与第一次总体规划的要求渐行渐远

up urban area was 131 km^2, with an average urban area per capita of 108 m^2 (XCUCRC, 2000: 106). According to China's First Five-year plan, as major inland development centers, 17 large-scale industrial projects were planned and built in Xi'an, and these were among the 156 projects supported by the then USSR (Zhang, 2002a: 12). Nonetheless, in a centrally planned economy, these tasks were assigned not to a team of planners working for the Xi'an local government but to a central government-organized planning team with Soviet technical assistance (Wang & Hague, 1992: 5).

With regard to the urban regeneration policies in the historic core, it was underscored in the plan that the historic core (or the CWA) should be of "enough utilization and basically no changes" should be made in order protect its historical capital features (Zhang, 1988: 36). As a result, the plan centered on the CWA and expanded towards the east, west, and south directions. New industrial developments were planned in the eastern and western parts of the city four kilometers from the CWA, and the southeast area was to be saved for future industrial development (Wang & Hague, 1992: 5). The location of the residential districts was intended to be between the CWA and new industrial development areas (SNU, 1988: 442). The southern suburb was planned to be the "cultural area" for residential, educational and research, and cultural use. To protect the Chang'an city ruins of the Han Dynasty and the Ruins of Daming Palace of the Tang Dynasty in the northern suburbs, no major development was proposed in these areas (Wang & Hague, 1992: 6). Nonetheless, the conservation of historic monuments did not play a significant role in the first master plan, when the overall urban development was central government-led and industrial development-oriented (XCUCRC, 2000; XMG & XSB, 1999).

4.2.1.2 Full utilization of the historic urban core and fragmented urban development (1958-1977)

Through the slogans "Great Leap Forward, Great Metropolis" and "building independent

(Zhang，1988：42）。1960年，由于苏联资金和技术支援的中断，中央政府没有足够的资金进行支援，相应地，西安大肆鼓励地方发展政策。市政府或区政府在其行政范围内兴建了不少地方工厂。随着中央政府权力下放政策的颁布和"独立自主、自力更生"的政治观念的倡导，西安一些区政府在历史核心区尽可能多地建立了街道作坊和小工厂。自那以后，早期规划的城市功能区内充斥了不少小型企业（Wang & Hague，1992：11）。根据"大跃进"时期的指示，以工业发展为主导的城市发展政策产生了很多问题，例如过度占用土地、规模超标、不平衡的标准和无授权的革新（Zhu & Wu，2003：542）。此外，过度的城市建设加重了旧城区的负担，使城区的交通系统条件恶化，破旧的基础设施和居住环境趋于恶化。1960年随着中央政府对三年城市规划的叫停，废除了西安市城市规划部门，并关闭了部分工厂企业（Wang & Hague，1992）。

"文化大革命"使整个中国的国家政治形势陷入了一种特殊状态中。这期间，西安的城市发展具有以下两个突出的特点。第一、城市发展和建设比较涣散。第二、西安城中的很多历史城区和文化古迹都遭到了很大程度的破坏。同时，没有一个统一部门控制公共投资，相反，不同部门独自制定规划（Wang & Hague，1992：14）。由于行政权力分散，土地公共所有权带来了很多问题，土地使用有以下三种方式：城市社区，农村和市政府。公共机构决定城市社区土地的使用；原村民安排农村和农业用地的使用；市政府则管理不在农民控制范围内的公共土地，主要包括道

and integrated industrial system" in 1958, many changes took place in Xi'an's urban development in terms of urban features, population capacity, development scopes, and planning layout. Many aspects had the tendency to break away from the requirements laid out in the first master plan (Zhang, 1988: 42). By 1960, with the end of the financial and technical aid from the USSR and without further funds available from the central government, Xi'an's local initiatives were encouraged. Many small factories were set up by the city government or urban district authorities within their administrative purview. With the decentralization policy and political ideology of self-dependence from the central government, some district governments in Xi'an set up local street workshops and small factories on any available sites in the historic core. From then on, the earlier planned urban functional areas became mixed up with small-scale developments (Wang & Hague, 1992: 11). In following the Great Leap Forward instructions, industrial development-oriented urban development policies created many problems such as occupying lands excessively, too grand scales, disproportionate standards, and unwarranted innovations (Zhu & Wu, 2003: 542). In addition, extensive urban constructions intensified the burdens to the old urban areas, exacerbating the conditions of the transportation systems within the areas, dilapidated infrastructures, and living environments. With the central government's instruction on the moratorium on city planning for three years in 1960, Xi'an's city planning department was abolished; small factories were shut down; and workers were forced to return to their rural homes (Wang & Hague, 1992).

During the Cultural Revolution, which brought about a different national political situation, Xi'an's urban development exhibited two salient features. First, urban development and construction was conducted in a rather fragmented manner. Second, many historic urban areas and heritage monuments in Xi'an underwent massive destruction. At the time, public investment in the city was not controlled by a unitary department. Rather, different functional departments made their own decisions

路和其他公共区域。因此，在"文化大革命"期间，西安城市发展产生了很多独立的、封闭式的社区（Zhang，2002a：13）。另一方面，在"文化大革命"期间，另一方面，无规划的任意建设破坏了西安很多历史庭院和遗产古迹（Zhang，2002a：13）。同样，大规模的工业发展也对历史遗迹产生了巨大影响。例如，一座铜线厂就建在了以青铜器著称的西周丰镐遗址上。铜线厂的建设对该历史遗迹造成了毁灭性的破坏（Zhang & Han，1981：16）。尽管如此，不少学者认为，虽然政治动乱时期打乱了西安的城市规划，并给其城市发展带来了很大的破坏，但是基于土地使用模式和1954年制定的第一次总体规划，西安仍是一个在新中国成立初期，如何实施当地城市规划的好例子（Wang & Hague，1992：16）。

1970年西安的国民生产总值是17.8亿，中心城区人口是188万。到1972年总建筑面积达131km²（XSB，2006：65）。图4-3和图4-4显示了更多关于西安国民生产总值和城区人口数量的信息[1]。由于"大跃进"后很多工人被遣回乡，"文化大革命"期间科技人员发送到农村进行改造，西安城区人口数量急剧下降。但是，总体而言，城区人口数量呈迅速上升趋势。

4.2.1.3 历史城区的现代化建设和多维度的城市更新策略（1978～20世纪80年代末）

1978年之后，西安城区被划定了五个主

separately (Wang & Hague, 1992: 14). With the fragmented administrative powers, public ownerships of land posed many problems, and land use was practiced in three ways: urban communities, rural areas, and the city government. The public institutions determined land use within the urban community walls; the original villagers determined land use within the village and the around agricultural land; and the city government controlled the public land withdrawn from the farmers mainly comprised of roads and other public areas. As a result, Xi'an's urban development during the Cultural Revolution produced many separate and independent closed-wall communities (Zhang, 2002a: 13). On the other hand, the random construction without planning during the Cultural Revolution destroyed many historic courtyards and heritage monuments in Xi'an (Zhang, 2002a: 13). Extensive industrial development likewise had substantial effects on many historical ruins. For example, a copper wire factory was built on the site of the Western Zhou Fenggao ruins, which are globally well known for their bronzeware. The construction of the copper wire factory caused devastating damage to this historic heritage site (Zhang & Han, 1981: 16). Nevertheless, academics argue that although the Great Leap Forward and Cultural Revolution disrupted urban planning and brought great damage to Xi'an's urban development, Xi'an is still a good example of how city planning was practiced in China in the early years of the Communist rule based on the land use pattern and the first master plan established in 1954 (Wang & Hague, 1992: 16).

Xi'an's national GDP in 1970 was 1.78 billion, and the central urban area population was about 1.88 million. By 1972, the total built urban areas reached 131 km² (XSB, 2006: 65). More information about the GDP and the urban area population in Xi'an[1] are presented in Figure 4-3 and Figure 4-4. Xi'an's urban area population

---

[1] 数据包括西安市行政范围内，1997年之后的临潼区和2002年之后的长安区的人口数据。这两个年份之后，先前的两个县成了西安城区的两个行政区（He，2006）。

[1] The population data include *Lintong* district after 1997 and *Chang'an* district after 2002, within Xi'an municipal administration periphery. After these two years, the former two counties became two administrative districts in Xi'an urban areas (He, 2006).

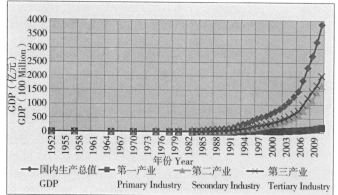

图 4-3：西安 GDP 的增长及构成
Figure 4-3: Xi'an GDP increase and structures
资料来源：(XSB, 2012)
Source: (XSB, 2012)

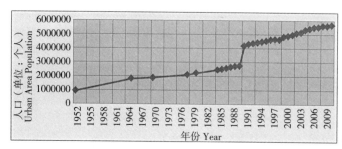

图 4-4：西安城区人口数量（1952—2011）
Figure 4-4: Xi'an urban area population (1952-2011)
资料来源：(XSB, 2012)
Source: (XSB, 2012)

要功能区：东郊的军工机械和纺织产业，西郊的电子工程，南郊的教育中心，北郊的文物和历史遗迹保护区，以及历史核心区被划定为行政、商业和居住区（Zhu & Wu, 2003：545-546）。随着1978年提出的"拨乱反正运动"和改革开放政策，西安明城区的城市发展与建设开始探讨如何将现代化的城市建设和古都特色以及文物古迹的保护相结合（Zhang, 1988：62）。

基于第一次总体规划的经验和教训，1981年西安市规划局制定了第二次总体规划（1980～2000），并于1983年得到了国务院的批准（Zhu & Wu, 2003：556）。该规划重新定义了西安的城市特色：是中国历史文化名城之一；陕西省的政治、经济和文化中心；以纺织和机械制造为主要产业的先进科学、文化和教育城市；以发展旅游产业保护城市

decreased because many workers were sent back to their rural homes after the Great Leap Forward, and technocrats were sent to the countryside to be rusticated during the Cultural Revolution. Nevertheless, the overall trend of the urban area population rapidly increased.

4.2.1.3 Modernization of the historic core and the multi-dimensional urban regeneration strategies (1978-late 1980s)

By 1978, Xi'an developed its urban areas mainly into five functional zones: military machinery and textile industries in the eastern suburban area, electronic engineering in the western suburban area, education centres in the southern suburban area, heritage and historical remains conservation zones in the northern suburban area, and the historic CWA that served the administrative, commercial, and residential functions (Zhu & Wu, 2003: 545-546). With the political ideology of bringing order out of chaos and adopting the open door policy (*bo luan fan zheng, gai ge kai fang*) in December 1978, urban development and construction within the CWA of Xi'an began to

历史特色（He, 2006 : 104 ; Wang & Hague, 1992 : 20。仔细对比第一和第二次总体规划, 我们会发现两者之间存在明显的连续性。两个规划都试图在发展制造业的基础上形成一种城市经济模式。1980 年增加了一项条款, 即强调旅游业的发展以及保护历史都城特色。此外, 该规划还提议将城区建筑面积从原来的 131 平方公里扩大到 162 平方公里, 降低人均土地使用面积, 由 1953 年时的 108 平方米降到 90 平方米（Zhang, 1988 : 33, 50）。图 4-5 显示了第二次总体规划中的土地使用规划。

根据之前的国家城市发展政策, 即"控制大城市的规模, 合理发展中型城市并严格发展小型城市", 第二次总体规划旨在重新调整西安土地利用模式。它把西安市划分为 10 个住宅区, 这些住宅区又被主要道路划分为 40～180 公顷大小的分区, 每个分区大约拥有 2 万人口。但是, 尽管该规划符合很多 20

explore how to incorporate modern structures with the conservation of the ancient capital city features and heritage monuments (Zhang, 1988: 62).

Based on the lessons and experiences of the first master plan, the second master plan (1980-2000) was proposed by the Xi'an City Planning Bureau (then called Xi'an Urban Planning Management Bureau) in 1981; it was approved by the State Council in 1983 (Zhu & Wu, 2003: 556). This plan redefined the urban features of Xi'an as one of the Chinese Historic and Cultural cities; the political, economic, and educational centre of Shaanxi province; a city of advanced sciences, culture, and education with textile and machinery manufacturing industries as the main sectors; and a city with a tourist trade orientation geared towards the protection of the city's historical features (He, 2006: 104; Wang & Hague, 1992: 20). A closer look at the first and second master plans shows that there was evident continuity between these plans. Both plans attempted to set up an urban economic form based on the manufacturing industry. The new item in the 1980 plan was the emphasis on the tourism industry and the proposal for the conservation of historic capital city features. Furthermore, the plan proposed the enlargement of the built-up urban

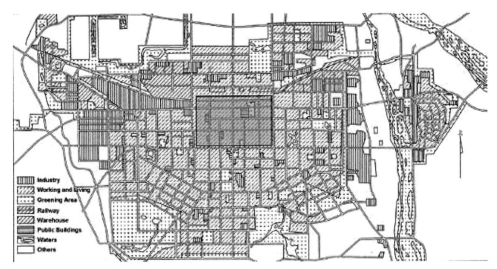

图 4-5：西安第二次总体规划（1980—2000）中的土地使用规划
Figure 4-5: Xi'an Second Master Plan (1980-2000)
资料来源：(XCUCRC, 2000 : 119)
Source: (XCUCRC, 2000: 119)

世纪80年代早期中央政府提出的国家政策，但是随着20世纪80年代全国范围内的政治权力下放和土地市场的出现，这些工程的实施显然是迎合了市场需要。20世纪70年代末的西安南大街就是一个很好的例子。由于过度拥挤和交通问题，这里的原住居民被重新安置；同时，为了与城市更新规划一致，街道两旁原有的传统建筑被拆除了。20世纪80年代末，该工程的结果是：拓宽了交通路线，增加了交通运输容量。另外，为了迎合日益繁荣的旅游业，街道两旁建立了大型国有超市和酒店（Wang & Hague, 1992：20—21）。这与第二次总体规划中规定的形成了鲜明的对比。因此，在历史遗产古迹和古都特色保护方面，西安城市规划极大地促进了旅游业的发展，改善了主要历史遗产古迹的物质环境，同时也修建了配套设施（Zhu & Wu, 2003：562）。同时，城墙区内的城市基础设施也得到了改善（XCUCRC, 2000：121）。

第一次总体规划后，明西安府城和当今的西安市政府都坐落于历史城墙区内。在第二次总体规划中提出城市更新规划采取多维发展途径，包括在不同地方采取保护、恢复和重建的方法。城市更新项目旨在将现代建筑和古都特色相结合来满足当地旅游业的需要（Zhu & Wu, 2003：548）。为提高古都特色，该规划要求对整个城墙区进行保护。在更新实践中，除了对一些现存古迹和历史街区进行了保护，还根据商业和旅游业的需要，在再开发项目中兴建了一些具有传统风格的建筑（XCUCRC, 2000：122）。在历史街区进行再开发活动的同时，规划要求重建项目要在建筑高度和风格方面和周围的传统建筑保

area from the previous 131 to 162 km^2 and the reduction of the average land use area per capita from the previous 108 m^2 in 1953 to 90 m^2 (Zhang, 1988: 33, 50). The land-use plan in the second master plan is shown in Figure 4-5.

According to the former national urban development policy, that is, "to control the size of large cities, develop medium cities in a rational manner, and rigorously develop small cities," the second master plan attempted to readjust Xi'an's urban land use pattern. It divided the city into 10 residential districts, which were separated by major roads into sub-districts of 40-180 hectares of land with about 20,000 people in each (Zhang, 1988: 52). Nevertheless, although this plan was developed to agree with the many national policies proposed by the central government in the early 1980s, the trend of political decentralization and the emerging land market in the whole of China in the late 1980s encouraged the implementation of projects that obviously catered to the market needs. A good example is the Xi'an South Street (*nan da jie*) project in the late 1970s. Faced with overcrowding and traffic problems, the original residents were relocated, and original traditional buildings on both sides of the street were demolished in line with the local urban renewal practices. The outcome of this project in the late 1980s included a much wider traffic line, increased capacity of transport channels, as well as large state-owned superstores and big hotels on both sides of the street, catering to the growing tourist industry (Wang & Hague, 1992: 20-21). This sharply contrasted with what had been laid out in the second master plan. Therefore, with regard to the conservation of historic heritage monuments and ancient capital features, Xi'an's urban plan strongly promoted the tourist industry, and many physical improvements of the major historic heritage monuments as well as supporting facilities were carried out (Zhu & Wu, 2003: 562). At the same time, the improvement of urban infrastructure took place within the CWA (XCUCRC, 2000: 121).

After the development of the CWA following the first master plan, it became the site where both the original government body of the Ming Dynasty (*ming Xi'an fu cheng*) and the current

持一致。例如，在 20 世纪 90 年代早期，北院门和书院门的商店保留了一些传统的住房标志性图案，并且大部分破旧的住宅都得到了重建（He, 2006：93）。

1982 年，国务院宣布西安成为第一批国家历史文化名城中的一座（Chan, 1999），有力地推动了西安城市发展中的城市保护力度。总体而言，20 世纪 80 年代，在城市保护和再开发项目的实践中，以及地方旅游业和商业发展方面，私人开发商的影响并不大，其影响仅限于一些小型服务企业部门。大部分城市再开发项目是由国家企业和政府机构来领导实施的（Wang & Hague, 1992：21）。不过，随着国家土地市场制度的出台，这种情况在 20 世纪 90 年代发生了很大变化。

4.2.1.4 以市场为导向、以房地产开发为主导的再开发过程（20 世纪 90 年代）

1984 年的城市经济体制改革以后，地方政府开始鼓励私营企业，特别是涉及海外投资的企业参与到地方经济建设中来。与此同时，国内对流动人口管治的放松，也大大促进了相关规划的实施（Wang, 2000：322）。如（图4-5）所示，1978 年后西安城区人口数迅速增长（He, 2006：105）。不过，这也给原本就已破败的城市基础设施带来了更大的挑战。随着越来越多的国外人士来到西安参观兵马俑，国际投资者敏锐地注意到了通过建设合资酒店来扩大业务的机遇。但是随着酒店开发需求的提高，为了迎合西安城墙区内新的重建项目，往往会对已有的保护政策妥协。最后，西安市政府为了这些再开发项目于 1988 年对第二次总体规划作了修改。结果是为了吸引

Xi'an municipal government was located. The urban regeneration plan in the second master plan proposed to adopt multi-dimensional development approaches including preservation, restoration, and redevelopment in different places. In order to serve the needs of the local tourism industry, urban regeneration projects were aimed to combine modern buildings with ancient capital features (Zhu & Wu, 2003: 548). To enhance the ancient capital features, the conservation plan required the conservation of the entire CWA. In practice, in addition to the conservation of several existing monuments and historic districts, large areas of redevelopment projects in traditional building styles were created according to the needs of the commercial and tourism industry (XCUCRC, 2000: 122). While conducting the redevelopment activities in the historic districts, the redeveloped projects were required to be coordinated with the surrounding preserved heritage buildings in terms of building heights and styles. For instance, in the shopping houses in Beiyuanmen and Shuyuanmen, some traditional housing representative motifs were conserved, and most of the dilapidated residential houses were redeveloped in the early 1990s (He, 2006: 93).

In 1982, Xi'an was declared member of the first batch of the Historic and Cultural Cities (*li shi wen hua ming cheng*) by the State Council (Chan, 1999), and this further propelled the conservation effort in Xi'an's urban development. Generally, in the practice of conducting local conservation and redevelopment projects and promoting local tourism industry and businesses in the 1980s, private developers contributed to only a very limited part of Xi'an's economy growth. Their contributions were limited to some small-scale service enterprises. Most of the urban redevelopment projects were led and carried out by public enterprises and agencies (Wang & Hague, 1992: 21). However, this situation significantly changed in the 1990s with the emergence of the land market in China.

4.2.1.4 Market-driven and property-led redevelopment process (decade of the 1990s)

Since the urban economic reform in 1984,

当地和国际投资，城墙区内实行了一系列的优惠减税政策（Wang，2000：323）。

1988年，随着中国城市土地改革的进行，引入了土地有偿使用制度，西安市政当局开始利用土地的公共所有权筹集城市发展所需资金（Yeh，2005：39；Tang & Liu，2005：202）。随着城市土地市场的出现，房地产开发商进一步参与到城墙区的再开发项目中。开发商支付的土地使用费和建设税成为市政府增加财政收入的一个来源（He，2002：50）。从1978年到1990年期间，房地产开发累计投资人民币1202.1亿元。但是，仅在1990年，投资就高达人民币922亿元（XSB，1993：134）。1992年8月，中央政府宣布西安为内陆开放城市，这给西安提供了沿海城市相同的机会和权力，来增加海外投资并授予其税收优惠政策。西安大力倡导建设三资公司，尤其是在新批准的高新区内（Wang，2000：323）。西安市政府颁布的优惠政策包括扩大海外投资领域和转变项目交易权（EBXY，1993b：49）。

很明显的一点是，西安的第二次总体规划不再符合当时西安的发展状况；因此，1994年底，西安市有关规划部门（西安市规划局和城市规划设计研究院）提出了第三次总体规划（1995～2010）。该规划在1999年获得了国务院的批准（Zhang，2002a：15）。图4-6显示了该规划中的土地利用规划情况。

第三次总体规划把西安市的城市特色重新定义为：一座世界历史文化名城，是中国一个集科研、高等教育、高科技产业为一体重要基地；是中国北方中西部最大的中心城市；是陕西省省会（Zhang，2002a：16）。该

private businesses have been encouraged by the local governments to participate in local economic growth activities, especially those involving overseas investment. With the relaxing control on the population movement within the country, it also had substantive implications on the planning practice (Wang, 2000: 322). As shown in Figure 4-5, Xi'an's urban area population has been increasing very rapidly since 1978 (He, 2006: 105). This posed many challenges to the already very dilapidated urban infrastructure. With more and more visitors from abroad coming to Xi'an to see the Emperor's Terra Cotta, international investors saw the opportunity to expand their businesses in building joint-venture hotels. However, when the demand for hotel developments increased, compromises were made to the established conservation policies to cater to the new redevelopment in Xi'an's CWA. In the end, Xi'an municipal authority altered the second master plan in 1988 to make way for these redevelopment projects. Accompanying this, a series of favorable tax-reduction policies were applied within the CWA to attract local and international investment (Wang, 2000: 323).

With the urban land reform in China in 1988, land use fees based on market demand were introduced, and Xi'an's local government took the opportunity of public ownership of land to raise funds for urban development (Yeh, 2005: 39; Tang & Liu, 2005: 202). With the emerging urban land market, property developers were encouraged further to get involved in the redevelopment within the CWA. This became a source for the municipal government to generate local revenues as developers needed to pay the land use fees as well as construction tax (He, 2002: 50). From 1978 to 1990, the accumulated investment in real estate development amounted to RMB120,210 million. Nevertheless, in the year 1990 alone, the investment reached 92,200 million (XSB, 1993: 134). In August 1992, Xi'an was declared an open city by the central government, and this provided Xi'an the same opportunities and power enjoyed by the coastal cities to attract overseas investment and grant tax concessions. Three investment (*san zi*) enterprises were advocated strongly in Xi'an,

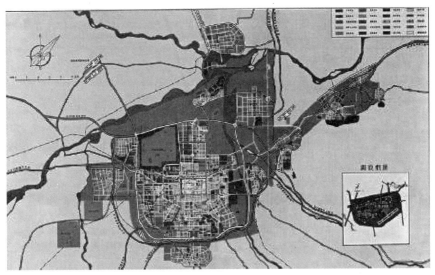

图 4-6：西安第三次总体规划（1995—2010）
Figure 4-6: Xi'an's third master plan (1995-2010)
资料来源：西安城市规划局，2008 年
Source: Xi'an Urban Planning Bureau, in 2008

规划的总目标是把西安发展成一所经济繁荣、功能齐全、环境友好和具有独特历史特色的美丽的现代化开放都市（Wang, 2000：325）。根据该规划，在 2010 年之前，城市区域面积要达到 275 平方公里，人口达到 320 万（XMG & XSB, 1999：112）。与之前的规划相比，纺织和制造业不再处于主导地位，主要突出强调西安的高新产业。因为西安的地理位置不适合大规模的棉花生产。随着市场和产业结构调整，西安纺织生产存在很多问题，在新规划中的主导产业转向高新技术产业。在西安南部和西部确定了新的高新区（Wang, 2000；XMG & XSB, 1999）。为了充分利用新获得的开放城市的地位，规划中高度重视贸易发展。距旧城区北部 5 公里处建立了经济技术开发区。市政府还提供了资金发展相应的基础实施（Wang, 2000：326）。

especially in its newly approved High-Tech Zone (Wang, 2000: 323). Privilege policies from the Xi'an municipal government included the enlargement of overseas investment fields and transition of project sanction rights (EBXY, 1993b: 49).

Clearly, the second master plan did not fit the reality of Xi'an anymore; thus, the third master plan (1995-2010) was proposed at the end of 1994 by the Xi'an planning authorities (Xi'an City Planning Bureau and Xi'an Urban Planning and Design Institute). This plan was approved by the State Council in 1999 (Zhang, 2002a: 15). Please refer to the land use plan in Figure 4-6.

In the third master plan, the features of Xi'an City was redefined as a world famous historic city, an important base for scientific research, higher education, and high-tech industries within China; the largest central city in the west and central part of northern China; and the capital of Shaanxi Province (Zhang, 2002a: 16). The overall aim of this plan was to develop Xi'an as an economically prosperous, functionally complete, environmentally healthy, and beautiful modern open city with its own unique historic features

城市更新政策主要特点

关于城墙区内的城市更新政策，第三次总体规划具有以下两个鲜明特征。一方面，西安城市保护规划中明确指出，作为保护地方历史城市特色的方法，要对历史古迹、古遗址以及整个历史城区进行保护（He，2006：112）。另一方面，为了保护历史都城的特色，在城墙区内重建传统城市景观。有学者认为，该政策的提出与地方政府对城市现代化观念的认识很相关（Wang，2000：328）。由于现有的规划与建设控制制度对再开发项目的约束力有限，所以规划和发展建设很大程度上以市场为主导。很多情况下，新建的建筑面积要比旧建筑的面积高出3～4倍。这样做的目的自然是为了从再开发项目中获得最大的经济回报。不过，这些不协调的现代建筑形式也进一步破坏了旧城区内的历史城区景观（Wang，2000：327）。

根据总体规划，各级政府基于土地有偿使用的政策积极进行再开发项目。房地产公司提供多种融资渠道，为再开发项目和城市基础设施建设提供必要资金。另外，这也是当地政府增加地方税收的重要途径（He，2006：87）。西安城区的房地产发展投资从1992年的4.33亿元增加到1998年的38.21亿元。1998年来自房地产行业的总税收高达11.95亿元。图4-7显示了房地产投资和财政收入的增长。1991年，西安地方政府选取了49块低洼地来进行更新项目，这标志着西安大规模的地方土地更新的开端。到2007年，其中47个项目已完成（EBXY，1993/2008）。但是，受市场驱动的影响，这些更新项目往往成为再开发项目和经济利益之间的置换。

(Wang, 2000: 325). According to the plan, the proposed urban area would reach 275 km^2 before 2010, with a population of 3.1 million (XMG & XSB, 1999: 112). Compared with the previous plan, the role of high-tech industries in Xi'an was stressed, and the textile and manufacturing industries in the previous two plans were not underlined anymore. This is because, geographically, Xi'an is not suitable for large-scale cotton production. With the open of market and industry restructuring, textile production in Xi'an became problematic, and the dominant industries in the new plan were to be high-tech based. The newly established High-Tech Zone was promoted in the southern and western regions (Wang, 2000; XMG & XSB, 1999). In order to take full advantage of the newly granted open-city status, the plan also cast light on trade development. An Economic and Technology Development Zone was proposed five kilometers north of the old town. The municipal government also provided funds to develop the corresponding basic infrastructure (Wang, 2000: 326).

Characteristics of the urban regeneration policies

The third master plan had two conspicuous characteristics with regard to urban regeneration policies within the CWA. On one hand, the conservation of heritage monuments, ancient ruins, and the entire CWA was explicitly laid out in the Xi'an urban conservation plan as a means of protecting local historic urban features (He, 2006: 112). On the other hand, local authorities tried to re-fabricate traditional urban scenes within the Xi'an CWA as a measure to "conserve" the historic capital features. Academics contend that this policy was generated by the local politicians' attitude towards urban modernization (Wang, 2000: 328). While the existing planning and development control system had very limited influence on these redevelopment projects, the planning and development processes became very market-oriented. Normally, the number of the newly built areas was three to four times more than the old buildings in terms of floor area. This was done mainly to maximize the financial returns from these redevelopment projects. As a result, the clash of building forms further disrupted the historic urban

图 4-7：房地产投资和财政收入
Figure 4-7: Investment and revenues from real estate development
来源：作者汇总自（EBXY，1993；XSB，2005）
Source: Author synthesized from (EBXY, 1993; XSB, 2005)

经过这些更新活动，西安历史城市的特色也在迅速消失（He，2002：50）。

4.2.1.5 城市空间重组与通过大规模的城市再开发来实施城市"保护"的目标（21世纪初期）

20世纪90年代末，西安城市建设中暴露的城市问题主要表现为：城市骨架尚未拉开、不平衡的城市土地利用情况、破旧的城市基础设施以及落后的环境设施，这些都导致西安历史古都的特色逐渐消失（Li，2005：106）。学者指出，以前的城市规划主要存在三个显著的城市问题。第一，以前的城市规划未能为西安第三产业的发展提供足够空间（He，2002：50）。第三产业所在西安地方经济所占的比例越来越大。（图4-4）显示了西安GDP的增长及从1952到2004年GDP的结构。第三产业占GDP的比例从1978年的23.39%增长到2004年的51.28%。西安第三产业主要有批发、零售、餐饮服务、银行和保险、房地产、教育、社会科学及其他行业。第二，基于历史核心区的规划重点强调保护和发展城墙区。今天，城墙区的历史核心区有多重功能，是城市政治、经济、文化和交通中心。这种境况给破旧的地方城市

scenes in the CWA (Wang, 2000: 327).

According to the master plan, different levels of the government were encouraged to carry out the redevelopment projects based on the paid land use rights policy. Various channels of funds were gained through real estate industries to provide the necessary funds for the redevelopment projects and basic urban infrastructure construction. Furthermore, it was a substantive way for local governments to generate local revenues (He, 2006: 87). The investment of real estate development in Xi'an urban areas increased from RMB0.433 billion in 1992 to RMB3.821 billion in 1998. The total revenue from the real estate industries in 1998 reached RMB1.195 billion. The increase in the real estate investment and the generated revenue are shown in Figure 4-7. In 1991, Xi'an's local authorities selected 49 blocks of low land to carry out the renewal projects, indicating the beginning of large-scale local land renewal in Xi'an. By 2007, 47 of these projects were finished (EBXY, 1993/2008). Nonetheless, in a market-driven situation, the renewal projects frequently turned out to be the exchange between urban redevelopment and economic profits. With such renewal activities, Xi'an's historic urban features began to disappear rapidly (He, 2002: 50).

4.2.1.5 Urban spatial restructuring and massive redevelopment-led "conservation" intentions (the early 2000s)

By the end of the 1990s, many urban

基础设施和历史城区环境的保护带来了压力(Chen et al., 2003：23)。第三，之前规划中不平衡的土地利用既和上述两个问题有关也和环境破坏有关。根据第三次总体规划，西安将其城市布局开发为一个中心城区，包括历史核心区和周围的11座卫星城镇（Zhang, 2002a：18）。这个特殊的城市布局的初衷是为了避免大规模的城市扩张。但是，由于缺乏一个有效的发展控制机制，卫星城并没有有效缓解历史核心区中人口拥挤带来的压力。事实上，这进一步加剧了历史城市中心的保护和发展之间的矛盾。结果，历史城市核心区被过度开发，南郊和北郊被大范围利用，而东郊和西郊却没得到有效开发。由于大规模的城市扩张，历史核心区陈旧的基础设施变的更加破败，城区环境被严重损害（He, 2003a, 2003b；Li, 2005；XCPB, 2007）。

1999年以来，中央政府制定了西部大开发的政策。因其是中国西北部的政治经济中心，西安在西部大开发工程中处于桥头堡的地位（Wang & Wei, 2003：305）。由于巨大的经济投资和快速城市化进程，大量人口开始涌入西安市区。到2003年，西安市的人口达到了510万，建筑面积约为405平方公里（SSB, 2007）。这些机遇和城市问题表明以前的城市发展模式已经不在适合当前市场主导的城市现状。

2004年底，西安地方规划局提出了第四次总体规划（2004—2020），并在2008年5月得到了国务院的批复（He, 2003b；SCC, 2008b）。该规划将西安的城市特色定义为：世界闻名的历史文化古都、旅游名城；中国重要的教育、科研、装备制造业、高新技术产

problems in Xi'an city development were mainly concerned with the existing inefficient urban structure, the imbalanced urban land use, the aged urban basic infrastructure, and environmental disruptions that have challenged and accounted for the disappearing historic capital features (Li, 2005: 106). Academics argued that three urban problems in the previous urban planning were outstanding. First, previous urban planning failed to provide enough space for the tertiary industry development in Xi'an (He, 2002: 50). In the composition of Xi'an's local economy, the tertiary industry has played a more and more significant role. Figure 4-4 shows the increase of Xi'an's GDP and the structure of GDP from 1952 to 2004. The ratio of the tertiary industry in the overall GDP increased from 23.39 percent in 1978 to 51.28 percent in 2004. In Xi'an, the local main tertiary industries include wholesale, retail, catering services, banking and insurance, real estate, education, social science, and other industries (XSB, 2005: 44). Second, previous historic core-based urban planning put great pressure on the conservation and development of the CWA. Today, the historic core of the CWA plays the multi-role of urban political, economic, cultural, and transportation centre. This situation put great pressure on the local aged urban infrastructures and the conservation of historic urban environments (Chen et al., 2003: 23). Third, the imbalanced urban land use in previous planning was related to the above two problems as well as to environmental disruptions. According to the third master plan, Xi'an developed its urban layout as a central urban area including the historic core and surrounding 11 satellite towns (Zhang, 2002a: 18). The initial rationale for this particular urban layout was to escape the massive urban sprawl. However, with the lack of capable development control mechanism, the satellite towns did not work efficiently to lessen the pressure caused by an overcrowded population in the historic core. In fact, it further intensified the contradictions between the conservation and redevelopment intentions in the historic core. As a result, the historic core was overdeveloped; the southern and northern suburban areas were extensively used,

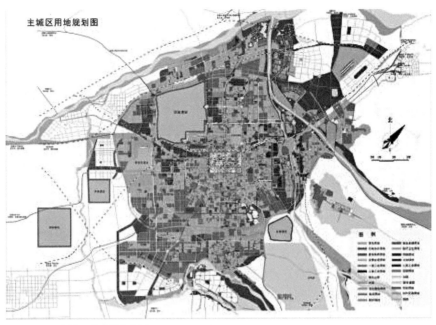

图4-8：西安第四次总体规划（2008—2020）
Figure 4-8: Xi'an Fourth Master Plan (2008-2020)
资料来源：(XCPB, 2005b)
Source: (XCPB, 2005b)

业基地和交通枢纽城市；新欧亚大陆桥中国段和西部及黄河中上游重要的中心城市；陕西省省会（He, 2004：13）。根据该规划，到2020年，西安中心（主）城区的人口将控制在528万人，建筑面积上限定为490km² (SCC, 2008b：1)。图4-8显示了第四次总体规划中的土地利用规划。

以市场为导向的规划和更新措施

随着我国西部大开发的机遇，西安所起的作用和城市地位得以在更广泛的区域经济范围内得到讨论，即西安的城市发展规模、城市空间分布、产业发展方向、城市交通以及城市保护问题。在解决现有的城市问题时，该规划着重强调两点。第一，该规划要符合市场经济的要求并对其产业布局做出调整。除了中心城区中现有的产业之外，要在附近

while the eastern and western suburban areas were not developed efficiently. Due to the massive urban sprawl, the aging basic infrastructure in the historic core was further challenged, and urban environments were seriously ruined (He, 2003a, 2003b; Li, 2005; XCPB, 2007).

Since 1999, the central government policy has advocated the development of the Chinese western regions. Considering its strategic location as the economic and political centre of northwest China, Xi'an was granted a leading role in the Western China Development Program (Wang & Wei, 2003: 305). With enormous economic investment and rapid urbanization speed, large numbers of population became drawn to the Xi'an urban areas. By 2003, the population in the Xi'an central urban area reached 5.10 million, while the built central urban area was about 405 km² (SSB, 2007). These opportunities and urban problems show that the previous urban development pattern did not fit the market-oriented urban reality in Xi'an anymore.

By the end of 2004, Xi'an's local planning

的卫星城建立大型产业。该规划还指出要营造良好的投资和空间环境来吸引大规模的装备制造业和其他知名企业（Zhang, Zhang, Feng, & Song, 2004）。第二，为保护历史核心区，规划提出重新调控旧城区内的土地利用现状，将从旧城区移出部分的城市功能，如旧城区的行政功能。为实现这一目标，该规划采用了所谓的九宫格局，该格局是根据周朝（公元前1046年）时的传统土地布局发展而来的（图4-9展现了九宫格局的功能结构规划示意图）。其他有关文献可见高娟和姜满年（Gao & Jiang, 2005），和红星（He, 2004），以及李红艳（Li, 2005）等学者的著作。

根据上述土地利用模式，旧城区将被保护重建为以旅游—商业为主的区域。作为总

authorities proposed its fourth master plan (2004-2020), which was approved by the State Council in May 2008 (He, 2003b; SCC, 2008b). This plan redefined Xi'an's urban features into those of a world famous historic city; a tourist city; an important base for scientific research, education, equipment manufacturing, and high-tech industries; a transportation junction city; the central city in western China, in the upper and middle reaches of Huanghe river, and in the Chinese section of Eurasia Land Bridge; and the capital of Shaanxi Province (He, 2004: 13). According to this plan, by 2020, the population in the central urban area would be restricted within 5.284 million, and the proposed built-up central urban area would be 490 km^2 (SCC, 2008b: 1). The land use plan is shown in Figure 4-8.

Market-oriented planning and regeneration policies

Obviously, because of the opportunity generated by the Western China Development Program, Xi'an was granted a chance to examine its role and position in a wider regional economic

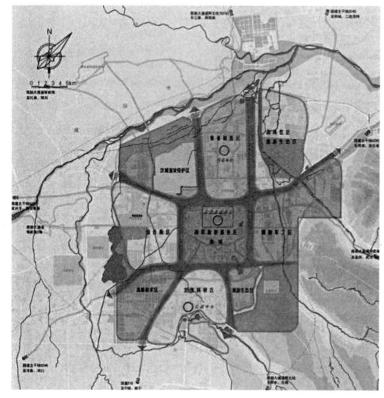

图4-9：西安中心城区的功能结构规划

Figure 4-9: The functional structure plan of Xi'an central urban area

资料来源：（XCPB，2005b：39）
Source: (XCPB, 2005b: 39)

体规划中的 29 个专项规划之一，西安唐皇城复兴规划和西安文化体系规划突出了该更新目标的主要特点。根据唐皇城复兴规划，我国杰出的文化和艺术将通过传统建筑设计项目和建筑形式表现出来。为实现这一目标，地方"保护"规划着重强调对现存历史文物古迹和历史街区进行保护，同时提议大范围的再开发项目应采用中国传统的建筑形式和景观（He，2004；XCPB，2005a，2005c）。

4.2.2 当前西安的城市保护条例

自 20 世纪 90 年代以来，西安市政当局制定了若干有关城市保护的法规，其中包括《西安城市建筑高度控制条例》（SCXPC，1993 年）和《西安历史文化名城保护条例》（SCXPC，2002 年）。这些规章都被新规划所采用。

西安城市建筑高度控制条例

为保护整体历史城市特色，1993 年西安市政府颁布了《西安城市建筑高度控制条例》。该条例提议在旧城区采用分区阶梯式来保护历史城市的轮廓线。该条例允许从城墙外以及其他标志性建筑为中心时，其他建筑高度可以逐渐增高。在城门、钟塔和鼓楼附近的区域规划建立传统风格的新房屋或其他建筑（SCXPC，1993；Wang，2000）。图 4-10 显示了旧城区中，从中心建筑（钟楼）到城墙之间的建筑高度控制线。但是，自 20 世纪 90 年代以来，迫于私营开发商的压力，城市发展已经从保护整个历史核心区转变为只保护了少数历史建筑和几条视线通廊，主要包括从市中心十字到四面的四个大门，或者到其他重要历史建筑的视线通廊（Wang，2000；

scale, that is, its urban development scale, urban space layout, industry development directions, urban transportation, and urban conservation. In addressing existing urban problems, this plan emphasized two perspectives. First, this plan catered to the requirements of the market-oriented economy and readjusted its industry layout. In addition to the existing industries in the central urban area, some new large industries would be introduced in nearby satellite towns. In order to attract large machinery manufacturing and renowned enterprises, the plan tried to create good investment and spatial environment (Zhang, Zhang, Feng, & Song, 2004). Second, to conserve the historic core, this plan proposed to restructure the functional land uses within the CWA and relocated some functions from this area, for example, the current administrative function. Accompanying this intention, the plan adopted a layout format known as *Jiu Gong Ge Ju*, which was developed based on the Chinese traditional land layout of the Zhou Dynasty (B.C. 1046) (see the functional structure plan of *Jiu Gong Ge Ju* in Figure 4-9). More discussions on this can be accessed at Gao and Jiang (2005), He (2004), and Li (2005).

According to this land use pattern, the CWA would be regenerated into a tourism-cum-commercial industries-based area. As one of the 29 special plans, Xi'an's Imperial Urban Regeneration of the Tang Dynasty and Xi'an City Cultural System plans would highlight this intention. According to the Plan of the Xi'an Imperial Urban Regeneration of the Tang Dynasty, outstanding Chinese literature and art would be exhibited through traditional planning schemes and building forms. To achieve this objective, the local "conservation" plan stressed the conservation of existing heritage monuments and historic precincts, while large scales of redevelopment projects in the Chinese traditional building forms and scenes were proposed (He, 2004, 2005; XCPB, 2005a, 2005c).

4.2.2 Existing urban conservation regulations in Xi'an

Since the 1990s, several regulations on urban conservation have been enacted by the Xi'an local authorities, including the Xi'an

图 4-10：1993 年旧城区中建筑高度控制线
Figure 4-10: Urban building-height control line within CWA in 1993
资料来源：（XCPB，2005a）
Source: (XCPB, 2005a)

图 4-11：2005 年旧城区的实际建筑高度和建筑轮廓线
Figure 4-11: Current situation of building-height and sky-line within CWA in 2005
资料来源：（XCPB，2005a）
Source: (XCPB, 2005a)

327）。图 4-11 展现了 2005 年旧城区实际的建筑高度和建筑轮廓线，表明了该条例的作用几乎完全失效。西安南大街的再开发项目是一个很好的例子，该街道早在 20 世纪 70 年代就开始进行再开发项目，经过 30 年的逐渐演变，这里几乎每栋建筑的高度都超过了该建筑高度控制线的要求（Zhang & Zhao，2003：29）。图 4-12 显示了 2005 年西安南大街从钟楼望向南门的效果。

随着中国 20 世纪 90 年代末颁布的有关历史街区保护的法律法规，西安当地也意识到，随着快速城市化，现有关于整体城市景观的保护法规并不能解决其城市保护问题，即不断消失的传统建筑和历史城市特色。在此背景下，西安市于 2002 年颁布了《西安历史文化名城保护条例》。

西安历史文化名城保护条例

该条例从三个层面描述了西安的城市保护策略。第一层，保护考古大遗址及其周围

Regulation on the Urban Building-Height Control (SCXPC, 1993) and Conservation Regulation on Xi'an Historic and Cultural City (SCXPC, 2002). These regulations were adopted by the new plan.

Xi'an regulation on the urban building-height control

To conserve its holistic historic urban features, the Xi'an municipal government announced in 1993 the Xi'an Regulation on the Urban Building-Height Control. Different building-height zones were proposed to preserve the historic urban skyline within the CWA. Building height was allowed to increase gradually from the City Wall and from other landmark buildings. New housing or other buildings in traditional style were planned in some areas around the gates and the Bell- and Drum-Towers (SCXPC, 1993; Wang, 2000). Figure 4-10 shows the section of the CWA from the urban centre—the Bell-Tower—and the building-height control line. However, since the 1990s, due to pressure from private developers, local development has shifted away from the policy of preserving the entire historic core to that of preserving only a few heritage buildings and several visual corridors running mainly from the central cross to the four gates and from each of the major heritage buildings (Wang, 2000: 327).

图 4-12：2005 年的西安南大街，从钟楼望向南门方向
Figure 4-12: Current view of Xi'an South Street (Note: Viewing from the Bell-tower toward the South Gate)

Figure 4-11 presents the actual building height and skyline within the CWA in 2005, which infers the total failure of this regulation. The redevelopment project of the Xi'an South Street is a good example. The redevelopment took place as early as the late 1970s. After three decades of gradual evolution, almost every building on this street today is much higher than the building-height control line (Zhang & Zhao, 2003: 29). Figure 4-12 shows the view of Xi'an South Street in 2005, as viewed from the Bell-Tower toward the South-gate.

With the promulgation of the conservation of historic districts in China in the late 1990s, Xi'an also realized that with rapid urbanization, existing regulations on holistic urban scenes could not solve its urban conservation problems, that is, disappearing traditional buildings and historic urban features. Against this background, the Conservation Regulation on Xi'an Historic and Cultural City was enacted in 2002.

Conservation regulation on Xi'an's historic and cultural city

This regulation depicts urban conservation strategies in Xi'an City in three levels. The first level of conservation involves the protection of archaeological sites and their precincts such as the Banpo Ruins (*ban po yi zhi*), Daming Palace Ruins (*da ming gong yi zhi*), and other sites. The second level includes the existing heritage monuments and their settings. These historic precincts express a holistic cultural significance such as the Wild Goose Pagoda (*da yan ta*), Eight Prophets Temple (*ba xian an*), and other areas. Surrounding environments are promoted to reflect and correspond with the existing heritage sites. The third level highlights the conservation of the city wall and the entire CWA. Comparatively, conservation strategies of the previous two levels are more explicit during the implementation because the historic monuments or archaeological sites physically remain. Currently, more discussions and controversies surround the conservation efforts at the third level (SCXPC, 2002: Article 28). The reality of urban conservation and development has witnessed the disappearance of many historic buildings and districts through the local

地段，如半坡遗址、大明宫遗址和其他遗址。第二层，保护现有的文物古迹及其周围环境。这些历史周围地区体现了整体的文化意义，例如大雁塔、八仙庵及其他地区。周围的环境要反映并符合现有的遗产古迹。第三层，保护城墙和整个城墙区。相比较而，由于历史文物遗迹和考古大遗址是以实物形式存在的，所以在实施过程中前两个层面的保护策略更明确。当前，更多的讨论和争议都是围绕第三层面的保护内容上（SCXPC，2002：第28条）。通过当地倾向于经济建设的再开发活动，在城市保护和发展实践中，很多历史建筑和街区都消失了。

通过对西安现有的保护法规的回顾，可以看到这些法规完全集中在对物质环境的保护，而没有关于历史环境中的地方原住民生活或传统活动的保护法规或条款。当前西安城市保护与发展过程中的一个实质性问题是，缺乏有效的规划和发展控制机制来监管当地再开发或保护项目的实施。例如，1993年，通过对西安城墙区内的传统建筑和历史遗迹的评估，西安市政当局[1]宣布了30个具有重大历史价值的建筑为历史建筑。但是，由于缺乏有效的保护管理和发展控制机制，到21世纪初时，这些历史建筑中70%以上的建筑已完全被拆除了，同时当地的原住居民也进行了重新安置。剩下的历史建筑也在城市更新过程中受到了不同程度的损坏（Bai, 2000：55; Zhang, Li, & Feng, 2007：217）。由于这些建筑未被划定为文物古迹，所以现有的国家级保护法律，如《文物保护法》对它们并不适用。同时，由于《西安历史文化名城保护条例》中没有针对地方传统生活的保护规定，加上也缺乏一个更加有效的城市保护和发展监管机制，近几年，在快速城市化和发展进程中，西安的历史街区也受到了很大影响（Zhang et al., 2007：215）。

4.2.3 西安的城市保护、发展控制和规划管理机制

城市保护和发展控制机制

自1978年改革开放政策实施以来，西安的城市规划和发展机构发生了很多变化。1979年废除了之前的西安城市建设局，成立了几个新机构，包括西安城市规划局、西安城市园林局以及西安市建设局（Zhang, 1988：71）。1979年，为了监督有效的城市规划，成立了几个专门部门，如西安城市规划局、西安房产管理局、西安文化遗产和园林局、西安公共事务局等。1980年，西安房产管理局更名为西安房地产管理局。1983年，西安基础建设委员会更名为西安城乡建设委员会，

---

[1] 西安市城市建设局、西安市城市规划局和西安市文物管理局。

economic-biased redevelopment activities.

The review of the existing conservation regulations in Xi'an shows that these regulations focus entirely on the physical environment, and there is no conservation regulation or items relating to the conservation of local indigenous lives or traditional activities in historic environments. Another substantive problem in the current urban conservation and development process in Xi'an is the lack of an effective planning and development control mechanism to monitor the implementation of local redevelopment or conservation projects. For instance, in 1993, after the assessment of traditional buildings and historic monuments within the Xi'an CWA, 30 historic buildings with high value were declared historic buildings by the Xi'an municipal authorities[1]. However, due to the lack of efficient conservation management and development control system, more than 70 percent of these buildings were totally demolished and original local residents were relocated. The rest of the 30 historic buildings were damaged in different degrees due to the local renewal programs (Bai, 2000: 55; Zhang, Li, & Feng, 2007: 217). Considering that these buildings are not heritage monuments, the available conservation laws at the national level for example, the Cultural Relics Conservation Act (*wen wu bao hu fa*), are not applicable to them. Nevertheless, due to the lack of conservation efforts for local traditional lives in the local Conservation Regulation on Xi'an Historic and Cultural City, which actually should be the indispensable parts in an integral conservation program, and the lack of an effective urban conservation and development management mechanism, the historic districts in Xi'an have been severely affected in the rapid local urbanization and development in recent years (Zhang et al., 2007: 215).

4.2.3 Urban conservation, development control, and management mechanism

Urban conservation and development control mechanism

Since the open door policy in 1978,

---

[1] Xi'an Urban Construction Bureau, Xi'an Urban Planning Bureau and Xi'an Cultural Heritage Bureau.

主要负责协调规划、交通运输、房地产行业、园林及其他方面事务。作为一个市政职能部门，它还负责城市基础设施和重要项目的建设与管理（XCUCRC, 2000：69）。一般而言，目前西安历史核心区的更新机制是以地方政府为主导、自上而下的实施与管理为特征。随着1988年土地改革的实施，西安所有城市土地归国家所有，并且在私人开发商需要进行项目时发行了土地使用许可证。西安国土资源局代表西安市政府管理西安的土地使用。所有县、区有他们自己的土地管理机构。根据陕西省土地资源局，不同级别的土地管理机构，基于其区域大小，行使权力也不同。1988年，西安国土资源局为土地使用许可证颁布了专门程序（Li, 2002：36）。

《中华人民共和国土地管理法》规定：下列建设用地，经县级以上人民政府依法批准，可以通过划拨方式取得：军事用地、城市基础设施用地、公益事业用地等，其他项目必须通过土地招标和拍卖的方式来取得土地使用权（SCNPC, 2004a：第54条）。为了使城市发展符合既定规划目标，1989年《城市规划法》制定了可自由支配的"一书两证"的方式来控制城区的发展（Ng & Wu, 1995：286）。《西安市城市规划管理条例》规定，所有土地的使用应该遵循西安总体规划和相应的各区规划和详细规划。参与其中的开发商，包括私人开发商和国家开发商，在建设之前，应该得到从西安市城市规划局（XUPB）发布的"一书两证"，指的是开发项目选址申请书（SSPDP）、建设用地规划许可证（PPCL）和建设工程许可证（PPCP）（Li, 2002：36）。开发项目选址方案划定一个区域的位置、大

many changes in Xi'an's urban planning and development institutions have taken place. In 1979, the previous Xi'an Urban Construction Bureau was abolished. Several new institutions were set up, including the Xi'an Urban Planning Bureau, Xi'an Urban Gardening Bureau, and Xi'an Municipal Construction Bureau (Zhang, 1988: 71). In 1979, several special departments were established to oversee a more effective urban planning, such as the Xi'an Urban Planning Bureau, Xi'an House Property Management Bureau, Xi'an Cultural Heritage and Gardening Bureau, Xi'an Public Affairs Bureau, and others. In 1980, the Xi'an House Property Management Bureau was renamed to Xi'an Real Estate Management Bureau (XREMB). In 1983, the Xi'an Basic Construction Committee was renamed to the Xi'an Committee of Urban and Rural Construction (XCURC). It is mainly responsible for the coordination among planning, transportation, real estate industries, gardening, and other related perspectives. As a municipal implementation agency, it is also responsible for the construction and management of municipal infrastructure and important projects (XCUCRC, 2000: 69). Generally, the current urban regeneration mechanism in Xi'an's historic core is characterized by local government-led top-down implementation and management. With the land reform in 1988, all urban lands in Xi'an have become state-owned, and Land Use Permit is issued when private developers need to conduct any projects. The Xi'an National Land and Resources Bureau (XNLRB) represents Xi'an's municipal government in administrating the land use in Xi'an. All counties and districts have their own land administrative agencies. According to the Land and Resources Bureau of the Shaanxi Province, different levels of land administrative agencies have different sanction rights based on the size of plots. In 1988, XNLRB set special procedures for granting the Land Use Permit (Li, 2002: 36).

The Land Administration Act of the P.R. China specified that except for several types of projects, namely, military purposes, municipal infrastructures, public welfare purposes, and other concerns that can gain the land use

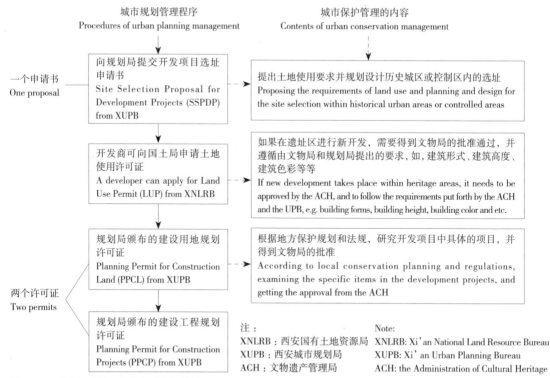

Figure 4-13: Procedures of urban planning management and contents of urban conservation management
资料来源：作者汇总自（Li, 1999：136–138；Wang et al., 1999：73）
Source: Author revised from (Li, 1999: 136–138; Wang et al., 1999: 73)

小和边界，然后开发商向西安国有土地资源局（XNLRB）申请土地使用许可证。在得到土地使用许可证之后，开发商可向西安市城市规划局申请建设用地规划许可证和建设工程许可证。图4-13显示了城市规划管理的过程。

上述过程展示了西安城市更新或再开发实践中城市规划管理的一般程序。考虑到当地的具体城市背景，还要遵循特殊地段的规划和建设管理程序。例如，西安历史街区中的地方居民大量自建活动，如鼓楼回民区和三学街区，则必须遵循当地的具体要求和有关程序。第一，当地房产公司或私人住户要向西安市城市规划

rights through allocation by the administrative governments above the county level, all the other projects need to undergo the land tender and auction processes to acquire the land use rights (SCNPC, 2004a: Article 54). To steer the development process to conform to the planning intentions, the 1989 City Planning Act stipulates the discretionary "one report and two permits" to control the development in China's urban areas (Ng & Wu, 1995: 286). According to the Xi'an Urban Planning Management Regulations, all land uses should follow the Xi'an master plan and corresponding district and detailed plans. All concerned developers, including both private and public developers, should get "one proposal and two permits" from the Xi'an Urban Planning Bureau (XUPB) before construction. This refers to the Site Selection Proposal for Development Projects (SSPDP), Planning Permit for

局提交建设申请书和施工的初步设计蓝图。如果审批通过,则可以进一步完善初步设计蓝图。第二,提交项目技术和实施蓝图,来申请施工许可证。第三,在研究了实施蓝图之后,西安市城市规划局发布执行证明,这意味着该项目应该在西安市城市规划局的监督检查下进行。最后,在交到相应的产权所有者之后,还要从西安市房地产管理局获得房产证(XCUCRC, 2000:132;XMHDPPO, 2003:76)。图 4-14 展示了整个过程。

Construction Land (PPCL), and Planning Permit for Construction Projects (PPCP) (Li, 2002: 36). In the SSPDP, the location, size, and boundary of a site are delineated. Developers will then apply for the Land Use Permit (LUP) from the Xi'an National Land and Resources Bureau (XNLRB). After the granting of the LUP, developers can apply for the PPCL and PPCP from XUPB. As underlined in the Xi'an Urban Planning Management Regulations, the purpose of issuing the permits is "to ensure that land use complies with the urban planning criteria and to protect those construction units that do comply with the legal requirements" (Li, 1999: 137). The procedure on urban planning management is shown in Figure 4-13.

This procedure illustrates the general process of urban planning management during the urban renewal or redevelopment practices in Xi'an. Given the local specific urban context, there are also other special planning and construction management procedures to follow. For instance, the many self-construction activities in the Xi'an historic districts by the local residents themselves, such as the Drum-Tower, Muslim District, and the Sanxuejie Historic District, have to follow explicit requirements and procedures. First, the local real estate office or individual tenants need to submit the application form together with the preliminary construction design blueprints to XUPB. If approved, the extended preliminary design may be carried out. Second, the project's technical and implementation blueprints should be submitted to apply for the implementation certificate. Third, after an investigation of the implementation blueprints, the implementation certificate is issued by the XUPB, and this entails that the project should be conducted within the supervision and inspection of the XUPB. Finally, after the handover to the corresponding property owners, it is necessary to obtain the house property certificate from the XREMB (XCUCRC, 2000: 132; XMHDPPO, 2003: 76). The entire process is presented in Figure 4-14.

As regards the urban conservation and development of historical urban areas and historic cities, it is under the responsibility of the local Urban Planning Bureau (*gui hua ju*), local Urban Construction Bureau (*cheng jian ju*), and

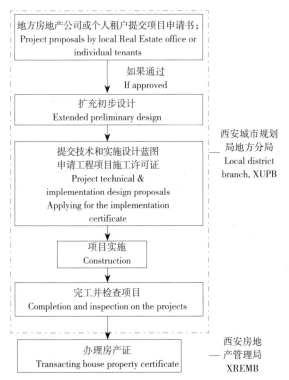

XREMB: Xi'an Real Estate Management Bureau;
XUPB: Xi'an Urban Planning Bureau.

图 4-14:西安历史街区中的城市规划和建设管理机制
Figure 4-14: Urban planning and construction management mechanism in Xi'an historic districts
资料来源:(XCUCRC, 2000:132;XMHDPPO, 2003:76)
Source: Author synthesized from (XCUCRC, 2000: 132; XMHDPPO, 2003: 76).

在当前我国的行政和管理体制下，城市规划局、城市建设局和文物局负责历史街区和历史城市的保护与建设工作。通常，在当地城市建设与更新过程中，文物局负责文物古迹的保护和管理，城市建设局负责历史建筑的保护和管理，而规划局负责当地历史建筑与历史城区的规划和管理。任何在历史建筑、历史城市或控制城区边界进行的城市规划或开发，都应由城市规划局、文物局和城建局共同负责（Wang et al., 1999：72）。图4-15展示了中国历史文化遗产保护行政主管机构体系（Wang et al., 1999：110）。

根据历史城区的有关保护和管理条例，在历史城区进行开发和建设活动时，私人开发商需要遵循一些具体的要求。例如，在遗

local Municipal Administration of Cultural Heritage (*wen wu ju*) under the current administration and management system of China. Normally, in the local urban development or regeneration processes, the Administration of Cultural Heritage is responsible for the conservation and management of heritage monuments, while the Urban Construction Bureau is responsible for the conservation and management of historical buildings. In addition, the Urban Planning Bureau is in charge of the planning and management of local historical buildings and urban areas. If there is any urban planning or development taking place along the boundary of historical buildings, historic cities, or controlled urban areas, it should be under the dual responsibility of the Urban Planning Bureau with either the Administration of Cultural Heritage or the Urban Construction Bureau (Wang et al., 1999: 72). The hierarchical administration and management system in the urban conservation and development in China is illustrated in Figure 4-15 (Wang et al., 1999: 110).

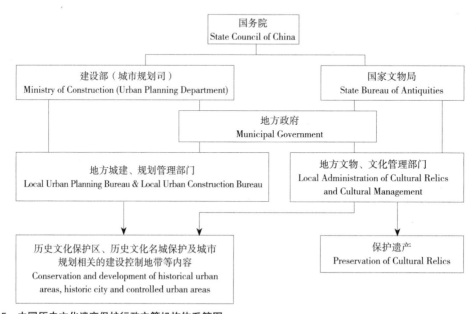

图4-15：中国历史文化遗产保护行政主管机构体系简图
Figure 4-15: Hierarchical administration and management system in the urban conservation and development in China
资料来源：（Wang, Ruan, & Wang, 1999：110）
Source: (Wang, Ruan, & Wang, 1999: 110)

址区开展的城市规划或开发活动要获得文物局的批示。另外，新开发或建设活动要遵循文物局和城市规划局制定的有关要求或标准，例如建筑形式、建筑高度、建筑色彩以及其他标准。在城市规划管理过程中，私人开发商要遵循的其他一些要求如（图4-15）所示。

城市管治模式

根据1989年《中华人民共和国城市规划法》第21条，我国的城市规划实行分级审批。"编制分区规划的城市的详细规划，除重要的详细规划由城市人民政府审批外，由城市人民政府城市规划行政主管部门审批。"（SCXPC，1990）。因此，包括城区内的每个建设项目的土地使用范围、建筑密度和建筑高度的控制指标、所有项目的城市布局在内的详细规划都在地方政府的监督下。另外，当地方政府积极干预地方企业活动时，政府和企业之间的关系就表现出和其他城市相似的特点。管理单位和企业，或者裁判和运动员的角色，有时变得难以区分。例如，在西安政府管理体制中，存在不少地方政府公司化现象。在再开发过程中，这种现象类似于西方国家的公私合营关系模式。根据程（Cheng）（1997）的说法，有些地方政府的权利过于集中，而在其他领域，权利则显得过于分散。例如，原本应该由相应政府部门拥有的管理职能，却被分散到各个单位。同时，原本属于企业的一些权利，却被政府部门接管了（Cheng，1997：512）。历史城区的保护过程还反映了地方政府倾向于和私营企业建立密切的合作关系趋势。王涛（Wang，2004：67）指出,考虑到历史街区保护中的经济权衡,

According to the regulations in the conservation and management of historical urban areas, private developers have to follow several requirements when the proposed development or construction projects are located within historical urban environments. For instance, if urban planning or development takes place within heritage areas, it needs to be approved by the Administration of Cultural Heritage. In addition, the new development or construction should follow the requirements or criteria set forth by the Administration of Cultural Heritage and the Urban Planning Bureau, for example, building forms, building height, building color, and other criteria. Some other requirements for private developers to follow during the processes of urban planning management are shown in Figure 4-15.

Mode of urban governance

According to article 21 of the 1989 City Planning Act, city planning in China is endorsed in different levels. "The detailed planning of cities with district planning is to be submitted to the planning administrative authorities of the city people's governments for examination and approval, with the exception of important detailed planning which should be submitted to the people's governments of the cities for examination and approval (SCXPC, 1990)." Hence, the detailed planning, which includes the land use scope for each construction project, control indexes for building density and building height, general layout of all items to be built within the urban areas, and other concerns are under the entire supervision of the local governments. Furthermore, the relationships between the Xi'an government and the enterprises also exhibit similar characteristics as of the other cities when local governments actively become involved in the enterprises' businesses. It is often difficult to cut clearly between the management units and enterprises or between the judge and the players. For example, in the Xi'an government management system, the condition of local-state-corporatism largely exists. In the redevelopment processes, it is similar to the mode of public-private-partnerships in Western countries. According to Cheng (1997), Xi'an local governments possess over-concentrated

很多地方政府在实施保护政策和追求经济效益之间徘徊，这往往造成既定保护政策和规划的搁置。这也突显了在地方城市的保护与开发过程中建立一个有效的发展监管机制的重要性。

上述讨论表明，在当前西安城市规划和更新体制中，几个市政机构（即，西安城市规划局、西安国土资源局、西安城乡建设委员会以及西安城市房地产管理局）在提出规划、管理土地使用权、协调并监控再开发或保护项目中起主导作用。值得注意的是，在这些过程中尚未纳入地方居民参与的环节。正如《中华人民共和国城市规划法》所指出的，城市规划一旦获批，应该由人民政府进行颁布。其目的是让公众更好地了解该规划，然后"参与"实施过程，并对规划进行"监督"（Ng & Wu, 1995: 288）。

宗教社团的管理

少数民族团体经常涉及专门的宗教事务。例如，所有回民都是穆斯林教徒，因此，对西安少数民族事务的管理，很大程度上是通过对宗教组织及其活动的管理来进行。西安民族事务委员会和宗教事务管理局的人员是一样的。这种情况在现行行政体系中被称作"一套人马，两块牌子"。在西安，最低级别的民族宗教事务委员会属于区级。例如，莲湖区政府和碑林区政府拥有各自的民族宗教事务委员会。关于宗教组织及其活动的管理，西安市的现行法律文件是2005年3月开始执行的《宗教事务条例》（下文简称《条例》）。《条例》规定，信教公民的集体宗教活动，一般应当在经登记的宗教活动场所内

powers in some areas, while in other areas the power is over-scattered. For instance, the management functions that should be under the corresponding institutions are actually scattered to various units. On the other hand, some rights that should belong to enterprises have been taken over by the government authorities (Cheng, 1997: 512). The tendency of local governments to build close relationships with private sectors to pursue economic interest is also reflected in the conservation of historic urban areas. Wang (2004: 67) pointed out that considering the economic tradeoffs in the conservation of historic districts, many local governments are confronted with the challenge of conducting conservation policies and pursuing maximum economic returns, which often puts aside the established conservation policies and regulations. This highlights the importance of an effective development control mechanism in local urban conservation and development processes.

The above discussion indicates that in the current Xi'an urban planning and regeneration mechanism, several municipal agencies (i.e., Xi'an Urban Planning Bureau, Xi'an National Land and Resources Bureau, Xi'an Committee of Urban and Rural Construction, and Xi'an Urban Real Estate Management Bureau) play a dominant role in proposing plans, managing land use rights, and coordinating and monitoring the implementation of redevelopment or conservation projects. Notably, in all these processes, there is no provision for the participation of local communities. As indicated in the City Planning Act, city plans, once approved, should be promulgated by the people's governments. The objective is to let the general public have a better understanding, and hence "participation," in the implementation and "monitoring" of the plans (Ng & Wu, 1995: 288).

Management of religious groups

Minor ethnic groups in Xi'an often relate to specific religious affairs. For instance, all Hui ethnic people are Muslims. Thus, the management of affairs of minor ethnic residents in Xi'an is largely conducted by managing

举行[1]（SCC，2005b：第13条）。而且，信教公民要建立相应的宗教管理组织，如穆斯林团体清真寺管理委员会。这些宗教管理组织中的成员要在地方信教公民中选举。这些成员应该在当地宗教活动场所所登记的政府管理单位进行登记（SCC，2005b：第16-17条）。在实践中，宗教管理组织负责人事、财务、安全、宗教和其他事务（SCC，2005b：第18条）。

根据地方行政管理部门的分配，当地信教居民及其组织也由相应社区居民委员会进行管理。例如，西安鼓楼回民区中的穆斯林居民受专门的清真寺管理委员会和社区居民委员会管理。因此，我国的民族和宗教团体往往由双重管理体系管理，每个管理体系负责不同的事务。相比地方社区居民委员会，当地居民和清真寺管理委员会的联系更加频繁。该组织负责与当地少数民族和信教居民更加密切相关的宗教事务。此外，在地方城市更新项目的实施过程中，有些因素也会影响地方宗教团体与地方政府之间的联系。这些都影响了当地宗教团体和政府部门之间联合型社会资本的积累。后面的案例分析进一步阐述了这个问题。

4.2.4 小结

通过对西安城市规划和发展建设的回顾，可以看到其历史核心区的城市发展与更新政策发生了巨大演变。1949年中华人民共和国

---

[1] 《条例》第十三条规定：筹备设立宗教活动场所，由宗教团体向拟设立的宗教活动场所所在地的县级人民政府宗教事务部门提出申请。宗教事务部门对设立其他固定宗教活动处所的，作出批准或者不予批准的决定。设区的区（县、市级）人民政府宗教事务部门负责该区宗教活动场所的管理（SCC，2005b）。

religious activities and organizations. The Xi'an ethnic affair committee and bureau of religious affairs comprise the same people. This situation is known as the "one group of crew with two administrative boards" (*yi tao ren ma, liang kuai pai zi*) in China's administration system. In Xi'an, the lowest hierarchy in the ethnic and religious affair committee is at the district level. For instance, the Xi'an Lianhu District Government and the Xi'an Beilin District Government have their respective ethnic and religious affair committees. Regarding the management of religious activities and organizations, the only legislative document in Xi'an City is the Regulation on Religious Affairs (*zong jiao shi wu tiao li*) (thereafter, the *Regulation*), which was enforced on March 1, 2005. The *Regulation* requires a local religious activity venue (*zong jiao huo dong chang suo*), such as mosques for Muslims, to be registered with the proper local governmental administration unit[1] (State Council of China, 2005b: Article 13). Moreover, religious residents are required to set up a corresponding religious administration organization, such as the Mosque Management Committee of Muslim groups. Staff in such religious administration organizations need to be elected from among local religious residents. The elected staff should also be registered at the same local governmental administration unit where the local religious activity venue is registered (State Council of China, 2005b: Article 16-17). In practice, the religious administration organization is responsible for personnel, finance, security, religion, and other affairs (State Council of China, 2005b: Article 18).

Local religious residents and organizations,

---

[1] As laid out in Article 13 of the *Regulation*: "If planning to establish a venue for religious activity in a district (county or city) within a directly administered municipality, an application may be put forth directly to the district (county or city) religious affairs bureau in that municipality. This religious affairs bureau can approve preparations to establish other fixed places for religious activity. District (county, city) religious affairs bureaus in the directly administered city are the offices in charge of managing the registration of venues for religious activity in the area" (State Council of China, 2005b).

成立之后，根据国家政策，西安的城市规划和发展偏向工业发展。在政治动荡的"文化大革命"时期，由于缺乏城市规划的指导，西安的城市发展和建设对历史环境造成了毁灭性的破坏。随着1978年中国政治改革的进行，尽管西安第二次总体规划还是比较强调以工业发展促进地方的经济增长，但已经提出了对西安历史古都特色的保护，并促进当地旅游业的发展。为了促进当地旅游业的发展，旧城区中既定的保护规划需要做出让步，实施了诸如酒店、饭店及其他服务行业项目。随着20世纪80年代末土地市场的出现，私人开发商积极地开展了旧城区的改造再开发活动。这些项目变成了土地使用权和经济利益的简单置换，也是地方政府增加其财政收入的重要途径。这种情况严重损害了旧城区的保护目标。考虑到以前城市发展模式所带来的种种城市问题，以及西部大开发所带来的宝贵机遇，拥有重要战略位置的西安在我国西部大开发过程中开始处于桥头堡的重要地位。西安市政府相应地于2004年提出了其最新的城市总体规划，为了适应经济市场的需求，决定对城市空间和土地使用方式进行重组。为了改善城市物质环境，并为资本投资创造机遇，第四次总体规划强调通过保护历史文物古迹以及开展大规模的再开发活动来更新修复传统的城市环境。西安城市再开发监管和控制机制表明，地方政府在制定规划、项目实施和管理过程中发挥出至关重要的作用。在当前的城市管治模式中，地方政府倾向于和私营部门建立密切的关系，并从更新项目中获取最大的经济回报。表4-1列出了西安不同时期总体规划和保护策略的特

according to local governmental administrative divisions, are also managed by corresponding Community Residents' Committee. For instance, the Muslim residents in Xi'an Drum Tower Muslim District are managed by both particular Mosque Management Committees and specific Community Residents' Committees. Therefore, ethnic and religious residential groups in China are often under double administration systems. Each administration system is responsible for different affairs. Comparatively, local ethnic residents tend to have more contacts with the religious administration organization, which is responsible for religious issues essential to local ethnic and religious residents, rather than the local Community Residents' Committee. Moreover, by implementing local urban regeneration projects, several factors further compromise the relationships between local religious groups and local government. These severely damage the bridging social capital between local religious groups and governmental agencies. This will be elaborated further in later case studies.

4.2.4 Summary

In the overview of urban planning and urban development in Xi'an, urban development and regeneration policies in its historic urban core are shown to have experienced substantive evolution. After the establishment of P.R.China in 1949, the orientation of the urban planning and development in Xi'an was partial to the industrial development to be in accordance with the national policy. During the Cultural Revolution, which was a period of political struggles, Xi'an's urban development and construction caused devastating damage to the historic environments due to the lack of direction in urban planning. With China's political reform in 1978, although industrial development was still emphasized as the means to promote local economic growth in Xi'an's second master plan, the conservation of Xi'an's historic capital features was put forth, and the tourism industry was promoted. In order to promote the local tourism industry, the established conservation plans in the

点。通过对西安城市规划和建设发展的历史回顾，人们可以看出，尽管原住民是地方历史住区保护中不可或缺的一部分，但他们目前尚缺少正式渠道可以参与到更新项目过程中。下一章，本研究将探讨有助于促进原住民的生活内容保护和历史街区更新的社会资本问题。

CWA gave way to such big projects as hotels, restaurants, and similar undertakings. With the emerging land-market in China in the late 1980s, private developers were encouraged to take part in urban redevelopment projects within the CWA. As a result, these projects became the trade-off between land use rights and economic returns. This turned out to be the significant way for local governments to generate their revenues. This situation severely compromised the conservation objectives within the CWA. The urban problems from previous urban development patterns and the opportunity of China's Western Development Program granted the strategically located Xi'an a leading role in the process. Consequently, Xi'an's municipal government proposed the latest master plan in 2004 to restructure its urban spaces and land uses to cater to an economic market. To improve the physical urban environments and create opportunities for capital investment, the fourth master plan emphasized the regeneration of traditional urban scenes through the conservation of limited heritage monuments and massive redevelopment activities. As indicated in Xi'an's urban development control and management mechanism, Xi'an's local governments played a critical role in making plans, projects implementation, and management. In the current mode of urban governance, local governments tend to build close relationships with private sectors in the pursuit of maximum economic returns from local regeneration projects. The characteristics of the master plans and conservation strategies in different periods in Xi'an are shown in Table 4-1. From the historical overview of Xi'an's urban planning and development, it is clear that the local community was not a part of the official regeneration plans, even if local indigenous residents should be an indispensable component of local historic residential districts. In the next part, this book will explore the issue of social capital in facilitating the conservation of local indigenous lives and the regeneration of historic districts in Xi'an.

表 4-1
Table 4-1

西安不同历史阶段的总体规划和保护策略的主要特点
Characteristics of the master plans and conservation strategies in different periods in Xi'an

时期 Period	项目 Item	内容 Contents	特点 Characteristics	城市发展中的问题 Problems in urban development	对策 Countermeasures
城市规划和保护, 1953~1972 Urban Planning and Conservation, 1953—1972		1) 道路系统产区传统的网格模式布局 2) 保护历史古迹不受城市建设的破坏 3) 在历史古迹和周围建设公园 4) 保护历史和重大遗产以提升城市内涵 1) road system adopting traditional grid-pattern layout 2) historical ruins and remains are protected from urban construction 3) parks being located at the precincts of historic monuments and remains 4) historic and significant heritages are conserved to enhance city essence	● 中央计划经济 ● 保护古长安不受城市建设的破坏的规划传统，采用唐朝网格道路模式，强调保护利用历史建筑。 ● centrally planned economy ● conserving the planning traditions of ancient Chang'an, absorbing the characteristics of grid-pattern road system of the Tang Dynasty, stressing the preservation and utilization of historic buildings	● 对历史城区附近的缓冲区关注不够 ● 历史遗迹周围不合理的城市建设 ● 侵犯并占用了公共绿地和公园 ● not enough attention given to the buffer zones in the historic vicinity ● inappropriate construction conducted around historic monuments ● encroaching and occupying public green space and parks	● 城市保护和经济发展的有机结合 ● 对新建筑制定的详细设计和建设控制定原则和条例 ● 为改善居民区的道路系统提供建议 ● 强化绿化建设 ● 专业机构和专家之间建立合作关系 ● organic combination of urban conservation with economic development ● principles and regulations in the detailed design and construction of new buildings ● suggestions for improving the branch-ways and road systems in residential districts ● intensifying the greenery construction ● setting up partnerships of specialist institutions with professionals
城市规划和保护, 1980~2000 Urban Planning and Conservation, 1980—2000		1) 保护体现唐朝繁荣景象的明城的整体城市布局，保护周、秦、汉和唐朝的历史遗迹 ● 通过划定区对历史遗产进行分类和归类 ● 保护整个城墙区 2) 遗产保护和利用相结合——建设遗迹公园和园林 3) 遗产保护和建设园林促进旅游产业相结合 ● 修复几个历史名胜古迹区 1) conserving the integral urban layout of Ming City, expressing the magnificent dimensions of Tang City, and protecting the historical ruins of Zhou, Qin, Han, and Tang dynasties ● classification and categorization of historic heritages by zone ● conserving the entire CWA 2) combining heritage conservation with utilization—building relic parks and gardens 3) combining heritage conservation and urban gardens with tourism industries ● restoring several historic landscapes resorts areas	● 从中央计划经济向市场经济的转型期 ● 明确西安城市特色和保护古都特色：保留道路系统的历史遗产的特点和对称的城市布局，并协调历史围区的建设工程。 ● 保护项目倾向旅游业的发展 ● transitional period from the centrally planned economy to the market-oriented one ● clarifying Xi'an urban features vs. conserving ancient capital characteristics: maintaining the characteristics of grid-pattern road system and the symmetric urban layout, and coordinating the construction in historical vicinities ● conservation programs geared towards the tourism industry	● 整个历史环境中的协调不够 ● 在一些标志性的历史遗迹周围没有足够的缓冲区 ● 破坏了城墙区内的空中轮廓线 ● 历史城市特色逐渐消失 ● not enough coordination within the entire historic environments ● not enough buffer zones around some iconic historic monuments ● destruction of the skyline within the CWA ● disappearing historic urban features	

续表
Continued

时期 Period	项目 Item			
	内容 Contents	特点 Characteristics	城市发展中的问题 Problems in urban development	对策 Countermeasures
城市规划和保护，1995～2010 Urban Planning and Conservation, 1995—2010	1) 保护体现明朝繁荣景象的整体城市布局，保护周、秦、汉和唐城模式和宏观环境角度的历史遗迹 • 在城市模式和宏观环境角度的历史城市特色 • 通过和唐朝著名的遗产古迹相结合，改善旅游景观 2) 城市保护、再开发利用和建设 • 协调城墙区以外和远离重要的历史遗产区，体现西安远郊区（例如，西郊的高新区） 21 世纪的现代化城市的城市特色 1) conserving the integral urban layout of Ming City, expressing the magnificent dimensions of Zhou, Qin, Han, and Tang Dynasties • conserving historic urban features in terms of urban pattern and macro environments • improving landscape tourism through linking the remarkable heritage sites of Tang dynasty 2) combining urban conservation, redevelopment, and utilization • coordinating urban conservation and construction • coordinating construction with historic environments within the CWA and around historic monuments • expressing Xi'an's urban features as a modern city of the 21st century outside the CWA and away from important historic heritage zones (e.g., the western high-tech zone)	• 市场经济 • 城市保护和更新规划中的全面考虑 • 强调保护和历史街区，并提出"把旧街区和新开发区结合"的城市发展策略 • market-oriented economy • comprehensive concerns in urban conservation and regeneration plans • stressing the conservation of the urban historic districts and proposing the urban development strategy of "Separating the Old Districts from the New Development"	• 整个历史城区的协调管理和控制力度不够 • 对历史环境造成了建设性的破坏 • 缺少绿化面积 • not enough coordination and controlling forces for the holistic urban environment • constructive destruction of historic environments • scarce greenery areas	• 城市保护和经济发展的有机结合 • 对新建筑式的详细设计和建设调控原则和条例为改善居民区的道路系统提供建议 • 强化绿化建设 • 建立机构和专家之间紧密关系 • organic combination of urban conservation with economic development • principles and regulations in the detailed design and construction of new buildings • suggestions for improving the branch-ways and road systems in residential districts • intensifying the greenery construction • setting up partnerships of specialist institutions with professional
城市规划和保护，2008～2020 Urban Planning and Conservation, 2008—2020	1) 规划并保护明城墙的整体城市布局和特色 2) 体现了唐朝长安城的繁荣景象，并提出对重大历史遗迹的保护规划 3) 提出对特定历史用地区，如四大古都遗址的保护规划 4) 控制城墙区的城市规模，并保护开发历史街区，包括附近的河流和湖泊 5) 保护西安市周围的水城系统，维护城市远景景观 6) 控制城墙区的建筑高度，体现城市轴线 7) 保护和开发唐朝网格式街道景观 1) planning and conserving the integral urban layout and features of Ming City 2) expressing the magnificent dimensions of Chang'an City of the Tang dynasty and proposing the conservation plan of important historic remains 3) proposing conservation plans on specific historic precincts, e.g., the Four Ancient Capital Ruins 4) controlling the urban scale of the CWA and conserving and developing the urban axis 5) conserving the water systems around Xi'an City including nearby rivers and lakes 6) controlling the building heights within the CWA and maintaining the urban vista corridor 7) conserving and developing the grid-pattern street layout	• 提出"强调保护和拯救濒危历史遗产"的原则，综合历史资源，提出功能分区的原则 • 提出"真实性和完整性相结合"的原则，继承传统布局，强调重点保护区 • 提出"保护和利用相结合"的原则，重建仿古建筑 • proposing the principle of "Stressing Conservation and Rescuing Endangered Historic Heritage," integrating historic resources and proposing the principle of functional zoning • proposing the principle of "Uniting Authenticity with Integrity;" inheriting traditional layouts and underscoring key conservation areas • proposing the principle of "Uniting Conservation with Utilization," redeveloping fake antique buildings		• 加强了宣传和立法程序 • 加强了不同部门间的协作，以及有关法规的实施 • 融合文物审批城市基础设施建设的职能 • 加强历史街区的保护与再利用 • intensifying the promulgation and legislation • intensifying the coordination of different sectors and the implementation of corresponding regulations • integrating the role of the Administration of Cultural Heritage in the sanctioning the process of infrastructure development • intensifying the conservation and utilization of historic district

资料来源：作者汇总自（He, 2002；XCPB, XACH, andXUPDI, 2005）
Source: Authorsynthesized from He (2002) and XCPB, XACH, and XUPDI (2005)

4.3 西安城市更新中的社会资本

上文对西安城市的发展规划与策略的历史回顾表明,在城市更新实践中,无论是再开发还是保护政策都强调物质环境的改善和经济发展。在保护地方非文化遗产方面,由于缺乏必要的法规,研究发现,在西安历史城市的更新过程中,地方原住民的生活内容与传统很多时候还未引起人们的足够重视。接下来,本文将探讨基于地方社区联系或邻里关系的社会资本,在促进社区参与到城市更新过程中所起的作用。本文将从网络、社会规范和信任的角度,来探讨社区成员之间的联系(聚合型社会资本)、同时也将探讨地方社区和其他机构之间的联系(联合型社会资本)。因此,根据在城市更新过程中社会资本建立的位置,可以在两个方面对其进行研究。聚合型社会资本与现存的社区居民之间的关系和联系有关。而联合型社会资本则与在城市规划和更新项目中的地方社区参与有关。鉴于传统商业活动在历史街区中的重大作用,本文针对地方经济形势有较大影响的商业网络关系进行了研究。

影响上述关系的因素有很多。根据对社会关系所在地的分类,本研究把这些因素归为两大类,即内部因素和外部因素。内部因素是指影响社区成员之间联系和关系的因素,主要是社会经济条件(例如,年龄、经济差距、就业类型、教育水平和宗教背景)、地方依附、宗教和民族问题以及其他条件。外部因素是指影响地方社区和其他政府机构之间关系的因素,主要是指社会经济条件(例如,非政府组织与其他社会组织)、更新策略、更新管

4.3 Social Capital in Xi'an's Urban Regeneration

The historical overview of the urban development plan and strategies in Xi'an shows that urban regeneration practices either in redevelopment or in conservation stress the improvement of the physical environment and economic development. With the lack of necessary regulations on the conservation of local intangible elements, the study finds that in the regeneration of the Xi'an historic city, local indigenous lives and traditional activities are largely overlooked. In the next section, this book will explore the role of social capital in facilitating community involvement in urban regeneration processes based on local community ties or connections. It will explore both the connections among community members (the binding social capital) and the connections between the local community and other agencies (the bridging social capital) in the forms of networks, social norms, and trust. Therefore, social capital can be examined in two aspects according to the locations where it is built during urban regeneration processes. The binding social capital is concerned with the existing communities' relationships and connections. Bridging social capital, on the other hand, relates to the participation of local communities in urban planning and regeneration projects. Considering that traditional business activities are also significant elements in local historic districts, the relationships in business networks that have a substantive impact on the local economic situation will also be explored.

Various factors influence these relationships. Following the above classification on where social relationships are located, this study divides these factors into two groups, namely, internal factors and external factors. Internal factors refer to those that affect the relationships and connections among community members, mainly the socio-economic conditions (e.g., age, economic disparity, types of employment, levels

治模式以及其他条件。需要提醒的是，这种对影响社会资本因素的分类也反映在前面第二章中对常见社会资本指标的讨论中。

4.3.1 西安城市中的社会资本

文献研究表明，直至最近几年，学者才开始对西安城市发展中的社会资本问题开展研究。有关研究可见《西安城市社区建设的理论与实践》（Cheng，2003）、《西安城市社会问题研究》（Shi，2004）以及《构建和谐西安》（Yu & Ju，2006）。20世纪90年代以来，随着中央计划经济体制逐渐向市场化的经济体制转变，社会结构的转型也给城市发展的各领域带来了巨大变化。在全球化和快速城镇化的背景下，中国社会结构变化的特征主要表现为，居住面积的不断扩张和城市人口密度的不断增长、人口迁移速度加快和人口组成更加多样化、下岗人数不断增加、人际关系和交流则渐趋冷漠和封闭、强调经济增长的同时却缺乏对社会问题的足够关注等内容（Song & Yuan，2003：11）。上述不少内容在西安市的城市发展过程中也有所体现。总体来说，目前针对西安社会问题的研究，在社会关系和社会认知方面，主要是通过公开采访与问卷调查的形式进行的[1]，其中大部分问题涉及地方社区居民间的信任、友谊、邻里生活状况以及其他问题（Shi，2004；Yang，2003；Yu & Ju，2006）。

of education, and ethnic background), place attachment, religious and ethnic issues, and other conditions. External factors refer to those that affect the relationships between the local community and other government agencies, mainly canvassing local socio-economic conditions (e.g., NGOs and other social organizations), regeneration strategies, mode of regeneration governance, and other conditions. Notably, this kind of classification of factors in social capital also refers back to the previously discussed common indicators of social capital in Chapter Two.

4.3.1 Social capital in Xi'an City

The literature review shows that social studies in the Xi'an's urban development did not attract academics' attention until recently. The related studies can be inferred from *Theories and Practices on Xi'an Urban Community Construction* (Cheng, 2003), *Research on the Urban Social Problems in Xi'an* (Shi, 2004), and *Constructing the Harmonious Xi'an* (Yu & Ju, 2006). With China's gradual transition from a centrally planned economy to a market-oriented one in the 1990s, numerous changes have manifested through social structural transformation. In the context of globalization and rapid urbanization, social structural transformation in China has been characterized by inhabitance area sprawl and increasing urban population density, speedy migrants flow and diversified population composition, gradually increasing lay-offs, aloof and closed human relationships and interactions, urban development emphasizing economic growth with little social concern, and other features (Song & Yuan, 2003: 11). Most of these aspects are also displayed in Xi'an's urban development process. Generally, current studies on Xi'an's social issues, in perspectives of social relationships and perceptions, are conducted through public interviews and questionnaire surveys[1], and most of the questions concern

---

[1] 关于西安快速城市发展中的城市社会问题研究，相应问卷可参见石英（Shi，2004：40），以及于小文和巨杭生（Yu & Ju，2006：203）的著作。

[1] The corresponding questionnaires on the urban social problems in Xi'an's rapid urban development may be found at Shi (2004: 40) and Yu and Ju (2006: 203).

4.3.1.1 社区关系与邻里联系：聚合型社会资本

目前已有的研究结果显示，西安市的社区居民邻里关系偏向于淡漠（Shi，2004；Song & Yuan, 2003）。正如石英（Shi，2004）描述西安城市中的社会问题时所指出的，不少人普遍认为现在人们的诚信度偏低。在缺乏足够信任的社区中，不少受访者表示，这种关系也会影响到他们的日常生活。于小文和巨杭生（Yu & Ju, 2006）针对人们对社区居民间的信任、友谊与交流的必要性进行了调查。调查结果显示，约0.5%的受访者认为这些内容没有必要；约14.1%的受访者认为这些内容无所谓；24.6%的受访者认为这些内容比较重要；58.5%的受访者认为这些内容非常必要（图4-16）（Yu & Ju, 2006：118）。上述调查结果表明，大部分居民还是认为社区居民间的信任和友好关系非常必要和重要。很多居民也指出良好的社区关系是建立在人们之间相互的理解、尊重、友谊、帮助和合作基础上的（Shi, 2004；Yu & Ju, 2006）。

研究还发现，公众对他们当前生活状态的满意度情况显示，大约5.7%的受访者表示非常满意；28.8%的受访者表示相对满意；43.8%的受访者认为，他们的生活状态不好不坏；14.8%的受访者表示相对不满意；近6.9%的受访者表示很不满意。（图4-17）显示了该调查结果。概括来看，近80%的受访者对他们的生活状态表示满意或一般。于小文和巨杭生认为，这种情况可以解释西安当前的社会状况是基本稳定的。不过，于小文和巨杭生进一步指出，约有20%的受访者对

social perceptions of the local community's trust, friendships, neighborhood living status, and other issues (Shi, 2004; Yang, 2003; Yu & Ju, 2006).

4.3.1.1 Community relationships and connections: The bonding social capital

Many studies define the situation of community relationships and connections in the current Xi'an City as aloof and indifferent (Shi, 2004; Song & Yuan, 2003). Just as Shi (2004) described the social problems in Xi'an City, many people held that one could not trust another too much these days. In the dearth of mutual trust, most respondents agreed that this kind of community relationship had affected their daily lives. Surveys on people's opinions on the necessity of social trust, friendship, and cooperation among community members were conducted. The survey result showed that only 0.5 percent of the respondents thought these issues were not necessary; about 14.1 percent thought they were average; 24.6 percent of the respondents believed they were relatively necessary; and 58.5 percent of the respondents thought the issues were very necessary (Figure 4-16) (Yu & Ju, 2006: 118). These survey results indicate that most residents in Xi'an still think that social trust and smooth relationships among the community members are necessary and important. Many residents expect good community relationships based on mutual understanding, respect, friendship, assistance, and cooperation (Shi, 2004; Yu & Ju, 2006).

Interestingly, it is noticed that the surveys on the public's satisfaction in their living status showed that about 5.7 percent of the respondents were very satisfied about it; 28.8 percent were relatively satisfied; 43.8 percent believed that their living status was average; 14.8 percent were relatively dissatisfied; and nearly 6.9 percent were very dissatisfied. These results are shown in Figure 4-17. In this survey, nearly 80 percent of the respondents expressed satisfaction or average opinions to their living status. Yu and Ju argued that this situation

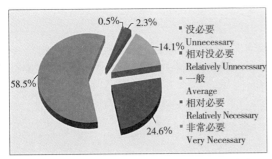

图 4-16：邻里间信任、友谊及交流的必要性的公众意见
Figure 4-16: Public opinions on the necessity of social trust, friendship and cooperation
资料来源：(Yu & Ju, 2006：118)
Source: (Yu & Ju, 2006: 118)

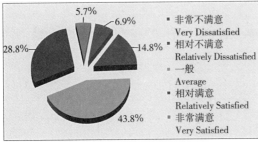

图 4-17：西安市城中居民对目前生活状态的满意度等级
Figure 4-17: Level of public satisfaction with current living status in Xi'an city
资料来源：(Yu & Ju, 2006：23)
Source: (Yu & Ju, 2006: 23)

他们的生活状态不满意，人们仍应关注市民中的社会阶层化和社会管理问题（Yu & Ju, 2006：23）。

影响聚合型社会资本的因素

根据当前已有研究（Cheng, 2003；Shi, 2004；Yu & Ju, 2006），影响西安城区居民关系的主要因素一般包括经济差距、就业类型、教育水平、地方依附性、宗教和种族等。通常情况下，受访者认为社会中不断扩大的经济差距是影响社区居民间关系的一个主要因素。另外，不同职业人员和不同身份、地位的人员会对社区居民间关系持不同的评价

could account for the current basic social stability in Xi'an. Nevertheless, Yu and Ju further pointed out that with about 20 percent of the respondents dissatisfied with their living status, one should be alert about social stratification and social management in Xi'an (Yu & Ju, 2006: 23).

Factors affecting the bonding social capital

According to the current available studies (Cheng, 2003; Shi, 2004; Yu & Ju, 2006), the key socio-economic factors affecting community relationships in Xi'an generally concern economic disparity, types of employment, levels of education, place attachment, and religion and ethnicity. In various studies, the issue of growing economic disparities in the society has been reported by the respondents to be one significant factor that undermines existing community relationships. In addition, members with different employments and members with different place identities often have different evaluations on community relationships. Yu and Ju's study found that many residents in the historic or old neighborhoods tended to have higher evaluation of local community relationships, while many temporary migrant workers in the city have very low evaluation (Yu & Ju, 2006: 68). The relation between the classification of employments and the community perceptions on social relationships is shown in Figure 4-18.

In addition, studies (Shi, 2004) have found that financial condition is a significant factor affecting community relationships. The survey found that, normally, people with low income (below RMB 500 per month) or high income (above RMB 1,500 per month) tend to give lower evaluation of social relationships than those people with moderate income. Shi explained that people with very low or high income might feel comparatively deprived of the normal social activities or interactions with the interest readjustment in various social strata in China's current transitional economy, which accordingly affect their perceptions of social relationships. Comparatively, those with

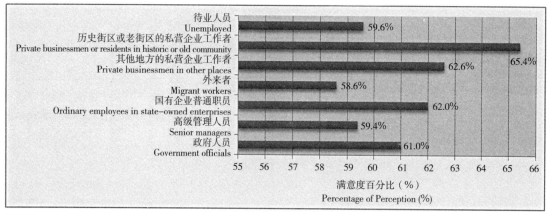

图 4-18：西安城市居民的职业类别及其对社会关系的认知
Figure 4-18: Classification of employments and their perceptions on social relationships in Xi'an
资料来源：(Yu & Ju, 2006: 68)
Source: (Yu & Ju, 2006: 68)

态度。于小文和巨杭生的研究还发现，历史街区或旧街区的居民一般对社区居民间的关系持较高评价，相比之下，城市中的临时工作者（如外来务工者）往往对社区关系的评价较低（Yu & Ju, 2006: 68）。（图 4-18）显示了居民的职业类别和其对邻里关系认知之间的关系。

有关研究（Shi, 2004）发现，居民的经济状况也是影响邻里关系的一个重要因素。通常情况下，与中等收入人群相比，低收入人群（低于 500 元／月）和高收入人群（高于 1500 元／月），对社区中的邻里关系评价较低。石英（Shi）解释说，在中国目前转型期经济体制下，随着不同社会阶层的利益再调整，低收入和高收入人群可能感到自己的正常社交活动或交流空间会受到较多影响，相应地也就影响到了他们对邻里关系的认知、评价。相比之下，中等收入人群受这个过程的影响较小（Shi, 2004: 6）。不过，本研究的重点并非城市发展中的社会问题。接下来，

moderate income feel less affected by this process (Shi, 2004: 6). Nevertheless, social problems in Xi'an's urban development are not the focus of the book. In the next, the book will accentuate on place attachment, historic urban fabric, and religion and ethnicity. These two factors closely relate to community relationships and local urban regeneration processes.

Place attachment and historic urban fabric

In exploring the relation between physical urban environments and social relationships, many academics (Brown & Perkins, 1992; Brown, Perkins, & Brown, 2003; Kasarda & Janowitz, 1974) argue that place attachment serves as the positive bond between residents and their socio-physical settings. Here, place attachment is understood as the "feelings of pride in the residential area and its appearance and a general sense of well-being" (Brown et al., 2003: 259; Harris, Werner, Brown, & Ingebritsen, 1995: 312). The interrelations of place attachment and social capital in the light of community relationships are elaborated in the next. Brown claimed that place attachments are nourished by the local residents' daily encounters with the environment and their neighbors, their seasonal celebrations, and other things (Brown et al., 2003: 259). Moreover, it was

研究将着重突出场所的依附性、历史城市肌理结构以及宗教和种族等因素的影响。这些因素也与邻里关系和城市更新过程有着密切联系。

在探讨城市物质环境和社会关系之间的关系中，不少学者（Brown & Perkins, 1992；Brown, Perkins, & Brown, 2003；Kasarda & Janowitz, 1974）也都认为地方依附性或者场所依附性在居民和周边物质空间之间发挥着积极的联系作用。在本研究中，地方依附性被理解为"居民区的一种自豪感和幸福感"（Brown et al., 2003：259；Harris, Werner, Brown, & Ingebritsen, 1995：312）。接下来，本文解释了地方依附和社区关系中社会资本之间的相互关系。布朗（Brown）认为地方依附性是当地居民与其周围环境以及邻居间的日常交流（包括定期的节日庆祝活动等）所产生的（Brown et al., 2003：259）。同时，有研究注意到不断增长的人口规模和人口密度以及居住时间长短也对社区居民的地方依附性和社区团结产生影响（Kasarda & Janowitz, 1974：338）。通常情况下，当人们之间的邻里关系较好时，人们也会拥有更强的社区认同感和场所依附性。有学者们认为地方依附性和邻里关系或社会资本的水平有着较大的联系。换句话说，"地方依附感较强的居民也拥有较高水平的（使用）社会资本"（Kleinhans, Priemus, & Engbersen, 2007：1083）。

本研究认为，在西安历史街区中，人们的地方依附性还与当地独特的历史城市（建筑）肌理相关。如果社区成员在历史街区中从事一些经济活动，居民也较容易和当地城

noticed that the increased population size and density in the communities and the length of residence have an impact on place attachment and social bonds in local communities (Kasarda & Janowitz, 1974: 338). Therefore, when people have better community relationships, they tend to have more community identity and stronger place attachment. Kleinhans et al. argued that place attachment has a remarkably strong association with community relationships or the level of social capital. In other words, "residents expressing a higher level of place attachment also report higher levels of (access to) social capital" (Kleinhans, Priemus, & Engbersen, 2007: 1083).

This book maintains that sense of place attachment in Xi'an's historic districts relates to the local specific historic urban fabric features. When community members undertake some economic businesses in Xi'an historic districts, these residents have established close relationship with the urban environments. They usually express stronger place attachment to the physical environment. This relates to the social capital in business settings. Xian is a historic and tourism city; thus, local commercial and business activities are attached to the local historic urban environments. When the majority of customers are local tourists, the business networks in historic districts become closely related to local historic urban fabric. Therefore, the social capital in local historic business settings may also be explored from the viewpoint of local historic urban fabric. Moreover, when the local historic urban fabric tends to bring opportunities for local businessmen and community members to conduct interactions and information sharing, it also contributes to the generation of social capital among community members. This argument is supported by Zhang (2002c) through the studies on the three historic districts of Xi'an, namely, Shuyuanmen, Beiyuanmen, and Defuxiang. Comparatively, given the religion and kinship bonds among the neighborhoods in the Beiyuanmen district, the Muslim district, Zhang noticed that there are more social interactions

市周边环境间建立起密切的关系。居民通常对城市的物质环境有着较强的地方依附性。这也与商业活动中的社会资本水平有关。西安是一个历史和旅游城市，因此，地方商业活动是依靠地方历史城市环境的。当大部分客户是地方游客时，历史街区的商业网络就和地方城市建筑的关系变得更加密切。此外，地方历史城市建筑给当地商人和社区居民进行交流和信息分享提供了机会，同时，这也有助于社区居民间社会资本的生成。值得一提的是，张凌（Zhang，2002c）在针对西安几块历史街区的研究当中（即书院门、北院门和德福巷），一定程度上也印证了本研究的上述论点。在北院门历史街区中，由于居民间具有的宗教和血缘联系，张凌注意到该历史街区的居民比其他历史街区中居民的日常交流更加频繁。交流活动一般涉及日常闲聊、下棋等活动。张凌指出，在北院门历史街区，"商业空间有生活化倾向"（Zhang，2002c：82）。基于其研究，张凌认为历史城市空间有助于促进当地社区居民间的互动与联系。另外，如果社区关系是建立在类似于宗教或血缘的亲密关系上时，也有助于社区居民间联系的增强和社会资本的积累。

历史城市空间对社区居民的互动和社区建设的意义也得到了其他学者的支持（Bai，2000；Chang，1999）。三学街区一项关于地方居民对社区关系的认知问卷调查表明，大约64%的受访者对地方庭院生活表示满意，只有24%的受访者表示不满意。而且，那些对地方传统庭院生活持乐观态度的人明确表示，和那些居住在高楼里面的人相比，生活在历史街区的最大好处就是易于邻里之间进

among the community members within this district than the other two districts. Such activities include casual chatting with others, playing chess, and other activities. Just as Zhang indicated, "there is a tendency for commercial space to be lifelike" in Beiyuanmen district (Zhang, 2002c: 82). Based on the study, Zhang (2002c) contended that the historic urban fabric contributes to the interaction and connections among local communities. Moreover, when the community relationships are built on such close ties as religion or kinships, these further reinforce the building of community connections and help increase local social capital.

The contribution of the historic urban fabric to community interactions and community building is supported by others (Bai, 2000; Chang, 1999). A questionnaire survey on the local community's perceptions on community relationships in the Sanxuejie historic district found that about 64 percent of the respondents were satisfied with the local courtyard life and only 24 percent of the respondents were dissatisfied. Moreover, those who gave positive comments on local traditional courtyard lives explicitly expressed that the best thing in living in historic districts was the easy access to neighborhood interactions and mutual assistance compared with that in high-rise apartments (Bai, 2000: 44). In exploring the relationship between traditional urban fabric and social relationships, some academics (Liu & He, 2004; Liu, 2004) further highlight the contribution of street scale in promoting mutual interactions. Through studying the correlation between spatial scale and human perception, Liu emphasized the street scale of the Xiyangshi lane in the Muslim District of Xi'an city, and remarked that the existing traditional narrow street produces a calm, accessible, and more interactive space (Liu, 2004: 14). Nevertheless, it is clear in Liu's study that there was no differentiation among the subjects, and therefore the mutual interactions could occur both among the local community members and between the local residents and tourists.

行互动和相互帮助（Bai，2000：44）。在探讨传统城市建筑和社会关系之间的关系中，有学者（Liu & He，2004；Liu，2004）强调了街区规模在促进互相交流中的作用。通过对空间规模和人类感知之间关系的研究，刘永望（liu）强调了西安鼓楼回民区西羊市巷道的街道规模，并认为现有的传统狭窄街道有助于形成一种平和、亲近、并更易于人们产生互动的空间（Liu，2004：14）。但是，很显然，在刘永望的研究没有进一步划分互动主体身份，因此，互动交流既可能发生在当地社区居民之间，也可能发生在居民和游客之间。

有关历史城市空间对社区互动的作用的探讨，也通过对西安历史街区中的社区网络的研究中进行开展。通过对西安正学街历史街区的研究，常海青（Chang，1999）确定指出历史街区为社区居民提供了多种社会网络。通常情况下，不同居民群体会形成不同的社会网络关系，而且居住较近或更便于交流的人们会分享彼此关系网，因此，居民间形成了相对明确的社会身份认同。例如，与其他公共设施相比，居民更常使用离家较近的公共设施，事实上，这为居民们碰面和交流也提供了更多机会，有助于提高社区的凝聚力和社区居民的认同感。从而，这些活动有助于在居民邻里之间生成更多的社会资本（Chang，1999：52）。

宗教和种族

正如上述提到的，在历史街区中，宗教和种族是影响社区居民关系的重要内容。2000年开展的第五次全国人口普查结果显示，西安市总人口达到了741万，其中，少数民族

The contribution of historic urban fabric to community interactions is also explored by studying the community networks in the historic district in Xi'an. By studying the Zhengxuejie historic district in Xi'an, Chang (1999) determined that historic districts tend to provide various social networks among local community members. Usually, different groups of people have various social networks, and people living nearer or together tend to share the same networks, therefore having relatively explicit community identities. For example, residents use the public facilities close to their houses more often than others, and this offers local residents more opportunities to meet and interact in favor of community cohesion and community identities. Hence, these activities play a part in creating local social capital among these nearby residents (Chang, 1999: 52).

Religion and ethnicity

As mentioned above, religion and ethnicity are also significant factors contributing to community relationships in historic districts. According to the latest fifth national demographic census in 2000, Xi'an's overall population was 7.41 million, among which the minor ethnic nationality reached 85,272. In addition, there were about 21,000 mobile minor ethnic populations (Yang, 2007: 11, 97). Among these minor ethnic people, Muslims account for 64,216 or 75.31 percent of the entire minor ethnic population of Xi'an (Yang, 2007: 123). Regarding the role of religion in facilitating cohesive community lives, according to Yang, one significant function of religion is to "construct, arrange, reiterate and intensify the fundamental social values." Yang categorized the role of religion in maintaining community solidarity and interrelationships among community members as the following five perspectives (Yang, 2007: 358):

1) providing a social structure model: Religion sets up the symbolic system to reflect the social structure, which also offers the conceptual basis for social structures.

常住人口有 85272 人。另外，还有 21000 流动的少数民族人口（Yang, 2007: 11, 97）。在这些少数民族中，回民约 64216 人，占到西安整个少数民族人口的 75%（Yang, 2007: 123）。关于宗教在促进社区凝聚力方面的作用，杨文炯（Yang）认为，宗教的一个重要作用就是"建立、整顿、重复并加强有关基本价值观"。杨文炯把宗教维护社区团结和社区成员之间相互关系的作用分为如下五个方面（Yang, 2007: 358）：

1）为社会结构提供模型：宗教通过一定的方式安排其象征体系，使之反映社会结构，并为社会结构提供理论依据。

2）提供共同的价值观：宗教的社会化功能就是把各种不同观点纳入同一价值结构。而且，形成共同的信仰意识和统一的行为规范，即通过信仰控制形成对人的行为的隐形支配。

3）缓和社群内部的冲突：宗教既是整合的力量，又是缓和社群内部冲突的力量。这一功能主要通过宗教仪式来实现。例如，在穆斯林社区中，主要通过礼拜或聚礼（回礼仪式）来解决内部冲突或个人分歧。

4）加强社会管理：每一种宗教都是解释是非行为的伦理体系，因此行为道德标准被赋予超自然的权威，从而形成对人的内在的自我约束，强化了社会管理。

5）凝聚社群：宗教的另一功能就是通过定期的仪式凝聚、整合社会群体，强化族群认同。例如，每周五的集会活动和两个年度庆典（开斋节和牺牲节）就发挥了这种作用。

因此，遵循共同的教义信条，有助于缓

2) providing the common values: It is one of the social functions of religion to include different viewpoints into the same value structure. Furthermore, it forms the common belief perceptions and behavior norms, which can invisibly dominate human activities through religious controls.

3) mitigating the conflicts and hatred among community members: Religion is the kind of force than can both converge opinions and mitigate the conflicts among community members. This function is accomplished mainly through religious ceremonies. For example, in Muslim communities, it is primarily through the salat or gathering ceremonies (*ju li, hui li yi shi*) that internal conflicts or individual disagreements are dealt with.

4) intensifying social management: Religious doctrines are ethic systems that differentiate the right from the wrong. When the behavior and ethic standards are bestowed with supernatural power, they form a kind of inward self-discipline and intensify the social management.

5) agglomerating communities: It is also a function of religion to unite and agglomerate communities to intensify group identities through regular ceremonies. For example, the gathering activities every Friday and the two annual festivals (the Eid-ul-Fitr (*kai zhai jie*) and the Eid-ul-Azha (*xi sheng jie*) or Sacrifice Festival) assume such a function.

Therefore, religion helps smooth the conflicts among its memberships following the common doctrines, and significantly contributes to the community trust, solidarity and place identity among them. All these contribute to build cohesive community relationships and to generate local bonding social capital.

4.3.1.2 Community participation in urban planning and regeneration process: The bridging social capital

Community participation in urban planning and local regeneration processes is seen, from the political viewpoint, as a significant

解成员之间的矛盾冲突，并有助于建立社区居民间的信任、团结和强化居民的地方认同感。这些内容都有助于增强社区的凝聚力，并生成当地聚合型的社会资本。

4.3.1.2 城市规划和更新过程中的社区参与：联合型社会资本

城市规划和更新过程中的社区参与是构建和谐社会的一个重要测量指标（Yu & Ju, 2006）。这也是公众了解当地城市更新政策的一个重要途径。在西安城市更新实践中，社区参与主要体现在将其融入规划、审批、实施和管理的环节当中（He, 2003a : 8-9）。通过把地方需求和历史街区的具体特征融入地方城市更新项目中，社区参与有助于弥补市场和政府决策的缺陷，并有助于推动政府更新政策的完善与实施，从而促进社区的建设（Yu & Ju, 2006 : 61）。

总体而言，在西安的地方更新实践中，社区参与的水平偏低。研究还发现，在城市更新过程中，居民参与的热情随着教育程度的不同而不同（Yu & Ju, 2006）。于小文和巨杭生的调查显示，那些积极参与更新过程的居民，分别是拥有小学和中学教育水平的受访者，但其中表示有兴趣参与城市更新的居民，也只是占到相应组别人数的 2.6% 和 3.0%。相比较下，拥有研究生或以上教育水平的高级知识分子，更加缺乏这种参与热情。该组受访者中，只有 1% 的人士表示有兴趣参与地方更新实践（Yu & Ju, 2006 : 83）。（图 4-19）显示了不同教育水平的居民分组以及相应组别中有兴趣参与城市更新项目的人口比例。

measure in building a Harmonious Society (Yu & Ju, 2006). It is also an important way for the public to know the government policies on local urban regeneration. In Xi'an's urban regeneration practices, community participation is advocated by the municipal government to be integrated in the planning, sanction, implementation, and management processes (He, 2003a: 8-9). By integrating the local needs and the characteristics of the historic districts in local urban regeneration programs, community participation helps to make up for the failure of market and local government policies and contributes to the smooth implementation of government regeneration policies and community building (Yu & Ju, 2006: 61).

Generally speaking, the level of community participation in local regeneration practices in Xi'an city is low. Studies find that participation enthusiasm in local regeneration processes varies significantly among the residents with different education levels (Yu & Ju, 2006). In Yu and Ju's survey, the people showing the most enthusiasm in participation, which were 2.6 percent and 3.0 percent of the respondents respectively, were usually those holding primary or secondary school levels. Comparatively, the respondents with the highest education levels of postgraduate or above showed the least enthusiasm of participation. In fact, only 1.0 percent of these groups' respondents were interested in participating in local regeneration practices (Yu & Ju, 2006: 83). The different education levels of residents and the proportion of their participation in local urban regeneration matters are shown in Figure 4-19.

This phenomenon shows that on the one hand, the concept of public participation is not accepted by the public yet, and many people maintain indifferent or detached towards public affairs. On the other hand, community or group lives do not constitute a significant part of the public's social lives. Especially, when the residents' lives have very few connections to local historic urban

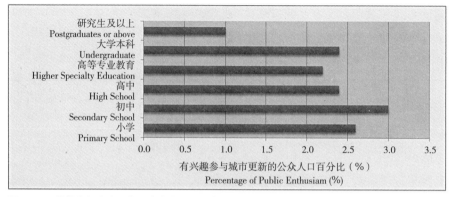

图 4-19：按教育程度分组的受访者中有兴趣参与城市更新的人口比例
Figure 4-19: Level of public enthusiasm in community participation
资料来源：（Yu & Ju，2006：83）
Source: (Yu & Ju, 2006: 83)

上述现象表明，一方面，社区参与的理念还没有被大众所了解和接受，不少人对公共事务持冷漠或远离的态度。另一方面，社区或群体生活也没有成为公众社会生活中的重要一部分。尤其是，当居民的生活内容与当地历史城市环境存在较少商业上或经济上的联系时，很多居民就看不到历史环境的改变对他们实际生活所产生的影响。同时，地方城市更新过程中社区参与程度不高，也与居民参与社区／社会组织的途径比较有限有关。

影响联合型社会资本的因素

社会组织

根据加拿大政策研究机构（PRI）的定义，社会经济指的是"一种多样的、不断演化的非政府组织的联合体，这种联合体可以在社区中进行物质产品与服务等内容的生产与分配。"通常，社会组织的任务是"为了满足大家的共同利益和以公共服务为目的"（PRI，2005c：1）。在西安，这样的社会组织既包括政府组织也包括非政府组织。政府资助的组

environments, in terms of businesses or economic connections, many residents can not see the impact of the changing historic environments on their actual lives. At the same time, the low level of community participation in the local regeneration process is also related to the rare participation opportunities in terms of community/social organizations.

Factors affecting the bridging social capital
Social organizations

According to the Policy Research Initiative (PRI, thereafter), social economy is defined as "a diverse and evolving combination of non-governmental organizations (NGOs) that have been producing and delivering goods and services in communities." Usually, the missions of social organizational are based on "a combination of common interest and public service objectives" (PRI, 2005c: 1). In Xi'an, such social organizations involve both the government-sponsored organizations and non-government organizations. The government-sponsored organizations encompass such institutions as labor and social security, social welfare and rescue, and other institutions. The non-government organizations include such associations as the Industrial and Commercial Trade Union of Beilin District, the Union of Women Association, and similar organizations (ECXBDR,

织机构包括劳动和社会保障、社会福利和救助和其他机构。非政府组织包括碑林区的工商业联合会、妇女联合会和其他组织（ECXBDR，2003：439）。还有一些政府资助的宗教社团，例如西安佛教协会、西安基督教青年会、西安伊斯兰教社团和其他宗教团体（ECXR，2006：94）。作为地方宗教团体的机构，这些社团得到了地方政府的认可和支持。对于中国的非政府组织，按照国家当前的管理条例，地方政府对这些组织进行负责。中国目前为此制定的国家相关法规包括1989年国务院颁布的《社会团体登记管理条例》。除了这部国家法规之外，还有一些地方性法规，西安区政府已经成立地方单位对现存的地方社会组织进行管理和整顿（ECXLDR，2001：542）。为了提高社会道德水平、形成良好的社会风气，以及更好地服务地方社区，一些非政府组织如社区学校、道德教育组织和其他类似的组织也已经成立了（Shi，2004：71）。不过，在西安，尽管有相当多的政府和非政府组织，但是相关文献表明，60%以上的地方居民普遍反映这些团体活动和老百姓的普通生活联系不够密切。一个鲜明的例子就是，地方居民很少有机会对他们真正关心的一些问题进行讨论，例如，居民重置和迁移规划。这不但影响到了地方社区和政府之间的信任，同时也影响了地方社区参与的热情（Yu & Ju，2006：75）。

政府行政等级体系和城市管理机构

石英（Shi）指出目前西安的城市发展当中尚存不少社会问题（表4-2），主要涉及地方政府的管理与服务、公民素质和社会道德、社会公平以及其他问题（Shi，2004）。研究认

2003: 439). There are also government-sponsored religious societies, for example, the Xi'an Buddhist Society, Xi'an Christian Society, Xi'an Islamic Society, and other religious groups (ECXR, 2006: 94). As institutes for local religious communities, these societies are acknowledged and supported by local governments. With regard to the non-government organizations, the local governments are usually in charge of these social organizations according to current Chinese regulations. In China today, the related national regulations for this purpose include the Regulation on the Registration and Management of Social Organizations by the State Council of 1989. Following this national regulation as well as some local ones, the Xi'an district governments have established local units to sort out and rectify many existing local social organizations (ECXLDR, 2001: 542). In order to improve social morality and form a good social ethos, some non-government organizations have been established for local communities such as community schools, morality educational organizations, and other similar endeavors (Shi, 2004: 71). Although there are quite a few government-sponsored and non-government organizations in Xi'an, literature indicates that more than 60 percent of the local residents believe that very few of them are closely related to the community building or to the local residents' own interest. As a sharp contrast, there are very few opportunities for the residents to participate in some urban matters they really care about, for instance, the residential relocation and displacement plans. This severely influences the trust between local community and government agencies and also affects local community's participation enthusiasm (Yu & Ju, 2006: 75).

Government administrative hierarchies and urban management institutions

In Xi'an at present, many serious social problems (shown in Table 4-2), as identified by Shi (2004), mainly concern government management and service, public qualities and social morality, social inequity, and other issues (Shi, 2004). One fundamental reason

公众认为较严重的城市问题　　　　　　　　　　　　　　　表 4-2
The most serious urban problem in Xi'an subjectively perceived by the public　　Table 4-2

最严重的城市问题 The most serious urban problems	认知的受访者人数 Respondents of the perceived	在全部受访者中的认知比例（%） Proportion of the perception to the overall respondents (%)	序号 Serial number
社会保障　Social security	205	22.5	1
政府管理和服务　Governmental management and service	153	16.8	2
环境卫生　Environmental sanitation	142	15.6	3
交通　Traffic	98	10.8	4
失业　Unemployment	79	8.7	5
公众素质和社会道德　Public quality and social morality	63	6.9	6
市场秩序和服务　Market order and service	41	4.5	7
经济增长和收入　Economic growth and income	39	4.3	8
教育　Education	21	2.3	9
医疗　Medical	14	1.5	10
社会不公平　Social inequity	13	1.4	11
农村 / 童工　Rural/child labor	9	1.0	12
人口　Population	7	0.8	13
老年人　Senior citizen	7	0.8	14
流浪儿童 / 乞丐　Homeless child/beggar	4	0.4	15
住房　Housing	3	0.3	16
其他问题　Other problems	13	1.4	
总计　Total		100	

资料来源：(Shi, 2004：23)
Source: (Shi, 2004: 23)

为，产生这些问题的一个重要原因是当前落后的政府管理体制。由于现有我国的政府管理机构还是在计划经济时期形成的，不少方面已很难适应目前市场经济体制下的城市更新现实 (Shi, 2004：69)。石英 (Shi) 认为，落后的城市管理体制往往引发许多社会问题，如社会不公平、社会道德水平低下、地方社区对政府的信任度等方面 (Shi, 2004：69)。足够信任与相互合作的匮乏会影响不同团体间联合型社会资本的积累，从而不利于地方历史环境中的更新进程 (Cao, 2005：42)。如上文所述，在西安的现行总规划中，当地的更新规划和发展策略更加侧重物质环境的改善与地方经济的发展 (Li & Yin, 2006：

for these problems is the lagging behind of the government management institutions. While the existing government management institutions were formed in the planned economic period, they do not fit the current market-oriented urban regeneration realities of Xi'an any more (Shi, 2004: 69). Shi contended that the current lag in the urban management institutions tends to breed many social problems such as social inequity, low level of social morality, and mistrust between the governments and local communities (Shi, 2004: 69). The lack of trust and cooperation between the communities and local authorities, which affects the accumulation of bridging social capital hitherto, is detrimental to the regeneration process of local historic environments (Cao, 2005: 42). In Xi'an's latest master plan, as discussed above, local regeneration plans and strategies focus on the

81)。在自上而下的规划和决策机制下，地方社区所起的作用微乎其微。这种背景下，社区居民也很少参与到历史街区的保护或更新过程中。

在我国，市级政府的行政等级体系一般包括"城区"和"街道办事处"。根据中华人民共和国宪法和地方组织法规，区政府的职能是区人民代表大会和区人民政府。不过，"城区"绝不是一个小的空间行政单位。通常，在市、区级的行政单位中会有成千上万的人口。因此，为了更好地进行社会管理，在区政府下面成立了街道办事处（Lin & Wang, 2002：87）。作为区政府的一个分支，街道办事处是"人民行使政治权利的基层机关"和"城市里级别最低的政府机关"（Editor, 1973：2）。在很大程度上，街道办事处代表政府，与社区居民保持着密切联系。同时，街道办事处也和居民委员会保持着密切关系。居民委员会是1956年由全国人民代表大会常务委员会颁布的法令下成立的。根据1989年通过并颁布的《中华人民共和国城市居民委员会组织法》，居民委员会作为城管居民的自治组织，具有如下主要功能（Lin & Wang, 2002：93）：

1）宣传党的路线、方针、政策，教育居民群众响应党和政府的号召，履行公民义务、遵守国家法令；

2）领导群众搞好治安保卫；

3）调节民事纠纷；

4）协助办理居民的公共福利事务；

5）向上级党政机关反映居民群众的意见和要求。

不过，尽管居委会的成员由地方社区居

improvement of physical environments and economic development (Li & Yin, 2006: 81). In a top-down planning and decision-making mechanism, local community plays a very weak role. Against this background, local residents are also rarely involved in the conservation or regeneration of historic districts.

In China, the local governments' hierarchies within the municipality administration generally consist of urban districts and sub-district offices. According to the Constitution of the P.R.China and Local Organization Regulations, the government authorities in urban districts are the district people's representative congress and serve as the district government. Nevertheless, the urban district is by no means a small spatial unit. Usually, several tens of thousands of people reside in such a unit. Therefore, for the purpose of social management, the sub-district office level operates under the district level (*jie dao ban shi chu*) (Lin & Wang, 2002: 87). As the substation of urban district government, the sub-district office is "the basic level organ of people's political power" and the "lowest level of government administration in the city" (Editor, 1973: 2). To a large extent, the sub-district office has a close relationship with the residents on behalf of the government. At the same time, the sub-district office is also closely associated with the residents' committee (*ju min wei yuan hui*), which was established under a decree issued by the Standing Committee of the National People's Congress in 1956. According to the Residents' Committee Organization Regulation of the P.R.China of 1989, the functions of the residents' committee, that is, the autonomous residential organization (Lin & Wang, 2002: 93), are as follows:

1) to propagate the Party's guidelines, policies, and politics, mobilize the public to respond to the calls of the Party and government, fulfill the public obligation, and obey the national decrees

2) to lead the public to carry out the security and defense works

3) to mediate disputes among residents

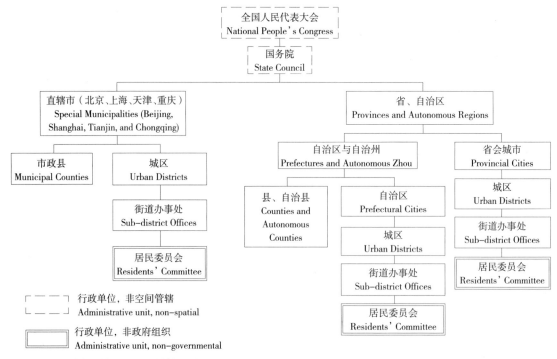

图 4-20：目前中国的空间和行政等级制度
Figure 4-20: Spatial and administrative hierarchy in current China
资料来源：作者汇总自（Deng, 2004：45；Ma, 1979：842；Wang & Hague, 1992：17）
Source: Author synthesized from (Deng, 2004: 45; Ma, 1979: 842; Wang & Hague, 1992: 17)

民组成，许多居民并没有把居委会当成社区的自治组织，对其缺乏重视（Dong, 2003：169）。尤其是当有些政策和策略在没有居委会成员参与的情况下就颁布、实施时，地方居民更是认为居委会是有名无实的。(图4-20)描述了当前中国的空间和行政机构等级体系。值得一提的是，西安市的行政体系中还建立了一种行政级别，称作社区居委会（Cheng, 2003），其行政级别低于区政府和街道办事办事处，但要高于上述居民委员会。该机构的具体形式有多种，包括：社区成员代表大会，行使决策权；社区居委会，行使执行权；社区协商委员会，讨论并协商相关事务，以及其他形式。然而，新旧行政机构之间经常存

4) to undertake public welfare for the residents
5) to relay the views and demands of the residents to the higher authorities

Nevertheless, although the residents' committee members are selected from the local communities, many local residents do not accept the residents' committee as their own organization and consider them unimportant (Dong, 2003: 169). When some government policies and strategies are implemented without the interference of the residents' committees, the local people often see this organization as a figurehead. According to the above descriptions, the spatial and administrative hierarchy in present China is illustrated in Figure 4-20. It is reported that the administrative system in Xi'an has created a new administrative level called the Community Residents' Committee (*she qu ju*

在一些矛盾和冲突，例如，管理责任、授权范围、人员分配、资金以及其他问题（XSSA, 2003：64）。需要注意的是，西安街道办事处的基本法规仍然延续1950年颁布的《城市街道办事处组织条例》。根据该条例，街道办事处在其行政范围内不具备协调功能，这个也不符合目前西安社区发展的实际需求（Song & Yuan, 2003：17）。同时，在当前市场主导的社区发展过程中，由于行政管理体系的落后，社区居委会和物业管理部门之间也存在着冲突（Li, 2003：172）。大部分冲突和经济利益的分配有关。例如，很多新建的社区是由房地产物业管理单位来管理的。通常情况下，这些物业管理单位关注经济回报而忽视社会问题。因此，这些物业管理单位经常照顾一些"高回报"的业务，如社区公共设施的建设和社区水、电、煤气、有线电视和其他公用失业的维护和服务。相比之下，社区居委会经常负责其他工作，例如，拉近党和群众的距离、宣传和贯彻党的方针、政策以及其他工作。社区居委会员工常常抱怨这些工作是"属于硬性任务，又费事费工，还没有费用可收"。这种情况也会影响到社区居委会的工作效率和积极性。人们戏称这一现象为"收钱的不管事，管事的没钱收"（Dong, 2003：169）。

因此，有学者认为西安城市建设管理机构中的一个根本问题是落后的政府管理体系，也不利于地方社区和政府机构之间关系的建立，从而影响联合型社会资本的生成（XSSA, 2003：67）。这也会影响到地方城市更新规划的顺利实施和有效管理。这会在后面的案例研究中进一步探讨。

wei hui) (Cheng, 2003). This administrative rank is lower than the urban district and sub-district office but higher than the previous residents' committee. The specific formats of this new institute vary including: Community Member Representative Congress (*she qu cheng yuan dai biao da hui*), which has the decision-making authority, Community Residents' Committee *(she qu ju wei hui)*, which is responsible for the implementation, Community Counseling Committee (*she qu xie shang wei yuan hui*), where businesses are discussed and negotiated, and other formats. Nevertheless, some conflicts often exist between the new and old administrative institutes, for example, management responsibility, authority scope, staff assignment, funds, and other concerns (XSSA, 2003: 64). It is noticed that the basic regulation in the sub-district office in Xi'an still remains the Organization Ordinance of the Urban Sub-district Office enacted in 1950. According to this Ordinance, the sub-district office does not have a coordination function in the units within its administrative periphery, and nonetheless this does not fit the current community development realities in Xi'an (Song & Yuan, 2003: 17). At the same time, in the current market-oriented community development, due to the lagging behind of the administration and management systems, the conflicts exist between the Community Residents' Committee and the Property Management Department (*wu ye guan li bu men*) (Li, 2003: 172). Most conflict surrounds the economic interest distribution. For instance, many newly built-up communities are managed by real estate property management units. Normally, these property management units focus on the economic returns and ignore other social concerns. Hence, property management units often take care of some "well-paid" businesses such as community public facilities and the maintenance and service of community water, electricity, gas, cable TV, and other utilities. Comparatively, the Community Residents' Committee often takes care of some jobs, such as bridging the Party and the public, propagating and carrying out the Party's guidelines and policies, and other jobs. These jobs are often complained by the Community Residents' Committee as "time-consuming and poorly paid". This situation severely affects the Community Residents' Committee's working efficiency and enthusiasm. Local people sarcastically call this situation "the paid does not care, and the one who cares is not paid" (Dong, 2003: 169).

Academics therefore contend that one fundamental problem in Xi'an's urban development management institution is the lagging behind of the government management system, which undermines the relationship-building and the bridging social capital-generating between local communities and government agencies (XSSA, 2003: 67). It also has substantial impacts on the smooth implementation and management of local urban regeneration plans. This will be further discussed in later case studies.

4.4　总结

本章回顾了城市更新政策和策略的转变，以及西安城市更新过程中影响社区关系和社区参与的社会资本问题。对西安城市发展与更新过程的回顾表明，西安的城市更新政策已经发生了若干变化。20世纪70年代之前，主要以工业发展为主导；到了20世纪80年代和20世纪90年代，主要以房产和再开发为主导；最后，20世纪90年代末以来，成为以城市再开发和保护的共同政策为特点。关于西安四次总体规划中的城市保护政策，城市保护规划的侧重点已经从20世纪80年代对文物古迹的保护转变为20世纪90年代的多层次、综合式保护。在最新的第四次总体规划中，西安市政府提出对城市土地利用进行重组，对传统城市景观进行翻新。根据该

4.4　Conclusion

This chapter reviews the transforming urban regeneration policies and strategies as well as the issue of social capital in relation to community relationships and to the involvement in the urban regeneration process of Xi'an. The overview of urban development and regeneration processes of Xi'an City shows that urban regeneration policies in Xi'an have transformed from industrial development-oriented before the 1970s to project-oriented and redevelopment-led in the 1980s and 1990s, and ultimately to the urban redevelopment-cum-conservation policies of the late 1990s. With regard to the urban conservation in Xi'an's four master plans, the focus of urban conservation plans has evolved from heritage monuments-centered in the 1980s to the multi-levels of conservation in the 1990s. In the latest or the fourth master plan, Xi'an's municipal government has proposed to restructure its urban land use and to re-fabricate traditional urban scenes. Following this policy,

项政策，整个历史城墙区内的城市更新过程采取的再开发和保护并进的策略，有利于地方旅游业和经济的发展。然而，相关文献表明，现有的保护政策和法规非常侧重于物质环境的保护。在地方更新过程中，一些非物质文化因素，如地方社区传统和原住民生活内容还未引起人们的足够重视。在这个过程中，地方原住民通常为了重建活动而经常搬迁。本研究认为这不利于历史街区的保护，也不利于综合更新目标的实现。

由于缺乏必要的参与式更新机制，本研究认为，在维护基于社区关系的地方原住民和促进社区参与到城市更新的过程中，社会资本发挥着重要作用。同时，本文探讨了影响社会资本积累的关键因素。基于社区居民关系及其联系的分析，本研究认为，社会资本有助于促进地方社区居民参与到城市更新过程中，有利于促进地方更新规划的顺利开展，有助于综合式保护目标的最终实现。不过，在此过程中，参与式的城市更新管理机制应当发挥着基础性的重要作用。

the entire historic CWA has been undergoing the urban regeneration process characterized by redevelopment-cum-conservation strategies in favor of the local tourism industry and economic development. Nevertheless, literature indicates that existing conservation policies and regulations in Xi'an focus much on the conservation of physical environments. Local intangible components such as the local communities and indigenous lives are overlooked in the local regeneration processes. During this process, the local original residents are often relocated to give way to the redevelopment activities. This book argues that this is detrimental both to the conservation of local historic districts and to the comprehensive regeneration aims.

With the lack of a necessary participatory regeneration mechanism, this book maintains that the local social capital plays a significant role in maintaining the local original residents based on community relationships and in facilitating community participation in the urban regeneration processes. At the same time, key factors affecting the accumulation of local social capital are examined. Based on community relationships and connections, this study contends that social capital contributes to the participation of local community in the urban regeneration process, the smooth implementation of local regeneration plans, and the comprehensive conservation objectives. Nevertheless, a facilitating and enabling participatory urban regeneration management mechanism is fundamentally important to this process.

第五章 具有较强社会资本的历史街区及其更新实践

CHAPTER 5　Urban Regeneration in Historic District with Social Capital based on Stronger Community Relationships

本研究在上一章中分析了转型期西安的城市规划、更新政策、城市管理机制以及城市发展当中各种形式的社会资本。接下来，本章将对西安一个具体的历史街区——西安鼓楼回民区——当中的城市更新实践和社会资本进行案例研究。首先，对回民区进行简单介绍，解释了街区的历史背景以及社会经济和行政管理特色。考虑到当地快速城市化进程，本章还探讨了在回民区进行城市更新的必要性。其次，探讨了西安地方政府主导的倾向于经济发展的城市更新政策。关于西安城市更新政策的转变，本研究注重分析了历史地段中的两个更新实践案例。其一是位于北院门历史街区，由地方政府主导的、通过重新建造传统城市风貌的"保护型"更新案例。该案例突显了参与式的城市更新和管理机制有助于基于社区的城市更新实践。其二是位于大麦市和洒金桥历史街区，以市场力驱动下的城市再开发和居民重置为特点的更新实践。这种倾向于经济发展的城市更新政策和以市场为主导的城市管治模式，对当地的街区历史和少数民族居民的生活产生影响，并可能影响社区社会资本的积累。随后，本研究还对回民区中的社会资本进行了探讨。作为一个少数民族聚居区，居民间的亲情和血缘关系有助于该区社会资本水平的提高。

In the previous chapter, Xi'an's transitional urban planning, regeneration policies, urban management mechanism, and various forms of social capital in urban development were examined. At this point, urban regeneration practices and the issue of social capital in a specific historic district, namely, the Xi'an Drum Tower Muslim District (DTMD), will be investigated. First, an introduction about DTMD is given, where its historical background as well as socio-economic and administrative features are explained. In the mean time, the need for urban regeneration in DTMD is explored while taking into consideration its rapid urban development. Second, Xi'an's local government-led pro-growth urban regeneration policies are explored. With regard to Xi'an's transforming urban regeneration policies, the book explores the regeneration practices in two historic quarters. The one is local government-led "conservation-oriented" regeneration plans through re-fabricating traditional urban scenes in Beiyuanmen street area, which underlines that a participatory urban regeneration and management mechanism contributes to the community-based regeneration process. The other is the market-driven redevelopment and residential relocation in the Damaishi and Sajinqiao streets area. These pro-growth urban regeneration policies and market-led mode of urban governance not only disrupts the conservation of local historic and religious residential lives, but also undermines the accumulation of community social capital. In addition, the issue

同时，本研究还考虑了影响当地社区社会资本的关键因素（涉及历史城市肌理以及居民对当地商业环境的依附性）、地方倾向于经济发展的城市更新政策和策略以及碎片化的城市管治模式。最后，基于对高水平社会资本的认识，本研究分析了在自建活动中的社区参与，并认为制度化的公众在地方城市更新过程中的参与机制，对综合式城市保护与更新目标的实现将起到基础性的重要作用。

of social capital in the DTMD is probed. As a Muslim residential quarter, kinship and blood ties strongly contribute to a high level of social capital in this district. At the same time, key factors affecting the community social capital (mainly the historic urban fabric and business setting attachment), local pro-growth urban regeneration policies and strategies, and fragmented urban governance are taken into account. Finally, based on the understanding of the high level of local social capital, community involvement in terms of self-construction activities is explored. It concludes by arguing that an institutionalized mechanism for public involvement in a local regeneration process is fundamental for carrying out comprehensive urban conservation and regeneration aims.

5.1 西安鼓楼回民区简介

5.1.1 历史背景

回民在西安聚居的悠久历史可以追溯到西汉年间（25 年～220 年）。丝绸之路在汉代发展起来以后，西部国家的人们逐渐开始东移，并在中国定居下来。随着时间的推移，阿拉伯人与汉族人开始长期共处，并互通婚姻（Hoyem，1989：2）。到了唐朝，皇帝李亨甚至借助了阿拉伯人的援助来平定安史之乱（755～763 年）。在成功镇压"安史之乱"之后，出于感恩，唐王授予很多阿拉伯官员和士兵在长安（今日的西安）长久居住的权利。而这些阿拉伯移民为了更好地了解长安人们的生活方式、礼仪和习俗，遂聚居在了回民区附近。这就是今天鼓楼回民区中"大学习巷"和"大学习清真寺"的最早来历（"大学习"本就意味着"广泛学习"的意思）（SMDPI et al.，1997：5）。到了元代

5.1 Introduction to the Xi'an Drum Tower Muslim District

5.1.1 Historical background

The history of Muslims living in Xi'an can be traced back to as early as the Han dynasty (A.D. 25-220). After developing the Silk Road, people from the West gradually moved eastwards and settled in China. Occasions such as Arabs mixing with the Han-Chinese people began to take place as time passed by (Hoyem, 1989: 2). During the Tang dynasty, then emperor *Li Heng* sought help from the Arabs in order to suppress the An-Shi Rebellion (A.D. 755-763). After successfully suppressing the rebellion, many Arab officers and soldiers were granted official permission to live in Xi'an (then Chang'an) in gratitude. To become acquainted with the Tang lifestyle, courtesies, and habits, the Arab migrants gathered around the Drum Tower Muslim District (DTMD). This was the origin of the *Daxuexi* lane as well as the *Daxuexi* mosque in DTMD (*Daxuexi* means "extensive study") (SMDPI et al., 1997: 5). During the Yuan dynasty (A.D. 1271-1368),

(1271～1368年），回族[1]被作为中国的穆斯林少数民族而长久地固定下来。他们主要居住在城墙之外，在唐代的西方郊区，他们在城墙内的西方市场以从事商业为生。但是在加入到对抗内蒙古人的战争之后，他们得到了新皇帝的信任，开始在城墙内的西方市场以北的两大街区中定居（Shi, 2008：418）。在清朝（1644～1911年），满族人再次把很多回族赶出城墙外。在抗日战争（1937～1945年）期间，许多穆斯林再次回到城市中心区（Hoyem, 1989；Shi, 2008）。如今，约有64000名穆斯林生活在西安，占整个城市少数民族居民的75.28%。在这些穆斯林居民中，30000多人居住生活在鼓楼回民区中（Yang, 2007：377）。

5.1.2 社会经济特征、行政管理体系以及城市更新的必要性

西安鼓楼回民区位于西安旧城区的中央位置，也是西安历史文化名城保护条例（SCXPC, 2002）和保护总体规划（XCPB, XACH, & XUPDI, 2005）中为数不多的完全被划定为历史街区的地段之一。回民区东起社会路，西至早慈巷；南起西大街，北至红埠街，整个街区面积约54公顷，拥有约60000人口，其中30000人为穆斯林居民（SCXPC, 2002；XMHDPPO, 2003）。鼓楼回民区在西安旧城区的位置见前图1-3所示。根据最新人口统计数据（即2000年第五次人口普查统计数据），

---

[1] 中国的回族：在中国少数民族中，与其他少数民相比，回族的人口更多，也更加分散。从历史的角度来看，回族是通过吸纳世界各地信仰伊斯兰教的人以及各种民族成分而形成的。穆斯林是伊斯兰教的信徒，穆斯林相信世间只有一位真神，即真主阿拉（Li, 2004）。

the Hui nationality[1] was consolidated as a Muslim minority nation of China. They lived mainly outside the city walls, in the western suburbs of Tang, and made their living largely as merchants at the West Marketplace inside the walls. However, by joining Han Chinese soldiers fight the Mongols, they obtained the new emperor's trust and began to settle down inside the city wall in two districts north of the West Marketplace (Shi, 2008: 418). During the Qing dynasty (1644-1911), the Manchurians forced many of them to move outside the city's walls again. Nevertheless, during the Chinese-Japanese war (1937-1945), many Muslims returned to the city center area once more (Hoyem, 1989; Shi, 2008). Today, there are approximately 64,000 Muslims living in Xi'an, making up 75.28% of the entire urban minority nationality residents. Among these Muslim residents, more than 30,000 live within the Xi'an DTMD (Yang, 2007: 377).

5.1.2 Socio-economic features, administration system, and the needs for urban regeneration

Xi'an DTMD is located at the very center of Xi'an CWA, and it is one of the few areas which has been entirely declared historic both in the Conservation Regulations on Xi'an Historic City (SCXPC, 2002) and in the Xi'an Conservation Master Plan (XCPB, XACH, & XUPDI, 2005). Ranging from Shehui Road in the east to Zaoci Lane in the west and from West Avenue in the south to Hongbu Street in the north, the whole DTMD is about 54 hectares, with a population of almost 60,000, of which 30,000 are Muslims (SCXPC, 2002; XMHDPPO, 2003). The location of this area is

---

[1] Hui nationality in China: In China's minor nationalities, Hui nationality is the one with more population and widely scattered compared with others. From historical perspective, Hui nationality came into being through absorbing many people believing in Islam all over the world, as well as the various national constituents. And a Muslim refers to an adherent of the religion of Islam, and Muslims believe that there is only one God, Allāh (Li, 2004).

回民区中的人口构成　　　表 5-1
Population composition of DTMD
Table 5-1

	男性（人） Male	女性（人） Female	总计（人） Total
穆斯林 Muslim	15005（50%）	14993（50%）	29998
汉族 Han Chinese	15572（52%）	14530（48%）	30102
总计 Total	30577	29523	60100

资料来源：2000 年西安市莲湖区第五次人口普查（SBLD, 2003）

Source: The Fifth Census of Lianhu District, Xi'an city, in 2000 (SBLD, 2003)

回民区中男女穆斯林性别比保持约 1：1。男性穆斯林居民总计 15005 人，女性穆斯林居民总计 14993 人。表 5-1 显示了穆斯林街区中人口构成的其他人口数据（SBLD, 2003）。

一般来说，穆斯林居民在中国形成了典型的"大分散、小聚居"式的居住特点。有学者认为这与他们的宗教信仰有关。由于穆斯林平时需要参加礼拜仪式，所以他们多居住在清真寺附近，也因此形成了特殊的穆斯林团体，例如"城市中的穆斯林街，或者乡村中的穆斯林村"（Ma, 1998b: 46）。这种居住模式也可以从西安市穆斯林居民的分布特点看出来，他们主要居住在三个历史街区（即新城区、莲湖区和碑林区），这几个区的穆斯林人口总数占到西安穆斯林居民总数的 80%（Yang, 2007: 377）。

在西安，教坊制度管理着穆斯林团体。教坊是一种围绕在清真寺周围的街坊，穆斯林教众和他们的家庭通常都生活在周围或附近。不论大小，教坊制度都是相互独立的。根据 1716 年颁布的《法典》，阿訇应该定期从教坊中的精英群体中选拔出。阿訇的职责是监督礼

shown in previous Figure 1-3. According to the latest available demographic data (i.e., the fifth census statistical data in 2000), the gender proportion remains approximately 1:1 between male and female Muslim residents in the DTMD. The male Muslim residents total 15,005, and the female Muslim residents total 14,993. Table 5-1 shows additional demographic data of the population composition of the DTMD (SBLD, 2003).

Generally, Muslim residents in China form the typical inhabitant tradition of "separating in large areas and gathering in small groups." Academics argue that this relates to their religious psychology. Out of their need to attend worship ceremonies, Muslims usually tend to live around mosques. Accordingly, they often form Muslim groups such as the "Muslim streets in cities or Muslim villages in countryside" (Ma, 1998b: 46). This kind of inhabitant mode can also be reflected in the distribution of Muslim residents in Xi'an city staying mainly in three districts (i.e., Xincheng, Lianhu, and Beilin), which together account for the 80% of the overall Muslim residents in Xi'an (Yang, 2007: 377).

In Xi'an, Muslim groups are governed by the *Jiao-Fang* system. A *Jiao-Fang* is a mosque-centered neighborhood, and the followers and their families usually live around or nearby. All *Jiao-Fang* systems are independent of one another whatever scales they may have. According to the code issued in 1716, an Imam should be chosen by the local elite board in each *Jiao-Fang* regularly. His duty is to oversee worship affairs as well as the teaching of students. Meanwhile, the local Hui elite board deals with many other matters such as resolving quarrels among the Muslims or the financial affairs of the mosque. In contrast with the Han society, the administration system in Muslim communities includes both governmental organizations and local mosque management committees, as illustrated in Figure 5-1 (Dong, 1995: 66-67).

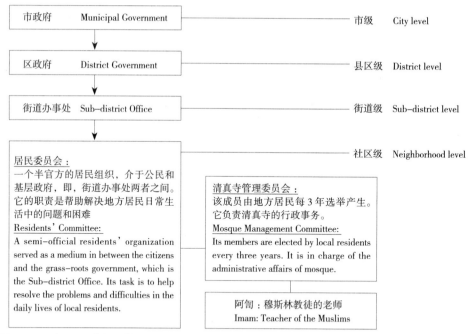

图5-1：西安回民区的行政管理体系
Figure 5-1: The Administration system in Xi'an DTMD
资料来源：(Dong, 1995：68)
Source: Author revised from (Dong, 1995: 68)

拜事务和向学徒们传授教义。与此同时，地方回族精英群体处理其他事务如解决穆斯林教徒之间的矛盾或清真寺的财政问题。与汉族社会不同，穆斯林社区中的行政管理体系既包括政府机构也包括地方穆斯林管理委员会，如图5-1所示（Dong, 1995: 66-67）。通过这种"双重渠道"社会管控结构，每个教坊中的回族精英群体管理清真寺，而地方政府管理居民委员会（Dong, 1995：74）。在西安回民区，教坊社区的空间布局在地方城市建筑中特点鲜明，被称为"七寺十三坊"。

当地城市肌理的空间特点

基于当地穆斯林居民的居住模式，鼓楼回民区的一个显著空间特点是"七寺十三坊"的空间格局（SMTL）。据有关文件记载，"七

This "double track" social control structure has the mosque controlled by the Hui elites in each *Jiao-Fang*, and the residents committee controlled by local government authorities (Dong, 1995: 74). In the Xi'an DTMD, the spatial features of *Jiao-Fang* are well revealed in the local urban fabric known as the "Seven Mosques and Thirteen Lanes."

Spatial features of local urban fabric

Based on the living pattern of local Muslim residents, one salient spatial feature of Xi'an DTMD is the "Seven Mosques and Thirteen Lanes (*qi si shi san fang*)" (SMTL). Documents show that the SMTL urban pattern came into existence at the end of the 18th century in the Qing dynasty (SMDPI et al., 1997: 7). Among the Seven Mosques, the Great Mosque is the most important religious building in Xi'an DTMD. Considered as a state-level heritage monument, most parts of the Great Mosque

寺十三坊"的城市格局是在18世纪末期的清朝开始形成（SMDPI et al., 1997：7）。七个清真寺中，化觉巷清真大寺是鼓楼回民区中最重要的宗教建筑。作为国家级遗产纪念碑，化觉巷清真大寺的主体部分是在明代建成的，在"文化大革命"中没有遭到严重破坏。今天，化觉巷清真大寺以其特有的传统建筑和周围的园林，成为西安一个主要的旅游景点，同时也是西安穆斯林值得骄傲和最神圣的地方（Hoyem, 1989：2）。十三巷指的是由十三条街道划分成的穆斯林居住区。图5-2显示了"七寺十三坊"格局的具体位置。在清朝末年的版图中，这些红点代表了七座清真寺的位置，彩色的线代表了主要的巷道。附近相同颜色的巷道构成一个穆斯林聚居区，生活在同一聚居区的居民，通常会去他们邻近的清真寺进行日常的祷告活动。同时，当地居民也是在这里来解决他们的个人或社会问题（Yang, 2007：255）。

"七寺十三坊"的城市格局突显了当地穆斯林街区的两个特征。一方面，"七寺十三坊"

was built in the Ming dynasty, escaping severe damages during the Cultural Revolution. Today, with its traditional architecture and surrounding gardens, the Great Mosque is a main tourist attraction of Xi'an, and the pride and the main sacred place of the Muslim people in Xi'an (Hoyem, 1989: 2). The Thirteen Lanes refer to local Muslim residential quarters delineated by 13 local streets. Specific locations of the pattern of SMTL are shown in Figure 5-2. In the map of the late Qing dynasty, the red dots indicate the locations of the Seven Mosques, and the colorful lines indicate local main lanes. The nearby lanes of the same color consist of one Muslim quarter. Normally, local residents living in the same quarter attend their nearby unique mosque to practice their daily praying activities. Moreover, this particular edifice is also the place where local residents tend to have their individual or social problems solved (Yang, 2007: 255).

The urban fabric of SMTL underlines two features of a local Muslim district. On the one hand, the SMTL provides traditional living space for local Muslims who live around a mosque in order to keep up with the regular congregation and praying activities. On the other hand, the SMTL is seen as an economic space by local residents. Doing businesses is the traditional way of making a living for many Muslim families. A survey indicates that about 70% of residents in Xi'an DTMD survive by engaging in retailing businesses (SMDPI et al., 1997: 70). Heavily supported by local needs and tourism industries, this area provides local businessmen with plenty of customers, while the courtyards serve as convenient working places. In fact, these business activities are so important to local Muslim families that they consider moving out from this area "unacceptable," because although moving away may result in better living conditions, it cannot make up for their economic losses since moving out means losing jobs (Li, 2004: 172; Wang, 2002c: 93; XMHDPPO, 2003: 10;

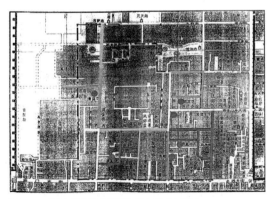

图5-2：西安鼓楼回民区的"七寺十三坊"
Figure 5-2: Seven Mosques and Thirteen Lanes (SMTL) in Xi'an DTMD
资料来源：（SMDPIetal., 1997：14）
Source: Author edited from (SMDPI et al., 1997: 14)

为生活在清真寺周围的穆斯林提供了一个居住空间，以便他们参加定期的集会和祷告活动。另一方面，当地居民把"七寺十三坊"看作一种经济发展空间。经商对很多穆斯林家庭来说是一种传统的谋生方式。调查显示，在鼓楼回民区，约有70%的居民依靠从事零售业来谋生（SMDPI et al., 1997: 70）。当地旺盛的需求以及繁荣的旅游业市场，为该地段的商业提供了源源不断的顾客，同时，这里的家庭院落也成为居民很方便的商业工作场所。事实上，当地的商业活动对穆斯林家庭来讲非常重要。因此，很多人都认为从该区中搬出去或被重新安置都是"无法接受的"。即便是搬出去可能会有更好的居住条件，但这还是无法弥补给他们带来的经济损失。对很多人来说，搬离这个地段可能就意味着失业（Li, 2004: 172; Wang, 2002c: 93; XMHDPPO, 2003: 10; Zhu, 1998: 257）。

城市更新的必要性

随着西安快速城市化的进程，历史街区也面临着诸多挑战与问题。通过2007年对莲湖区建设局的实地考察与访问，本研究概括出回民区发展面临的挑战可分为如下外部因素和内部因素。

外部因素

随着西安城市更新政策的逐步实施，旧城区的南大街、北大街和东大街的再开发与更新项目相继在20世纪80年代开始进行。相比之下，西大街的更新过程显得落后（XMPG, 2005）。2008年，随着西安第四次总体规划开始实施的契机，整个旧城区都被划定为商业旅游区，考虑到西大街所处的重要地理位置，总体规划将其建设成一条商业街。

Zhu, 1998: 257).

Needs for urban regeneration

With rapid urban development, there are many challenges and problems facing this historic district. Throughout field observation and interview to several working staffs in Lianhu District Construction Bureau in 2007, generally, the urban development challenges in the DTMD may be divided into external and internal perspectives.

External factors

With the gradual urban renewal policies in Xi'an, redevelopment and renewal projects within the South, North, and East Streets have started since the 1980s. Comparatively, West Street lags behind (XMPG, 2005). With the implementation of the Xi'an fourth master plan in 2008, the entire CWA will transform into a commercial and tourism zone and, considering its important geographic location, West Street into a commercial lane. After four years of intensive redevelopment from 2001 to 2005, including the massive elimination of residential buildings, West Street endeavors to be the First Western Gold Street (*xi bu di yi jin jie*), which will hopefully serve as a source of wealth for the entire Xi'an City (Sun, 2007). Accordingly, a state-of-the-art transportation system is necessary in order to vitalize the businesses in West Street. The original transportation system often results in traffic jams, seriously affecting the operation of many businesses here (Sun, 2007). Businesses within DTMD close to the West Street area are also severely affected. Local professionals contend that business operations in the Muslim district can hardly get out of its current predicament when there is no substantive renewal on the transportation system (Zhu, 1998: 258). Seeing this, Lianhu District Construction Bureau, since early 2000, proposed to widen Damaishi and Sajinqiao Streets to pave the way for the proposed enhancement of transportation in this historic district. Nevertheless, in order to carry out

从2001年到2005年，经过四年紧锣密鼓的再开发建设，包括拆除了大量的住宅楼，西大街成了西部第一金街，并有望为整个西安市创造财富（Sun, 2007）。为了振兴西大街的经济，建设一个先进的交通运输体系非常必要。不过，原有交通运输体系经常造成交通堵塞，严重影响了当地许多企业的正常运营（Sun, 2007）。事实上，回民区靠近西大街的企业也受到了较严重的影响。有专家认为，如果交通运输体系没有实质性的更新，那么穆斯林街区中的企业运营仍很难摆脱目前的困境（Zhu, 1998：258）。鉴于此，自2000年初开始，莲湖区建设局就提出了拓宽大麦市和洒金桥街道的建议，这为改善该历史街区的交通运输系统的计划迈出了坚实的一步。但是，根据地方政府的规划，这个项目的实施需要重新安置涉及的原住穆斯林居民，从而引发了许多争论。

内部因素

除了上述因素之外，城市更新还面临着伴随人口激增而来的许多地方性挑战（例如，破败的物质环境、破旧的基础设施和杂乱的建筑物）。根据分别于1964年、1982年、1990年和2000年对回民区进行的四次人口普查结果，图5-3显示了穆斯林街区中迅速增加的人口。这些少数民族中，90%以上是穆斯林。例如，2000年第五次人口普查结果显示，在34988个少数民族的人中，穆斯林居民人数达32318人，或者说占少数民族总数的92%（SBLD, 2003：28, 45）。急剧增加的城市人口给当地城市物质环境带来了巨大的压力。统计数据表明在西安，人均住宅面积快速增加。这种情况在西安回民区中也有所

this project, many original Muslim residents need to be relocated according to the local government's plan, thus arousing many discussions.

Internal factors

In addition to the above factors, there are also many local challenges confronting urban regeneration (e.g., dilapidated physical environments, rundown infrastructures, and cluttered constructions) that go with rapidly increasing population. This swift increase of urban population in the Muslim district is shown in Figure 5-3, according to four demographic censuses in DTMD in 1964, 1982, 1990, and 2000, respectively. Among these minority ethnic people, more than 90% are Muslim. For instance, in the fifth census of 2000, among the 34,988 minority ethnic people, there were 32,318 Muslim residents, or 92% of the total ethnic minority population (SBLD, 2003: 28, 45). With the increasing urban population, there is an urgent pressure on local urban physical environments. Statistics show that the floor space of residential houses per person in Xi'an has been increasing very fast. This situation is also reflected in Xi'an DTMD, and as a result, plenty of living space has been built here.

As reflected from the interviewees in Lianhu District Construction Bureau in 2007, another issue concerning changing local urban space relates to the transforming property rights. The introduction of land reform at the end of the 1980s has resulted in the redistribution of the original land use rights and the formation of a corresponding urban fabric. For example, there were 23 landowners in the Muslim district in the 1950s. Later, with the implementation of central governmental policies by the end of the 1990s, this area was subdivided into 85 plots (Li, 2002a: 54). Previously, one family owned and populated its own courtyard with the traditional, regional organization of houses and spaces. Now there are about 10 to 15 families in one courtyard

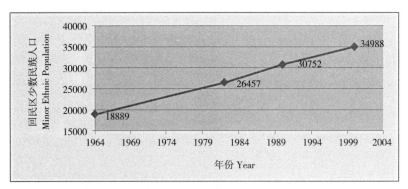

图 5-3：回民区中的少数民族人口
Figure 5-3: Minor ethnic population in DTMD
资料来源：(SBLD, 2003)
Source: Author synthesized from (SBLD, 2003)

体现，结果，这里开始建设大量的住宅。

2007 年莲湖区建设局的调查反映出，与当地城市空间的变化密切相关的另一个问题涉及建筑产权的转变。20 世纪 80 年代进行的土地改革，造成了原有土地使用权的再次分配，以及相应城市肌理结构的形成。例如，20 世纪 50 年代，穆斯林街区中有 23 个土地所有者。随着 20 世纪 90 年代末中央政府政策的实施，穆斯林街区被划分成了 85 块（Li, 2002a：54）。以前，一个家庭对自家庭院拥有使用权，该院落具有传统的、地域组织特色的房屋与空间。而现如今，在这个被大规模利用的地段中，一个庭院就可能居住了 10 到 15 户家庭（Hoyem, 1989：3）。这种情形也加剧了目前城市环境中拥挤不堪的状况。作者在 2007 年走访了北院门街道办公室的工作人员以及大麦市和洒金桥地段的居民代表，总结出回民区现存的问题主要涉及以下几方面。

1) 居民混乱无序的自建行为

为了改善居住环境，不少居民随意在自家房屋上增建楼层，或在公共空地上构建新建筑

in the densely exploited sites (Hoyem, 1989: 3). These situations worsened the huddling congestions in current urban environments. DTMD was identified during the author's interviews with a few working people in the Beiyuanmen Street Office and local residents in Damaishi and Sajinqiao area in 2007, as including the following aspects.

1) Chaotic self-construction activities by local residents

To improve living areas, many residents randomly add more stories on top of their own buildings or construct other new buildings in public areas. Such activities often ignore existing urban planning and conservation regulations. As a result, many traditional courtyards have been destructed and replaced with simple three- to four-storey brick-made buildings. Characterized by crude construction, poor quality, and simple decoration (Cao, 2005: 6), these tentative buildings form a sharp contrast with the overall historic district fabric (shown in Figure 5-4) (Li, 2002a: 39; Xiao, 2004: 60). In addition, these self-construction activities have been causing such problems as infringement on public space, waste of space, and poor lighting and ventilation (XMHDPPO, 2003: 11). The main reason for these self-construction activities by local residents is due to their actual needs for more housing space and residential areas

图 5-4：回民区的居民自建现象
Figure 5-4: Local self-constructions in the DTMD
图片来源：作者摄于 2008 年
Source: Author in 2008

物。这些行为经常违背现行的城市规划和保护方案与法规。结果，许多传统庭院都被拆毁，取而代之的是用砖建成的 3～4 层简易楼房。这些尝试性的构筑物结构粗糙、质量低下、装修简单（Cao, 2005：6），与整个历史街区的整体建筑风貌形成鲜明对比（见图 5-4）(Li, 2002a：39；Xiao, 2004：60)。另外，这些自建活动也带来不少问题，例如，侵占公共空间、空间浪费以及照明和通风条件差（XMHDPPO, 2003：11）。调查发现，当地居民进行这些自建活动的主要原因是，一方面，由于不断增加的家庭成员，有些家庭确实需要更大的居住空间和面积。另一方面，有些居民期望最终在与地方政府洽谈拆迁补偿方案时，能够从"拆一还一"的政府政策中获得更多补偿（Dong, 1995：94；Wang, 2002c：95）。不过，很多受访居民向作者强调，第一个原因对很多家庭来讲占主要原因。

with the increasing family members. Another reason for local self-construction activities is that, when some residents may eventually give way to the redevelopment plan promoted by local government, they expect to get more compensations according to the government policy of "demolishing one and returning one" (Dong, 1995: 94; Wang, 2002c: 95). Many local residents opine that the previous reason is the dominant one.

2) Environmental disruption accompanying flourishing business activities

In Xi'an DTMD, about 70% of local residents have been conducting private businesses (SMDPI et al., 1997: 70), and almost every family is doing businesses here. Generally, local Muslim businesses include catering, butchery, antiques, and curios, among many others (Li, 2004: 172-174), which normally take place at shops' entrances along narrow streets. The areas behind the shops serve as the place for butchery and preparation (e.g., slaughtering animals and boiling meat). Some retailers and vendors even set up their stalls in more popular streets like

图 5-5：回民区街道上的零售活动
Figure 5-5: Retailing activities in the street of DTMD
图片来源：(A)(Li, 2002a: 88)；(B) 作者摄于 2006 年
Source: (A): (Li, 2002a: 88) reproduced by courtesy of Hongyan Li; (B): Author in 2006

2) 繁华的商业活动引发的环境破坏

在西安回民区中，大约 70% 的居民从事着私营企业 (SMDPI et al., 1997：70)，从家庭来看，几乎每户家庭都在做地方生意。一般而言，地方穆斯林企业包括餐饮、肉食店、古董店、古玩店以及其他，这些生意通常就在狭窄的街道两旁的商店的入口处进行。这些商店后面的区域是用来屠宰和准备的（例如，屠宰动物和煮肉）。一些零售商和商贩甚至在一些更繁华的街道建立自己的商铺，如在北院门或西羊市街道（图 5-5），尤其是当他们商店的位置不易发现或比较偏远的时候。作为西安城墙区内最繁华的市场之一，自 20 世纪 90 年代以来，这些商业活动在为地方商人带来巨大利润的同时，也产生了许多交通和环境问题 (Zhu, 1998：258)。

3) 基础设施落后、街区环境破败

当地街区中很多环境问题，如街区狭窄道路上的交通拥堵和废气排放问题，都与落

Beiyuanmen or Xiyangshi Street (Figure 5-5), especially when their shops' positions are hidden from public view or are relatively remote. As one of the busiest marketplaces within Xi'an CWA, it brings great fortune to local businessmen, but it has also generated many traffic and environmental problems ever since the 1990s (Zhu, 1998: 258).

3) Lack of basic infrastructure and rundown district environment

Many environmental problems such as traffic jams and exhaust fumes in the narrow district roads abound with the lagging district environment management system. Moreover, some residents and private property owners are accustomed to discharging used or contaminated water directly to the public sewage system, which often results in the blockage of sewage pipes (Cao, 2005: 16) (Figure 5-6). These problems seriously affect the reputation and value of this historic residential district. Most local residents express their expectation of having their local living environment improved but are unwilling to relocate primarily because of their religious beliefs and their economic lives, which are connected to local tourism settings.

后的环境管理系统有关。有些居民和营业主习惯于将使用过的污水直接排放到路面上的公共雨水道系统中,经常导致路面下水道被堵塞(Cao, 2005：16)(图5-6)。这些问题严重影响了该历史街区的知名度与价值。同时,大多数当地居民期望他们的居住环境能够得到改善,但却不愿意搬迁,这主要与居民的宗教信仰以及很多家庭依靠地方旅游业来进行经济生活有关。

图5-6：回民区大麦市街的现状
Figure 5–6: The situation of current Damaishi street of DTMD
图片来源：作者摄于2007年
Source: Author in 2007

5.2 鼓楼回民区的更新实践

根据西安不同时期的总体规划方案,回民区的更新实践有很多不同的特点。作者于2007年在莲湖区建设局进行的访谈中,受访者指出,当地的城市更新以整治为主要特点,包括城市保护与再开发两方面内容。一方面,根据西安的保护总体规划,为了迎合当地旅游业的发展,城市更新规划倾向于保护与重建传统城市景观。另一方面,当地的城市更新以大规模的房地产再开发为特色,在此期间,出现了大量的房屋拆迁和居民重置现象。通过改善城市的物质环境,可以吸引私营企业,为当地的经济发展创造更多机会。概括来讲,自20世纪80年代初期以来,鼓楼回民区的更新实践可划分为两个阶段。

1) 零星的居民自助式房屋整修和建设(1978年～20世纪80年代末)

自20世纪80年代以来,随着居民改善住房条件的经济能力不断提高,地方居民开始依靠自己的力量来整修自家破败的住房。不过,这些活动还仅限于在原有住房上增加

5.2 Urban Regeneration Practices in Xi'an DTMD

According to Xi'an master plans set in different periods, the regeneration practices in DTMD also have many embedded characteristics. As indicated by the interviewees in Lianhu District Construction Bureau in 2007, local urban regeneration approach, featured by urban renovation work (*cheng shi zheng zhi*), comprises both urban conservation and redevelopment. On the one hand, according to the Xi'an conservation master plan, local regeneration plans tend to conserve and reconstruct traditional urban scenes catering to the local tourism industry. On the other hand, local regeneration plans feature large scales of property-led redevelopment, during which massive housing removal and residential relocation would have to take place. By improving the physical environment, more opportunities to attract private businesses for local economic development are created. Since the early 1980s, there have been roughly two stages for regeneration practices in Xi'an DTMD (Cao, 2005; Xiao, 2004).

1) Sporadic self-help refurbishment and constructions (from 1978 to the late 1980s)

With their increasing economic capability to improve their housing conditions since

几层临时建筑。有些原住房主开始拆毁旧房屋，并在原址建造两层砖盖的住房。不过，由于经济条件所限，这种建筑还较少，当时在回民区也仅占18%左右。

2) 地方政府主导的大规模再开发（20 世纪90年代～至今）

自20世纪90年代以来，西安市政府提出了进行大规模城市再开发的政策。随着20世纪90年代初期，西安第三次总体规划的提出，一种新型城市发展结构也应运而生。旧城区实施的城市更新政策提出了进行房地产主导的城市再开发实践，与对著名的历史文物古迹进行保护的双重要求。20世纪90年代末，当地政府进一步提出了第四次总体规划，其中除了大规模的房地产主导的再开发项目之外，该规划提出对整个旧城区进行"保护"。对整个历史古城进行"保护"的规划宣传，突出了通过促进当地旅游业的发展，来吸引私人投资并促进地方经济发展的发展政策。

5.2.1 地方政府主导和促进地方经济发展的更新实践（自20世纪90年代）

为了改善当地的城市物质环境，促进经济发展，自20世纪90年代早期以来，西安旧城区内开展了大规模的房地产主导的城市再开发活动。20世纪90年代早期，在地方市政府提出再开发政策之后，随之而来的是当地居民大规模私人住房自建的快速繁荣期，这种现象在20世纪90年代几乎遍布整个回民区。(图5-7)显示了鼓楼回民区东部的房屋建设时期。

事实上，不少居民家庭甚至修建了许多不必要的房屋。究其原因是，根据1989年颁布的《西安城市住房拆迁管理办法》，以及当

the early 1980s, local residents began to refurbish dilapidated housing on their own. Nevertheless, those activities were limited to partial additions of more tentative floors. Some original housing owners began to replace their old housing with two storey brick-made apartments in situ. Nevertheless, considering the economic limitations, the numbers of constructed buildings at this time were few, accounting for only 18% in the DTMD.

2) Massive government-led redevelopment (from the early 1990s to the present)

Since the 1990s, the Xi'an municipal government has proposed the massive scale of urban redevelopment policies. With the proposal of the third master plan at the beginning of the 1990s, a new urban development structure in Xi'an was put forth. Urban regeneration policies within the CWA have promoted the property-led redevelopment practices together with the conservation of outstanding heritage monuments. By the end of the 1990s, with the proposal of the fourth master plan, urban "conservation" in the entire CWA had been put forward in addition to massive property-led redevelopment. By propagating the overall urban "conservation" plan, the intention of promoting local tourism industry and attracting private investments to boost local economy development was underlined.

5.2.1 Local government-led pro-growth urban regeneration plans since the early 1990s

To improve local physical environments and promote economic development, large scales of property-led urban redevelopment activities have taken place within Xi'an CWA since the early 1990s. When the local municipal government put forth the redevelopment policies in the early 1990s, it was accompanied by an immediate boom of private houses construction by local residents nearly throughout the DTMD in the 1990s.

图 5-7：回民区东部的房屋建设时期
Figure 5-7: The building periods in the east section of the Muslim district
资料来源：(XMHDPPO，2003：24)
Source: Author edited from (XMHDPPO, 2003: 24)

地政府"拆一还一"的再开发政策，在政府主导的再开发过程中，当居民被迁移或重置后，受影响的地方业主可以获得相应的补偿(SCXPC，2003：第18条)。为了得到更多的补偿，很多居民都开始迅速建造或增加他们的住房面积。

拆一还一

这个政策意味着在住房重建项目中，作为对被拆除老房子的补偿，发展商会置换给原住居民与被拆老房子同等面积标准的新住房。但是，这并非完全免费。一个人可以以低于其成本的价格购买一个新住房。很多时候，有些家庭希望得到更多的住房。原则上，在收到他们应得的补偿部分后，每户家庭仍能以成本价购买更多的住房面积，来达到国家制定的居住标准(人均居住面积为8平方米，约合人均建筑面积12平方米)。另外，如果在此基础上，有人还想得到更多的住房面积，就必须以市场价格进行购买。2005年时，当

The building periods in the east section of the Muslim district are shown in Figure 5-7.

In fact, many residents have even built many unnecessarily large houses. The main reason was, according to the *Xi'an Urban Housing Removal Management Method* enacted in 1989 and the local governmental redevelopment policy of "demolishing one and returning one," that affected local property owners in the government-led redevelopment processes could get corresponding compensations when they are displaced or relocated (SCXPC, 2003: Article 18). To get more compensation, many residents built or increased their housing areas hastily.

"Demolishing one and returning one" (chai yi huan yi)

This policy means that in housing redevelopment projects, the developer will return the original residents to the same area with a new apartment as compensation for demolishing their old houses. Nevertheless, it is not free. One can get a new apartment in the same area at a price lower than its cost price. Very often, some families want more. The principle is, after receiving their compensation, each family can buy extra areas at the cost price to meet the national living standard ($8m^2$ of living area per person, approximately equaling $12m^2$ of building area per person). Further, if one wants even more areas, he has to buy it at the market price, which in Xi'an in 2005 was two to three times higher than the cost price (Dong, 1995: 88).

Financial consideration, housing removal and residential resettlement

At the end of 1991, the Xi'an municipal government issued the first 49 renewal blocks in Xi'an's rundown urban areas, which were all taken over by various housing and real estate companies. The strategies in these projects were simple and straightforward: housing removal and residential relocation (Dong, 1995: 89). These were the cheapest and simplest ways of conducting renewal projects. From the perspective of the local

地的住宅市场价格要比成本价高出2～3倍（Dong, 1995：88）。

财政因素、房屋拆迁和住宅重置

1991年末，西安市政府公布了第一个针对西安旧城区中49块地段的更新决策，随后，有关更新项目全被住房和房地产公司所接管了。这些项目的策略也相对明确：房屋拆迁和居民重置（Dong, 1995：89）。这是进行更新项目时最便宜、最简单的一种方式。从地方政府和私人发展商的角度来看，考虑到出租土地使用权可以获得巨大的经济效益，他们也乐于通过调整土地性质，把住宅用地调整为商业—住宅综合体，以此来提升原有土地价值。原住居民则通常被重置于郊区（Dong, 1995：90）。但是，伴随这个目标而来的，往往是居民、地方政府和私人开发商之间尖锐的冲突。事实上，这种做法也不利于地方政府提出的保护规划的最终落实。例如，根据有关保护规划（XCPB, XACH, & XUPDI, 1994），西安旧城区的人口数量希望进一步得到控制或者减少。但是为了获得更大的经济效益，私人开发商则会希望通过建设更多楼层来吸引更多的人，因而西安旧城区内的实际人口数目不降反升。而对很多地方居民而言，他们认为当地最严重的问题还不是他们的住房问题，而是街区缺乏必要的城市基础设施，以及社会服务、文化机构和高效的管理机制。同时，随着很多居民自建房屋的不断增加，私人开发商往往需要支付巨额补偿费和再开发成本，使得回民区变得越来越"贵"，因此，当地政府主导的更新规划也显得很难进行。值得一提的是，无论是在自上而下的当地政府主导的再开发项目过

government and private developers, a land's value can rise by changing its function from residential to commercial-cum-residential, considering the enormous profits from leased-out land use rights. Original residents are usually resettled in sub-urban areas (Dong, 1995: 90). Nevertheless, along with these intentions are sharp conflicts among the residents, local governments, and private developers. This intention is also detrimental to conservation plans proposed by local governments. For instance, according to the conservation plan (XCPB, XACH, & XUPDI, 1994), the population within Xi'an CWA should be reduced. However, in order to get more profits, all private developers wish to build more floor areas to attract more people, while the urban population within Xi'an CWA is itself growing continuously. For many local residents, the most serious problems within the district are not their houses, but the lack of basic urban infrastructures, social services, cultural institutions, and efficient management. In addition, with more unnecessary private structures built, DTMD became very "expensive" for private developers to compensate and redevelop, making local government-led regeneration plans even more difficult to pursue. Noteworthily, whether it is the top-down government-led redevelopment or the self-help constructions by local residents, Huis tend to actively participate in local regeneration plans and prefer to take the principal role in it (Dong, 1995: 94). The involvement of Huis in the local regeneration process also relates to the urban "conservation" policy within the CWA. When the regulation on the building height control within the CWA was enacted in 1993 (SCXPC, 1993), the maximum building area was also "limited" to some extent. This actually undermined the invested interest of private sectors. Furthermore, considering the cohesive community neighborhoods with a great deal of social capital, the involvement of

程中，还是在地方居民自发的自建活动过程中，当地居民都倾向于积极参与到更新规划的实施环节中，并且乐于发挥主导作用（Dong，1995：94）。回族在地方更新过程中的参与也和旧城区实行的有关城市"保护"政策相关。1993年颁布的关于控制旧城区中建筑高度的规定（SCXPC，1993），使得新建建筑面积也被"控制"在一定范围内。这实际上影响了私营部门的投资利益。此外，考虑到拥有大量社会资本的凝聚力较强的社区生活，当地居民通过自建活动而参与到更新规划中，也可以被认为是一种当地居民自发的城市更新规划方式。

地方政府主导、通过重建传统城市景观进行"保护型"的更新规划

在当地各种更新实践中，北院门历史街区自1991年以来实施的人文景观规划项目和回民区自1997年以来实行的保护项目，在探索该历史街区的更新实践中是最为显著的。其中，北院门项目侧重于关注传统街道景观的再开发与修缮，而回民区的保护项目则侧重于探讨大规模住宅区的保护与再开发。1993年西安回民区的更新项目被西安市建设局作为一个试点项目批复开展，也是希望借此来寻求一个对整个历史街区都可行的更新模式。自那以后，不同的部门机构，如，西安市规划局、西安市城市规划设计研究院、大清真寺以及学院机构（西安交通大学和挪威理工学院）都参与到对更新模式的探索中来（XMHDPPO，2003：2）。

<u>政府主导的北院门历史街区人文景观保护项目</u>

在西安，北院门历史街区有着复杂的功能

local residents in regeneration plans in terms of self-help constructions was not only an autonomous activity but also a way out of local regeneration plans.

Local government-led "conservation-oriented" regeneration plans through re-fabricating traditional urban scenes

Among various local regeneration practices, the Cultural Landscape Project of Beiyuanmen Street since 1991 and the Muslim Historic District Protection Project since 1997 are the most conspicuous in exploring the regeneration of this historic district. The Beiyuanmen Street project focused much on the redevelopment and improvement of traditional streetscapes, while the Muslim Historic District Protection Project has explored the conservation and redevelopment of massive residential areas more. In 1993, the regeneration project of Xi'an DTMD was approved by the Xi'an Municipal Construction Bureau as a pilot research project, wishing to seek a feasible regeneration model for the whole historic core area. Since then, various agencies such as the Xi'an City Planning Bureau, the Xi'an Urban Planning and Design Institute, the Great Mosque, and academic institutes (Xi'an Jiaotong University and Norwegian Institute of Technology) have been involved (XMHDPPO, 2003: 2).

<u>Government-led "Beiyuanmen Street Cultural Landscape Project"</u>

In Xi'an, Beiyuanmen Street has a very complex character. On the one hand, it is a monumental axis between the Drum Tower and the Xi'an Municipal Government Building. On the other hand, influence from the Muslim district through the business activities and the activities of the courtyards and of the neighboring streets are more evident here compared with other streets (Hoyem, 1989: 3). Facing the derelict street condition, in 1991 the Xi'an municipal government decided to renew it, focusing mainly on the buildings and courtyards flanking Beiyuanmen Street. The

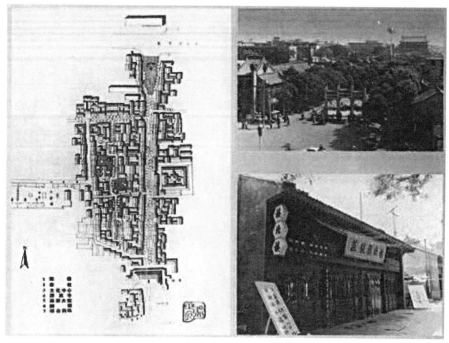

图 5-8：西安市政府关于北院门街道的更新方案
Figure 5-8: Official proposal on Beiyuanmen street renewal
资料来源：2005 年当地政府的宣传资料
Source: Local official promulgation materials in 2005

特点。一方面，它是鼓楼和西安市政府大楼的轴线。另一方面，和其他历史街区相比，穆斯林街区通过商业活动的影响力更为明显，庭院和街道的活动也更加显著 (Hoyem, 1989: 3)。面对破败的街道状况，1991 年，西安市政府决定对其进行更新，主要是针对那些北院门街道两边的建筑和庭院。这些重建的建筑采用了明清时期传统的住宅建筑外观（图 5-8）。在 20 世纪 90 年代末之前，政府没有在这一街区中再开展其他更新活动（Xiao, 2004：59）。

为了迎合当地旅游业的发展，西安第四次总体规划强调对传统城市特色的保护 (XCPB, XACH, & XUPDI, 2005:19)。相应地，2005 年，政府提出了西安北院门街道人文景观规划方案。（图 5-9）显示了西安市城市规

reconstructed buildings adopted the traditional housing facade of the Ming or Qing dynasty (shown in Figure 5-8). No further official activities were done in this area until the end of the 1990s (Xiao, 2004: 59).

The Xi'an fourth master plan stresses the conservation of traditional urban features to cater to the local tourism industry (XCPB, XACH, & XUPDI, 2005: 19). Accordingly, the Cultural Landscape Planning of Xi'an Beiyuanmen Street was proposed in 2005. Figure 5-9 illustrates the plans proposed by the Xi'an City Planning and Design Institute, which evidently promote the local tourism industry by exhibiting historic buildings, traditional courtyards, the Great Mosque, and so forth, and re-fabricating streetscapes like the Beiyuanmen and Xiyangshi traditional tourist streets.

During the implementation of the local

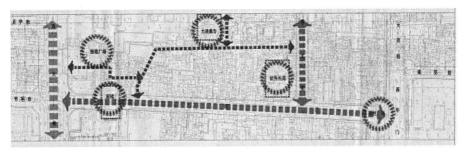

(A)

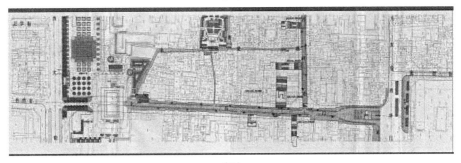

(B)

图 5-9：旅游路线规划（A）；总体规划（B）
Figure 5-9: Tourist Route Planning (A) and General Planning (B)
资料来源：(XCPDI, 2005)
Source: Author edited from (XCPDI, 2005)

划设计院所提出的规划方案，主要通过展示历史建筑、传统庭院、清真大寺等已有历史建筑物，以及重新建造北院门和西羊市的传统旅游街道景观，这些都有利于促进当地旅游业的发展。

在当地城市更新规划的实施过程中，各级政府机构对其控制与管理。例如，北院门人文景观街道管理委员会负责北院门传统街道的更新和维护，而西大街管理委员会负责管理西大街的发展。2006 年开展景观规划项目时，为了避免由于各级机构职责不明而带来的工作低效率，莲湖区政府专门成立了钟鼓楼广场管理委员会和北院门人文景观街道管理委员会。委员会的主要职责之一就是通过改善物质环境来吸引更多的投资（XLDG，

regeneration plan, control and management were under the supervision of various government agencies. For example, the renewal and maintenance of the Beiyuanmen traditional street was the responsibility of the Management Committee of Beiyuanmen Cultural Landscape Street, while the development of neighboring West Street was under the regulation of the Management Committee of the West Street. While carrying out the landscape planning project in 2006, the Lianhu district government set up the Management Committee of Bell-Drum Tower Square and Beiyuanmen Cultural Landscape Street to avoid inefficiency caused by ambiguous duties of various agencies. One of the Committee's main functions was to attract more investments by improving the physical environment (XLDG, 2005: 1). Similarly, specific governmental agencies were set up in nearby projects. For instance, the Management

2005：1）。类似地，在临近的项目中也成立了专门的政府机构。例如，2005 年当地区政府成立了西大街管理委员会，其目的就是促进街道振兴，以此吸引私人投资（XLDG，2005：2）。成立这些专门委员会的初衷是，减少项目管理中的各个机构职责的不明确性，不过，这种分散的管理体系在实践中反而可能使分工更加不明确，结果，就导致一些地方的开发活动完全处于无人监管的状态。为了应对这种情况，地方政府于第二年即 2006 年，成立了一个类似机构，即，历史城区保护与更新领导小组，由其来负责各个社区更新项目（XUPDI，2007）。

政府主导的"回民历史街区保护与更新项目"

西安回民区的另一项更新尝试就是地方政府主导的"回民历史街区保护与更新项目"。该项目自 1997 年开始实施。作为中国国家科学委员会和挪威发展合作机构共同正式签署的九份文件之一，自 1997 年以来，政府就已经在回民区选定一处居民区来开展城市更新项目（XMHDPPO，2003：2）。选定的区域总面积约 4 公顷，范围包括：南起化觉巷、北至西羊市；西起广济街、东至北院门（图 5-10）。

该试点项目提出要实现的三个目标（Hoyem，1989：4；Li，2002a：40）：

1）改善当地居民的居住质量；

2）保护清真大寺周围的环境；

3）促进地方经济增长，修复／复原保存完好的庭院和建筑。

在学术界的倡议下，地方政府在实施更新项目的过程中采取了相对综合的措施（Hoyem，1989：4）。更新项目中的多方面利

Committee of the West Street was established by local district government in 2005 to promote street revitalization and to attract private investments (XLDG, 2005: 2). The original intention of setting up these special committees was to reduce ambiguity in project management, but with its fragmented management system, it often turned out to be even more ambiguous in practice, resulting in situations where some local development activities had no management at all. In response, the local district government set up a similar institution just a year later in 2006, the Conservation and Renewal Leading Team of Historic Urban Areas to take charge of the various community renewal projects (XUPDI, 2007).

Government-led "Muslim Historic District Protection Project (MHDPP)"

Another regeneration intention in DTMD is the local government-led "Muslim Historic District Protection Project (MHDPP)," which has been in implementation since 1997. As one of the officially signed nine cooperation contracts between the National Science Committee of China and the Norwegian Agency for Development Cooperation (NORAD), urban regeneration practices in a selected residential quarter have been conducted since 1997 (XMHDPPO, 2003: 2). The total selected quarter is about four hectares within the following borders: to the south, Huajuexiang; to the west, Guangji Street; to the north, Xiyangshi Street; and to the east, Beiyuanmen Street (Figure 5-10).

As a pilot project, it was proposed with three objectives (Hoyem, 1989: 4; Li, 2002a: 40):

1) to improve the dwelling quality of local residents;

2) to safeguard the environmental setting of the Great Mosque; and

3) to stimulate local economic growth and to restore well-preserved courtyards and buildings.

Advocated by academics, the local municipal government has taken a comparatively

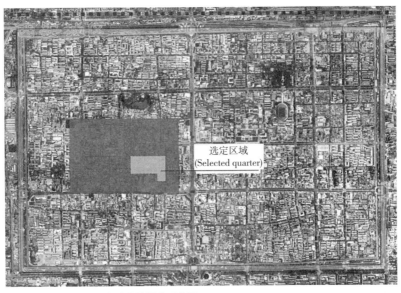

图 5-10：西安回民区中选定的居住区范围
Figure 5-10: Selected residential quarter within Xi'an DTMD

益相关者，包括当地社区、地方政府以及专家学者所起的作用，都进行了考虑。霍伊姆（Hoyem）认为这些利益相关人士所起到的作用可以从以下几方面看出（Hoyem, 1989：4）：

1) 通过社区参与的方式，来决定居民的需求和愿望的优先级别，并且充分利用地方居民的人力物力资源进行发展；

2) 通过市政府的登记和规划，利用街区和地方的潜能，来改善现有住宅结构；

3) 在专业工程师的协助下，为当地居民营造一个良好的生活环境。

根据有关协议，在该更新项目中，西安市政府提供基础设施建设所需资金，并负责临时办公室（即回民区项目办公室）员工的薪资；而挪威发展合作机构需提供经认定的传统建筑修复所需资金；地方居民需要支付自家房屋的更新费用（Li, 2002a：40）。在这样一个结构化的基金体制下，该项目由隶属

integrated approach in implementing the regeneration projects (Hoyem, 1989: 4). The roles of various stakeholders including local communities, authorities, and professionals were taken into consideration. Their contributions, as Hoyem contends, cover the following (Hoyem, 1989: 4):

1) to define the priority of the needs and wishes of the tenants, and to develop and take advantage of the human and material resources of the inhabitants by means of community participation;

2) to improve the existing housing fabric by using the potential of the district and the location through registration and planning with municipal authorities; and

3) to create a good living environment for local inhabitants with professional architects' work.

According to the agreement, during the regeneration practices, the Xi'an municipal government will provide the fund for infrastructure and the salaries for the staff working in a temporary working office (i.e.,

于西安房地产第二分局的开发公司负责开展（XMHDPPO，2003：56）。尽管促进地方经济发展是更新项目的重要目标之一，但是在该项目工作机制下的更新实践并未完全以获得最大的经济回报为目标（XSB，2005b）。值得注意的是，地方社区也参与到了规划和实施的过程中，通过各种利益相关者的合作，改善庭院以及住房的物质环境的具体规划目的得到了实现（Hoyem，1989：4）。

到了2002年末时，选定区域中的试点项目完美落幕。化觉巷的125号庭院、西羊市街道的77好庭院以及北院门街道的144号街道都得到了妥善修复。图5-11展示了修复后的庭院以及新建的实验性住房的情况（XMHDPPO，2003：51）。

修复后的化觉巷125号院落给人留下了深刻的印象，并于2002年被授予了联合国教科文组织的文化遗产保护奖。这项保护项目的成功和它独特的更新机制和机构设置密

Muslim District Project Office, etc.), NORAD will provide the fund for the restoration of traditional buildings; and local residents will need to pay for the renewal costs of their own houses (Li, 2002a: 40). With this structured funding system, this project conducted by development companies belonged to the Second Branch of the Xi'an Real Estate Bureau (XMHDPPO, 2003: 56). Even though promoting local economic development is one of its ultimate objectives, the regeneration project in such a working mechanism does not aim at maximizing the economic returns (XSB, 2005b). Notably, local communities were engaged in the processes of planning and implementations as well. Through the collaboration of various stakeholders, some detailed plans for improving the courtyards and housings were worked out (Hoyem, 1989: 4).

By the end of 2002, the experimental projects in the selected quarter came to an end. Courtyard nos. 125, 77, and 144 in Huajue Lane, Xiyangshi Street, and Beiyuanmen Street, respectively, were restored (XMHDPPO, 2003: 56). The conditions of the restored courtyards and the newly built experimental model houses are illustrated in Figure 5-11 (XMHDPPO, 2003: 51).

（A）

（B）

图5-11：（A）修复后的高家院（北院门144号）；（B）新建的实验性住房模型
Figure 5-11: (A) Family Gao courtyard (no. 144 in Beiyuanmen Street); (B) Newly built experimental model apartment
资料来源：（XMHDPPO，2003：51）
Source: (XMHDPPO, 2003: 51)

不可分。正如联合国教科文组织（UNESCO，2007）指出的，在整个更新过程中展示的公众参与是非常值得关注的。项目组在一开始就向社区居民分发调查问卷来寻求他们最初的参与。另外，他们还参与了前期规划、规划、决策和实施的过程。这种机制极大地促进了各种利益相关者，包括地方专家、学者、政府官员和社区居民之间的合作（Leaf & Hou, 2006：574），从而加强了社区联系（UNESCO, 2007：235）。不过，人们也注意到，这种地方社区参与的工作机制是由研究学者所提出和最终决策的，他们认为"公众参与是整个街区保护项目中的一个重要组成部分"（XMHDPPO, 2003：63）。

更新规划和开发控制机制

在回民历史街区保护与更新项目进行的同时，成立了一个特殊的规划和开发控制机制来执行提议的规划（图5-12）。这个机制的一个主要特色就是在居民参与了更新过程。正如上面所提到的，各种利益相关者，尤其是地方居民都在前期规划、决策和实施的阶段中有所参与。为了处理项目管理中的不确定性和动态因素，在北院门街道144号庭院成立一个回民区项目办公室，这对地方居民而言是件值得高兴的事，因为这为他们进行咨询提供了方便。为了更好地了解地方意见和要求，街道委员会事先举行了多次会议，在这个会议中，社区代表、受保护庭院中的居民、项目办公室人员以及地方官员都有参加。事实上，在更新项目实施过程中，社区居民也表现出了极大的兴趣和热情。他们的热情参与对更新项目的顺利实施具有建设性的作用（XMHDPPO, 2003：63）。当项目进

So impressive was Courtyard no. 125 in Huajue Lane that it was presented the 2002 UNESCO Cultural Heritage Protection Award. The success of this conservation project has much to do with its special regeneration mechanism and institutional setup. As pointed out by UNESCO (2007), public engagement exhibited throughout the entire processes was noteworthy. Questionnaires were distributed at the very beginning to local residents to seek their initial inputs. Moreover, they were also engaged in the pre-planning, planning, decision-making, and implementation stages. This mechanism significantly contributed to the cooperation among various stakeholders, including local professionals, academics, government officials, and the general public (Leaf & Hou, 2006: 574), therefore strengthening community bonds (UNESCO, 2007: 235). However, it is noticed that the mechanism of engaging local communities was proposed and decided by the research academics, who assume that "public involvement and participation is an important part of the whole district protection project" (XMHDPPO, 2003: 63).

Regeneration planning and development control mechanisms

While conducting the MHDPP, a special planning and development control mechanism was set up to implement the proposed plans (Figure 5-12). A major feature of this mechanism was the involvement of local residents in the regeneration process. As mentioned above, various stakeholders, especially local residents, were integrated from the pre-planning to the decision-making and implementation stages. To deal with the many uncertainties and dynamics of project management, an office was established at Courtyard no. 144 in Beiyuanmen Street, much to the delight of local residents as easy access to consultation became possible then. To better understand local opinions and requirements, several meetings at the Sub-district Committee were organized beforehand,

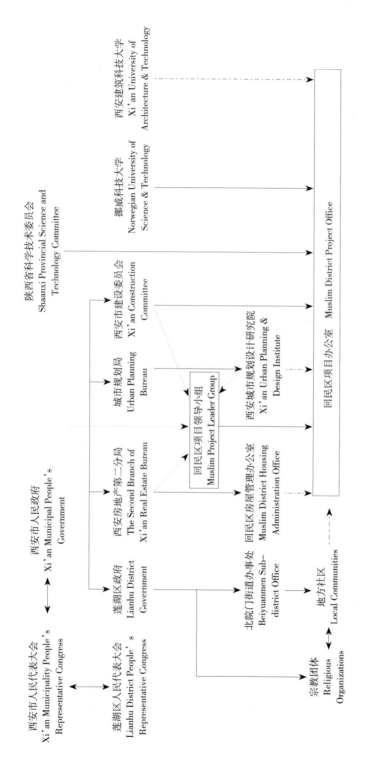

图 5-12：所选回民区项目的管理组织机构图

Figure 5-12: Institutional set-up of the selected Muslim district project

资料来源：(Li，2002：41)

Source: (Li, 2002, p. 41)

行的时候，项目办公室人员还不得不面对多层次的政府机构。这里介绍其中几个关键机构的主要责任范围。西安城市规划局拥有规划、管理和监督项目实施的权力（XCUCRC，2000：132）。根据西安城市规划管理条例，在历史城区内，任何详细的规划和城市建设都必须首先获得西安城市规划局的批复。同时，各级人民代表大会负责对有关项目进行监督（SCXPC，2005：第10条）。由于回民区属于莲湖区政府的管辖范围，所以当地的回民街区住房管理办公室，负责管理住房的产权问题。随着各种利益相关者的参与（即专家、地方居民以及项目工作人员），在更新过程中，产生了很多争论和冲突（Li，2002a：41）。不过，这个项目的最终更新结果在各个方面都取得了进步和改善：地方经济取得了发展、当地的物质环境得到了改善、历史城市建筑和原住民生活内容得到了保护。

大麦市和洒金桥的市场主导下的再开发项目与居民重置

随着快速城市化进程，经济发展在西安的城市开发政策中一直发挥着非常重要的作用（XCPDI，2002：90）。这点在回民区的大麦市和洒金桥的更新案例中得到了很好体现。所提项目位于穆斯林街区的西部，两条街道贯穿回民区南北，在庙后街处相连，当地居民称该交叉十字为"洒金桥十字"，两条街道总长约1100米（图5-13）。

大麦市和洒金桥大街的历史可以追溯至唐代。651年，伊斯兰教传入西安（当时的长安）。经过一千多年的发展，该地区成为远近闻名的穆斯林聚居区。通过经营多家地方家族产业，这里也成了独具穆斯林特

during which community representatives, tenants of the protection courtyards, members from the Project Office, and local officials were all invited. Indeed, during the implementation, local residents showed plenty of interest and enthusiasm. They were engaged in the projects keenly and constructively (XMHDPPO, 2003: 63). When the projects were carried out, the Project Office members had to face multiple governmental agencies. The responsibility scopes of several critical agencies were introduced here. Xi'an Urban Planning Bureau possessed the unique power of planning, managing, and supervising its implementation (XCUCRC, 2000: 132). According to Xi'an Urban Planning Management Regulations, any detailed planning and urban constructions within the historical urban areas need to get the approval from the Xi'an Urban Planning Bureau, while different levels of Representative Congresses also take the responsibility of supervision (SCXPC, 2005: Article 10). As the Muslim district is within the jurisdiction of the local Lianhu District government, the local Muslim District Housing Administration Office manages the property issues. With the participation of the various stakeholders (i.e., professionals, local residents, and project team members), many discussions and conflicts showed up in the regeneration processes. It took a long time and great efforts for them to reach a consensus (Li, 2002a: 41). Nevertheless, as mentioned above, the regeneration outcomes of this project were multi-dimensionally rewarding: local economic growth, improvement of local physical environment, and conservation of local historic urban fabric and indigenous lives.

Market-driven redevelopment and residential relocation in the Damaishi and Sajinqiao areas

With the trend of rapid urbanization, economic consideration has been playing a very significant part in Xi'an urban development policies (XCPDI, 2002: 90). This was well revealed in the Damaishi and

图 5-13：回民区大麦市和洒金桥位置以及当地更新建议
Figure 5-13: Damaishi & Sajinqiao area and officially proposed widened streets in DTMD

色的商业街区。虽然自 20 世纪 90 年代以来，西安市政府就提出了关于大麦市和洒金桥地段的再开发规划（SCXPC, 2002；XCPB et al., 2005），但是直至 2005 年西安第四次总体规划提出之前，一直没有取得实质性进展。第四次总体规划中，地方政府打算在回民区整治其交通系统。根据地方政府提出的再开发规划，为了解决当地的交通问题，以及减少对西大街商业的影响，决定对贯穿大麦市和洒金桥大街的两条街道进行拓宽整治（Hoyem, 2002：386）。图 5-13 的蓝色区域显示了打算拓宽的街道。随着道路的拓宽，现有的低层住宅也会被多层的现代化建筑群所取代，而项目涉及的居民会被安置到旧城区以外的地方。图 5-14 展示了所提规划方案。

当地再开发方案的社会影响

自从 1998 年大麦市和 2005 年洒金桥开展更新活动以来，不少原住居民进行了搬

Sajinqiao Streets area project within DTMD. Located at the western part of the Muslim district, this area runs from south to north, connecting with each other at Miaohou Street, an intersection which local residents named the "Sajinqiao crossroad" (*sa jin qiao shi zi*). The two streets together are about 1,100 meters long (Figure 5-13).

The history of the Damaishi and Sajinqiao Streets dates back to the Tang dynasty, when Islam was introduced into Xi'an (then Chang'an) in A.D. 651. After more than 1,000 years of development, this area is now well known as a Muslim residential quarter serving Muslim flavors through many local family-based businesses (SMDPI et al., 1997: 18). Even though the redevelopment plans of the Damaishi and Sajinqiao Streets area have been proposed by the local government ever since the late 1990s (SCXPC, 2002; XCPB et al., 2005), it did not flourish until the proposal of the Xi'an fourth master plan in 2005, during which the local government planned to redevelop the transportation system

酒金桥街景　　　　酒金桥景观
西安市酒金桥大街规划设计

酒金桥规划

图 5-14：酒金桥规划方案
Figure 5-14: Proposed plans in Sajinqiao street area
资料来源：2005 年当地政府的宣传资料
Source: Local official promulgation materials in 2005

迁，当地原有的私营企业也受到不同程度的影响。值得注意的是，在有关规划的制定与实施过程中，当地居民的意见并未得到慎重考虑（XMG，2005a：4），而很多时候只是被通知要求遵循既定规划的指示。这种情况下，本研究认为再开发规划方案不但会改变已有历史城市的肌理结构，而且会影响原住居民的生活，如传统商业活动和社区邻里关系。

自从 2005 年地方区政府宣布再开发规划方案以来，关于所有权的补偿问题一直存在着争议。今天，回民区的房屋产权大致分为三类：1）私人土地使用权；2）公共土地使用权（房地产第二分局代表西安市政府管理有关土地及其使用权。该机构还负责产权的维护）；3）租赁房屋（主要有三种类型：用于居住的公共租赁权；用于公司或政府单位的公共租赁权；用于商业的私人租赁权）（Li，2002a：63）。在全部产权中，私人产权占

within DTMD. According to the government-proposed redevelopment plan, two streets passing through the Damaishi and Sajinqiao Streets area will be widened to avert traffic problems and its influences on the businesses in West Street (Hoyem, 2002: 386). They are shown in the blue belts in Figure 5-13. With this, existing low-level houses will be replaced with multi-level modern complexes, and residents concerned will be relocated outside the CWA. Figure 5-14 presents an illustration of the proposed plan.

Social impacts of local redevelopment plan

Since launching the housing removal activities in the Damaishi area in 1998 and in the Sajinqiao area in 2005, many local residents have been displaced and original private businesses have been seriously affected. Notably, in the plan-making and implementation processes, they were not taken into consideration (XMG, 2005a: 4), but were just required to follow the plans announced by local government. The local redevelopment plan will not only change the existing historic urban fabric, but will also affect local indigenous lives in terms of traditional business activities and community neighborhoods (resources provided by local residents).

Since the local district government began to promulgate the redevelopment plan in 2005, the issue of property compensation has been very contentious. Today, the property rights in DTMD consist of three categories: i) private land use rights; ii) public land use rights (The Second Branch of Real Estate Bureau represents the Xi'an municipal government in administrating the plots and the properties on the plots. The agency is also responsible for the maintenance of the properties.); and iii) rental houses (three main types: public rental property for residential usage; public rental property for company or governmental units; and private rental house for businesses) (Li, 2002a: 63). Private property right occupies 75.2% of total local properties; public property, 23.0%, and

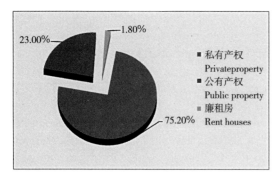

图 5-15：回民区中各种产权的比例

Figure 5-15: Ratio of the various property rights in the DTMD

资料来源：(SMDPI, XUAT, UTPRI, RERI, & RUDRI, 1997 : 65)
Source: (SMDPI, XUAT, UTPRI, RERI, & RUDRI, 1997: 65)

2005 年 3 月，西安旧城区拆迁补偿费与商品房之间的价格对比 表 5-2
Prices comparison between removal compensation and market-price of local commodity housing within Xi'an city-wall area, March 2005 Table 5-2

产权类型 Property Categories	价格类型 Price Types	政府给的拆迁补偿 （元 / 平方米） Removal compensation from government (RMB yuan/M^2)	当地二手房的价格 （元 / 平方米） Local second-hand apartment price (RMB yuan/M^2)	当地商品房的市场价格 （元 / 平方米） Market-price of local commodity housing (RMB yuan/M^2)
私有住宅 Private-owned residential housing		1680 ~ 2180	3000 ~ 3500	4000 ~ 5000
私有商业住房 Private-owned business housing		3800 ~ 4000	约 10000	约 30000（包括 47% 的公摊） About 30,000 (including 47% public apportion)
公有住宅 Public-owned residential housing		800 ~ 1000		
公有商业住宅 Public-owned business housing		约 2000 About 2,000		

资料来源：2007 年，莲湖区建设局
Source: Lianhu District Construction Bureau, in 2007

75.2%；公共产权占 23.0%；租赁房屋占 1.8%（图 5-15）。这意味着回民区的大部分产权属于私人所有。因此，为了体现土地使用权的合理性，地方区政府颁布一个合理的补偿方案显得尤为重要（表 5-2）。不过，地方政府公布的补偿方案引起了当地居民的不满，居民认为他们原有住房的补偿价格与当地住房的市场价格之间存在较大差距。

作者 2008 年对大麦市和洒金桥的居民进行的面访显示，在当地再开发过程中，当地居民非常关心他们植根于该历史街区的商业谋生方式以及他们的宗教生活可能会受到地方

rental houses, 1.8% (SMDPI et al., 1997: 65) (Figure 5-15). This means most properties in the DTMD are privately owned. Therefore, a proper compensation plan by the local district government is necessary to appropriate land use rights (Table 5-2). However, even this plan is surrounded by enormous discontent among local residents. They believe that the compensation is far less than the market price.

The author's interview to local residents in Damaishi and Sajinqiao area, in 2008, reveals that local Muslim residents' priority concerns in the redevelopment process are their original economic ways of making a living, which are rooted in this historic district, and having their religious lives disrupted by the

政府所提议规划方案的较大影响。在采访过程中，不少居民表示，如果他们接受了公布的补偿方案，那么他们可能无法承担将来原址建设的新住宅；这样一来，不少人可能别无选择，只能寻找偏远地段的价格更低廉的住处。同时，这也意味着他们可能需要离开回民社区。被重置之后，受影响的部分居民还有可能会失去他们原来基于历史街区旅游业的谋生经营方式。考虑到上述原因，当地穆斯林居民表示，较难接受有关再开发和房屋补偿方案。

5.3 鼓楼回民区更新实践中的社会资本

本研究主要从两个层面来探讨鼓楼回民区的社会资本问题，包括聚合型社会资本（或者说是社区居民之间的关系）和联合型社会资本（或者说是地方社区与其他机构之间的关系＜例如，地方政府、房地产开发商等＞）。影响这两种社会资本水平和生成的因素有很多，本研究根据研究案例的特征，大致也将其分为两类：第一类因素影响聚合型社会资本，包括亲属关系、宗教和种族以及场所依附性和当地的商业环境；第二类因素影响联合型社会资本，包括当地倾向于经济发展的城市更新政策和策略以及碎片化的城市管治模式。

5.3.1 鼓楼回民区的社会资本

据已有研究调查（Hoyem，1989），总体而言，当地穆斯林居民之间保持着良好的邻里关系，由此，居民之间相应的社会资本水平也会较高。早在1989年，当地研究小组就对邻里

residential relocation plan proposed by the local government. During the interview, many residents complained that if they agreed to the compensation plan, they would not be able to afford the new apartments upon its completion in the original community, leaving them no choice but to go for other cheaper and remote living areas. This means they need to move out of their original Muslim community. This plan also means that after being relocated, most affected residents may lose their ways of making a living through local historic district-based businesses. Therefore, considering these two reasons, local Muslim residents can hardly accept the redevelopment and compensation plans.

5.3 Social Capital in the Regeneration of Xi'an DTMD

In Xi'an DTMD, social capital is seen in various forms including the bonding social capital, or the relationships among local residents, and the bridging social capital, or the relationships between local communities and other agencies (e.g., local governments, property developers, etc.). Several factors affect the level and the generation of the two kinds of social capital. Roughly, these factors may be divided into two sets. The first group of factors, influencing the bonding social capital, comprises kinship relationships, religion and ethnicity, and place attachment and local business setting. The second group, influencing the bridging social capital, includes local pro-growth urban regeneration policies and strategies, fragmented urban governance, and lack of official policy toward ethnic minority in urban regeneration.

5.3.1 Social capital within Xi'an DTMD

Survey (Hoyem, 1989) reveals that, generally, local Muslim communities maintain very good relationships and the corresponding level of social capital among them is also higher. Early in 1989, local research groups

之间的信任和友谊情况开展了现场调查，总共分发了 497 份调查问卷，最终收集到了 193 份，1133 名居民参加了调查 (Hoyem, 1989: 5–13)。好几个问题都涉及地方居民之间的相互信任和关系。在回答问题"你和邻居们之间的关系如何？"时，193 位受访者中，有 80.4% 的居民表示关系很好，18.3% 的居民表示和邻居之间没什么联系，而 1.3% 的居民表示关系不好（图 5–16）。在回答问题"和你生活联系密切的最好的朋友居住在哪里？"时，如下是 125 个受访者的回答：30.4% 的住户说他们都住在同一个院落里；5.6% 的住户表示他们居住在同一楼层上；11.2% 的住户表示他们的庭院挨着；28.8% 的住户说他们住在同一胡同中；21.6% 的住户说他们在同一工作单位；2.4% 的住户说他们居住在其他地方（图 5–17）。

conducted field surveys on trust and friendships among local residents. A total of 497 questionnaires were distributed and 193 were collected covering 1,133 local residents (Hoyem, 1989: 5-13). Several questions were related to mutual trust and relationships among local residents. Regarding the question, *"How is the relationship between your neighbors and you?"* 80.4% indicated very good; 18.3% had no contact; and 1.3% rated negative, among the 193 respondents (Figure 5-16). The following were the answers of 125 households to the question *"Where do your best friends with whom you have much contact live?"*: 30.4% remarked that they resided in the same courtyard; 5.6% resided on the same storey; 11.2% were in the surrounding courtyards; 28.8% were in the same alley; 21.6% were in the same working unit; and 2.4% resided somewhere else (Figure 5-17).

The result means that most local residents have a relatively good relationship with their

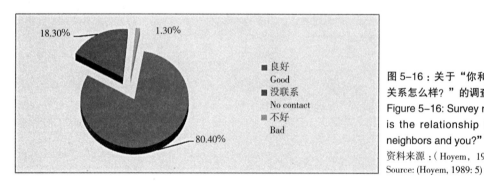

图 5–16：关于"你和邻居们之间的关系怎么样？"的调查结果
Figure 5-16: Survey regarding "How is the relationship between your neighbors and you?"
资料来源：(Hoyem, 1989：5)
Source: (Hoyem, 1989: 5)

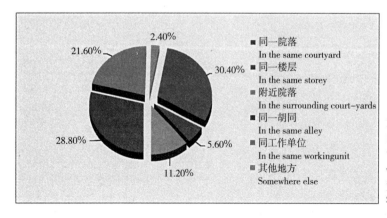

图 5–17：关于"和你联系密切的最好的朋友住在哪里"的调查结果
Figure 5-17: Survey regarding "Where do your best friends with whom you have much contact live?"
资料来源：(Hoyem, 1989：5)
Source: (Hoyem, 1989: 5)

调查结果显示，大部分地方的居民和邻居之间都保持着相对较好的关系，而且他们经常见面的地方离他们居住或工作的地方比较接近（即同一所庭院、同一胡同和同一工作单位）。在某种程度上，地方社区居民之间这种紧密的关系也解释了为什么很多居民在被重置后会选择搬回到他们原来的住处。与当地穆斯林居民之间的密切关系形成鲜明对比的是，当地以经济发展为主导的更新政策和规划可能影响当地居民和政府机构之间达成的互信关系。例如，在笔者2008年进行的现场调查中，对于问题"你有多信任政府官员？"的回答中，只有12.5%的受访者表示对政府官员比较信任，50%的受访者表示持中立态度，大约37.5%的受访者直接表示对政府官员的不信任。由于缺乏必要的相互信任，在地方的再开发过程中也容易产生冲突。这也破坏了地方居民积极参与和配合的热情。在当地居民自发建设的建筑案例中，很多居民不愿意遵守地方政府制定的指导方针，这都会阻碍当地更新规划的顺利实施。

影响聚合型社会资本的因素
亲属关系

穆斯林居民之间密切的邻里关系，在很大程度上与其婚姻习俗有关，并从而使当地的邻里关系以血缘关系为基础。根据伊斯兰教义，穆斯林结婚时其配偶应该也是穆斯林。这种婚姻关系成为穆斯林社区里最密切的居民间的互动模式（Ma, 1998a：331）。同时，这也是维护和表达居民民族认同感的一种方式。俗话说的好，"穆斯林中十之八九是亲戚"。这种关系促进了居民之间频繁的互动和密切关系网络的建立。基于对西安穆斯林街区的

neighborhood, and the places they usually meet their friends are the areas close to where they live or work (i.e., the same courtyard, the same alley, and their same working units). To some extent, these kinds of close relationships among local community members account for the situation when many resettled residents during a local residential relocation process chose to move back to their original area. As a sharp contrast to the intimate relationships among local Muslim residents, the local government-led economic development-biased regeneration plan largely spoils the mutual trust and relationships between local residents and government agencies. During the field survey in 2008, regarding the question "*How much do you trust government officials?*," only 12.5% of respondents partially expressed trust in government officials, 50% had average opinions, and about 37.5% explicitly conveyed distrust. Without the necessary mutual trust, many conflicts arise during the local redevelopment process, undermining local residents' active involvement and cooperation. In self-constructed buildings, for instance, many residents are reluctant to adhere to guidelines set by the local government, severely undermining the smooth implementation of local regeneration plans.

Factors affecting the bonding social capital
Kinship relationships

The intimate neighborhood relationships among local Muslim residents have much to do with their marriage custom and thus formed kinship relationships. According to the Islamic doctrine, a Muslim can only get married to another Muslim. This kind of marriage network becomes the most intimate interaction mode for Muslim communities (Ma, 1998a: 331). It is also one way of maintaining and expressing their national identity. Just as the folk words say, "nine of ten Muslims are relatives." This kind of relationship promotes frequent mutual interactions and intimacy. Based on the surveys in Xi'an DTMD, Yang (2007: 266)

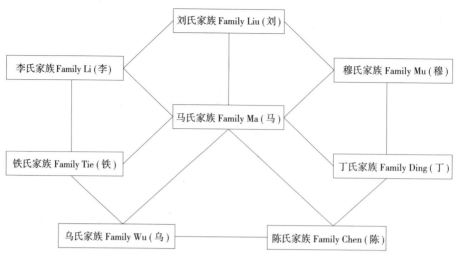

图 5-18：西安穆斯林街区中主要家族之间的婚姻网络
Figure 5-18: Marriage network of the main surname families in Xi'an DTMD
资料来源：(Yang, 2007：266)
Source: (Yang, 2007: 266)

调查，杨文炯（Yang, 2007：266）围绕马氏家族绘制了一幅关系网络图（图5-18）。从该图可看出，马氏家族和街区里很多其他大姓家族都存在婚姻关系，如，铁氏家族、刘氏家族和李氏家族。据当地住户反映，有些家庭可以保持这种亲戚关系或亲属关系长达三至五代。这种情况表明，回民区不仅是具有相同民族和宗教信仰的社区，而且还是一个庞大的亲戚网络存在地。正是这种血缘关系和亲情让这个社区变得如此紧密团结。

除了亲情关系之外，另一个促进居民达成紧密社区关系的原因与当地宗教信仰的团体生活模式有关。这种群体生活模式进一步加强了穆斯林的民族认同感，也促进了他们之间的日常交流（下面会进一步阐述）。一些俗语生动形象地描述了当地紧密的邻里关系，例如"世上回民是一家"、"回民之间自然亲切友好"等（Ma, 1998b：47）。这些因素都有助于当地穆斯林居民之间形成良好的社区关系。

depicts a relative network surrounding the families with the surname Ma (Figure 5-18). This diagram shows that the Ma family has marriage relationships with many other big families in this district, such as Tie, Liu, and Li. According to local residents, some families could maintain this kind of relative and kinship ties for three to five generations. This situation shows that DTMD is not just a community of the same nationality or religion, but a big network of relatives. It is the ties of blood and emotion that tightly build this community.

In addition to kinship ties, another factor contributing to local intimate community relationships relates to the local group living pattern based on religious necessities. This kind of group living pattern further strengthens the Muslims' national identity and mutual interactions (further elaborated in the next section). Local close neighborhood relationships are vividly expressed through such sayings as "all Huis in the world are families," "natural geniality among Huis," and so on (Ma, 1998b: 47). All these aspects contribute to good community relationships among local Muslim residents.

作者在2008年的实地调研也注意到了这种密切关系。在回答问题"你和邻居之间的关系如何？"时，60户受访家庭中有15户，或者说25%的受访者的回答都是"非常好"，剩下45户家庭，或者说75%的受访者回答是"还可以"。同时，几乎所有的受访者都提到，他们很怀念以前（指的是20世纪60年代和20世纪70年代）的老邻居、那些逝去的岁月、过去美好的时光以及当地和睦相处的生活。与此同时，关于他们之间的相互信任和友谊问题，作者问及了如下几个问题：

"你觉得在和人相处中，这儿的大多数人可以信任吗？"；"对下面的表述你是如何看待的？同意？还是不同意？：在你需要帮助的时候，社区里的大多数人都很乐意伸出援助之手；在你的社区中，人人都需警惕或某人想要利用你"；"如果社区的一个项目对你没有直接好处，但是对社区中的其他很多人都有好处，那么你会对这个项目投入时间或金钱吗？"

对于"在你需要帮助的时候，社区里的大多数人都很乐意伸出援助之手"的表述，在60位受访者中，有37位，也就是62%的受访者选择"非常同意"，其他23位，也就是38%的受访者选择"不大同意"。这表明当地社区中的大多数人还是乐意帮助那些需要帮助的人的。同时，对于"在你的社区中，人人都需警惕或某人想要利用你"的表述，在60位受访者中，有30位，也就是50%的受访者选择"不大同意"；7位，也就是12%的受访

This kind of relationship has been echoed in the author's 2008 field survey. Regarding the question *"How do you describe your relationship with your neighbors?"* 15 of the 60 households, or 25% of the respondents, described it as "very good," and the other 45, or 75% of the respondents, described it as "good." At the same time, nearly all the respondents remarked that they cherished their old neighborhoods, their childhoods, past time, and harmonious lives in earlier times (referring to the 1960s and 1970s). At the same time, regarding mutual trust and friendships, the following questions were asked:

"Would you say that most people here can be trusted in dealing with people?"; "Do you agree or disagree with the following statements?: Most people of your community are willing to help if you need it; One has to be alert or someone is likely to take advantage of you in your community"; "If a community project does not directly benefit you but has benefits for many others in the community, would you contribute time or money to the project?"

Concerning the statement that "Most people of your community are willing to help in your need", 37 of the 60 respondents, that is 62%, select "agree strongly", and 23 persons, that is 38%, select "agree partially". These indicate that most people in the local neighborhoods are willing to help the needy. At the same time, concerning the statement that "One has to be alert or someone is likely to take advantage of you", 30 of the 60 respondents, that is 50%, select "agree partially"; 7 persons, that is 12%, select "neither agree nor disagree"; 8 persons, that is 13%, select "disagree partially"; and 15 persons, that is 25%, select "disagree strongly". This implies that although many people acknowledge the impacts of current China's transitional economy to their community relationships, most residents still show propensity to trust their neighbors. This

者选择"既不赞成也不反对";8位,也就是13%的受访者选择"不大同意";另外15位,也是就25%的受访者,选择"非常不同意"。这表明了虽然很多人承认当前中国转型期的经济体制对他们社区关系的影响,但是大多数居民仍然选择相信他们的邻居。这在他们对群体利益问题的态度中得到了进一步的证实。例如,关于问题"如果社区的一个项目对你没有直接好处,但是对社区中的其他很多人都有好处,那么你会对这个项目投入时间或金钱吗?",受访的60位居民都表示愿意为社区项目投入金钱或时间,即使他们不是直接受益人。

宗教和种族

在西安回民区中,宗教信仰在促进当地居民之间形成紧密的社区关系和凝聚力方面,发挥了很重要的作用。基于宗教文化,当地居民间形成了明确的社会行为规范和社区交流方式,这对社区中社会资本的积累有着实质性的作用。总体而言,在西安回民区中,宗教社区生活和群体生活与五方面内容紧密相关(Yang, 2007: 364)。第一,与当地"坊"的居住模式有关,这是一个类似街区的中国传统城市居住单位。通常情况下,地方居民通过"坊"来进行组织。在穆斯林街区中,以围绕清真寺居住为特色(Dong, 1995: 66)。第二,与当地教育制度有关。在当地清真寺,向穆斯林教徒传授伊斯兰教义和文化有助于地方宗教文化的传播和继承。第三,回族成员或穆斯林居民之间特有的婚姻关系,有助于通过血缘关系和亲属关系来形成文化整体性和民族认同感。第四,当地的经济活动具有独特的世袭和垄断特色,通过这种活动,由种族群体形成的家族企业建立了自己的认同感。第五,清真寺作为当地宗

has been further proved by their attitudes to the issues for group interests. For instance, as regards the question "If a community project does not directly benefit you but has benefits for many others in the community, would you contribute time or money to the project?", all 60 respondents show willingness to contribute time or money to community projects, even if they do not benefit directly from them.

Religion and ethnicity

In the Xi'an DTMD, religion plays a significant role in contributing to the intimate community relationships and cohesion among local residents. Based on the religious culture, it forms explicit local social norms and modes of community interactions, which substantially contribute to the level of local community social capital. Generally, in the Xi'an DTMD, the religious community lives and group-living activities are closely related to five perspectives (Yang, 2007: 364). First, it relates to the local residential pattern of *Fang*, the traditional urban unit in China close to a city block. Usually, local residents are organized by it, and in a Muslim district, it is featured by living around a mosque (Dong, 1995: 66). Second, it relates to the local education system. The education of Islamic religion and culture to Muslim pupils in local mosques contributes to the spread and continuity of the local religious culture. Third, the exclusive marriage among Hui ethnic members or Muslim residents contributes to the construction of a cultural integrity and an ethnic group identity through blood ties and kinships. Four, the family-based business made up of the ethnic group establishes its identity through local economic activities with special hereditary and monopoly features. Finally, mosques play substantive roles in local Muslim lives as the center of the local religious community. They also form a self-management structure through local management units (i.e., Xuedongxianglao Society (*xue dong xiang lao hui*), Shetou Society (*she tou hui*), or Mosque Management Association. With mosques as the dominant

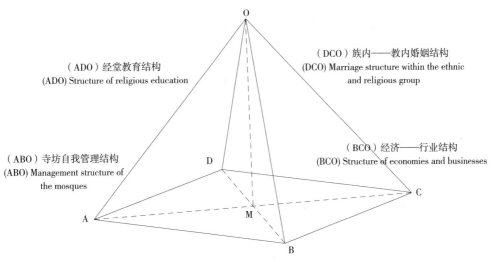

图 5-19：回族社区五维一体的社会和宗教结构图
Figure 5-19: Social and religious structure integrating the five dimensions of Muslim community
注：A、B、C、D 代表社区单位（家庭或个人）；O 代表共同信仰的信合点；M 代表地缘结构的中心——清真寺
Note: A, B, C and D representing community units (families or individuals); O representing the integration of common belief; M representing the centre of local geographic structure–the Mosques.
资料来源：（Yang，2007：366）
Source: (Yang, 2007: 366)

教社区中心，对穆斯林的生活产生了巨大的作用。他们通过当地的组织单位（即学东乡老会、社头会）或寺管会形成了自我管理组织。清真寺在其中发挥着主导作用，当地回族社区社会行为规范的五个方面彼此紧密相关。共同的信仰和民族认同感是形成当地宗教文化的内在因素。上述五方面内容互相影响，相互补充，共同为文化整体性和种族认同感的形成发挥了重要作用。基于上述五方面内容之间的相互关系及其作用，图 5-19 展示了回族社区生活行为规范中，存在较多内部沟通和彼此互惠的特点（Yang，2007：364）。

对穆斯林社区而言，清真寺是文化传承、宗教仪式和地方交流的中心。除了宗教功能之外，清真寺还是当地穆斯林居民举办婚礼、葬礼仪式以及解决家庭或邻里之间矛盾的场所。此外，这里还是青少年接受传统教育的

element, these five perspectives of local Muslim social norms are closely interrelated to one another. Their common belief and ethnic group identity are the tacit elements that form the local religious culture. These five perspectives have been reciprocally and mutually supplementary, playing the role of constructing cultural integrity and ethnicity. Based on the inter-relationships among these five perspectives, the Muslim living norm featured by the reciprocal and interactive social structure is illustrated in Figure 5-19 (Yang, 2007: 364).

In a Muslim community, the mosque is the center for cultural inheritance, religious ceremony, as well as local interactions. In addition to its religious function, a mosque is the place for local Muslim residents to have their marriages and funeral ceremonies and where conflicts in families or among neighborhoods are resolved. Moreover, it is the place where the youth have their traditional education. As to their local religious lives, the Muslims normally have five regular praying activities everyday and

地方。关于他们的宗教生活，穆斯林通常每天会定期做五个祷告仪式，每周五去清真寺参加一个集会仪式。另外，在清真寺，他们还有两个年度节日——开斋节和牺牲节（Li, 2004：83-84）。这些定期的日常礼拜和仪式为当地社区成员见面和交流提供了很多机会。2007年作者对大麦市和洒金桥地区的穆斯林居民的采访和问卷调查中显示，几乎所有的受访者都表示宗教团体活动是他们生活中最重要的一部分。此外，他们还指出居民之间的互动不仅表现在参加一个清真寺的活动，而且表现在参加不同的清真寺活动。杨文炯在西安鼓楼穆斯林历史街区中对145名穆斯林居民就他们参加清真寺活动作了一项调查。结果显示，32.9%的居民每天去清真寺进行五次祷告；59.4%每周五去清真寺参加主麻日和尔德节[1]，例如开斋节和牺牲节；7%的居民只在尔德节的时候去清真寺；只有0.7%的居民从来不去清真寺（Yang, 2007：253）。在这些居民中，98.6%的居民表示他们和邻居之间联系密切（Yang, 2007：259）。这一结果表明广泛的社交活动是围绕清真寺中的宗教活动来进行的（Yang, 2007：253-258）。这种宗教社会结构促进了穆斯林居民之间的相互理解、相互交流、增强了民族认同感。马（Ma）表示这种民族认同感成了他们内部凝聚力的形成打下了基础（Ma, 1998b：92）。穆斯林的社会和宗教行为规范对居民的共同信仰和世界观的形成具有很大影响，并极大促进了居民之间密切关系网的形成以及高水平社会资本的生成。

one gathering rite every Friday in the mosque. Furthermore, they have two annual festivals, the Eid-ul-Fitr (*kai zhai jie*) and the Eid-ul-Azha (*xi sheng jie*) or Sacrifice Festival, in the mosque (Li, 2004: 83-84). The regular daily worship and ceremonies provide many opportunities for local community members to meet and interact. Throughout the author's interview and questionnaire surveys to local Muslim residents in Damaishi and Sajinqiao area in 2007, nearly all respondents indicated that religious group activities constitute the most important part in their lives. Further, they pointed out that these interactions exist among the attendees not only in one mosque, but also among different mosques. Yang conducted a survey on 145 Muslim residents about their attendance to mosques in the Xi'an DTMD. It shows that 32.9% of the residents go to mosques to pray five times a day; 59.4% go to mosques every Friday for the Eid Greetings (*Zhu Ma Ri*) and the Eid Mubarak (*Er De Jie*)[1], for example the Eid-ul-Fitr and the Eid-ul-Azha; 7% only go to mosques on the occasions of the Eid Mubarak; and only 0.7% never go to mosques (Yang, 2007: 253). Among these residents, 98.6% remark that they have frequent contacts with their neighbors (Yang, 2007: 259). This result indicates that extensive social interactions have been conducted surrounding religious activities in the mosques (Yang, 2007: 253-258). This religious social structure promotes mutual understanding, group identities, and interactions among Muslim neighborhoods. Ma remarks that this kind of national identity serves as the basis for their internal cohesion (Ma, 1998b: 92). These Muslim social and religious norms contribute to the formation of the common beliefs and world outlook among local residents, and significantly help to build the intimate relationships and the generation of higher level of social capital among them.

---

[1] 尔德节表示接受祝福的节日（Yang, 2007：253）。

[1] The Eid Mubarak means blessed festival (Yang, 2007: 253).

场所依附性和当地的商业环境

西安回民区中高水平的社会资本存量也反映了当地居民的地方依附性,这既和上面所提到的宗教因素有关,也和当地家族商业活动有关。从城市的物质环境来看,低层住房、小巷和露天广场为邻里之间的日常交流提供了机会。这些交流互动既可能发生在一些公共场所(对很多穆斯林居民而言,即便不是祷告时间,很多人也愿意去清真寺;而对于很多当地的汉民而言,很多人更喜欢待在自家庭院或巷子里),也可能发生在居民家中,尤其是当居民去拜访朋友或朋友来拜访他们的时候。当地的居民表示,他们经常在自家周围或方便去的公共场所进行他们的日常公共活动,要么在室内或庭院中,要么在小巷里。这种情况表明历史城市或历史街区的建筑往往为居民邻里之间的联系和交往提供了很多便利。

位于西安的旅游和商业中心或者说位于城市的市场集中区域,该历史街区实际上将当地的居住建筑和商业活动紧密结合在一起。因此,很多临街道而建的住宅被改造成住宅—商业一体化的商住建筑,以更好地满足当地商业活动的需求。正如前面所提到的,西安回民区中约有70%的回民是通过家族零售业的方式来维持其家庭生计(SMDPI et al., 1997:70)。当地传统的城市模式(即住宅—商业一体化的建筑)为这些商业活动的进行提供了必要场所。通常情况下,这些商住建筑面朝街道的房屋被作为零售业的店面(例如,销售牛肉和羊肉),而后面的房屋则被用来加工肉食。另外,基于当地的宗教和民族特色,穆斯林社区中生成了高水平的社会资本。该历史街区被进一步灌输了这些社会和

Place attachment and local business setting

The higher level of social capital in DTMD also reflects local residents' place attachment, which has much to do with both the abovementioned religious issue and the local family-based business activities. The low-level houses, alleys, and public squares make the neighborhood's daily interactions possible from the perspective of an urban physical environment. Such interactions might take place either in some public areas (for local Muslim residents, many people prefer to go to mosques even when it is not praying time; for local Han Chinese, many people would like to stay in the courtyards or in the alleys;), or in residents' homes when they are visiting/visited by their friends. Local residents indicate that they often have their daily activities close to their houses' surroundings or the easily accessed public areas: indoors, in the courtyards, or in the alley. This situation infers that historic urban fabric bears rich neighborhood connections and interactions.

Located at the tourist and business core of Xi'an or the concentrated urban marketplace area, this historic district actually ties local residential places with business activities. Accordingly, many residential buildings facing the streets were transformed into residential-cum-commercial compounds to serve local business necessities. As mentioned earlier, about 70% of residents in Xi'an DTMD survive through family-based retailing businesses (SMDPI et al., 1997: 70). Local traditional urban pattern (i.e., the residential-cum-commercial compound) provides the space needed to conduct these businesses. Usually, houses facing the public street become shopfronts for retailing businesses (e.g., selling beef and mutton), while the spaces behind are used to process meat. In addition, based on local religious and ethnic features, a high level of social capital is generated among the Muslim community. The historic district has been further imbued with these social and cultural features,

文化特色，这有助于吸引游客的注意。

大量的游客和旅游业的发展为当地带来持续的商业收入，这也体现在该街区繁荣的生意和昌盛的旅游景象上。事实上，西安城墙区内的游客数量就从1992年的401559人次增加到了2006年的867273人次，而相应的商业收入也从9951万元增加到了153356万元。当地大多商业活动都属于传统活动，并与穆斯林的宗教和回族文化相关，例如餐饮、肉食店、传统饰品和古玩（Li，2004：172-174）。图5-20和图5-21展示了持续增长的游客数量以及旅游商业收入。

which contributes to tourist attraction.

The vibrant business situation of this historic district has been reflected in the large number of local tourists and increasing commercial income from tourism. The tourists' number within Xi'an CWA has increased from 401,559 in 1992 to 867,273 in 2006, and the corresponding commercial income has increased from RMB99.51 million yuan to RMB1,533.56 million yuan. Most of the local businesses are traditional and related to the Muslim religion and the Huis ethnic culture, such as catering, butchery, antiques, and curios (Li, 2004: 172-174). Increasing local tourists' numbers and commercial income are reflected in Figures 5-20 and 21.

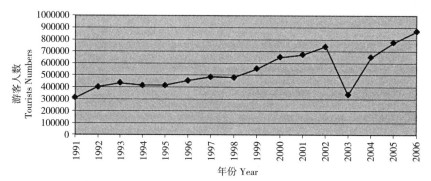

图5-20：西安城墙区内持续增加的游客数量（1991-2006）
Figure 5-20: Increasing tourists' numbers within Xi'an CWA (1991-2006)
资料来源：（XSB，1993~2007）
Source: Author synthesized from (XSB, 1993 ~ 2007)

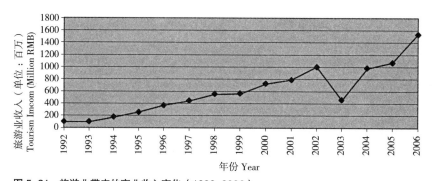

图5-21：旅游业带来的商业收入变化（1992-2006）
Figure 5-21: Increasing commercial income from tourism industry (1992-2006)
资料来源：（XSB，1993~2007）
Source: Author synthesized from (XSB, 1993 ~ 2007)

鉴于当地充满活力的商业形势，当地大多数穆斯林住户在经济上也非常依赖于该历史地段，也更进一步加强了他们对当地的依附感。2007年在对当地进行再开发时，受访的住户强调说，如果他们为了地方政府倡导的再开发方案，而接受被重置的话，可能不少人都会失去他们的家庭经济来源，所以，很自然地，这些居民都很反对有关规划方案。因此，对于当地的再开发活动对他们私营商业和宗教生活所产生的影响，不少住户也表示出担忧和不情愿。

影响联合型社会资本的因素

当地倾向于经济发展的城市更新政策和策略

上述关于回民区自20世纪90年代以来的城市更新政策和策略的讨论表明，当地的更新实践很侧重于重组城市空间以及为经济发展来创造更多的机会。为了迎合当地旅游业的发展和吸引更多的商业活动，当地区政府提出了以房地产主导的再开发方式。房屋拆迁和居民重置是这一更新策略的典型特征。值得一提的是，该策略中较少关注到当地社区的实际需要和要求，本研究认为这很容易导致破坏已有的传统商业模式和宗教社区的生活，并容易恶化居民和当地政府之间所达成的信任关系。作者在2008年的现场调研中，关于问题"影响你与邻居之间关系的最重要因素？"，60位受访者中，其中有50位，也就是87%的受访者表示当时的房屋拆迁和重置活动是影响他们邻里关系最大的一个因素。这些受影响的居民抱怨说，地方政府的再开发政策没有考虑他们的宗教传统和日常交流活动。这些受访者表示，补偿和再开发政策

Considering the vibrant local business situation, most local Muslims are economically dependent on this historic area, further strengthening their place attachment. As stressed by the interviewed local residents in the redevelopment area in 2007, when they are resettled in order to give way to local redevelopment plans led by local governments, many of them will lose their source of income and, logically, they resist the plan. Thus, these people have expressed severe discontent on the impacts of local redevelopment activities on their private businesses and religious lives.

Factors affecting the bridging social capital

Local pro-growth urban regeneration policies and strategies

Previous discussions on urban regeneration policies and strategies in Xi'an DTMD since the 1990s show that local regeneration practices focus much on restructuring spatial urban areas and creating more opportunities for economic development. In catering to the local tourism industry and attracting more business activities, a property-led redevelopment approach has been proposed by the local district government. Housing removal and residential relocation are typical features of this scheme. Very little attention, if any, was given to the local community's actual needs and requirements, resulting in the disruption of existing local traditional businesses and religious community lives and the further deterioration of trust between local residents and the district government. During the author's field survey in 2008, regarding the question "*What affects the relationships between you and your neighborhoods most?*", 52 out of the 60 respondents, that is 87%, remarked that the then housing removal and relocation activities were the biggest factors affecting their neighborhoods relationships. These affected residents complained that the local government's redevelopment plan did not consider their religious traditions and daily interactions at all. According to these

已经损害到了他们的利益。因此，他们很反对这个规划的实施，事实上，这也严重阻碍了地方政府所提更新方案的顺利开展。实际上，为了调查在大麦市街道和洒金桥街道中受更新规划影响的居民和业主的意见，作者于2008年在该地段进行现场调研时受访者们就向作者提供了，当地居民代表在2005年时曾作过的一次调查结果。当地居民曾对该区域的三种再开发方案进行了对比：第一个是北院门街道实行的方案，即直接在临街商店后面建立现代化住宅；第二个是地方居民所提出的方案，即在该地段里原有的工作单位搬离后，在原址兴建现代化的住宅和购物中心；第三个是当地政府所提方案，即对需要搬迁到其他地方的住户进行补偿，再实行综合的再开发规划。调查结果显示，93%的受访者赞成第一个方案，7%的受访者同意第二个方案，没有人选择第三个方案。

碎片化的城市管治模式

除了缺乏制度化的更新机制之外，回民区更新实践的另一个显著的特点就是碎片化的城市管治模式。(图5-2)所示的西安回民区的行政体系结构中，穆斯林居民既受当地居民委员会的管治也受到清真寺管理委员会的管治。在面对住宅拆迁和再开发政策时，穆斯林居民往往倾向于清真寺管理委员会的管治，以此来保护他们传统的宗教生活和商业活动。尽管居民委员会是协调居民和地方政府之间关系的半官方机构，但当居民选择清真寺管理委员会作为其管理机构时，居民委员会就常常无法推进其有些事务。另外，尽管该地区的遗产保护问题是由西安城市规划局和文物局来管理，但是除了当地文物古

87% of respondents, their interest had been infringed upon by the compensation and redevelopment plan. Their severe opposition to its implementation had caused the local government's proposed procedures to severely lag behind. In fact, in order to investigate the opinions of concerned residents and business owners regarding the renewal plans in the Damaishi and Sajinqiao Streets area, as provided by the interviewees during the author's field study in this area in 2008, a comprehensive survey was conducted by local residents' representatives in 2005. Three redevelopment proposals for this area were compared: the first was the one conducted in Beiyuanmen Street, which built modern houses immediately behind the shop fronts; the second one was proposed by local residents, which built modern houses and shopping malls in the original sites of moving out work-units; and the third one was proposed by the local government, which compensated and relocated the residents to suburban areas and conducted the local government's comprehensive redevelopment plan. The survey result shows that 93% of the respondents favored the first choice, 7% approved the second, and none chose the third.

Fragmented urban governance

In addition to the lack of an institutionalized regeneration mechanism, one outstanding feature of the regeneration practices in Xi'an DTMD is the fragmented urban governance. As shown in Figure 5-2 of the Administration Structure of Xi'an DTMD, the Muslim residents are organized both by local residents committee and by the mosques management committees. Facing local residential relocation and redevelopment policies, Muslim residents tend to be organized by mosques and strive to protect their traditional religious lives and local traditional businesses. Even though the residents' committee officially serves as the agency that coordinates relationships between residents and the local government, very

迹外的任何规划和再开发问题，就受其他各区政府机构的管辖了（例如，西安城市规划局莲湖区分局和西安房地产管理局）。当历史地段受到这么多不同级别的不同政府机构来管理时，有时很难界定具体某一问题到底属于哪个政府部门来监管。

5.3.2 当地居民在城市更新中的参与：自建活动

西安历史城市的更新过程中，一个显著特征就是存在大量的居民自建活动，这在很多历史街区的更新实践中都有所体现，如正学街（Chang, 1999）、三学街（Zhang, Li, & Feng, 2007）、七贤庄（Yu, 2003）以及回民区（Wang, 2002c；XMHDPPO, 2003）。事实上，据2008年受访的当地居民代表们声称，这种持续的居民家庭自建活动就从来没有停止过。为了满足当地不断增长的人口和房屋功能改变的需要，很多家庭都进行了诸如建设新住宅或增加楼层的活动。这一点在其他学者的研究中也多有提及（Dong, 1995：94, 95）。一般来说，自建活动有两大优势。

1）传统文化的继承与更新：有学者认为（Wang, 2002c：94），在自建活动中，规划师、建造者和使用者形成三位一体的情形。新建住房的功能、形式、特点和费用都由使用者自己决定，因此，这些新建的住房能够最大限度上满足使用者的需要，并最能反映当地的生活传统和特点。

2）保留原有街道—庭院式住宅格局的同时也维护了邻里关系：考虑到现有房屋的产权问题，当地的自建活动主要都在居民自家庭院内进行。因此，原有的城市格局几乎没

often they fail to perform this function when the former opts to be organized by mosques management committees. In addition, even though the heritage conservation issues in this area are managed by the Xi'an Urban Planning Bureau and the Xi'an Administration of Cultural Heritage, any planning and redevelopment concerns, except for local heritage monuments, are under the jurisdiction of various district governmental agencies (e.g., the Lianhu District Branch of the Xi'an Urban Planning Bureau and the Xi'an Real Estate Management Bureau). When this historical urban area is managed by different governmental agencies of different levels, job descriptions become very ambiguous to define.

5.3.2 Community involvement in local regeneration processes: local self-construction activities

In the renewal of Xi'an historic urban areas, the exhaustive self-construction activities are an outstanding feature, which may be seen in the regeneration practices of many historic districts such as Zhengxuejie (Chang, 1999), Sanxuejie (Zhang, Li, & Feng, 2007), Qixianzhuang (Yu, 2003), and DTMD (Wang, 2002c; XMHDPPO, 2003). In fact, according to local residents' representatives interviewed by the author in Xi'an DTMD in 2008, the continuous family self-construction activities have never stopped ever since it came into being. To meet the requirements of the increasing population and the functional alteration, many families conduct activities such as constructing new buildings or adding more floors. This point is echoed in other studies (Dong, 1995: 94, 95). Generally, there are two advantages in self-construction activities.

1) The inheritance and transformation of the traditional culture: as Wang (2002: 94) states, in self-construction, the planner, the builder, and the user are three-in-one. The function, form, characteristics, and cost are decided by the user himself, and therefore,

有大的变化。类似建立新的邻里关系和环境认同感这样的问题也只会在完全新建的地区才发生（XMHDPPO，2003：11）。

随着地方再开发过程中出现的各种利益冲突，居民自建活动实际上已成为改善当地物质环境的一种有效方式。不过，人们也不应忽视自建活动所产生的问题，而这些问题也会对当地社区邻里关系产生影响。在1989年对回民区的调查中，81.5%的受访者曾解释道，在居民自建活动中，如果邻居占用了过多的空间就会产生冲突；13.6%的受访者认为他们的矛盾来自小孩的争吵；只有4.9%的受访者认为他们的矛盾是由经济问题所产生（Hoyem，1989：5）（图5-22）。在1991年的调查中，大约70%的邻里争执案例都和邻里之间的房屋边界问题有关（Dong，1995：95）。这表明，居民自建活动虽然能够改善当地的城市物质环境，但也有可能削弱当地的邻里关系，并影响到其社会资本的积累。本研究认为，这需要对居民的自建活动颁布必要的法规，并建立相应的政府管理机制。

实际上，当地的区政府已着手制定了管理地方自建活动的工作方法和流程（图5-23）。

the newly constructed buildings can meet the user's needs most and reflect his living tradition best.

2) The original street-courtyard urban fabric as well as the neighborhood relationships remain: associated with the existing property rights, self-construction activities take place mainly within family boundaries. Hence, there are very little changes in its original urban fabric. Problems such as building new neighborhood relationships and environment identities occur only in totally new-built areas (XMHDPPO, 2003: 11).

With various conflicts of interests emerging in the local redevelopment process, self-construction actually becomes an active way for the physical environment to be improved. Nevertheless, the problems accompanying self-construction activities should not be ignored, as these also have an impact on community relationships. In a 1989 survey, 81.5% of respondents explained that conflicts take place when neighbors take excessive space in self-construction activities; 13.6% thought that their contradictions resulted from petty children's quarrel; and only 4.9% of respondents thought that it was because of economic problems (Hoyem, 1989: 5) (Figure 5-22). In a 1991 survey of DTMD, about 70% of the police cases related to a local neighbor's conflicts concerning property boundaries (Dong, 1995: 95). This indicates that self-construction activities as a way of improving local physical environments may weaken local neighborhoods' relationships and the accumulation of social capital. This thereby called forth the necessary regulations and governmental management mechanisms on local self-construction activities.

The district government has developed a working procedure governing local self-construction activities (Figure 5-23). The local district government plays a dominant role in development control and management. If proposed self-construction projects have been

图5-22：关于"你与邻居之间产生矛盾的主要原因？"的调查结果

Figure 5-22: "What is the main reason of having contradictions between you and your neighbors?"

资料来源：（Hoyem，1989：5）
Source: (Hoyem, 1989: 5)

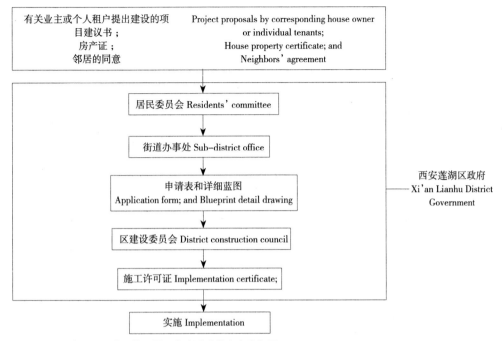

图 5-23：西安回民区中，指导居民自建活动的官方流程图
Figure 5-23: Official procedure for the residents in Xi'an DTMD to conduct self-constructions
资料来源：(XMHDPPO，2003：19)
Source: Author edited from (XMHDPPO, 2003: 19)

区政府在当地的开发、控制和管理中起到了主导作用。如果居民提出的自建项目被区级规划部门所批准，那么区建设委员会将会发布实施证明来监管整个建设过程。通过这个程序，当地居民从开始到实施阶段都会参与到项目当中。但是现实中，调查小组在 2002 年对当地穆斯林居住环境的调查结果显示，只有不到一半的家庭会按照官方制定的程序来开展自建活动，这也意味着约超过一半的建设活动是不合法的，在一开始，居民的建设活动也就没有得到法律的允许（XMHDPPO，2003：19）。大部分居民只是关心保护他们自己的空间不被邻居所侵犯。为了获得更多的生活区域，很多地方可以看到侵占公共空间的非法建设活动（XMHDPPO，2003：19）。

approved by the planning department at the district level, the district construction council will issue the implementation certificate and monitor the construction processes. Through this, local residents are able to participate from the very beginning until the implementation period. In reality, however, less than half the households have gone through the official procedure, according to the survey conducted by a research team on local Muslim living conditions in 2002. This means more than half the existing self-constructions are illegal and that no legitimate consent to their construction activities was granted from the very beginning (XMHDPPO, 2003: 19). Most residents just care about protecting their own space from being invaded by their neighbors. To acquire more living areas, illegal activities such as nibbling public space can be seen in many places (XMHDPPO, 2003: 19).

一般来说，地方居民之所以不愿遵循政府制定的有关程序,主要有两个原因。一方面,迫于巨大的人口压力,很多建设方案都倾向于侵占或啃噬公共空间,如公共场所,或在庭院和街道之上进行房屋空间的扩建。公共场所的侵占是指在建设住房时把前墙推至公共空间内；而空间的侵占指的是通过把第二层或更高的楼层向大街上空出挑50～100cm的做法,以此来啃噬公共空间（Dong，1995：95）。这些非法做法的蓝图如果提交给有关部门,也一定不会得到官方批准,因此,不少居民都不遵循官方制定的工作流程。另一方面,这种现象也与地方建设委员会不力的管理体系有关。地区建设委员会往往只关注旅游街区的环境或能够带来经济利益的项目。那些"后街背巷"的自建活动不会对当地的旅游街区景观产生直接影响。同时，在当前中国强决策和弱效率的城市管理体系下,很少政府部门会关心这种"后街背巷"地段的建设活动。因此,在繁华的旅游街道背后就建造了许多非法建筑。

5.4 总结

本章探讨了西安回民历史街区更新过程中的社会资本问题。作为一个穆斯林居住区,回民区显示出西安市政府在西安明城区内推行的城市更新政策,突出了城市保护的同时也兼顾城市再开发。这些更新政策在北院门街道和大麦市街道的实践过程中已得到具体体现。为了探讨地方政府主导的"保护型"更新规划,本文对北院门街道地区的更新项目进行了分析,该项目强调参与式的城市更

Generally, two reasons account for the residents' disobedience to the official set-up procedure. On the one hand, with the high population pressure, many construction plans tend to invade or nibble public space like ground floors or extend spatially in the courtyard as well as upon the street. The ground-floor technique means moving the front wall onto the public space while building a new house; the spatial way means nibbling the public space by protruding the second or upper floors upon the public street from 50-100 cm (Dong, 1995: 95). These illegal blueprints can never get officially approved if submitted to the corresponding departments, hence the disobedience. On the other hand, this phenomenon relates to the poor management system of the local district construction council, who quite often just pays attention to tourist streets environments or projects which can bring economic interest. Self-construction activities within the "minor lanes or the backyards" (*hou jie bei xiang*), do not form a direct challenge to local tourist street scenes. Additionally, in current China's discretionary authoritarian decision-making and inefficient urban management system, no government department would even care about it. Thus, many illegal buildings have been built behind glorious tourist streets.

5.4 Conclusion

This chapter explores urban regeneration processes and the issue of social capital in a historic district—the DTMD. As a Muslim residential area, Xi'an DTMD showcases the urban regeneration policies of the Xi'an municipal government within the CWA, stressing both urban conservation and redevelopment. These urban regeneration policies have been shown through the practices in Beiyuanmen Street and Damaishi and Sajinqiao Streets. In exploring local government-led "conservation-oriented" regeneration plans, the book explores the

新和管理机制有助于推进以社区为基础的更新过程。在这个过程中，当地的历史建筑得到了保护，破败的城市环境也得到了改善，原住民社区被保留了下来。相比之下，大麦市和洒金桥街道的更新实践，在市场经济主导下，其特色是推行城市再开发和居民重置的更新策略。当地倾向经济发展的更新政策和以市场为导向的城市管治模式密不可分。在当地城市更新过程中采取的大规模的居民重置政策，不仅影响到了对地方历史和宗教居民生活的保护，而且也损害了当地社会资本的积累。

在穆斯林住区，地方居民一般都保持着良好的社区关系，并拥有基于亲情关系的较高水平的社会资本、共同的宗教信仰以及民族特色。另外，穆斯林社区的日常宗教生活和集体活动模式有助于社会行为规范的确立，这些也有利于高水平社会资本的生成。同时，本文还对影响社会资本积累的几个因素进行了探讨。关于影响社区聚合型社会资本的因素，本文认为，历史城市的建筑和商业背景在居民的地方依附性方面，对社区社会资本有一定影响。但是，在倾向经济发展的城市更新政策和策略中，社会资本却容易受到损害。当前碎片化的城市管治模式和居民与地方政府之间缺乏足够的信任，都阻碍了社区社会资本的积累。

regeneration project in the Beiyuanmen street area, which underlines that a participatory urban regeneration and management mechanism contributes to the community-based regeneration process. In this process, local historic buildings have been conserved, dilapidated urban environments have been improved, and original communities have been maintained. In contrast, the regeneration practices in Damaishi and Sajinqiao Streets featured urban redevelopment and residential relocation within the market-oriented economy. Local pro-growth urban regeneration policies relate closely to the market-led mode of urban governance. Massive residential relocation policies adopted in a local regeneration process not only disrupt the conservation of local historic and religious residential lives, but also undermine the accumulation of community social capital.

In a Muslim residential district, local residents generally have very good community relationships and a high level of social capital based on kinship ties and common religious and ethnic characteristics. In addition, a Muslim community's daily religious life and group living pattern contribute to the establishment of social norms, all geared toward the generation of a high level of social capital. At the same time, the book explores several factors affecting local social capital. Regarding the factors affecting the cohesive community relationships, the book argues that local historic urban fabric and business setting have an impact on community social capital in terms of place attachment. Nevertheless, community social capital tends to get compromised in pro-growth urban regeneration policies and strategies. The current fragmented urban governance and the lack of trust between local residents and the district government compromise the accumulation of community social capital.

第六章 具有转型期社会资本典型特征的历史街区及其更新实践

CHAPTER 6　Urban Regeneration in a Historic District with Social Capital based on the Transformation of Community Relationships

通过对西安回民区的城市更新政策和实践以及社会资本的探讨，上一章着重分析了基于宗教和地方旅游环境的高水平的社会资本，以及更新过程中当地居民的参与对地方更新项目的重大意义。基于另一个汉族历史街区，即西安三学街历史街区的调研，本章将进一步探讨社会资本在城市更新过程中的作用。作为汉民生活区，西安三学街历史街区的社会资本与回民区的情况不大相同。由于没有促进高水平社会资本形成的宗教和民族因素，西安三学街历史街区的社区关系更多地反映出转型期中国城市发展中的社会资本现状及问题。

首先，本章简要介绍了西安三学街区，包括其历史背景以及几个当地历史建筑的现状。这些历史遗迹有助于人们了解当地"保护"传统城市景观的更新政策。其次，研究了该地段转型期的城市开发政策和策略。自20世纪80年代至今，三学街区的更新实践大致可分为三个时期：20世纪80年代地方政府主导的，以社区为基础的城市复兴政策；20世纪90年代，地方政府主导的倾向于经济发展的城市再开发政策；自2001年以来的地方政府主导的，保护和再开发并进的城市重组政策。同时，本章还探讨了该地段进行真正意义上的城市更新的必要性。为了研

By exploring urban regeneration policies and practices, as well as the local social capital in Xi'an DTMD, the previous chapter examined the significant contribution of the higher level of social capital based on religion and the local tourism business setting, and the involvement of local Muslim residents in regeneration processes on the outcomes of local regeneration plans. Based on another historic Han Chinese district, Xi'an SHD, this chapter further explores the role of social capital in local urban regeneration processes. As a Han Chinese residential community, Xi'an SHD has a local social capital which is very different from that of DTMD. Without the salient factors of religion and ethnicity contributing to a high level of social capital, the community relationships and connections in Xi'an SHD explicitly reflect the condition of social capital in transitional urban China.

First, this chapter briefly introduces the historic district of Xi'an SHD, including its historical background and the presence of several local historical buildings. These historic physical remnants help us understand the local regeneration policy of "conserving" traditional urban scenes. Second, the transitional urban development policies and strategies in this area are examined. Roughly, from the late 1980s to the present, the regeneration practices in this area have been divided into three periods: local government-led and community-based urban revitalization in the 1980s; local government-led, pro-growth urban redevelopment in the 1990s; and local government-led, conservation-cum-redevelopment urban restructuring

究三学街区自 2001 年以来的更新政策和实践，本章主要研究了两个具体案例：自 2001 年以来，三学街区历史核心区的保护和再开发、自 2004 年以来，大吉昌巷区的再开发。接下来，本章探讨了三学街区中低水平社会资本的问题。通过对外来务工者的涌入、当地破败的居住条件以及脆弱的社区地方依附感的探讨，研究了造成这种低水平社会资本的关键因素。最后，研究认为，由于社会资本水平的低下，尽管这种生活方式对老年人来讲有着特殊意义，但很多年轻人都已经放弃了之前传统的庭院生活方式。另外，随着该历史街区原住民的搬迁，类似原住民集体活动这样的地方非物质文化遗产也容易受到破坏。本章最后总结：当地城市更新政策与策略和脆弱的社区关系密切相关；低水平的社会资本造成了在当地更新规划中社区参与的欠缺，并进一步威胁到整体更新和综合保护目标的实现。

6.1 西安三学街历史街区简介（下文简称三学街区）

6.1.1 历史背景与空间特征

三学街区是西安明城区内的另一个历史居住区，它与回民区最大的不同是该街区的居民以汉族居民为主。西安历史文化名城保护规划（XCPB, XACH, & XUPDI, 2005）和西安历史文化名城保护条例都指出，三

since 2001. At the same time, the needs for genuine urban regeneration in this area are also explored. In examining the regeneration policies and practices in Xi'an SHD since 2001, this chapter focuses on two specific cases: the conservation and redevelopment of the historic core of SHD since 2001 and the redevelopment of the Dajichang alley site since 2004. The chapter further probes into the issue of the low level of social capital within SHD. Several key factors affecting this low level of social capital are studied, along with an examination of the influx of migrant workers, local dilapidated living conditions, and weak communities' place attachment. Finally, we argue that with a low level of social capital, previous traditional courtyard living styles have been challenged and abandoned by many young workers, although they remain to be significant especially to local senior residents. In addition, with the loss of the original residents of this historic district, the conservation of local intangible heritage in terms of the original residents' group activities has been undermined. This chapter concludes by emphasizing that local urban regeneration policies and strategies relate to weak community relationships, and amid the lack of community involvement in local regeneration plans due to the low level of social capital, comprehensive urban regeneration and integrated urban conservation outcomes have been further undermined.

6.1 Introduction to the Sanxuejie Historic District (hereafter, SHD)

6.1.1 Historical background and spatial features

The Sanxuejie Historic District (hereafter, SHD) is the other historic residential district within the Xi'an City wall area, and its major difference from DTMD is that its local residents are all Han Chinese. As laid out in the Xi'an Urban Conservation Plan of the Historic City (XCPB, XACH, & XUPDI, 2005) and in the

三学街历史街区中的人口构成
Population composition of SHD

表 6-1
Table 6-1

	2000 年　In 2000			2008 年　In 2008		
	永久居民　Permanent residents			永久居民　Permanent residents		
	男性（人）Male	女性（人）Female	总数（人）Sub-total	男性（人）Male	女性（人）Female	总数（人）Sub-total
三学街社区（或三学街区的核心区）Sanxuejie community (or historic core of SHD)	2302（54%）	1961（46%）	4263	2521（53%）	2235（47%）	4756
书院门社区　Shuyuanmen community	2010（54%）	1713（46%）	3723	2113（53%）	1873（47%）	3986
总计　Total	4312	3674	7986	4634	4108	8742

资料来源：2009 年 3 月，碑林区公安局提供
Source: From the police station in Beilin district in March 2009

街区东起开通巷、西至南大街、北起东木头市、南至城墙（SCXPC, 2002）。三学街区在西安明城区的地理位置见前图 1-3 所示。该街区占地面积共约 36.5hm² (XUPDI, 2001: 2)，2000 年这里的永久居民人数约 7900 人。据 2009 年碑林区公安局提供的数据，至 2008 年，永久居民总人数增至约 8700 人。（表 6-1）显示了三学街区中的人口构成情况。

现有文献（Ju, 2005；XCUCRC, 2000；XUPDI, 2001；Yuan, 2008）表明，该历史街区是从唐朝逐渐形成的"一庙三学"演变而来的，"一庙三学"指的是：孔子庙、西安府学、长安县学以及咸宁县学。此外，几个其他当地历史古迹，如、卧龙寺、关中学院和华塔寺共同构成了该区的历史环境。（图 6-1）所示西安地图中，黄色区域代表了三学街区的位置，并标示出了该区中的重要历史遗迹位置。

"一庙三学"的城市模式
孔庙和碑林

自唐朝开元二十七年（739 年），孔子被封为文宣帝以来，中国很多地方，为了纪念

Xi'an Conservation Regulations of the Historic City, SHD ranges from the Kaitong Lane in the east to the South Street in the west, and from the East Mutoushi Street in the north to the city-wall in the south (SCXPC, 2002). See previous Figure 1-3 for the location of SHD within Xi'an CWA. The total area of the SHD is approximately 36.5 hectares (XUPDI, 2001: 2), with a population of approximately 7,900 permanent residents in 2000. Total permanent residents rose to approximately 8,700 in 2008, according to the statistical data from the local police station in 2009. Table 6-1 shows the population composition of SHD.

Available literature (Ju, 2005; XCUCRC, 2000; XUPDI, 2001; Yuan, 2008) shows that this historic district has derived its pattern from one temple and three schools (*Yi Miao San Xue*) that were gradually formed since the Tang dynasty: the Confucius Temple, the Xi'an Prefecture School (*Fu Xue*), the Chang'an County School (*Chang'an Xian Xue*), and the Xianning County School (*Xian Ning Xian Xue*). Moreover, several other local historic sites such as the Dragon-crouching Temple (*Wo Long Si*), Guanzhong School, and the multi-colored pagoda temple (*Hua Ta Si*) together contribute to its historic environment. Figure 6-1 shows the location of SHD in a yellow square and the locations of historic monuments within this district in the current Xi'an map.

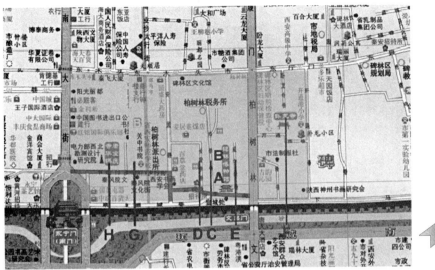

图 6-1：三学街历史街区中的孔庙（A），碑林（B），西安府学（C），长安县学（D），咸宁县学（E），卧龙寺（F），关中学院（G）华塔寺（H）

Figure 6-1: Locations of Xi'an Confucius temple (A), Beilin (B), Xi'an prefecture school (C), Chang'an county school (D), Xianning county school (E), Dragon-crouching temple (F), Guanzhong school (G) and Multi-colored-pagoda temple (H) within the SHD in current Xi'an map

孔子已经建立了孔子庙（XUPDI，2001：4）。很多研究（SACH，1998：23；Wu，1979：265；Zhao，1992：253）表明，西安孔庙是第一所孔庙，建于1080年，即北宋元丰三年（ECXBDR，2003a：685）。但是，西安孔庙现有的大部分建筑都是在明清时期建立的（Zhao，1992：253）。西安孔庙的正北就是碑林，碑林始建于1087年，即北宋元祐二年，是为了保护唐朝的开城石经（837年）以及石台孝经（公元745年）（ECXBDR，2003a：687；SACH，1998：26）。明朝正统年间（1436～1449年）扩建了整个西安孔庙和西安碑林建筑群，并且整个区域自1961年被列为国家级遗产纪念碑。作为世界最大的雕刻石头的收集地，1993年，该区被名为西安碑林博物馆（Zhang，2002e：1）。

碑林博物馆覆盖面积3.19万平方米，展

Historic urban pattern of "one temple and three schools (Yi Miao San Xue)"

The Xi'an Confucius Temple and Xi'an Beilin (the Stele Forest)

Since Confucius was named as Wen Xuan King in A.D. 739 (the 27th year of Kai Yuan of Tang dynasty), Confucian temples as spatial structures in memory of Confucius have been built in many places in China (XUPDI, 2001: 4). Many studies (SACH, 1998: 23; Wu, 1979: 265; Zhao, 1992: 253) indicate that the Xi'an Confucius Temple was first constructed in A.D. 1080, the third year of the Yuanfeng period during the North Song dynasty (ECXBDR, 2003a: 685). Nonetheless, most parts of existing structures within the Xi'an Confucius temple have been built during the Ming and Qing dynasties (Zhao, 1992: 253). Just to the north of the Xi'an Confucius Temple locates Xi'an Beilin (Stele Forest), which was first constructed in A.D. 1087, the second year of the Yuan You period of the North Song dynasty, for the preservation of Kai Cheng Stone Scriptures (*Kai Cheng Shi*

览面积大约 4900 平方米，馆区由孔庙、碑林、石刻艺术室三部分组成（Zhang，2002e：2）。现有馆藏文物 11000 余件，其中包括自各个朝代著名书法家的作品。一些石碑记录了和其他国家进行文化交流和往来的历史事件，因此，这些石碑在记录中国的社会和文化的演变过程中起了很大的作用（Ma，1985：162）。但是，西安碑林博物馆面临的一个现实问题就是碑多地少。据已经出版的资料（Lu，2000：181；XSFM，2000：1；Zhang，2002b：2）显示，博物馆内的石碑有 2500 多个，但是目前的展厅只能展览大约 1100 件，这就意味着一半以上的石碑不能展列，这就有可能削弱了未被展出的石碑的社会、文化和经济价值。为了解决这个问题，西安市政府租了博物馆周围的土地，即，三学街区的历史核心区来进一步拓展博物馆的空间。但是，由于这片潜在的地区位于地方碑林街区的行政边界中，所以这片土地的产权变得不明确。这种所谓的产权问题，引起了西安碑林博物馆、地方区政府和地方居民之间，就更新目的，产生的冲突。图 6-1（A & B）显示了西安西安碑林博物馆的位置。

西安府学

西安府学是明代当地县政府成立的一所中学。根据中国古代的等级教育体系[1]，这所中学在 1949 年中华人民共和国成立之前是西安最好的教育机构（Yuan，2008）。最初，西安府学位于府学巷，实际上是唐代皇家祠堂的位置。公元 904 年，太学从城墙外移到

---

[1] 中国旧时期（1911 年辛亥革命之前），学校主要分为两大类：官学和私塾，官学又分为三个级别：国子监、府学或州学以及县学（Yuan，2008）。

Jing) (A.D. 837) and the Stone Scripture on Worthy Progeny (*Shi Tai Xiao Jing*) (A.D. 745) of the Tang dynasty (ECXBDR, 2003a: 687; SACH, 1998: 26). The entire spatial area of the Xi'an Confucius Temple and Xi'an Beilin was enlarged during the Zheng Tong period (1436-1449) of the Ming dynasty, and the whole area was classified as a heritage monument at a national level since 1961. As the world's largest collection of inscribed stones, it was named as the Xi'an Stele Forest Museum in 1993 (Zhang, 2002e: 1).

Covering a ground area of 31.9 thousand square meters, this museum has an exhibition area of about 4,900 square meters, and it consists of three sessions: the temple of Confucius, the Stele Forest, and the Halls for Carved Stones (Zhang, 2002e: 2). Nowadays, on display are over 1,000 stones, among which are the works of famous calligraphers from various dynasties. Some stones record important historical events and cultural contracts and interactions with other countries, and therefore, these stones have a significant role in the social and cultural evolution of China (Ma, 1985: 162). Nevertheless, a practical problem facing the Xi'an Stele Forest Museum is the contradiction between its exuberant storage and limited exhibition areas. Published sources (Lu, 2000: 181; XSFM, 2000: 1; Zhang, 2002b: 2) suggest that there are more than 2,500 stone tablets stored in the museum, but current exhibition rooms can only display around 1,100 pieces. This means that more than half of the total number of stones could not be displayed, and this potentially undermines the social, cultural, and economic values of the unexhibited tablets. To address this, the Xi'an municipal government reserved the land around it, that is, the historic core area of SHD, for the future development of the museum. Nevertheless, since this potential area is located within the administrative boundary of the local Beilin District, the rights for this property became ambiguous. The said rights issue spurred conflicts regarding regeneration intentions among people in the Xi'an Stele Forest

了今天的府学巷的位置。各个朝代不同级别的学校都坐落在这里，如，宋朝的金京兆府学、金代的经济府学、元朝的奉园路学以及明清时期的西安府学（XUPDI，2001：4）（图6-1（C））。

长安县学

长安县学是明朝长安县政府建立的一所县级中学。起初，它坐落于西安市西门外，在明清时期也称为西安府城（XCUCRC，2000：35）；在1370年，即洪武年间三年，它被重新安置在西安城内的西街。最后，在1473年，即成化年间九年，它被重新安置在西安府学的西侧，就是现在的长安学巷（EBCCR，1999：610；Zhang & Dong，1967：471）（图6-1（D））。

咸宁县学

咸宁县学，是明朝初年成立于咸宁县的一所县级中学。最初，这所学校坐落于咸宁县西部。1471年，即明朝成化七年，它被重新安置在西安府学的东边，也就是现在咸宁学巷的位置（EBCCR，1999：610；Ju，2005：29）（见图6-1（E））。

光绪三十年（1905年），清政府宣布废除科举制度（科举考试）。随后，这三所教育中心，即西安府学、长安县学和咸宁县学，丧失了它们之前的教育功能，成了历史文物（XUPDI，2001：5）。由于其文化和人文环境，这个地方吸引了很多人前来购买房屋和土地，例如，著名书法家于柔仁先生。20世纪80年代以后，随着快速城市化的进程，之前的三所学府都被住宅和零售商店所取代，它们的名字还是延续了原来学校的名字（Zhu，2006：52）。今天，这个地区内，之前学校留下的遗迹已经很少

Museum, the local district government, and also local residents. The site of the Xi'an Stele Forest Museum is shown in Figure 6-1 (A & B).

The Xi'an Prefecture School (Fu Xue)

The Xi'an Prefecture School, or Xi'an Fuxue, was a middle school established by the Xi'an local prefecture government during the Ming dynasty. According to the hierarchical education system in ancient China[1], this middle school offered the highest level of education in Xi'an City before P.R. China was established in 1949 (Yuan, 2008). Originally, Xi'an Prefecture School was located at Fuxue Lane, which was actually the location of the Imperial Ancestral Temple of the Tang dynasty. In the year 904, the imperial school or grand school (*Taixue*) was relocated to the location of today's Fuxue Lane from the outside city wall. The schools of different dynasties at different levels were all situated here, such as Jingzhao Fuxue in the Song dynasty, Jingji Fuxue in the Jin dynasty, Fengyuanlu Xue in the Yuan dynasty, and Xi'an Fuxue in the Ming and Qing dynasties (XUPDI, 2001: 4) (shown in Figure 6-1 (C)).

The Chang'an County School (Chang'an Xian Xue)

The Chang'an County School was a middle school established by the Chang'an County government at the county level during the Ming dynasty. Originally, it was located outside the west gate of Xi'an City, also called Xi'an Fucheng during the Ming and Qing dynasties (XCUCRC, 2000: 35); in 1370, the third year of the Hongwu period, it was relocated at the west street within Xi'an City. Then finally, in 1473, the ninth year of the Chenghua period, it was relocated at the western side of the Xi'an Prefecture School, the location of today's Chang'anxue Lane (EBCCR, 1999: 610; Zhang

---

[1] In ancient China (before Xinhai revolution in 1911), schools were divided into two categories: public schools (Guanxue) and private ones (Sishu), and within the public schools three levels were created: the grand school of central government (Guozijian), the prefecture schools of regional government (Fuxue or Zhouxue) and the county schools (Xianxue) (Yuan, 2008).

图 6-2：长安学巷中，以前长安县学遗留下来的二道门
Figure 6-2: Two remaining gates of previous Chang'an county school in the Chang'anxue lane

了，其中一个就是长安县学的第二个大门（图6-2）。这是古代考试的入口，是古代唯一的考试的地方。而如今，它只不过是一个荒废的结构。

三学街区的其他历史建筑

卧龙寺

位于现今三学街区内柏树林的东边，卧龙寺最初建立于汉朝灵帝时（168～189年），隋朝时称"福应禅院"（ECXBDR，2003a：667）。直到北宋年间，这座寺庙才更名为卧龙寺（Ju，2005：30；XUPDI，2001）。今天，卧龙寺占地面积约15000平方米（ECXBDR，2003a：667），已被列入省级文物保护单位，是当地朝圣者的宗教场所，也是旅游胜地（SACH，1998：25）（图6-1（F））。

关中书院和书院门

关中书院始建于明朝1609年，它是知名学者冯从吾讲学的地方（XUPDI，2001：6；Zhao，1992：295）。关中书院是明清时期教学质量最好、级别最高的官学。清朝废除科举制度之后，1906年，关中书院改建为

& Dong, 1967: 471) (shown in Figure 6-1 (D)).

The Xianning County School (Xian Ning Xian Xue)

The Xianning County School, a middle school at the county level, was established in Xianning County during the beginning of the Ming dynasty. Originally, the school was situated at the western side of Xianning County. In 1471, the seventh year of the Chenghua period of the Ming dynasty, it was relocated to the eastern side of the Xi'an Prefecture School, where the Xianningxue Lane is situated today (EBCCR, 1999: 610; Ju, 2005: 29) (shown in Figure 6-1 (E)).

By 1905, the 31st year of Guangxu, the Qing government announced the abolition of the imperial examination system (*Keju Kaoshi*). Following this, the three education centers, namely, the Xi'an Prefecture School, the Chang'an County School, and the Xianning County School, became historic relics without their previous educational functions (XUPDI, 2001: 5). Due to its cultural and literati environment, the placed has attracted many people to buy houses and lands within the area, such as Mr. Yu Youren, a well-known calligrapher. After the 1980s, with rapid urban development, these previous three schools have all been replaced by residential buildings and retail shops, the names of which were taken from their original school names (Zhu, 2006: 52). Today, only very limited relics relating to the previous schools remain in this area, one of which is the second gate of the Chang'an County school (shown in Figure 6-2). This is the entrance to the old examination area, a solemn area for examinees, during the ancient period. At present, it only exists as a deserted structure.

Other historical buildings within SHD

The Dragon-crouching Temple (Wo Long Si)

Located at the east side of today's Baishulin Street within SHD, the Dragon-crouching Temple was first built during the reign of a Ling emperor (168-189) of the Han dynasty, and it has the original name of Fu Ying Chan Si (ECXBDR,

陕西第一师范学堂，中华人民共和国成立之初，将其改为陕西师范学院。今天，作为市级文化遗产，这里既是西安小学以及附属小学的位置，占地面积约3.4万平方米（SACH，1998：24）（图6-1（G））。

由于关中书院名闻天下，其前面的街道命名为书院门（或学习经典的学院大门），这条街道早在1609年就存在了（ECXBDR，2003a：512）。今天，书院门东至柏树林大街，西至西安南大街，长约570米，书院门街道连接了三学街区内的所有历史古迹（EBXY，1993：308）。考虑到书院门显著的历史和文化特色，该地区自20世纪80年代末以来，已经被开发为一个著名的旅游景点，这里囊括了丰富的文化内涵，例如书法、绘画以及文房四宝（笔、墨、纸、砚）（XCUCRC，2000：163）（图6-1中的紫色部分）。接下来将会进一步讨论这条街道发展演变过程。

华塔寺

紧邻关中书院的另一个历史遗址就是华塔寺。华塔寺始建于隋朝人寿年间（601～604年）（ECXBDR，2003a：669），而寺中的宝塔是在唐朝开成年间（Ju，2005：30）建造的。宝塔之所以被称为华塔是因为它是由各种颜色的砖块建成的。但是，清朝末年，除了华塔之外的其他建筑都遭到了严重的破坏，或被摧毁了（SACH，1998：21）。今天，只有明代景泰二年（1451年）重新修建的宝塔保留下来，而且已经在1957年被列为省级遗产古迹（XCUCRC，2000：607）（图6-1（H））。

随着西安市的快速城市化进程，三学街区同样面临着保护的必要性。自2000年以来，西安市政府和碑林区政府提出了在保护当地

2003a: 667). The temple was not named as Dragon-crouching Temple until the North Song dynasty (Ju, 2005: 30; XUPDI, 2001). Today, covering an area of about 15,000 square meters (ECXBDR, 2003a: 667), the temple has been classified as a heritage monument at the provincial level and serves as a religious place for local pilgrims and also a scenic spot for tourists (SACH, 1998: 25) (shown in Figure 6-1 (F)).

The Guanzhong School and Shuyuanmen (academy gate of classical learning)

The Guanzhong School was first built in 1609 during the Ming dynasty as a venue for then well-known scholar Feng Congwu's lectures (XUPDI, 2001: 6; Zhao, 1992: 295). The school was considered as the best and also the highest level of Guan school (*guan xue*) during the Ming and Qing dynasties. With the abolition of the imperial examination system during the Qing dynasty, Guanzhong School was converted to the First Normal School of Shaanxi (*Shaanxi diyi shifan xuetang*) in 1906 and the Normal School of Shaanxi Province at the beginning of the republic of China, respectively. Today, as a heritage monument at the municipal level (ECXBDR, 2003a: 672), it is the location of the Xi'an Normal School as well as the Attached Primary School, spanning an area of about 34 thousand square meters (SACH, 1998: 24) (shown in Figure 6-1 (G)).

Owing to the presence of the well-known Guanzhong School, the street in front of it was named Shuyuanmen (or the academy gate of classical learning) which was in existence since 1609 (ECXBDR, 2003a: 512). Today, starting from Baishulin Street in the east to Xi'an South Street in the west, Shuyuanmen is about 570 meters long, through which the historic sites within SHD are all connected (EBXY, 1993: 308). Considering its conspicuous historic and cultural features, this area has been re-developed into a famous tourist street since the late 1980s, which exhibits and sells exuberant cultural contents such as calligraphies, paintings, as well as the

历史古迹和环境的同时，对破旧的地区进行更新。在不同时期，地方政府对三学街区的不同地段提出并实施了这些规划，包括2002年三学街区的保护和复兴规划，2003年南门至文昌门的控制细则，以及2004年顺城巷的保护和更新规划（顺城巷规划），等等。值得注意的是，地方政府在实施促进地方经济发展以及房屋拆迁和补偿的更新政策时，很多原住民都倾向于离开这个街区。这与前文中的穆斯林街区的情况形成了鲜明对比。随着原住民逐渐搬离该街区，三学街区的邻里关系和社交网络也受到了影响，而且历史街区应有的"灵魂"因素也逐渐消逝了。

four treasures of the study (writing brush, ink stick, ink slab, and paper) (XCUCRC, 2000: 163) (shown in the purple belt in Figure 6-1). More discussions on the evolution of this street will be made in the succeeding discussions.

<u>The Multi-colored Pagoda Temple (Huata Si)</u>

Another historic site adjacent to Guanzhong School is the Multi-colored Pagoda Temple (*Huata Si*). The temple was first built during the Renshou period (A.D. 601-604) of the Sui dynasty (ECXBDR, 2003a: 669), while the pagoda within the temple was built during Kaicheng period of the Tang dynasty (Ju, 2005: 30). The pagoda obtained the name of *Huata* or Multi-colored Pagoda because of the use of bricks in various colors. Nonetheless, the temple and all other buildings except for ***Huata*** were damaged and destroyed at the end of the Qing dynasty (SACH, 1998: 21. Today, only the reconstructed pagoda at the second year of the Jingtai period of the Ming dynasty exists, which has been classified as a heritage monument at the provincial level in 1957 (XCUCRC, 2000: 607) (shown in Figure 6-1 (H)).

With the rapid urban development and urbanization of Xi'an City, this historic district also faces the necessity of preservation. Since 2000, the Xi'an municipal government and the local Beilin District government have proposed to revitalize the run-down areas while conserving local historical monuments and environments. These plans have been proposed and implemented by the local district government in various areas of SHD in different periods, including the Conservation and Renewal Plan of Sanxuejie Historic District in 2002, the Controlled Detailing Plan of the Southern-gate-Wenchang-gate Area in 2003, the Conservation and Renewal Plan of Shuncheng alleys (around-city-wall alleys) in 2004, and so on. Notably, despite the local district government's pro-growth as well as rehousing and compensation policies, many original residents chose to leave this area. This forms a sharp contrast with the situation in the Muslim district previously discussed. With the moving-out of local indigenous residents, the relationships among neighborhoods and networks have been disrupted, and the "soul" of the area as a historic district also began to vanish.

6.2 三学街区的更新政策与更新实践

三学街区的更新政策和策略是根据不同历史时期，当地特定的更新需求而提出的。随着1949年中华人民共和国的成立，当地政府提出了改善大范围的破败城区的需求，城市更新主要集中在改善基础设施方面，例如修建了新的沥青马路，以取代之前的土路（XCUCRC，2000：163）。自20世纪80年代以来，为了改善书院门的居住环境，并进一步促进当地的经济发展，该地段被提议重建为文化旅游街。重建之后的街区，为了与其历史地位相一致，街道两旁都修建了传统风格的建筑物（ECXBDR，2003a：277）。与此同时，三学街或"一庙三学"地段，即府学巷、长安学巷和咸宁学巷[1]，被开发成了传统居民区。随着西安市在1992年被划定为内陆开放城市，碑林区政府也加快了当地发展的速度。1999年，随着中国西部大开发计划的提出，为了适应更具竞争性的环境，西安各级政府开始调整他们的城市政策（Gui，2000，2001）。2004年，随着西安第四次总体规划的颁布，碑林区政府提出了改善城市环境的更新政策，其目的是通过保护三学街区的历史环境能够吸引更多的投资（EBXY，2006b：335）。

6.2 Transitional Urban Regeneration Policies and Practices in Xi'an SHD

Generally, urban regeneration policies and strategies in SHD relate to some specific local regeneration needs in different phases. With the establishment of P.R.China in 1949 and the need to enhance large-scale run-down areas, urban revitalization mainly focused on the improvement of fundamental infrastructures, such as replacing previous earth roads with asphalt ones (XCUCRC, 2000: 163). Since the 1980s, in order to improve the physical living environment of Shuyuanmen and its local economic development, this street was proposed to be redeveloped into a cultural tourist street. After its redevelopment, it has been flanked by traditional-styled constructions in tune with its historic position (ECXBDR, 2003a: 277). In the meantime, the Sanxuejie Street area or the area of the "one temple and three schools," that is, the Fuxue Lane, the Chang'anxue Lane, and the Xianningxue Lane[1], has been redeveloped into a traditional residential quarter. With Xi'an City named as an in-land open city (*nei lu dui wai kai fang cheng shi*) in 1992, the local Beilin District government also increased its local redevelopment pace. In 1999, with the proposal of China's Western Development Program, Xi'an's various levels of governments actively began to adjust their urban policies in order to cater for a more competitive environment (Gui, 2000, 2001). With the promulgation of the Xi'an fourth master plan in 2004, the Beilin District government proposed its regeneration policy of improving its urban

---

[1] 府学巷是前西安府学的所在地，长安学巷是前长安县学的所在地，咸宁学巷是前咸宁县学的所在地。

[1] Fuxue lane is the area of previous Xi'an prefecture school, Chang'anxue lane is the area of previous Chang'an county school, and Xianningxue lane is the area of previous Xianning county school.

6.2.1 地方政府主导和基于社区的城市修复阶段（20世纪80年代末～20世纪90年代初）

根据1980年通过的西安第二次总体规划，西安城市规划管理局于1989年制定了西安中心城区规划，其目的是通过展示当地的历史文化价值——如文物古迹、特色建筑以及传统购物街——来促进当地旅游业的发展。正如规划中所述，要从各方面入手来构建一个旅游网络，促进西安历史、传统以及当地的城市特色的恢复（XUPEMB, 1989）。在这个背景下，三学街区进行了大规模的城市复兴实践，其特点是保护历史古迹，修复破败的城市环境。类似于把三学街建设成"一条古香古色的旅游文化街"的概念，就是那时提出的（XCUCRC, 2000：163）。在1990年更新还没进行前，三学街区中很多房子破败不堪，基础设施也很残破。面对这种情况，碑林区政府提出将书院门改造为商业街，重点是再现当地的历史文化风貌。这也是西安作为古城展示其历史文化的第一条商业步行街（EBXY, 1993a：308）。

有关改造项目于1990年10月开工，遵循碑林区政府提出的"统一规划、统一管理、统一实施"的方针。该项目的一大特色是必要的经费由当地居民提供（EBXY, 1993a：308）。该项目中，书院门街道的598户原住居民家庭提供了共约1500万元人民币（ECXBDR, 2003a：277）。虽说书院门街道的规划是由当地区政府实施的，但是街道后面的房屋修复工作则是由当地住户自己来开展。这样一来，大多数原住民也得以保留在了这个地区。1991年10月，该项目顺利竣工，书

environment in order to attract more investments through the "conservation" of SHD's historic urban environment (EBXY, 2006b: 335).

6.2.1 Local government-led and community-based urban rehabilitation (late 1980s to early 1990s)

Based on the approved second master plan in Xi'an in 1980, the Xi'an Urban Planning and Management Bureau worked out the Xi'an Central Urban District Plan in 1989, which aimed to promote the local tourism industry through the exhibition of local historical and cultural values, such as local heritage monuments, residential housings, and traditional shopping streets. Through organizing these perspectives to form a tourism network, as indicated in the plan, Xi'an's ancient history, tradition, and local urban features were promoted (XUPEMB, 1989). Against this background, large-scale urban revitalization practices within SHD commenced, which were characterized by the conservation of historical monuments and rehabilitation of dilapidated urban environments. The concept of developing Sanxuejie Street within SHD into a "tourist and cultural street in ancient style" was proposed at that time (XCUCRC, 2000: 163). Before its renewal in 1990, most houses in SHD were very dilapidated, along with the derelict basic infrastructures. Confronting these situations, the Beilin District government proposed to redevelop Shuyuanmen into a commercial street focusing on exhibiting its cultural and historical characteristics, which was actually the first street in Xi'an that exclusively exhibited its culture as an ancient capital city (EBXY, 1993a: 308).

The project started in October 1990, with the policy of "unified planning, unified management, and unified implementation" by the Beilin District government. One characteristic of this project was that necessary funding was raised by local residents (EBXY, 1993a: 308). In this project, a total of 15 million (RMB) was raised from 598 original households along the street (ECXBDR, 2003a: 277). While the project

1992 年碑林区政府在房屋更新中的收入和投资　表 6-2
Beilin district's revenues and investment in housing renewal in 1992　Table 6-2

	单位（百万元） Unit (million RMB)
碑林地区的收入 Beilin district revenues	99.63
碑林地区的花费 Beilin district expenses	99.63
碑林区政府在房屋更新中的投资 Investment in housing renewal by Beilin district government	4.87

资料来源：（XSB，1993：31）
Source: (XSB, 1993: 31)

院门街道成为西安第一条古文化街（EBXY，1993a：308）。

当地碑林区政府之所以采取以社区为基础的复兴策略，在很大程度上与当时地方政府有限的财政预算有关。据官方统计数据表明，1990 年，西安市政府对城市房屋更新的总投资总计 5367 万元人民币，其中只有 78 万元用在区级或县级的房屋项目上（XSB，1990：202）。有限的财政预算给地方政府造成很大压力。表 6-2 显示了 1992 年碑林区政府在城市更新项目上的收入和支出。

相对于碑林区政府有限的财政预算而言，由地方居民自己提供房屋更新的资金能够起到实质性的作用。尽管在以不同方式收集房屋更新资金，以及由地方居民自己施行更新工程具有很多好处，但是，需要注意的是，这同时也带来了许多不利影响。例如，历史街区中很多居民自建的房屋质量和整体风貌都得不到控制，不少房屋不符合相关规定。这种情况在 20 世纪 90 年代随着快速城市化的进程而更加突出。为了对西安城墙区内的这种历史城市景观加以控制，西安市政府在

along Shuyuanmen was conducted by the local district government, the revitalization work on housing structures behind the street was carried out by respective households themselves. This way, most original residents have been retained in the area. With the completion of the project in October 1991, Shuyuanmen became the first ancient cultural street (*gu wen hua jie*) in Xi'an city (EBXY, 1993a: 308).

The reason why local Beilin District government promoted the community-based revitalization strategy had much to do with the limited budget of the local government at that time. According to available official statistical data, the total investment of the government in urban housing renewal projects in 1990 was 53.67 million (RMB), of which only 780 thousand was spent for housing projects within the district or county jurisdiction (XSB, 1990: 202). This gave much pressure to the local government owing to its limited budget. Beilin District's revenues and expenses on urban renewal in 1992 are shown in Table 6-2.

As compared to the limited budget of the Beilin District government, the necessary funds for housing renewal by the local residents themselves turned out to be substantial. Although there are benefits in raising the housing renewal funds in various ways and conducting the rehabilitation projects by local residents themselves, it was noticed that this also had many disadvantages. For example, many of the self-constructed structures within the historic district are of poor quality and were conforming to no particular regulation. This situation worsened in the late 1990s with the rapid urbanization process. In order to control the historic urban scene within Xi'an CWA, the Xi'an Regulations on Urban Building-Height Control was enacted by the Xi'an municipal government in 1993 (SCXPC, 1993: 376), which stipulated that the building height within SHD should not exceed nine meters. This regulation set the ceiling for many local three- to four-storey buildings which are serving the needs

1993年颁布了《西安市控制市区建筑高度的规定》(SCXPC, 1993: 376), 该规定明确了三学街区中的建筑高度不应超过9米。由于当地人口的迅速增长, 很多地方都建造了3～4层的建筑物, 所以该条例为这种情况设置了界限。不过, 仍然有很多自建房屋在侵占更多的土地或公共空间, 这使得当地的居住环境变得异常拥挤, 并带来不少火灾隐患。需要注意的是, 缺乏专门的法律法规针对当地居民的自建活动进行监管, 也是造成这种问题的一个原因。除了缺少必要的法规, 当地适于监管的管理机制也严重影响了已经颁布法规的有效实施。这种地方性的疏于监管控制的低质量自建活动, 在该地段的大吉昌巷表现尤为突出 (Liu, 2006a: 81)。作为西安顺城巷 (城墙周围的巷道) 的一部分, 大吉昌巷在2004年被列入顺城巷保护和更新规划的市级城市更新计划。

6.2.2 地方政府主导和促进地方经济发展的房地产再开发阶段(20世纪90年代)

在20世纪90年代, 碑林区的城市再开发表现出以市场为导向的更新政策的特点。随着1986年国家土地改革以及1988年住房制度的改革, 房地产企业也参与到了碑林街区的城市再开发项目中。到1993年, 碑林街区已经成立了28家房地产公司。碑林区政府最初在低洼土地的更新项目中积极利用这些开发公司。随着房地产开发公司的参与, 城市更新进程取得了长足进展,建立了新型的城市面貌。因此, 区政府一如既往地鼓励这种再开发模式, 例如, 在1992年, 瓦窑村的更新项目中 (EBXY, 1994: 298), 重建了大约8.4万平方米的新

of the rapidly increasing local population. Nevertheless, many self-constructed buildings tended to occupy as much land or seized public spaces, which resulted in the very crowded living environment and some hidden fire hazards. It is noticed that with regard to the self-construction activities of the local residents in this area, there remains to be no specific laws or regulations to guide these. In addition to the lack of necessary regulations, there has also been a loose development control and management mechanism, which severely undermines the effectiveness of the already enacted regulations. The poor quality of the local uncontrolled self-construction activities was reflected prominently in the Dajichang alley site between Shuyuanmen Street and the city wall (Liu, 2006a: 81). As part of the Xi'an Shuncheng alleys (around-city-wall alleys), Dajichang alley site was included in the municipal urban renewal scheme of the Conservation and Renewal Plan of the Shuncheng alleys (around-city-wall alleys) in 2004.

6.2.2 Local government-led pro-growth urban redevelopment (during the 1990s)

In the 1990s, urban development in Beilin District exposed the characteristics of the market-oriented urban policy of Xi'an City. With the national land reform in 1986 and the housing reform in 1988, real estate companies also began to be involved in the urban redevelopment of Beilin District. By 1993, 28 real estate companies had entered the Beilin District. The Beilin District government actively made use of these development industries initially in the renewal projects of the low-lying lands. With the involvement of real estate companies, urban renewal progressed, and new cityscapes were developed. Therefore, the district government enthusiastically encouraged this kind of redevelopment mode. For instance, during the renewal project of *Wayao* village in 1992 (EBXY, 1994: 298), about 84 thousand square meters of new construction areas were redeveloped, and 216 households were relocated (Zhang, 1993:

型建筑区，216 户家庭被重置（Zhang，1993：307）。之后在 1993 年德福巷历史街区的更新过程中，重建了大约 8 万平方米，1295 户家庭被重置（EBXY，1994：298）。

以经济增长为导向，房地产主导的更新政策

1992 年西安被国务院划为 11 个内地开放城市之一，相应地，当地的发展规划进一步促进私人开发商在项目中的参与。西安市政府和区政府也从私营部门获得更多的动力和机会来吸引投资（XMG，1993：1）。这从碑林区迅速增长的街道经济[1]也可以看出来。从 1985 年到 1993 年，碑林区的街道经济总收入从 6690 万元人民币增长到 8.77 亿元人民币，总利润从 423 万元人民币增长到 5065 万元人民币，区财政收入从 300 万元人民币增长到 4100 万人民币。这期间碑林区的企业数从 1985 年的 68 个增加到 1993 年的 1856 个（Ji，1994：371）。碑林历史街区中，三学街所在的柏树林街道办事处，依靠当地的历史特色和城市文化特色，其街道经济自 20 世纪 90 年代以来也取得了很大的进步。到 1994 年，在柏树林街道办事处管辖的范围内又新增了 130 多家企业。1994 年柏树林街道办事处的经济收入 1.28 亿元人民币，街道财政收入 1027 万元，街道利润 1165 万元，其街道经济在陕西省街道中名列第一（Ma，1995：270）。柏树林街道经济的快速增长和当地的再开发项目密不可分。到 1994 年，柏树林街道超过 30% 的行政区域已经被拆除并进行了重建（EBXY，2000b：296）。图 6-3 显示出 1999 年

---

[1] 街道经济是指街道行政级别的经济结构形式及其发展情况。

307). Then during the renewal of *Defuxiang* historic district in 1993, about 80 thousand square meters were redeveloped, and 1,295 households were relocated (EBXY, 1994: 298).

Economic growth-oriented and property-led regeneration policies

With Xi'an's ratification as one of the 11 inland open cities by the State Council in 1992, the involvement of private developers in local development plans was further promoted. Local municipal and district governments earned more impetus and opportunities to attract investments from private sectors (XMG, 1993: 1). This was well illustrated from the rapid growth of the street economy (*jie dao jing ji*)[1] in Beilin District. From 1985 to 1993, the general income of the street economy increased from 66.9 to 877.0 million (RMB), the overall profit increased from 4.23 to 50.65 million (RMB), and the district revenue increased from 3.00 to 41.00 million (RMB). The number of enterprises entering Beilin District during this period increased from 68 to 1,856 (Ji, 1994: 371). Within Beilin District, the Baishulin Street office or Sub-district, where SHD is located, also made great progresses in developing its street economy since the early 1990s, based on local historic and urban cultural features. By 1994, more than 130 new enterprises were set up within the administrative boundary of Baishulin Street office. With a economic income of 128 million (RMB), a street revenue of 10.27 million (RMB), and a street profit of 11.65 million (RMB) in 1994, the street economy within Baishulin Street office first among all other streets in Shaanxi Province (Ma, 1995: 270). This kind of rapid growth of the street economy within Baishulin Street was closely related to that of local urban redevelopment projects. By 1994, more than 30% of the administrative areas of Baishulin Street had been demolished and redeveloped (EBXY, 2000b: 296). Figure 6-3 shows the increasing tax revenues through city

---

[1] Street economy means the economic structure and development at the administration level of Street or Sub–district.

到2005年，碑林区通过城市建设和保护项目所增加的财政税收。

受企业投资和当地基于文化的城市再开发项目带来的巨大经济效益的驱使，碑林区政府强调城市开发政策为"一方面抓城市再开发，一方面抓城市管理"，目标是将街区建设成一个设备齐全、设施现代化、经济繁荣的现代化街区。结合这一点，区政府也强调了私人开发商在地方经济发展中的巨大作用（EBXY，1997：282）。事实上，在整个城市中，对城市建设和城市土地使用所征收的税收是增加地方财政收入的重要方式。图6-4显示了从1998年到2005年，西安市城市建设带来的财政税收。总体而言，20世纪90年代，西安市政府和房地产开发商在对城市住房的基础建设和重建方面的投资增长巨大，相比之下，对房屋修复的投资却较少（图6-5）。

在20世纪90年代中期，西安第三次总体规划提出了城市保护的三个层次（1995～2020）。根据第三次总体规划中的保护政策，三学街被列为历史街区，要求在保护

construction and maintenance projects in Beilin District from 1999 to 2005.

Encouraged by the great economic profit via entrepreneurial investments and local culture-based urban redevelopment projects, the Beilin District government stressed its urban regeneration policy as "urban redevelopment on one hand and urban management on the other," with the target of constructing a modern district with comprehensive equipment, modern facilities, and a prosperous economy. In this regard, the district government stressed the significant role of private developers in local economic development (EBXY, 1997: 282). Actually, within the whole city, the levying of taxes on city construction and the use of urban land were substantive ways to earn local revenues. Figure 6-4 shows the increase in Xi'an municipal revenues from urban construction from 1998 to 2005. Generally, throughout the 1990s, the investment in urban residential buildings in terms of fundamental constructions and redevelopment by the municipal government and real estate developers increased immensely. Comparatively, the investment in housing rehabilitation was very little, as shown in Figure 6-5.

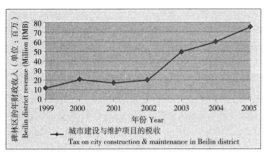

图6-3：碑林区通过城市建设和保护项目带来的财政税收
Figure 6-3: Tax revenues through city construction and maintenance in Beilin district
资料来源：(XSB, 2000～2006)
Source: Author synthesized from (XSB, 2000～2006)

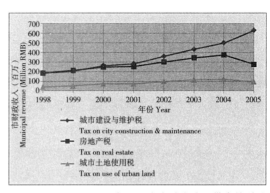

图6-4：1998～2005年，西安市城市建设带来的财政税收
Figure 6-4: Xi'an municipal revenues from urban constructions from 1998 to 2005
资料来源：(XSB, 1999~2006)
Source: Author synthesized from (XSB, 1999～2006)

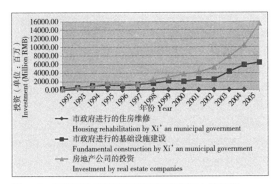

图 6-5：不同机构在城市更新和建设项目中的投资

Figure 6-5: Investment in urban renewal and constructions by different agencies

注：基础建设指的是一些新的或扩大项目，如，工厂、矿山、学校、医院等，这些项目的兴建是为了扩大生产的规模或项目的利益（XSB，2002：127）

Note: Fundamental construction refers to the new or extension projects, e.g. factories, mines, schools, hospitals and etc, which serve the purpose of extending or enlarging the production capacity or project benefits (XSB, 2002: 127).

资料来源：(XSB，2000~2006)

Source: Author synthesized from (XSB, 2000 ~ 2006)

城市历史景观的同时，也要保护当地文物古迹，例如，西安碑林博物馆、关中书院，等等。

6.2.3 地方政府主导的城市保护与房地产再开发并行的城市空间重组（2001年~至今）

在作者 2007 年的现场调研中，碑林区建设局的工作人员表示，在 2000 之后，三学街开展了两个更新项目：2001 年以来的三学街历史核心区的保护和再开发；2004 年以来的大吉昌巷再开发。在一定程度上，这些项目的特点都是自上而下的地方政府主导的和私人开发商参与的再开发项目，并导致了大量的居民重置和城市绅士化现象的产生，这些在大吉昌巷项目中表现尤为明显。本研究认为，在这些再开发项目中，由于居民的重置，三学街中的邻里关系以及社会资本的积累也会受到影响。接下来将对三学街保护与更新

In the mid-1990s, the three tiers of urban conservation policies were put forth with Xi'an's third master plan (1995-2020). According to the conservation policies in this third master plan, SHD was identified as one of the historic urban areas, where the conservation of an ancient historic urban scene was required together with the preservation of local heritage monuments such as Xi'an Stele Forest Museum, Guanzhong School, and so on.

6.2.3 Local government-led conservation-cum-redevelopment urban restructuring (from 2001 to present)

As indicated by the working people in Beilin District Construction Bureau during the author's field study in 2007, two regeneration projects have been conducted within SHD after 2000: the conservation and redevelopment of the historic core of SHD since 2001 and the redevelopment of the Dajichang alley site since 2004. Largely, these projects have been characterized by top-down local government-led and private developer-involved redevelopment, which have resulted in massive residential relocation and urban gentrification, especially in the project of the Dajichang alley site. This book argues that with these urban redevelopment practices featured by residential relocation, local neighborhood relationships and the accumulation of local social capital within SHD have been severely disrupted. The following sections discuss the existing social capital condition within SHD.

Needs and opportunities for urban regeneration

Evolving socio-economic factors

Since the 1992 declaration of Xi'an as an inland-open-city, local pro-growth urban policies attempted to attract as many investments as possible, during which the Xi'an municipal government forcefully promoted the development of tertiary

项目中涉及的社会资本问题进行讨论。

城市更新的必要性与机遇

不断演变的社会经济因素

西安自1992年被封为内地开放城市以来，当地倾向于经济发展的城市政策试图尽可能多地吸引投资，在此过程中，西安市政府有力地推动了第三产业，如，商业、贸易和旅游业的发展。这些政策也为小型零售企业提供了就业机会（EBXY，2000a：365）。结果，这些城市政策就吸引了更多的外来务工者涌入西安市区，尤其是自2000年以来，碑林区的人口也大幅增长（图6-6）。图6-7显示出

industries like commerce, trade, and tourism. These policies also provided employment opportunities to small-scale retail businesses (EBXY, 2000a: 365). As a result, these urban policies absorbed many migrant workers flooding into Xi'an urban areas, and the population in the Beilin district increased enormously, especially since 2000, as seen in Figure 6-6. This situation of immense migration to Beilin district is shown in Figure 6-7.

Another reason that may account for the flooding of mobile population from nearby counties or villages into Xi'an urban areas relates to the persistence of high levels of unemployment since the late 1990s in China

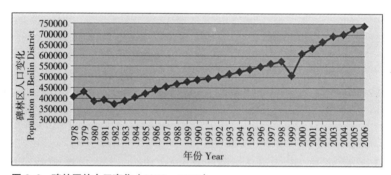

图6-6：碑林区的人口变化（1978—2006）
Figure 6-6: Population changes in Beilin district (1978–2006)
资料来源：（ECXBDR，2003b；XSB，1990~2007）
Source: Author synthesized from (ECXBDR, 2003b; XSB, 1990–2007)

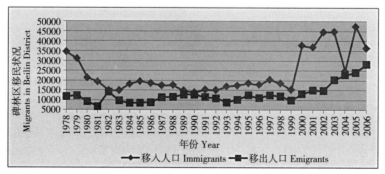

图6-7：碑林区移民变化情况（1978—2006）
Figure 6-7: Migrants' changes in Beilin district (1978–2006)
资料来源：（ECXBDR，2003b；XSB，1990~2007）
Source: Author synthesized from (ECXBDR, 2003b; XSB, 1990–2007)

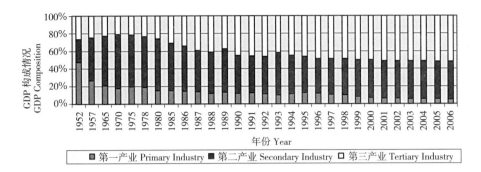

图 6-8：西安 GDP 的构成（1952 ~ 2006 年）
Figure 6-8: Composition of Xi'an GDP (1952-2006)
资料来源：(XSB, 2007)
Source: Author synthesized from (XSB, 2007)

碑林区大量移民潮的情形。

另一个导致附近县城或村庄的流动人口涌入西安城区的可能原因是，中国自 20 世纪 90 年代以来居高不下的下岗失业率[1]（Li, 2007a；Liu, 1998；Shi, 2000）。由于第三产业是西安经济来源的重要组成部分，很多人，尤其是不少无业人员，利用这个时机开始在零售业、餐馆或服务行业从事工作。这点可以从西安市持续增长的工人数量和当地第三产业的经济产出看出来。图 6-8 显示出西安 GDP 的结构变化，图 6-9 显示了西安城市就业人口的构成。位于西安市旅游和商业中心的三学街区自然也吸引了大量的外来务工人员。

三学街区的很多拥有房屋产权的居民，通过出租房间或者甚至整个房子给那些流动

(Li, 2007a; Liu, 1998; Shi, 2000)[1]. With the tertiary industry accounting for a significant portion of Xi'an's economic output, many people, especially those who were laid off, took advantage of the available opportunities and took up such jobs in retailing, restaurants, or service businesses. This can be observed in the increasing number of workers and economic output in the tertiary industry in Xi'an. The evolving composition of Xi'an's GDP (1952-2006) is shown in Figure 6-8, and the demographic composition of Xi'an's urban employment is shown in Figure 6-9. Located within the Xi'an tourist and business centers, the SHD area also attracts a lot of migrant workers.

Many residents in SHD who own property rights take a chance at making money by renting out rooms or even their whole houses

---

[1] 李志宁称，中国的下岗失业问题早在 1994 年就开始出现，到 20 世纪 90 年代末，尤其是在 1997 年、1998 年和 1999 年，失业问题开始成为影响中国经济发展的重要问题（Li, 2007b：5）。关于中国的失业情况，有学者指出，到 1997 年中国的失业总人数超过了 2000 万人（Liu, 1998：4）。这个数字在当年占到全国工作总人数的 20%。其中 70% 的失业人口后来又成功地再就业（Shi, 2000：1-11）。很多这些再就业人口涌入城市范围，并从事起小型商业活动。

[1] According to Li, the unemployment situation in China began to turn up as early as 1994, and by the end of the 1990s, especially in 1997, 1998 and 1999, the unemployment issue became a crucial issue baffling Chinese economic development (Li, 2007b: 5). Regarding the unemployment situation in China, some academics saliently pointed out that by 1997 the total unemployed in China amounted to more than 20 million people (Liu, 1998: 4). This figure accounted for 20% of the entire working people in China that year. Among these unemployed, about 70% later successfully realized their re-employment (Shi, 2000: 1-11). Most of these re-employed people poured into urban areas and took up small scale businesses.

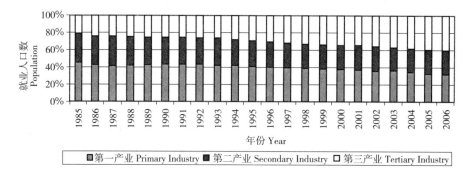

图 6-9：西安就业人口的构成（1985 ~ 2006 年）
Figure 6-9: Demographic composition of Xi'an employment population (1985-2006)
资料来源：（XSB, 2007）
Source: Author synthesized from (XSB, 2007)

人员，尤其是当这些原住居民还有其他地方可以待或居住的情况下，例如他们单位的房子或在单位附近的房子等[1]。通过这种途径获得的经济回报为很多三学街区的居民带来了好处，尤其在当地居民也面临失业的情况时。为了取得更多的居住面积，很多家庭已经另外兴建了不少住房。大量的自建活动以及缺少高效的城市管理机制，严重破坏了三学街中原有的本地建筑和历史城市的结构肌理。

当地破败的居住环境

西安历史城区里很多地方的基础设施都非常破旧，如（图6-10）所示。流动人口的大量涌入也给现有的陈旧基础设施增加了更多的沉重负担，尤其是在西安快速城市化的背景下。面对当地破败的居住条件，很多居民非常渴望开展城市更新项目。例如，2004年的一项涉及大吉昌巷的更新项目调研中，将近90%的受访者渴望尽快开展城市更新活动，如图6-11所示。类似于大吉昌巷破败的

to the mobile population, especially when they have other places at which to stay or live, such as houses in their Danwei or work places[1]. This way of obtaining economic returns is beneficial to many local residents within SHD, especially because local residents themselves may also be part of the group of people who have been laid off. In order to have more residential areas, many households have built extra housing areas. These activities of massive self-construction and the lack of an efficient urban management system heavily disrupted the original vernacular buildings and historic urban fabric in SHD.

Local dilapidated living environments

In many parts of Xi'an's historic areas, the remaining urban infrastructures were decrepit, as shown in Figure 6-10. The sudden influx of a mobile population added a heavy burden to the existing facilities, especially since rapid urbanization also existed. Facing dilapidated living conditions, many residents showed a strong desire for local urban renewal projects. For instance, in a project involving the Dajichang alley site in 2004, nearly 90% of the local residents expressed their desire for urgent urban renewal, as shown in Figure

---

[1] 该信息是作者在2009年的实地调研中，由当地居民提供。

[1] Information provided by local residents during the author's field study in 2008.

图 6-10：三学街历史街区的居住建筑
Figure 6-10: Residential buildings within SHD
资料来源：2007 年碑林区建设局
Source: Construction Bureau of Beilin District, in 2007

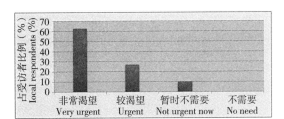

图 6-11：当地居民对城市更新活动的意见调查
Figure 6-11: Survey of the opinions on urban renewal activities among local residents
资料来源：(Zhu，2006：57)
Source: Author edited from (Zhu, 2006: 57)

城市基础设施情况，同样也存在于三学街的其他地方，如，咸宁学巷、长安学巷以及府学巷（Zhu，2006：57）。

在作者对当地居民就城市更新项目的意见调查中，40 位受访者中，有 28 位或者说 70% 的受访者表达出他们对更新项目的极其热切的关注和期望。他们希望通过更新项目来改善他们的居住环境和当地的公共设施。当问及"应该如何改善当地的城市环境？"时，40 位受访者中，其中有 28 位或者说 70% 的人期望由地方政府来引导拆除现有的建筑，并重建新房

6-11. The problem of dilapidated urban infrastructures, similar to the conditions found in the Dajichang alley site, also exist in other locations within SHD, such as the Xianningxue Lane, the Chang'anxue Lane, and the Fuxue Lane (Zhu, 2006: 57).

During the author's survey of local residents' opinions on urban renewal projects, 28 of the 40 respondents, or 70%, expressed their earnest concerns and expectations of local renewal projects, through which they hoped their living conditions and local public facilities would improve. At the same time, when asked the question "*How should local urban environments be improved?*", 28 of the 40 respondents, or 70%, expected the local government to conduct renewal activities through demolishing existing buildings and rebuilding new apartments, 4 of the 40 respondents, or 10%, thought they should be done by rehabilitating existing buildings by themselves, and 6 respondents, or 15%, thought their houses can remain the same except for minor repairs. These findings are shown in Figure 6-12. Another investigation in 2000 echoed local residents' opinions on urban renewal methods. During the survey, 40% of the 50 respondents opined

图 6-12：关于问题"应该如何改善当地的城市环境？"的调查结果
Figure 6-12: Surveys on "How should local urban environments be improved?"
资料来源：作者，2008 年
Source: Author, in 2008

来开展更新活动，另有 4 位或者说 10% 的人认为应该由他们自己通过修复现有建筑来改善，剩下 6 位或者说 15% 的受访者认为他们的房子可以保持原样，仅仅进行小修就可以。图 6-12 显示了上述调查结果。2000 年进行的另一项调查反映出当地居民对城市更新方法的意见。在调查的 50 位受访者中，40% 的人认为更新活动应由地方政府来引导进行，28% 的人认为应由居民自己来完成（Bai，2000：45）。上述调查结果表明大部分居民更喜欢通过地方政府主导的再开发项目来改善他们的居住环境。进一步的调研发现许多居民不在乎，甚至希望被重新安置到郊区的高层住房中。

2001 年以来三学街历史核心区的保护与再开发

为了更新三学街破败的居住环境，2001 年，西安市政府提出了《西安市三学街历史文化街区保护规划》(XUPDI, 2001)。在规划中，三学街历史核心区指的是"一庙三学"的历史地段，如图 6-13 (A) 所示。三学街区历史核心区东起柏树林大街，西至安居巷，北起东木头市，南至城墙。整个区域面积大约

that renewal should be done by the local government, and 28% of the respondents believed that repairs should be undertaken by themselves (Bai, 2000: 45). This showed that most residents prefer the improvement of their living environments through redevelopment projects led by the local government, and they do not care for or even wish to be relocated to high-rise apartment buildings in sub-urban areas.

The conservation and redevelopment of the historic core of SHD from 2001

In an attempt to renew run-down living environments within SHD, the Xi'an municipal government proposed the Conservation and Renewal Plan of the Traditional Historic District within SHD in 2001 (XUPDI, 2001). This move had been promoted by urban development opportunities since the late 1990s. In this plan, the historic core of SHD referred to the historic setting of "one temple and three schools" (*Yi Miao San Xue*) as shown in Figure 6-13 (A). The historic core of SHD started from Baishulin Street in the east to Anjv Lane in the west, and from the East Mutoushi Street in the north to the City-wall in the south. The whole area was about 11.68 hectares and was inhabited by 1,420 households (Zhang, Li, & Feng, 2007: 216). The permanent residents in this area in 2008 included 4,756 people, and this number

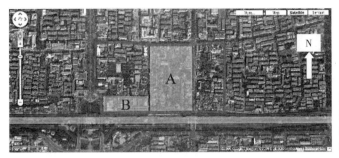

图 6-13：(A) 历史核心区位置；(B) 三学街历史街区中的大吉昌巷区域
Figure 6-13: (A) Locations of the historic core; and (B) Dajichang-alley area within SHD

11.68 公顷，有 1420 户住户（Zhang，Li，& Feng，2007：216）。2008 年，该区域的永久居民人口为 4756 人，其中不包括流动人口[1]。

该区域的更新面临的挑战主要来自城市空间和社会两个方面。已有基础设施和公共设施老化比较严重，例如，城市排污系统和人行道已经十多年——或者说自 1991 年书院门文化商业街成立以来——就没有再维修过。此外，该地区中不少基础设施都不完善，很多住户家中一直没有洗手间（Zhang et al.，2007）。由于城市缺乏排水或排污系统，所以很难在每家每户都装设用水量大的卫生间。此外，地方居民开展了大量的自建活动和移民潮，这给本来就已十分脆弱的城市基础设施系统和居住环境增加了额外负担和破坏。同时，保护珍贵的历史遗产建筑方面也面临着巨大的困难。由于倾向于经济发展的城市再开发策略，20 世纪 90 年代，很多珍贵的历史遗产建筑都遭到了破坏或摧毁。例如，20 世纪 30 年代，东木头市大街的 108 号院落最初是原来一位国民党将领的故居。即便早在 1993 年，该院落就被认定为一处历史遗址了（Bai，2000：56），而且，当地居民也被严禁对该院落里的建筑进行自行改建，但院落还是逐渐

[1] 2008 年的人口资料由碑林区公安局提供。

did not include the mobile population[1].

The regeneration challenges in this area came mainly from spatial and social perspectives. Some infrastructures and public facilities, such as the sewage system and pavements, had not been maintained for more than a decade, or ever since the establishment of the Shuyuanmen Cultural and Commercial Street in 1991. In addition, many basic facilities like private washrooms for many families had been lacking in this area, (Zhang et al., 2007). Due to the lack of a water or sewage system, it was difficult to introduce private washrooms to every household. Furthermore, given the large number of self-building activities by local residents and migrant flooding in the area, extra menaces and damages were added to the already fragile infrastructure system and living environment. Great difficulty in conserving precious heritage buildings was also experienced. Many precious historic buildings were damaged or destroyed in the 1990s due to economic-biased urban redevelopment strategies. For instance, Courtyard No. 108 in East Mutoushi Street was originally the household of a Kuomintang officer in the 1930s. Even though this courtyard has been claimed as a historic site as early as 1993 (Bai, 2000: 56), and local residents have not been allowed to perform self-constructions to the buildings of this courtyard, it gradually decayed and became severely damaged. The lack of necessary funds and an effective conservation management system from the local government made the destruction of this

[1] The population data in 2008 is obtained from local police station.

衰败并后来受到了严重的破坏。地方政府缺少必要的资金以及有效的保护管理机制，在一定程度上也加速了该历史建筑物的破坏。这个地段中的其他 30 多处类似建筑，虽然在 1993 年都被认定为历史建筑，但不少也遭遇了同样的命运（Zhang et al., 2007：217）。

更新规划及影响因素

● 来自旅游业的经济因素

遵照第三次总体规划（1995～2020）中保护规划的规定，在地方政府的领导下，三学街开展了相应的更新与保护项目，其目的是将西安整体历史城市肌理的保护和促进经济发展为导向的房地产再开发相结合。根据这个目标，西安旧城区内所有新建建筑和建筑环境都应展现古都风貌（He, 2006：112；XCPB, XACH, & XUPDI, 1994：11）。正如前面章节所提及，西安第四次总体规划（2004～2020）保留并强化了这些城市建设政策，保护与再开发并进的更新策略也体现在西安唐皇城复兴的概念规划中（XCPB, 2005a）。作为西安市总体更新规划的一部分，三学街历史核心区更新实践的每个方面也反映出总体更新政策的特点。根据第三次总体规划，三学街的更新项目不仅要促进对历史文化环境的保护，而且应该促进西安城市经济的发展，提高社会生活水平（XUPDI, 2001：8）。经济发展在保护与更新规划中的一个重要问题，相应的保护目标如下：

"以保护西安碑林博物馆为中心，继承和保护关中地区民居中的传统和本土住宅，通过提供现代化的生活服务设施来改善居住环境（XUPDI, 2001：8）。"

historic landmark progress. The same situation happened to 30 other similar constructions within the area, all claimed as historic buildings in 1993 (Zhang et al., 2007: 217).

Local regeneration plan and influencing factors

● Economic consideration of tourism industry

Following the conservation plan in the Xi'an third master plan (1995-2020), the renewal and conservation intentions in SHD, led by the local district government, have been featured by integrating the conservation of Xi'an's integral historic urban pattern with property-led redevelopment. According to this goal, all new constructions and built-environments within Xi'an CWA should display the cityscape of an ancient historic capital city (*gu cheng feng mao*) (He, 2006: 112; XCPB, XACH, & XUPDI, 1994: 11). As indicated in an earlier chapter, these urban policies were retained and strengthened in the Xi'an fourth master plan (2004-2020), and these urban conservation-cum-redevelopment regeneration strategies were shown in the Concept Plan of the Xi'an Imperial Urban Regeneration of the Tang Dynasty (XCPB, 2005a). As part of the big regeneration picture of Xi'an City, the regeneration practices in the historic core of SHD reflected this policy in every aspect. According to the third master plan, the regeneration project in SHD should not only promote the preservation of the historic and cultural environment but also contribute to the development of Xi'an's urban economy and social lives (XUPDI, 2001: 8). Economic development was a major concern in this regeneration plan, and the conservation target was proposed as follows:

> Centering the preservation of Xi'an Stele Forest Museum, the traditional vernacular residential buildings in the style of Guanzhong regional dwellings should be inherited and preserved, and the living environment should be ameliorated with the provision of modern living and servicing facilities (XUPDI, 2001: 8).

该保护目标强调了对当地文物古迹以及本地传统的住宅和街道风貌的保护。不过，作为历史居住区，当地原住民的传统和生活内容则不在保护之列。为了促进地方经济发展，该规划将居住区的土地划分为三种利用类型：(1) 历史建筑保护区（如西安碑林博物馆、关中书院以及卧龙寺）；(2) 重建商业服务区（如三学街、书院门以及安居巷附近地段）；(3) 部分建筑保护和重建区（主要是长安学巷和府学巷）(XUPDI, 2001：11)。根据对土地使用的重新规划，很多住宅区转变为商业用地。图6-14和图6-15展示了当地

This conservation target underlined the preservation of local heritage monuments and the "traditional" streetscapes of vernacular residential buildings. Unfortunately, as a historic residential area, local indigenous people were not included. In order to promote local economic development, the plan restructured local residential areas into three types of land uses: (1) the conservation of monument buildings (such as the Xi'an Stele Forest Museum, Guanzhong School, and the Dragon-Crouching Temple); (2) the redevelopment of commercial and service areas (such as those surrounding Sanxuejie Street, Shuyuanmen Street, and Anjv Lane); and (3) the conservation and redevelopment of partial vernacular

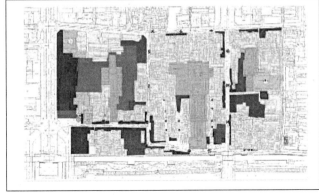

图 6-14：三学街历史街区现有土地使用情况

Figure 6-14: Existing land use condition in SHD

资料来源：2007年西安城市规划局

Source: Xi'an Urban Planning Bureau, in 2007

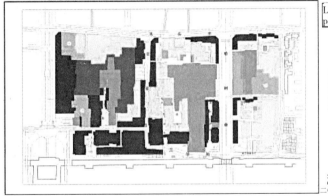

图 6-15：三学街历史街区的土地利用规划

Figure 6-15: Land use planning in SHD

资料来源：2007年西安城市规划局

Source: Xi'an Urban Planning Bureau, in 2007

土地利用在规划前后的变化。长安学巷和府学巷的部分地段已经转变为商业地段，而整个咸宁学巷完全转变为商业用途。附近的大吉昌巷也突出了这种情况。这在下一节中将会进一步讨论。

- 政策权力的碎片化与房屋产权的冲突

根据当地提出的规划方案，住宅的修复工程既可以由地方政府来进行，也可以由居民自己进行。但是，如果由居民自己进行修复，相应的地方管理部门，即西安市规划局需要对实施规划进行修正与监督（XUPDI，2001：15）。另外，即使当地居民能通过自己的能力来改善他们的居住环境，房屋的修复提议也应有助于当地历史城市风貌的恢复。基于地方居民的财政状况，一般较少居民有能力达到这一要求。2005年当地住户平均每人花费在住宅自建的费用为71.89元，这比当地每平方米的平均建设费用要低很多，2005年每平方米的平均建设费是 2000～2500 元（表6-3）。此外，由于缺少必要的基础设施，如排水系统和室内卫生间，几乎所有受访者都明确表示他们期望在该街区施行由政府主导的全面的更新项目。

正如前文所提，西安碑林博物馆是历朝石碑的储藏地，是当地历史核心区的一个重要元素。目前，博物馆面临一个挑战是其庞大的石碑存储量和非常有限的展出面积之间的矛盾。2008 年，据受访的府学巷三学街居民委员会的工作人员称，西安城市保护发展规划要求三学街历史核心区的发展应该考虑对西安碑林博物馆的保护及其未来的发展。这意味着周边居民区的再开发项目，在平衡西安碑林博物馆的土地利用需求的同

buildings (mainly Chang'anxue Lane and Fuxue Lane) (XUPDI, 2001: 11). According to this re-categorization of local land uses, many residential areas have been converted to commercial use. Figures 6-14 and 6-15 illustrate the contrasting land use changes before and after the plan. It shows that many residential areas have been transformed into commercial use. Parts of Chang'anxue Lane and Fuxue Lane have been converted into commercial areas, while the whole of Xianningxue Lane has been transformed from residential to commercial use. This situation is also conspicuously reflected in the adjacent case of the Dajichang alley site, which will be further discussed in the next section.

- Fragmentation of political power and conflicts over property rights

According to the local proposed plan, the revitalization of local residential buildings could be conducted either by the local government or by respective residents themselves. However, if it is conducted by local residents, the implementation plan needs to be ratified and monitored by the corresponding local management department, which is the Xi'an City Planning Bureau in this case (XUPDI, 2001: 15). Furthermore, even though local residents could improve their living environment by themselves, the requirement is that their revitalization proposals need to contribute to local historic cityscapes. Based on local residents' financial standing, very few can actually meet this requirement. The average expense on house self-construction per person in this area is 71.89 yuan (RMB), and this is much lower than the local average construction price per square meters, which is 2,000-2,500 yuan (RMB), in 2005 (shown in Table 6-3). In addition, with regard to the lack of necessary basic infrastructures, such as the water system and indoor private washrooms, almost all respondents expressed clearly that they look forward to a government-led overall renewal project in this area.

2005 年碑林区居民的年人均收入、居民在购房 / 建房上的花费以及每平方米建筑费用之间的对比　　表 6-3
Comparison of the average income per person per year, their expenditures on house purchasing/construction, and the construction price per square meters in Beilin District in 2005　　Table 6-3

			单位（元 / 人民币） Unit (Yuan RMB)
每年每人的平均收入 Average income per person per year	工资和补贴	Wages and subsidies	6708.06
	其他劳动所得	Other fees related to labor	218.22
	总计	Sub-total	6926.28
每年每人在购房 / 建房的平均花费 Average expenses on house purchasing/construction per person per year	购房	House purchasing	650.46
	建房	House construction	71.89
	总计	Sub-total	722.35
每平方米的平均建设费（市场价） Average construction price/m² (at market-price)			2000–2500
每平方米平均商品房费用（市场价） Average commodity house price/m² (at market-price)			4000–5000

资料来源：(XSB，2006)
Source: (XSB, 2006)

时，还要对传统的街道风貌进行改善（XUPDI，2001：10）。尽管西安市政府和文化局主张优先为西安碑林博物馆的需求来再开发这片土地，但是，行政优先权并没有为其带来金融优先权。2007 年碑林区建设局的工作人员声称，鉴于其有限的财政预算，西安碑林博物馆目前还没有能力单独重建整个历史核心区。而且，根据市政府有关规定，禁止吸纳私营开发商来对该区域进行再开发，以免可能对博物馆保护造成不利。不过，由于在历史核心区的产权纷争，当地破旧的居住环境并有没多大改善。作者在 2008 年现场调查时注意到，很多居民都希望哪怕两个单位中的任何一个，即西安碑林博物馆或碑林区政府，出面来改善或重建他们的居住环境。这些受访者还明确表示他们希望尽快搬迁到现代住房甚至搬到郊区去居住。

当地更新规划的实施结果

根据当地保护和再开发并进的规划要求，相关的基础设施和商业设施由当地国有建筑

As mentioned earlier, the Xi'an Stele Forest Museum, considering this is where the stone tablets of various dynasties are stored, is the essential element in the local historic precinct. One challenge facing this Museum, however, is the contradiction between its vast storage and very limited exhibition areas. According to the interviewed working staffs in Sanxuejie Residents' Committee in Fuxue Lane in 2008, Xi'an Urban Conservation and Development Plan requires that the development of the historic core of SHD to consider the conservation and future development of the Xi'an Stele Forest Museum. This means the redevelopment of the surrounding residential areas needs to improve the traditional streetscape while balancing the land-use needs of the Xi'an Stele Forest Museum (XUPDI, 2001: 10). Although the Xi'an municipal government and the Xi'an Cultural Heritage Bureau offered to prioritize the redevelopment of this piece of land for the Xi'an Stele Forest Museum, however, administrative priority did not bring with it the required financial priority. Given the limited budget, as mentioned by the working people in Beilin District Construction Bureau in 2007, the Xi'an Stele Forest Museum could not afford to redevelop the whole historic core area

图 6-16：书院门古文化街风貌
Figure 6-16: Street scene of Shuyuanmen ancient cultural street

alone for the time being. Moreover, according to municipal regulation, the redevelopment of this area by attracting private developers is not allowed because doing so might not be in the best interest of conserving the Museum. Nonetheless, as a result of the conflicts over the property rights of this historic core area, there was no progress with regard to the local dilapidated residential environment. As of the author's field study in 2008, many local residents expressed their concerns in having their living environment improved or redeveloped by any of the two units, that is, the Xi'an Stele Forest Museum or the local district government. These respondents also stated explicitly that they wish to be resettled as soon as possible into modern apartments even in sub-urban areas.

Outcomes of the local regeneration plan

According to the local conservation-cum-redevelopment plan, the related urban infrastructures and commercial facilities have been constructed by a local state-owned construction company, the Xi'an Urban Renewal and Construction Company Ltd, which has been set up under the administrative control of the Xi'an City Real-estate Bureau (XMG, 2008a). The overall regeneration project started in 2004 as a part of Xi'an's Concept Plan (Liu, 2006a: 81; SBBD, 2005: 1). With the relocation of concerned residents, the commercial areas and some basic infrastructures including pavements, road lighting, and sewage systems have been completed in 2007. The situation of the completed commercial and pedestrian street is shown in Figure 6-16.

The tourist and commercial businesses brought substantial economic returns to Beilin District. In 2004, there were more than 8.2 million tourists in Beilin District. The retailing and service business income accounted for 72% of the total Beilin District GDP. The total retailing business income of Beilin District in 2004 was 11,704 million RMB (SBBD, 2005: 1), and by 2005, this figure reached 13,355 million RMB, which was 14.6% more than that in 2004 (XMG, 2006: 335).

公司兴建，即由西安市房产局监管下建立的西安城市改造建设有限公司（XMG，2008a）。作为西安唐皇城复兴规划的一部分，该更新工程于 2004 年开始施工（Liu，2006a：81；SBBD，2005：1）。随着有关居民的重置，商业区以及基础设施，包括马路、路灯、排水系统等都在 2007 年先后完工。图 6-16 显示了建好后的商业区和步行街。

旅游业和商业给碑林区带来了巨大的经济回报。2004 年碑林区的旅游人数超过 820 万，零售业和服务业的收入占到整个碑林区 GDP 的 72%。2004 年碑林区零售业总收

图 6-17：碑林区的零售业总收入（1980 年 ~ 2005 年）
Figure 6-17: Total retailing business income in Beilin district (1980-2005)
资料来源：（ECXBDR, 2003b: 251 ; XMG, 1993, 1994-2006）
Source: Author synthesized from (ECXBDR, 2003b: 251; XMG, 1993, 1994 ~ 2006)

入为 117.04 亿元人民币（SBBD, 2005: 1），到 2005 年时就达到 133.55 亿元人民币，比 2004 年高出 14.6%（XMG, 2006: 335）。（图 6-17）显示了 1980 年到 2005 年期间，碑林区旅游业带来的迅速增长的零售业收入。

当地旅游业的发展带动了经济的快速增长，相比之下，居民区的修复工作则远远落后了（Zhang et al., 2007），这和上文提到的碑林区政府和碑林博物馆就碑林博物馆周边居住区的权属问题所产生的行政冲突有关。而且，随着更新计划的实施，很多原住民为了当地旅游业和商业的发展，而陆续搬迁到了其他地方。根据移民安置规划，大约 30% 的原住居民仍将保留在该区域。但是，考虑到再开发之后他们店面的产权潜在的升值空间，大多数居民把他们的房子都出租给商人使用，而他们自己则选择到其他地方居住。近 50% 受影响的原住民被重置到旧城区外的新开发区域中，而 20% 的居民选择搬到他们工作单位的房子或其他地区的商品房[1]。

乐居场区域位于西安城墙外的东南角，占地面积大约 23.45 公顷。最初，这里是一个

The rapidly increasing retailing business income from the tourism industry in Beilin District from 1980 to 2005 is shown in Figure 6-17.

As compared to the successful economic growth of local tourism industries, the revitalization work pertaining to residential areas has lagged behind (Zhang et al., 2007), which has much to do with the abovementioned administrative conflicts, between the local district government and the Xi'an Stele Forest Museum, over the property rights of the residential areas surrounding the Xi'an Stele Forest Museum. Moreover, with the implementation of the regeneration plan, many original residents have been resettled to give way to the local tourism and commerce development. According to the resettlement plan, about 30% of the original residents will remain in the area. However, considering the high values of their property as potential shop-fronts after the redevelopment process, most residents have rented out their houses to businessmen, while they chose to live in other places. Nearly 50% of the concerned original residents have been relocated to the newly redeveloped area of Lejv course out of CWA, while about 20% moved into the houses of their work units or into commodity apartments in other areas[1].

Lejv course, with an area of about 23.45

---

[1] 2007 年作者的实地研究中，从碑林区建设局所获数据。

[1] Data from the Construction Bureau of Beilin District, during the author's field study in 2007.

城中村。经历了 1987 年和 1994 年的转型发展阶段之后，这个地方被重建为居住区。自 20 世纪 90 年代以来，碑林区政府在这里已经开展了很多再开发项目。如今，这里已成为安置三学街很多原住居民的主要地区（XMG，2008a：1）。值得注意的是，随着三学街原住居民的重置，一些传统的社团和社区也受到了影响。例如，当地的古乐社已经解散了。尽管政府成立了一些"专业"团队以官方的形式来表演古乐，但是这是对保护地方非物质文化遗产的误解。当地土生土长的音乐"表演者和作曲家"已经消失了（Huashangnews，2004；Xinhuanews，2004）。

2004 年以来大吉昌巷以房地产开发为主导的更新项目

大吉昌巷（图 6-18（B））北起书院门大街、南至城墙角下，坐落于三学街历史核心区的西南部。占地约 35000 平方米，2003 年这里约有 285 户家庭（其中大约 1246 名原住民，900 多名外来务工者）（EBXY，2006：337；Ju，2005：40，44；Zhu，2006：54）。2008 年，资料显示整个书院门社区[1]的永久居民总人数是 3986 人，需要注意的是该数字未包括当地流动人口[2]。尽管大吉昌巷与上述历史核心区都处在三学街区当中，但是这两个地方却受不同的街道办事处管辖。三学街历史核心区属于柏树林街道办公室的行政管辖范围内，而大吉昌巷属于南大街街道办公室的管辖（2002 年 9 月以前叫做南院门街道办公室）（XMG，2003：357）。因此，尽管整个更新规

---

[1] 书院门社区位于三学街区边界处，北邻东木头市，南邻城墙，东邻乐居巷，西邻南大街。

[2] 2008 年的人口资料由碑林区公安局提供。

hectares, is located at the southeast corner out of the city-wall. Originally, it was an urban village area. After the two transition periods in 1987 and 1994, respectively, this area has been redeveloped into residential districts afterwards. Since the late 1990s, the Beilin District government has conducted many redevelopment projects here, and today, it has become the main area for resettling many original residents of the SHD area (XMG, 2008a: 1). Notably, with the resettlement of the original residents in SHD, some local traditional societies and communities have been disturbed. For instance, memberships to a local ancient music society have been disbanded. Although the local government has already set up some "professional" teams to perform ancient music in an official way, these were obvious misinterpretations of the conservation of the local intangible heritage. Local indigenous music "performers and composers" have disappeared (Huashangnews, 2004; Xinhuanews, 2004).

Property-led redevelopment of the Dajichang alley site since 2004

Starting from Shuyuanmen Street in the north and bounded by the city wall in the south, the Dajichang alley site (shown in Figure 6-16 (B)) is located at the southwest portion of the historic core of SHD. Within an area of about 35,000 square meters, nearly 285 households (about 1,246 original residents and more than 900 migrant workers) resided in this area in 2003 (EBXY, 2006: 337; Ju, 2005: 40, 44; Zhu, 2006: 54). Available data for the year 2008 show that the total number of permanent residents within the whole Shuyuanmen community[1] is 3,986 persons, but note that this figure does not include the local mobile population[2]. Although the Dajichang alley site and the above-

---

[1] Shuyuanmen community locates within the SHD boundaries, with the East Mutoushi Street in the north and the city-wall in the south, with Anjv lane in the east and South Street in the west.

[2] The population data in 2008 is obtained from local police station.

划是由西安市政府提出的，但是这两个地段的详细规划和具体项目的管理却由两个不同的行政部门来执行（XUPDI，2001：15）。

自 1905 年废除了科举制度以来，大吉昌巷一直是一个历史居住区。但是，如今几乎没有什么遗留证据可以证明这个地方曾经是一个历史住区（Zhu，2006：52）。1991 年书院门大街被重建为旅游商业街时，大吉昌巷并没有在重建计划之内。主要是因为它地处所谓的"后街背巷"，地理位置不利于吸引商业投资（Liu，2006a：81）。20 世纪 90 年代以来，由于当地居民大量无节制的自建行为，大吉昌巷充斥了一大批低质量的临时性建筑[1]，带来许多火灾隐患问题，也严重影响了当地居民的生活（ECXBDR，2003b：209）。由于该区人口激增以及缺乏有效的管理机制，之前传统的庭院已经退化成散乱的大场棚，而且人口密集。同时，当地的基础设施非常破败，如狭窄的道路以及恶劣的排水系统，从当时的物质环境几乎很难辨别出以前的庭院。刘临安（Liu）认为，唯一能证明这个地区曾经是历史住区的证据就是以前的街道和巷道格局，还有以前建设环境的规模（Liu，2006a：82）。

2005 年朱文龙（Zhu）的实地调研中，关于 21 世纪 20 年代早期当地破败的居住条件，可以从当地城市基础设施的角度，从三个方面进行分析。第一是缺乏供水和排污系统。整个地区，只有一个公共自来水水源，所以家庭只能自己打水自己排污水。第二是破旧的道路状况。这个地区唯一的一条道路仅两米宽。很多居民抱怨说这条街太窄了，在道路高峰期都需要排队。

---

[1] 在中国，这些低质量的临时性建筑被称为棚户区，属于最低级别的住宅建筑物。

mentioned historic core are both located within SHD, these two sites are under two respective administrative street offices of Beilin District. The historic core of SHD is within the administrative boundary of Baishulin Street office, while the Dajichang alley site belongs to the Nandajie Street Office (called Nanyuanmen Street Office before September 2002 (XMG, 2003: 357)). Accordingly, even though the entire regeneration plan has been proposed by the local municipal government, the implementation of the detailed plan and the management of specific projects of the plans in these two sites have been conducted by two different administrative departments (XUPDI, 2001: 15).

The Dajichang alley area has been a historic residential area ever since the abolition of the imperial examination system (*Keju Kaoshi*) in 1905. Today, however, there are very few physical evidences that can prove this place was once a historic area (Zhu, 2006: 52). When Shuyuanmen Street was redeveloped into a tourist and commercial street in 1991, this site was not included in the scheme at that time. It was mainly because its geographical location was characterized by so-called "minor lanes or backyards" (*hou jie bei xiang*), which implies that economically speaking, the place was not ideal or attractive for commercial businesses (Liu, 2006a: 81). With the tremendous uncontrolled self-construction activities by local residents since the 1990s, the area had been filled with a lot of temporary constructions of poor quality[1], which involved many fire-hazard problems and which seriously affected the lives of local residents (ECXBDR, 2003b: 209). Due to the rapidly increasing population and the lack of an effective management system of the area, previous traditional courtyards have been degenerated into messy big compounds inhabited by many households. Together with local dilapidated infrastructures such as narrow roads and the

---

[1] In China, these temporary constructions of poor quality are named as Peng Hu Qu, which belongs to the least level of residential buildings in China.

由于恶劣的排水系统，这种情况在雨天会更糟。第三是狭窄的公路。一些车辆，如消防车和救护车都不能通过，这给当地带来了生命和火灾隐患。很多居民抱怨说由于担心火灾隐患，他们都不敢使用压力锅（Zhu，2006：56—57）。

除了上述朱文龙（Zhu，2006）的实地研究中体现的这些关于当地破败的物质环境问题之外，作者在2008年采访了搬迁到乐居大场棚的地方居民，他们说大吉昌巷的社区居民之间几乎没有交流和信任，所以社区缺乏凝聚力。这种现象产生的主要原因是，大部分地方居民虽然居住在这里，却在其他地方工作。对他们来说，他们和大吉昌巷之间的联系很少。实际上，很多受访居民承认他们更愿意和他们工作中的人建立关系网。通过对搬迁到乐居大场棚的居民进行调查研究，作者发现30位受访者中，有13位（大约43.3%）表示他们在大吉昌巷区一个朋友都没有，而其中9位受访者（约30%）表示他们只有1～2个朋友。这表明，在再开发项目实施之前，三学街区中73%以上的原住居民可能最多只有两个朋友。

另一个严重影响地方居民之间关系的因素是20世纪90年代末以来大量外来务工者的涌入。统计数据显示，大吉昌巷区中的大部分居民都是个体商人，他们的生活方式包括制造业、经商以及生活（Liu，2006a：82）。实地研究发现，2001年，123所当地商店中（Zhang，2002c：31），只有50家是由原住民来经营的（Ju，2005：40）。因此，通过对大吉昌巷区中的285户家庭的对比（Ju，2005：40），发现很少有地方居民在这里工作，230户当地家庭，约占家庭总数的81%，他们

poor sewage system, it is even difficult to identify the previous courtyards from the existing physical environment. The only physical remains or evidence of the site as a historic residential area, as Liu argues, are the patterns of previous streets and alleys, as well as the scales of previous built environments (Liu, 2006a: 82).

During Zhu's field study in 2005, the local dilapidated living conditions could be illustrated in three aspects in relation to local urban infrastructures in the early 2000s. The first was the lack of water supply and drainage system. In the entire area, there was only one public tap water source, and all households had to carry their drinking water in and their contaminated water out by themselves. The second was the dilapidated road condition. The only road within this area was only about two meters wide. Many residents complained that the street was so narrow that they needed to queue up to pass by during busy hours. This situation would become even worse during rainy days due to the poor drainage system. The third concerned the narrow public road system. It was impossible for some vehicles such as fire trucks and ambulance to pass through, and this posed potential life and fire hazards. Many residents complained that because of fear of fire hazards, they could not use pressure cookers (Zhu, 2006: 56-57).

Aside from problems concerning the local dilapidated physical environment as identified in Zhu's field observation (Zhu, 2006), the relocated local residents in the Lejv course interviewed by the author in 2008 stressed that the relationships among the communities in the Dajichang alley area lacked cohesiveness as manifested by few interactions and little trust. The problem had much to do with the situation that most local residents lived there but worked elsewhere. To those residents, there was little connection with the Dajichang alley site. In fact, many interviewed residents admitted that they tended to build their interaction networks within their workplaces.

只是在这里居住，而在其他地方经营个体商店。在大吉昌巷区，大多数地方居民都是个人业主。随着20世纪90年代末以来大量外来务工者涌入西安城区，很多个体业主把他们的房子租赁给了这些外来人员，而他们则居住在工作单位的房子里。2002年，大吉昌巷区的约42%的居民是外来务工人员，他们主要从事各种零售业（Ju，2005：40；Zhu，2006：54）。随着大量外来务工者的迁入，而大量原住居民的搬出，这也影响了当地社区居民之间的关系，损害了当地社区凝聚力的形成。鉴于这种情况，当地原住居民难以对该区域维持社会经济依附感，而且，很多外来务工者很难有归属感。因此，地方居民之间的关系和联系是非常微弱的。此外，由于很多居民优先考虑自己的利益而忽视公共利益，很多人为活动侵犯并破坏了当地的公共环境。因此，常常造成邻里之间关系紧张，并进一步恶化了该区居民之间的关系。

当地更新规划和影响因素对经济发展与旅游业的考虑

面对各种挑战，西安市规划局于2004年提出了更新规划。同年，随着西安第四次总体规划和概念规划[1]的提出，在概念规划的指引下，西安市政府提出将顺城巷（城墙周边的巷道）建设成一个旅游观光景区。作为顺城巷的一部分，大吉昌巷被选为试点工程，以便为西安顺城巷的更新项目提供一些实际经验（Liu，2006a：82）。为了完成这个任务，碑林区政府在2004年提出了大吉昌巷更新规划的指导方针：

---

[1] 西安概念规划：唐皇城复兴规划。

When the author surveyed those relocated residents in the Lejv course, it was find that 13 out of 30 respondents (about 43.3%) indicated that they did not have any friends within the Dajichang alley area, while 9 out of 30 respondents (about 30%) indicated that they had only one to two friends there. This means that more than 73% of the original residents do not have more than two friends in the SHD area before the redevelopment project.

Another issue significantly undermined the relationships among local residents related to the evidence that lots of migrant workers poured into the Dajichang alley area since the late 1990s. Statistics show that most residents within the Dajichang alley area were private businessmen, and their lifestyle involved manufacturing, running businesses, and living within the area (Liu, 2006a: 82). The field survey showed that among 123 local business shops (Zhang, 2002c: 31), only about 50 had been run by original residents in 2001 (Ju, 2005: 40). Therefore, compared with the 285 households in the Dajichang alley site (Ju, 2005: 40), very few local residents actually worked here, and more than 230 local families, or about 81% of the total households, just resided here while they worked as private businessmen in other places. In the Dajichang alley area, most local residents were private property owners. With the large number of migrant workers entering the Xi'an urban areas since the late 1990s, many owners of private properties rented out their houses to temporary migrants and lived in the houses of their work units. In 2002, about 42% of the residents in the Dajichang alley area were migrant workers who were mainly involved in many retail business activities (Ju, 2005: 40; Zhu, 2006: 54). With many migrant workers moving in and many original residents moving out of the area, it undermined the connections among local neighborhoods and the building of local cohesive communities. Given this situation, it was difficult to maintain social and economic attachment to the area

"该重建项目是西安古都更新规划中的一个重要组成部分；是保护历史城市和更新传统文化的一个实验性项目；是进行顺城巷更新的一个试点工程；是构建和谐社会的一个项目。通过这个更新项目，重塑历史城市风貌，改善当地环境，提高地方居民谋生手段，改变当地落后的社会现状（Zhu，2006：57）。"

在实施这个以经济发展为主线的更新规划的同时，地方市委、市政府也提出了从物质文化遗产和非物质文化遗产两个方面来保护该区域。既然没有物质遗迹可以成为遗产建筑保护政策提出物质文化遗产应该指的是原始街道的模式以及原始庭院和传统风貌的规模，非物质文化遗产指的是商业活动、当地传统的家庭商业以及传统的生活方式。新的建筑区应该考虑其周围的历史古迹，其建筑风格继承传统，以便和古迹相协调[1]（Liu，2006a：83；Wang，2002b：12-13）。（图 6-18）显示了所提出的再开发规划方案。

当地更新规划的结果

由于当地开发公司和西安房地产管理局第一分局的共同参与（XMG，2008b：第 6 条），整个再开发项目于 2005 年顺利竣工。项目过程中共有 266 户家庭被重置，新建区域面积扩展到 20000 多 m²。在此期间，开发公司负责建筑的拆迁和建设，碑林区政府提供实施政策以及基础设施建设的资金，例如，碑林

---

[1] 按照书院门街道两边的建筑风格，大吉昌巷的新建筑也采取了关中住宅建筑的传统风格，以和周围的历史遗迹，即西安碑林博物馆、关中书院和城墙，取得协调（Liu，2006：83）。

for local original residents, and moreover, it was difficult to have a sense of ownership for many temporary migrants. Therefore, the relationships and the connections among local residents were very weak. Moreover, when many residents prioritized their self-interest and ignored the public interest, there were many man-made activities that invaded and damaged the local public environment. Accordingly, this often resulted in much tension among neighborhoods and further deteriorated the relationships among the area's inhabitants.

Local regeneration plan and influencing factors Economic consideration of tourism industry

Facing these various urban challenges, a local regeneration plan was proposed by the Xi'an Urban Planning Bureau in 2004, the same year Xi'an's fourth masterplan and the Concept Plan[1] were also put forward. Guided by the Concept Plan, Xi'an's local municipal government decided to redevelop the Shuncheng alleys (around-city-wall alleys) into a tourist and sight-seeing destination. The Dajichang alley site, as part of the Shuncheng alleys, was chosen as the pilot project site in order to provide some practical experience to the Xi'an Shuncheng alleys regeneration project (Liu, 2006a: 82). Following this mission, the guiding principle of the regeneration plan in the Dajichang alley site was proposed by the Beilin local district government in 2004:

This redevelopment project is one substantive part of Xi'an's regeneration plan of the imperial capital; is one experiment of the conservation of historic city and the regeneration of traditional culture; is one pilot project to get through Shuncheng alleys; and is one project to build harmonious society. Through this renewal project, the historic cityscape will be remolded, local environment will be improved,

---

[1] Xi'an's Concept Plan: The Concept Plan of the Regeneration of the Tang Imperial City.

图 6-18：顺城巷规划方案的第五立面
Figure 6-18: The fifth façade planning of the Around-city-wall Lane
资料来源：(XCPB, 2005b：73)
Source: (XCPB, 2005b: 73)

规划前后的土地利用情况对比　表 6-4
Comparison of land use before and after the plan
Table 6-4

土地利用类型 Type of land uses	面积（m²） Areas (m²)		比例（%） Percentage (%)	
	之前 Before	之后 After	之前 Before	之后 After
办公 Office	5790	0	16	0
商业 Business	9610	17200	27	48
住宅 Residential	13380（三级）[1] 13,380 (grade III)[1]	3280（一级） 3,280 (grade I)	38	9
绿化 Greening	0	3850	0	11
公园 Parking	0	350	0	1
道路 Road	6790	10890	19	31
总计 Total	35570	35570	100	100

资料来源：(Ju, 2005：44)
Source: (Ju, 2005: 44)

博物馆外围东西广场的建设 (EBXY, 2006：337)。项目完成后，该区域大部分地段，也从以前的住宅用地转变为商业旅游用地。表 6-4

---

[1] 根据《城市用地分类与规划建设用地标准》，一级住宅用地是指基础设施完善、布局合理、环境良好的地区，而三级住宅用地是指基础设施欠缺、布局不完整、环境较差的地区（DURCEP, 1991：2）。

local residents' means of making a living will be upgraded and local laggard social situations will be changed (Zhu, 2006: 57).

While conducting this pro-growth regeneration plan, the local municipal government also proposed to conserve this area from both tangible and intangible perspectives. Since no physical remains had been identified as heritage buildings, the conservation policy proposed that tangible heritage should refer to the patterns of original streets and the scale of original courtyards and traditional cityscapes, while intangible heritage should refer to the commercial activities, local traditional household industries, and traditional living styles. The newly built areas should consider the surrounding historical monuments, be in harmony with them, and be built in traditional building styles [2] (Liu, 2006a: 83; Wang, 2002b: 12-13). The proposed redevelopment plan is shown in Figure 6-18.

Outcomes of the local regeneration plan

Through the involvement of local development industries and the first branch of the Xi'an real estate management bureau (XMG, 2008b: Article 6), the entire redevelopment project was completed in 2005. A total of 266 households were relocated, and newly constructed areas spanned more than 20,000 square meters. During this time, development companies were responsible for the demolition and construction of buildings, while the Beilin district government provided the

---

[1] According to *Chengshi yongdi fenlei guihua jianshe yongdi biaozhun* (Classification standard of urban land planning and construction), grade I of residential land refers to the area with complete infrastructures, good layout and environment, while grade III of residential land refers to the area with average infrastructure, incomplete layout and common environment (DURCEP, 1991: 2).

[2] Following the building styles flanking Shuyuanmen street, the new buildings in Dajichang-alley area also took the style of traditional Guanzhong residential buildings, which tried to be in harmony with the surrounding heritage monuments, i.e. the Xi'an Stele Forest Museum, the Guanzhong school and the City-wall (Liu, 2006: 83).

显示出项目完工前后土地利用情况的变化。

该区新建了很多具有传统风格的庭院，这些庭院也被区政府看做顺城巷的"成功保护"模式（XCPB，2005b）。随着居住环境等级由三级上升为一级，新建居民楼的价格也开始飙升，这让很多居民都承受不起。不少新建建筑都试图体现当地的精英文化，一些历史名人——如于右任和吴三达——都曾在这里居住（Zhu，2006：74）。为了保护这些精英文化，当地重建了一些高质量的庭院民居，并生产和销售有关商品。尽管这也被看作是保护当地非物质文化遗产的一部分，但其目的更多的是赚取商业利润。同时，由于很多原住民的搬迁，古乐团体及其成员也已

图6-19：再开发项目完成后的大吉昌巷

Figure 6-19: Current conditions in Dajichang-alley after redevelopment

implementation policies and invested on local basic infrastructure constructions, such as the east and west squares outside of the Museum (EBXY, 2006: 337). After the project, large parts of this area were converted from previous residential to current commercial-cum-tourist use. Table 6-4 shows the comparison of land uses before and after the project.

Many new courtyards in traditional styles were built in this area, and these were seen as a "successful conservation" project of the Shuncheng alleys by the local district government (XCPB, 2005b). In addition, the upgrading of the residential environments from grade III to grade I resulted in very few local residents who could afford the high prices of the new residential buildings. Many of the buildings in this area were thus used for the purposes of elite culture. In local history, a number of famous people, such as Yu Youren and Wu Sanda, resided here (Zhu, 2006: 74). In order to protect this elite culture, some residential courtyards of high quality were redeveloped to produce and sell commercial items. While this was seen as a part of the local plan for conserving local intangible heritage, these activities in fact targeted the earning of commercial profits. Some business types and commercial activities, however, have been sustained, although the ancient music society and its members were dismantled due to the resettlement of many original residents. The current conditions of this area after redevelopment are shown in Figure 6-19.

According to the local residential resettlement plan, residents living in public houses and those in privately owned houses measuring less than 60 square meters were to be resettled to the Lejv course area. Other private property owners, mainly businessmen, were to be resettled in situ. After the local redevelopment, an estimated 80 original business households were moved back to their original locations in order to continue the previous commercial activities and scenes in this

经解散，但原有一些商业形式和商业活动被保留了下来。图 6-19 展示了重建后的状况。

　　根据当地的居民重置计划，居住在公共场所的居民以及个人住房面积不足 60 平方米的居民需要搬迁到乐居场。其他个体业主，主要是商人，则考虑原地安置。当地的再开发项目完工后，共约有 80 户原有商家重新回到了他们原来的地方，来继续以前的商业活动。搬回来的大部分住户都能继续从事他们的商业活动，包括出售传统画作、书法、工艺品以及旅游纪念品（Zhu，2006：76）。在保护当地商业形式和商业活动方面，这种做法完全符合地方政府在重建规划中制定的规则。

　　开发项目完成后，约有 50% 的原住民搬迁到了乐居场附近[1]。在 2007 年的实地调研中，共有 30 名受访者，其中 23 名原住居民，约占总参与者的 76.7%，表示出他们对现有居住条件的不满，原因各不相同。23 名原住居民中的 12 名，约占 52.2%，抱怨说他们直到 2007 年，也就是他们原来的房子拆除 3 年后，才得以搬进他们的新家。而在三年的时间里，他们不得不花很多钱在其他地方租房子过渡。而且，他们现在的的房子空间很小，房屋设计也不美观。例如，有些房屋的客厅小且没有窗户。23 名原住居民中，有 10 名，约占 43.5%，抱怨说，和以前大吉昌巷相比，现在居住的地方不方便购物。另外，23 名原住居民中，有 5 名，约占 21.7%，抱怨说，没有电梯，上下楼较困难，这种情况对老人尤其构成问题。在调研中，当被问及"对你的老街坊，你有什么想念的吗？"时，23 位受访者中，有 15 位，约占

---

[1] 该数据是 2007 年作者调研时，由碑林区建设局的受访工作人员所提供。

area. Most of those who moved back were able to continue their business activities, including selling traditional paintings, calligraphy, handicrafts, and tourist souvenirs, among others (Zhu, 2006: 76). In terms of conserving local business types and commercial activities, this move successfully met local government rules laid out in the proposed redevelopment plan.

After the project was completed, about 50% of the original residents resettled in the Lejv course area[1]. During a field study in 2007 in which 30 respondents participated, 23 original residents, or about 76.7% of the participants, said they were not satisfied with their current living situations for various reasons. Twelve out of the 23 original residents, about 52.2%, complained that they were unable to move into their settlement houses until 2007, almost 3 years after their original houses had been demolished. During this three-year period, they had to pay large sums of money to rent houses in other places. In addition, their current resettlement houses were too small, and the house types were not well designed. For example, some houses had very small sitting rooms without windows. Ten out of the 23 original residents, about 43.5%, complained that the house locations made shopping an inconvenience, as compared to the previous Dajichang alley area. Moreover, 5 out of the 23 original residents, about 21.7%, complained that the absence of elevators made coming down from higher floors difficult. This was especially felt by families with senior members. During the survey, the question "*Do you miss anything in your old neighborhood?*" was asked. Interestingly, 15 of the 23 original residents, or 65.2%, stated that they missed certain aspects of their previous lives in the Dajichang alley area, such as the easy access to neighborhoods' assistance and daily casual interactions. At the same time, when asked

---

[1] This data is obtained from interviewing staffs in Construction Bureau of Beilin District in 2007.

65.2%，都表示他们很怀念以前大吉昌巷的生活，如方便找邻居帮忙、串门聊天。同时，当被问及他们和新邻居之间的关系时，受访者中有14位，约占60.9%，说他们搬到高层楼房之后，很少与新邻居碰面。调查结果显示，在原住民搬离他们原来的住处后，他们也不太方便使用城市的公共设施。另外，他们需要在新环境中重新建立邻里关系。调查结果表明了两方面的内容：(1) 当地居民希望所在的社区具有较强的凝聚力；(2) 在政府主导的大吉昌巷更新过程中，他们的可选择空间非常有限。

6.3 三学街历史街区更新中的社会资本

通过对三学街更新规划和实践的探讨，可看出当地居民较少被吸纳在更新方案的决策环节中。这种现象可以从两个方面来理解。一方面，在我国当前的城市更新机制中，社区参与还未达到完善的参与过程。正如前面章节所讨论的，由于中国城市规划进程中尚缺乏实际的公众"参与"机制（MC，1990：61-62；SCNPC，1990：第28条；SCNPC，2008：第26条），很多更新项目是通过公私合作的融资模式来实施的。居民对更新规划和再开发项目的了解，往往是通过地方政府对有关项目的官方宣传来获得。另一方面，本研究认为，居民之间松散的社区关系或者低水平的社会资本也不利于社区居民自主参与到城市更新过程中。面对当地破败的居住环境，大多数居民都把政府主导的再开发规划看作是改善他们居住环境的唯一出路。

about their relationships with neighbors in the new environment, 14 of the respondents, or about 60.9%, answered that they seldom met with their new neighbors after they moved into the high-rise buildings. The survey results showed that when these original residents left their original areas, they also gave up easy access to some public facilities within the downtown area. In addition, they needed to re-establish their neighborhoods' networks in the new environment. The survey results also revealed two things: (1) the local residents' wishes to live in a cohesive community and (2) the reality that they had very few choices during the government-led redevelopment processes in the Dajichang alley area.

6.3 Social Capital in the Regeneration of Xi'an SHD

Throughout the discussions on urban regeneration plans and practices in Xi'an SHD, local residents were excluded from the regeneration processes. This can be understood from two perspectives. On one hand, in China's current urban regeneration mechanism, community participation is not yet a genuine participatory process. As discussed in an earlier chapter, because of the lack of a genuine public "participation" mechanism in China's urban planning and development process (MC, 1990: 61-62; SCNPC, 1990: Article 28; SCNPC, 2008: Article 26), the regeneration projects have been implemented mainly through quasi-public-private-partnerships. Local residents actually learned about the regeneration plans and redevelopment projects only through the official promulgation of corresponding projects by local governments. On the other hand, this book argues that loose community relationships or low levels of social capital among local residents contribute very little to the local community's autonomous involvement in regeneration processes. Facing local dilapidated living environments, most local residents see

在接下来的章节中，本文将重点讨论当地社会资本的特点。同时，明确几个影响当地社会资本水平的关键因素，其中包括：流动人口、简陋的居住环境、居住环境的依附性以及倾向于经济发展的城市更新政策。

6.3.1 三学街历史街区的社会资本

不少学者，如乐国安（Yue，2002：218-247）和张云武（Zhang，2008：202）都认为自20世纪80年代以来，在中国转型期经济中，人际关系已经由以前的基于友谊的人际关系，逐渐转变为当前基于物质利益的人际关系，其特点是商品化、追求物质利益以及自我中心主义。相应地，这对社会资本的水平也产生了一定的影响。在三学街案例中，大部分居民之间的关系比较松散，不过，很多居民也表示良好的邻里关系对他们的生活很重要。在2008年进行的一项实地调研中，当40位受访者被问及"你和邻居之间的关系怎么样？"时，其中6位受访者，占15%，表示他们和邻居之间的关系非常好；另有17位受访者，或者说42.5%的人，认为他们和邻居之间的关系很好；11位受访者，或者说27.5%的人，感觉他们和邻居之间的关系一般；只有6位受访者，或者说15%的人，表示他们和邻居之间的关系不好。图6-20显示了调查结果。

结果显示，约57.5%的受访者都认为当地的邻里关系总体上是好的。但是，需要注意的是，17位受访者在表示他们邻里关系良好的同时，也指出了这样一个事实，即他们的关系往往只停留在"面子"层次上。进一步探讨还发现现有的邻里关系往往是有问题的。例如，当问及"当你向邻居借钱的时候

government-led redevelopment plans as their only choice to improve their living conditions.

In the following sections, this book will focus on the discussion of the features of local social capital. Also, several key factors affecting the level of local social capital are identified. These include the floating population, derelict living conditions, the level of place attachment, and local economic growth-biased urban regeneration policies.

6.3.1 Social capital within Xi'an SHD

Several academics such as Yue (2002: 218-247) and Zhang (2008: 202) argue that during China's transitional economy since the 1980s, human relationships have transformed from the previous friendship-based to the current material interest-based one, and are characterized by commoditization, the pursuit of material benefits, and ego-centricity. This accordingly imposes an impact on the level of local social capital. In the case of SHD, most local residents have loose relationships among them, and yet, many residents also express that good relationships among neighbors are important to their lives. In a field survey conducted in 2008, the question "*How do you describe your relationship with your neighborhoods?*" was asked among 40 respondents. Six of them, or 15% of the respondents, said they had very good relationships with their neighbors; 17 of them, or 42.5% of the respondents, thought that their relationships with their neighbors were good; 11 of them, or 27.5% of the respondents, felt they had average relationships with their neighbors; and only about 6 of them, or 15% of the respondents, stated that their relationships with their neighbors were bad. This result is illustrated in Figure 6-20.

The survey indicates that about 57.5% of the respondents assume that local relationships among neighbors are generally good. However, it was noted that 17 respondents who stated that their local relationships were good also pointed to the fact that most of the time these relationships

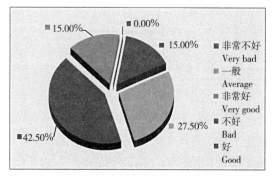

图 6-20：关于"你和邻居之间的关系怎么样？"的调查结果

Figure 6-20: Survey on "How do you describe your relationship with your neighborhoods?"

remain only on a "face" level or, in short, superficial. Upon further examination, existing neighbors' relationships tended to be questionable. For instance, when asked *How do you find it when you need to borrow money from neighborhoods?*", only six out of the 40 respondents, or about 15%, thought it was possible, while 34 out of 40, about 85%, thought it difficult to borrow money from neighbors. The 34 respondents were asked to whom they usually turn when they were in urgent need of money. Eleven of the 34 respondents, or 32.4%, indicated that they had never borrowed money from neighbors, and the remaining 23 respondents, about 67.6%, indicated that they could only borrow money from relatives or very close friends in other places. These results are shown in Figure 6-21.

会怎么样？"时，40 位受访者中只有 6 位，即约 15% 的受访者表示邻居愿意借给钱；其中 34 位，即约 85% 的受访者认为很难向邻居借钱。当问这 34 位受访者当他们急需用钱的时候，通常向谁借时，其中 11 位，即 32.4% 的受访者，表示他们从没有向邻居借过钱，其余的 23 位，约 67.6% 的受访者表示他们只能向远房亲戚或非常要好的朋友借钱。（图 6-21）表示出了这些调查结果。

受访者关于地方再开发项目的意见也反映了这种社区关系的特点。面对当地破败的

The situation of local community relationships is also reflected in their opinions towards local redevelopment projects. In the face of local dilapidated living conditions, as mentioned earlier and as shown in Figure 6-15, 28 out of the 40 respondents, about 70%, indicated that they supported local government-led urban renewal plans, and they were happy to be resettled in high-rise apartments in sub-urban areas. These remarks and mindset, on the one hand, show their earnest demand to see their living conditions improved. On the other hand, it also revealed

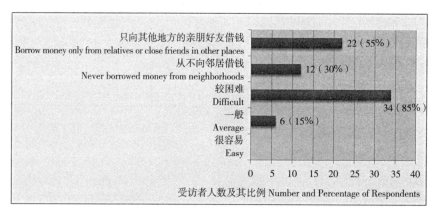

图 6-21：关于"当你向邻居借钱的时候会怎么样？"的调查结果
Figure 6-21: Survey on "How do you find it when you need to borrow money from neighborhoods?"

居住条件,正如之前所提到的以及(图6-15)所示,40位受访者中的28位,即约70%的受访者表示他们支持地方政府主导的城市更新规划,而且很乐意搬到郊区的高层住宅中。这些表述和心态,一方面体现了他们希望自己的居住环境得到改善的迫切愿望;另一方面,也表明了如果这意味着失去原有社区的话,很多居民宁愿不选择更好的居住环境。有些学者认为这种人际关系具有临时性、表面性和功利性的特点(Han,1995;Yue et al.,2002;Zhang,2008:202)。

图6-26和图6-27显示的综合调查结果表明,尽管其他因素影响了当地关系,居民仍然渴望和邻居之间保持良好的关系。于小文和巨杭生在2006年的调研也突出了这种现象。图4-19和图4-20显示,西安的大多数居民,约有83.1%的受访者认为,良好的邻里关系在生活中是非常必要和重要的,即使他们对社会信任和友谊的满意度并不高。只有约20.4%的受访者对当前的社会信任和友谊状况表示满意(Yu & Ju,2006:113,118)。

保持良好的邻里关系对一些老年人而言尤其重要。这主要是因为在老年人的日常生活中,邻居可以帮助他们解决很多麻烦琐碎的问题[1]。从表6-5可以看出,当地居民中约17.8%的人年龄在60岁以上。他们大多退休在家,并远离了子女。调查发现,大多数老年受访者表示他们的邻居对他们很重要,尽管如今人们不愿意干涉其他人的事务。有一些熟悉的邻居在他们周围也是很重要的,尤其是在他们急需帮助的时候,例如生病或不

that many residents would rather not choose a better living environment if it meant losing their original neighborhoods. Some academics view this kind of human relationship as a temporary, superficial, and utilitarian relationship (Han, 1995; Yue et al., 2002; Zhang, 2008: 202).

A combination of the survey results in Figures 6-26 and 6-27 shows that local residents still expressed their propensity towards having good relationships with their neighbors, even though other factors had an impact on local relationships. These results echo the findings by Yu and Ju in 2006, shown in Figures 4-19 and 4-20, that most residents in Xi'an, about 83.1% of the respondents, think that good relationships among neighbors are necessary and important to their lives, even though the level of their satisfaction on social trust and friendships was not very high. Only about 20.4% of the respondents think that current social trust and friendships are satisfactory (Yu & Ju, 2006: 113, 118).

Keeping good neighborhood relationships is particularly important to some senior residents. This is mainly because to many senior residents, the mutual assistance from close neighborhoods could help them overcome many petty problems in their daily lives[1]. Table 6-5 indicates that about 17.8%

当地居民的年龄构成 表6-5
Composition of local residential ages Table 6-5

年龄 Age	人数 Number of residents	比例 Percentage
12岁以下 Below 12	473	12.3%
12~17	362	9.4%
18~60	2324	60.5%
60岁以上 Above 60	683	17.8%

资料来源:(Bai,2000:43)
Source: (Bai, 2000: 43)

---

[1] 这种情况在当地老年人中尤为突出。不少老年人在此独居,但他们的子女并不在身边。

[1] This is especially true among those senior residents who live along in this area while their children live in their own houses in other places.

小心摔倒的时候。这表明建立并维持良好的邻里关系对老年群体更加重要。

影响当地社会资本的因素

在三学街案例中，本研究确定了几个影响社会资本水平的因素：(1) 杂乱的居住环境；(2) 大量流动人口造成的脆弱的地方依附感和社区之间的自我中心主义倾向。作者在2008年的实地调研中，当问及"什么对你和邻居之间的关系影响最大？"时，共有40名受访者，其中23名，或57.5%的受访者表示破坏当地邻里关系的主要因素是自私和自我中心主义的倾向；另外17名，或42.5%的受访者认为杂乱的居住环境是造成邻居之间冲突和紧张关系的原因。图6-22显示了调查结果。本研究认为，事实上，这些问题是相互联系的。根据作者在2008年采访的三学街居民，由于一些人的自我中心主义，他们往往为了自己的利益而破坏公共环境。另外，由于频繁的搬迁以及对该处缺乏地方依附感，一些居民往往对公共的居住环境造成破坏。受访的居民表示，这种情况尤其在临时居住的外来务工者群体中表现更加突出。

of the local residents were over 60 years old. Many of them are retired and live separately from their children. During the survey, most of these senior respondents stated that their neighborhoods were very important to them, even though nowadays, people do not like to intervene into others people's businesses. It was still very important to have some familiar neighbors around them, especially when they are in dire need of help, such as when they get ill or accidentally fall. This indicates that the building and maintenance of good relationships among neighbors are significantly important to local senior groups.

Factors affecting local social capital

In the case of SHD, several issues have been identified as affecting the level of local social capital: (1) the messy residential environment and (2) the weak place attachment and ego-centricity tendency among the communities, with massive migrant populations in this area. During the author's field survey in 2008, the question "*What affects the relationships between you and your neighborhoods most?*" was asked. Twenty three out of 40 respondents, or 57.5%, indicated that the main factor that undermined the relationships among local neighborhoods is the trend of selfishness and ego-centricity, while 17 out of the 40 respondents, or 42.5%, assumed that the messy living environment accounts for the conflicts and tensions among neighborhoods. The results of this survey are shown in Figure 6-22. This book contends that these issues actually intertwine with one another. According to the local residents interviewed by the author in SHD in 2008, some people tend to destroy the public environment for their own interests because of their egocentricity. In addition, due to the frequent movement and lack of place attachment to this area, some residents tend to damage their public living environment. This is true especially among temporary migrant workers, according to the interviewed local residents.

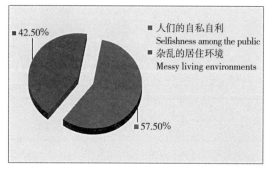

图6-22：关于"什么对你和邻居之间的关系影响最大"的调查结果

Figure 6-22: Survey on "What affects the relationships between you and your neighborhoods most?"

图 6-23：关于"当地历史建筑环境对你和邻居之间的关系产生了怎样的影响？"的调查结果

Figure 6-23: Survey on "How does local built-environment affect your relationships with your neighborhoods?"

在调查中，当被问及"当地历史建筑环境对你和邻居之间的关系产生了怎样的影响？"时，40 名受访者中的 17 名，或 42.5% 的受访者表示该区域吸引了很多临时工的到来，他们的工作日程对原住居民的正常生活产生了很大的影响。图 6-23 显示了这些调查结果。受访者表示，很多新来的住户或者外来务工者临时居住在这里，有些人不遵守基本规则。他们损坏了公共的生活环境，并造成邻里之间关系紧张。例如，一些受访的居民抱怨，由于缺乏私人卫生间，有些人甚至在公共场所或在狭窄的道路上随地大小便。这都严重影响了邻里间的团结关系。

基于以上讨论，本文认为影响当地社会资本的主要因素有三方面：(1) 流动人口；(2) 破旧的居住环境；(3) 脆弱的地方归属感

流动人口

随着这个地区外来务工者和临时租户的涌入，以及他们的频繁搬迁，当地居民发现很难和临时住户建立亲密的关系。在作者的调查中，当问及 70 名受访者"你和移民邻居之间的关系怎么样？"时，其中 55 名受访者，

In the survey, the question "*How does the local historic built-environment impact your relationships with your neighborhoods?*" was asked. Seventeen out of the 40 respondents, or 42.5%, indicated that the area attracted many temporary workers whose work schedules have a big impact on local original residents' normal life. These results are shown in Figure 6-23. According to the respondents, many newcomers or migrant workers temporarily lived in the area, and many of them did not obey basic regulations. They often spoiled their common living environments and caused many tensions among neighbors. For instance, some interviewed local residents complained that because of the lack of private toilets, some people often discarded their excrements in public spaces or on the narrow roads at will. These occurrences had a severe impact on the cohesive relationships among neighborhoods.

Based on the above discussions, this book identified three factors that had an impact on local social capital: (1) massive migrant population, (2) local derelict living environments, and (3) weak place attachment.

Massive migrant population

With the influx of migrant workers and temporary tenants in the area, local residents found it difficult to build close relationships when

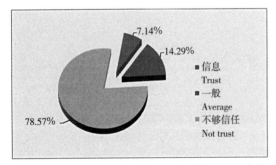

图 6–24：关于"你和外来邻居之间的关系怎么样？"的调查结果

Figure 6–24: Survey on "How do you find your relationships with local migrant neighbors?"

their neighbors moved frequently. During the author's survey, the question "*How do you find your relationships with local migrant neighbors?*" was asked of 70 respondents. Fifty five out of the 70 respondents, about 78.6%, indicated that they did not trust strangers, and it was difficult to build lasting or deep relationships with temporary tenants. Ten of the respondents, or about 14.3%, indicated that their relationships were average, and only 5 of the respondents, or about 7.1%, indicated that they trusted their migrant neighbors. These results are shown in Figure 6-24. As a result, local residents tended to build shallow relationships with their migrant neighbors or only maintained their relationships on a superficial level (Han, 1995). This kind of loose relationship among community members has worsened with local top-down redevelopment plans of relocating many of the original residents.

Derelict living environments

The existing living structures in SHD are generally dilapidated due to local residents' wide-spread self-construction activities and the conservation and infrastructure constructions that are lagging behind. As a result of the lack of proper development controls and management, most existing buildings within SHD were simply one- and two-storey residential buildings in bad quality and were lacking in necessary indoor amenities like private tap water tap sources and wash rooms. Many of them were built by the residents themselves in recent years. In addition, the road and sewage systems were severely decayed due to the lack of regular maintenance work. These problems in the physical environment have seriously affected local residents' lives (Ju, 2005: 31). Many people chose to move out of this area for new apartments in gated communities when possible. Some people chose to live in the houses of their workplaces and rented out their houses in SHD to migrant workers. Thus, local derelict living environments pose great challenges to the keeping of original residents and their connections, and undermine the accumulation of local social capital.

大约 78.6%，表示他们不信任陌生人，很难和临时住户建立长久或深入的关系。其中只有 10 名受访者，约 7.1%，表示信任他们的移民邻居。图 6–24 显示出这个调查结果。因此，当地居民和临时住户之间的关系往往是一种浅层的或表层的关系（Han，1995）。随着当地自上而下的再开发规划中，大量原住居民的迁移，这种松散的社区邻里关系就更加恶化了。

破旧的居住环境

由于当地居民大规模的自建活动以及落后的基础设施保护和建设，三学街现有的居住环境普遍比较破败。由于缺乏合理的开发控制和管理机制，三学街大部分现有建筑只是 1～2 层的居民楼，并且这些居民楼的质量整体低下，缺少必要的室内便利设施，如私家自来水以及卫生间。最近几年，居民自己建设了不少这种楼房。另外，由于缺少定期的维修，当地道路和排水系统也严重老化了。这些物质环境中存在的问题严重影响到了当地居民的生活（Ju，2005：31）。如果有可能，人们都会选择搬离这个地方，到封闭式小区的新住房中去。另有些人会选择到工作单位分配的房中居住，把他们三学街的房子租赁给外来务工者。这样，当地破旧的居住环境给保留住原有居民以及他们的社区关系网也带来不少挑战，同时也破坏了当地社会资本的积累。

脆弱的地方归属感

鉴于当地破败的居住环境，很多原住居民选择搬离这个地方。2008 年，三学街的居民中约有 85.7% 的人是新来者，例如当再开发项目完成后，吸引到大吉昌巷的新居民，或者移民租户，如三学街历史核心区的临时租户。三学街历史核心区自愿搬出去的原住居民中，一些人在他们的工作单位有住房，一些人已经在现代化的封闭式小区中购买了商品房。当他们离开这里的时候，很多房东会把房子租赁给临时租户。三学街中的各种产权状况使很多人在搬出去后都能将这里的房子租赁出去。2000 年，据现有数据显示，大约 52% 的房屋产权是由个人拥有，约有 39% 的产权是国家所有[1]，约有 9% 的产权属于一些当地工作单位所有，譬如柏树林街道公安局和西北测绘设计研究院（Bai，2000：44）。

随着快速城市化的进程，很多人倾向于和他们日常生活中接触的人们建立"工具性"的网络联系（Yue et al.，2002：255）。随着城市里单位制度的逐渐解体（Chai，2009），很多人会在原来的居住区内居住，而在其他地方工作。目前很多城市都存在这种居住区和工作区分离的状况。三学街也存在这种情况。统计数据表明，三学街中 89% 的居民在

---

[1] 该物业由西安房地产管理局来监管。

Weak place attachment

Given such dilapidated living environments, many original residents moved out of the area. In 2008, about 85.7% of the residents existing within SHD were newcomers, such as the newcomers in the Dajichang alley area after the local redevelopment project, or migrant tenants, such as the temporary tenants in the historic core area of SHD. Among the original residents in the historic core of SHD who voluntarily had moved out, some of them had available houses within their workplaces, while some of them had purchased new houses in modern gated-communities. When they left the area, many house owners rented out their houses to temporary tenants. Various property rights within SHD made it possible for many people to rent out their houses after they moved out. According to available data in 2000, about 52% of local property rights were privately owned, about 39% were publicly owned[1], and about 9% belonged to some local work-units like the Baishulin Street Police Office and the Institute of Northwest Survey and Design (Bai, 2000: 44).

With rapid urbanization, many people tended to build instrumental networks and connections with the members they related to in their daily lives (Yue et al., 2002: 255). When the Danwei system in urban China was gradually dismantled (Chai, 2009), many people only dwelled in their original residential area while working in other places. The separation between living and workplaces is commonly seen in many Chinese cities. It is also the case in the SHD. Statistics show that 89% of the residents in SHD had worked in places other than within SHD. Among these working residents, 61% of them were ordinary workers in public or private companies, and 28% of them were relatively senior employees, for example, professionals and cadres in state-owned work units. Only about 11% of the residents were engaged in

---

[1] The estates belong to the property of Xi'an Real-estate Bureau.

其他地方工作。在这些工作居民中，61%的人是国企或私企中的普通工人，28%的居民是高级雇员，例如，国企工作单位的专家和干部，只有11%的居民在三学街区中从事个体零售业。此外，这11%的居民中的大多数是外来务工者，从事地方零售业和服务业。这意味着89%的地方居民在本地区中没有工作联系或雇佣关系。因此，除了在这拥有个人产权之外，大多数人对这个历史街区没有其他的依附感。这意味着很多居民对该地区有很少的经济和社会依附感，这就破坏了地方居民之间的联系，阻碍了高水平社会资本的生成。

6.3.2 较低水平的社会资本对城市更新的影响

研究认为三学街中较低水平的社会信任和邻里关系影响了城市更新中社会经济发展综合目标的实现。除了上述对当地邻里关系以及老年居民的正常生活产生影响之外，本研究从两方面进一步分析较低水平的社会资本对当地城市更新过程的影响：(1) 对传统庭院生活的影响；(2) 对保护当地原住民的集体活动的影响。

对传统庭院生活的影响：很多当地居民放弃了庭院生活

尽管庭院生活有利于居民的相互帮助和相互交流，同时也能促进社会资本的积累，但是，很多居民，尤其是在这里居住却在其他地方工作的居民，更加倾向于住在现代化住宅或者由私人开发商开发的商品房中。基于2000年对三学街中的50名居民进行的调查，下图显示了当地居民对庭院生活的意见。

private retailing businesses in the SHD area (Bai, 2000: 43; ECXBDR, 2003b: 108). In addition, most of the 11% of the residents were migrant workers, engaged in local retailing and service businesses. This means that about 89% of local residents do not have working ties or employment relationships within the area. Thus, most people do not have other attachments to that historic area except having privately owned properties there. This implies that many residents have very little economic or social attachment to the area, which undermines the connections among local residents and the generation of higher level of social capital among them.

6.3.2 Impacts of the lower level of social capital to urban regeneration

This book argues that the lower levels of community trust and relationships in SHD undermine the comprehensive urban regeneration aims of socio-cultural development. In addition to the impacts to local neighborhoods' interactions and to the normal lives of senior residents as mentioned above, this book will further explore the impacts to local urban regeneration processes from two perspectives: (1) impacts to traditional courtyard lives, and (2) impacts to the conservation of local indigenous residents' group activities.

Impacts to traditional courtyard lives: Courtyard lives are given up by many local residents (mainly young workers)

Even though courtyard living is conducive to local residents' mutual assistance and interaction, as well as the accrual of local social capital, many residents, especially those residing in the area while working in other places, show a propensity to live in modern apartments or commodity houses developed by private developers. Based on a 2000 survey of 50 local residents within SHD, the following figures illustrate local residents' opinions towards courtyard life. Figure 6-25 shows the level of satisfaction of local residents with courtyard life. Meanwhile, Figure 6-26 reveals local residents'

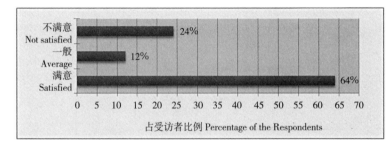

图 6-25：三学街居民对庭院生活的满意度调查
Figure 6-25: Survey on the level of satisfaction towards courtyard life among the residents within SHD
资料来源：(Bai, 2000：44)
Source: (Bai, 2000: 44)

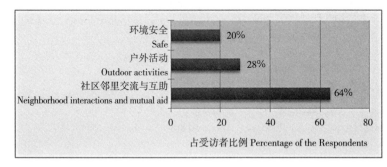

图 6-26：三学街居民认为庭院生活的优势
Figure 6-26: Advantages of courtyard life identified by local residents within SHD
资料来源：(Bai, 2000：44)
Source: (Bai, 2000: 44)

图 6-25 显示出居民对庭院生活的满意度；图 6-26 显示出居民对庭院生活优势的意见。本研究认为这两个调查结果是相互关联的。当居民意识到庭院生活的优势时，他们往往也会表现出更多的满意度。

从图 6-32 可以看出，大约 64% 的受访者对庭院生活表示满意，另外几乎同样多的人相信庭院生活有利于邻里互动和互惠互利的。这种情况在另一项调查研究中也得到了证实。当受访者被问及"在你需要帮助时，你会求助于谁？"时，50% 以上的受访者表示他们更多的是从同一庭院的邻居的帮助，而不是附近的亲戚。图 6-27 显示了该调查结果。但是，同时这些受访者也表示，邻居仅仅是在特定情况下帮助他们，比如，在生病的时候。当涉及经济问题时，比如说借钱，他们通常就会更依赖他们的亲戚或者好朋友。这些受访者称，向邻居借钱很困难。

opinions on the advantages of courtyard life. This book argues that these two figures are interrelated. When local residents identified the many advantages of local courtyard living, they also tended to show corresponding satisfaction.

Figure 6-32 shows that about 64% of the respondents were satisfied with courtyard living, and almost the same number of people believed that courtyard life was conducive to neighborhood interaction and mutual reciprocity (*hu hui hu li*). This situation had been proven yet again in a survey when the respondents were asked "*Whom will you turn to when you are in need of help?*" More than 50% of the respondents remarked that they received more assistance from neighbors in one courtyard than from nearby relatives. This is shown in Figure 6-27. Nevertheless, these respondents also indicated that neighbors' help is usually limited to certain conditions, such as being ill. When it came to economic issues like borrowing money, however, they normally relied more on their relatives or close friends. According to these respondents, it was very difficult to borrow money from neighbors. This finding is illustrated in previous Figure 6-21.

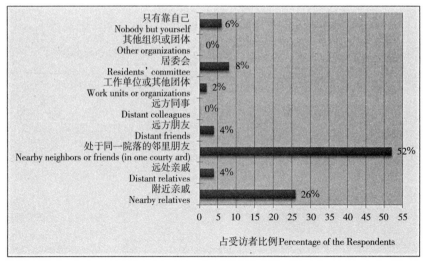

图 6-27：关于问题"在需要帮助时，你会求助于谁？"的调查结果
Figure 6-27: Survey on "Whom will you turn to for help when you are in need of help?"
资料来源：(Bai, 2000：44)
Source: (Bai, 2000: 44)

前图 6-21 显示了调查结果。这个结果证明当地的邻里关系通常很薄弱，或者说他们的关系仅停留在表面。这在张云武的研究中（2008：202，203）也有所体现，随着快速城市化的发展，中国社会的社会关系往往更具临时性、表面性和功利性。另外，人们经常依赖亲属关系或血缘关系，尤其是在他们遇到紧急情况时。

尽管调查中 64% 的受访者认为庭院生活有利于他们的日常生活，如促进相互交流。邻里之间互惠互利，但是很多人也指出，庭院生活也带来了诸多不便，如，公共场所中不断积累的垃圾以及拥挤（图 6-28）。外来务工者的涌入加剧了这种情况。在 2008 年的一项实地研究中，当问及"传统的庭院生活对你和邻居之间的关系有何影响？"，40 位受访者中，其中 17 位，即 42.5% 的受访者表示，该区中很多移民、商贩以及流浪者

This result further proves that local neighborhood relationships generally remained shallow, or their relationships just remained on a superficial level. It also echoed the observation of Zhang (2008: 202, 203) that social relationships in Chinese societies tended to be more temporary, superficial, and utilitarian with rapid urbanization. In addition, it was often the relationships based on kinships and blood ties that people can turn to especially in critical moments.

Even though 64% of the respondents in the survey identified benefits to their daily lives, such as the improvement of mutual interactions and neighborhood reciprocities of local courtyard life, many also pointed out that courtyard life carried with it many disadvantages, such as the proliferation of messy and crowded public spaces (shown in Figure 6-28). This situation worsened with the influx of migrant residents. During the field survey in 2008, when asked the question "*How does the historic courtyard life impact on your relationships with your neighborhoods?*", 17 out of the 40 respondents, or 42.5%, stated that many migrants, private

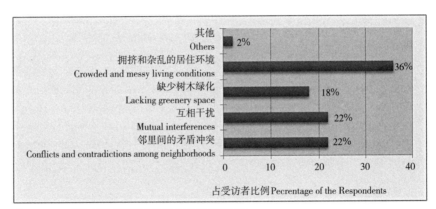

图 6-28：三学街原住居民认为庭院生活带来的不便
Figure 6-28: Disadvantages of courtyard life identified by original residents within SHD
资料来源：(Bai, 2000：44)
Source: (Bai, 2000: 44)

使居住环境变得又脏又乱，不安全，而且犯罪率不断增加。当问及"当地城市环境应该如何改善？"时，40位受访者中，其中28位，即70%的受访者明确表示，他们支持政府主导的再开发活动，拆除破旧的庭院，建立新的住房；相比之下，只有4位，即10%的受访者赞成通过自己修复现有的建筑来改善居住环境；其中6位，即15%的受访者认为他们的房屋可以保持原样，只进行一下小修就可以了。图6-12显示了这些调查结果。这种情况表明，即使在当地不断变化的社会经济条件下，尤其是基于松散的社区关系之上的低水平社会资本，传统的庭院生活能够为地方居民带来很多便利，但是为了改善当地的居住环境，地方居民还是倾向于遵循区政府主导的再开发规划，放弃传统的庭院生活。

对保护当地原住民集体生活的影响

2008年，作者在对三学街的实地研究中，受访的当地居民以及三学街居民委员会的工作人员都提及，当地居民了解并参与城市更

vendors, and idlers in this area caused messy living environments, making them untidy, noisy, unsafe, and increasing the incidence of petty theft. When asked the question "*How should local urban environments be improved?*", 28 out of 40 respondents, or 70%, explicitly indicated that they supported government-led redevelopment activities to replace dilapidated courtyard houses with new apartments. Comparatively, only four respondents, or 10%, supported the improvement of their living conditions by way of rehabilitating existing buildings by themselves. Six respondents, or 15%, thought their houses could remain the same except for some minor repairs. These results are shown in Figure 6-12. This situation reveals that even though traditional courtyard life has many benefits considering local rapidly changing socio-economic conditions, especially the low level of social capital based on loose community relationships, there is a tendency that traditional courtyard living can be given up by local residents following district government-led redevelopment plans in order to improve local living environments.

Impacts to the conservation of local original residents' group activities

During the author's field study in SHD in 2008, it was mentioned by both the interviewed local residents and the working

新过程的唯一方式就是通过区政府对再开发项目和居民重置规划的宣传。但是，由于再开发规划会对当地社区人们的日常生活产生直接影响，尤其是对那些急切渴望改善居住环境的人们，因此，很多受访居民对参与再开发规划就表现出了极大的热情。作者在2008年的实地调研中，当被问及"如果地方政府就地方更新规划组织一些研习会或居民大会，你会参加吗？"，大多数受访者都表达出对参与这些活动的极大热情。29位受访者中有23位，即79.3%的受访者都表示愿意参加由地方政府组织的公众会议；其余6位，即20.7%的受访者表示除非是强制性的，否则就不参加。不过，值得注意的是，这29位受访者同时也表明他们愿意参加当地的更新规划，因为他们最关心的是他们自己的财产和利益。正如大多数受访者所指出的，他们想参加更新规划的原因是他们不想让地方政府的更新实践损害到他们自己的利益。那些希望被重置的居民则最关心补偿方式和金额；而那些希望留下来的居民，主要是个体商人，除了关心赔偿之外，他们还关心重建

staffs in Sanxuejie Residents' Committee, that, regarding the means for local residents to be aware of and participate in local regeneration progresses, there are no other ways except through the promulgation of the redevelopment projects and residential relocation plan by the local district government. Nevertheless, since local redevelopment plan directly influences the local community's daily lives, especially for those eager to have their living environments improved, many interviewed residents show a strong enthusiasm for participating in the redevelopment plan. During the author's field survey in 2008, when asked the question "*If the local government organizes some workshops or residents' meeting concerning local renewal plans, would you attend?*", most respondents expressed great interest in getting involved in these activities. Among the 29 respondents, 23 of them, or 79.3%, stated that they would attend the public meetings organized by local governments. Six respondents, or 20.7%, indicated that they would not attend unless it was compulsory. It is noteworthy to mention, however, that these 29 respondents also expressed explicitly that they would like to attend such local renewal plans because they cared most about their own property and interests. As pointed out by most respondents, the reason they would like to attend is because they did not want their

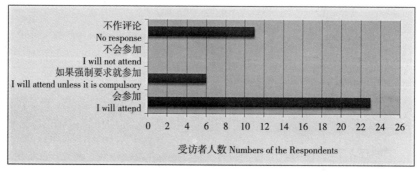

图 6-29：当地居民对参加官方会议的意见调查结果
Figure 6-29: Local residents' opinions of attending official meetings
Source: Author, in 2008

后他们商铺的区位。图 6-29 显示了这些调研结果。

调查结果显示，居民对参加当地的更新过程表现出很大的热情。但是在现实中，政府并没有提供这样的官方会议。另外，由于当地居民社区关系网和联系的薄弱，居民几乎没有动力来进行自我组织，并自发地参与到当地更新规划当中，来追求他们的集体利益。由于社区关系网的脆弱，聚合型社会资本水平也偏低，很多居民都表示愿意接受地方政府提出的重置方案，而放弃现有的社区生活。如图 6-15 所示，约 70% 的受访者期望通过政府主导的再开发项目来改善他们的居住环境。研究认为，随着原有社区的解体，当地传统的集体活动和珍贵的非物质文化遗产也会随之消失。一些传统活动需要通过当地居民团体来发起，例如古代长安的铜器社和古乐社。

铜器社

2007 年咸宁学巷受访的一些居民提到，自从该区域在清朝末年逐渐演变为居民区以来，古代长安铜器社就一直存在于三学街历史核心区。现如今，只有很少的资深音乐人掌握演奏铜锣和鼓的技艺，例如东仓乐器社的赵庚辰、李石根和李明忠等人 (Chinanews, 2005)。通常，古乐乐谱是记录在纸上的，但是演奏的技艺则需要向高级工艺人来请教。如今很少有人再对这项技艺感兴趣，而这种非物质文化遗产的保护也面临着很大的挑战 (Huashangnews, 2004)。

古乐社

大吉昌巷一个类似的居民协会就是古乐社，它是西安现存为数不多的古乐团体中的

own interests to be infringed upon by local redevelopment practices. Those wishing to be relocated cared most about the compensation types and amounts, and those wishing to stay, mainly private businessmen, cared about the positions of their shop-fronts after redevelopment, in addition to the compensation. The survey results are illustrated in Figure 6-29.

This survey reveals that local residents are enthusiastic to get involved in local regeneration processes. In reality, however, there are no such official meetings organized by the government. In addition, because of weak community relationships and connections, there is little drive for local residents to self-organize and autonomously get involved in local regeneration processes in order to pursue their collective benefits. Closely related with the effects of weak community ties and the low levels of bonding social capital, many local residents are willing to accept the relocation plans proposed by local governments and give up existing community neighborhoods. As shown in Figure 6-15, about 70% of the respondents expected to have their living conditions improved through government-led redevelopment projects. This book argues that with the disruption of original neighborhoods, local traditional group activities and valuable intangible heritages will also disappear. Some of the local traditional events need to be organized through local residents' groups, such as the ancient Chang'an brass instrument society (tong qi she) and the ancient music society (gu yue she).

The Ancient Chang'an Brass Instrument Society (tong qi she)

As referred to by some local residents interviewed in Xianningxue Lane in 2007, the Ancient Chang'an Brass Instrument Society (tong qi she) has existed within the historic core area of SHD since the area gradually became a residential area in the late Qing Dynasty. Today, the skills necessary to perform with gongs and drums have been mastered by very few senior musicians, such as Zhao Gengchen, Li Shigen, and Li Mingzhong of

一个[1]。这一当地古老的民间音乐传统可以追溯到明朝。今天，大部分有记录的古乐只有年长的民间工匠能够演奏。这种音乐已经成为这里非物质文化遗产中的一个重要组成部分，而且很多工匠在工作的时候仍然喜欢哼唱这些曲调（Zhu，2006：52）。居民作为这种非物质文化遗产的载体，他们既是表演家也是作曲家（Huashangnews，2004）。因此，居民的搬迁会严重影响大吉昌巷的古乐社团的保护与传承。很多当地的工匠已经搬迁到其他地区，这些也是当地非物质文化遗产的一大损失（Renminnews，2006；Xinhuanews，2004）。

作者在2007年对咸宁学巷的实地研究发现，地方居民的集体活动不仅是当地宝贵的非物质文化遗产，而且也是他们聚在一起，相互交流的宝贵机会。在对三学街的实地调研中，40名受访者中的15位，或者37.5%的受访者表示这些活动实际上也是他们仅有的集体活动内容。因此，当地原住居民的搬迁可能会使这些基于社区的活动完全消失。

the Dongcang instrument society (Chinanews, 2005). Normally, ancient music is inscribed on papers, but the skills necessary to perform them need to be learned from senior craftsmen. When few people are interested in learning these skills, the conservation of this type of intangible heritage also becomes a big challenge (Huashangnews, 2004).

The Ancient Music Society (gu yue she)

A similar residential association within the Dajichang alley site is the Ancient Music Society (gu yue she), one of the few remaining ancient music societies in Xi'an[1]. According to Zhu (2006), this tradition of local ancient folk music may be traced back to the Ming Dynasty. Today, much of the recorded ancient music can only be performed by senior folk craftsmen. This music has become a significant component of the intangible cultural heritage in the area, and many local craftsmen still like to sing them when they are working (Zhu, 2006: 52). Local residents, as the carriers of this type of intangible heritage, are not only performers but also composers (Huashangnews, 2004). The Dajichang ancient music society has been badly disrupted because of the resettlement of local residents. Many local craftsmen have been resettled to other areas, and it is also a big loss to the local intangible cultural heritage (Chinanews, 2005; Renminnews, 2006; Xinhuanews, 2004).

The author's field study in Xianningxue Lane in 2007 finds that these group activities of local residents serve not only as local valuable intangible heritage but also as one of the very rare opportunities for local residents to come together and interact. During the field survey in the historic core in SHD, 15 out of the 40 respondents, or 37.5%, stated that these events were actually their only public activities together. The relocation of local original residents may result in the complete disappearance of these community-based activities.

---

[1] 西安现存的古乐社包括：何家营、大吉昌巷、南集贤、东仓、西仓、端履门和城隍庙（Xinhuanews，2004）。

[1] Today, the existing ancient music societies in Xi'an include Hejiaoying, Dajichang, Nanjixian, Dongcang, Xicang, Duanlvmen and Chenghuangmiao (Xinhuanews, 2004).

6.4 总结

本章探讨了三学街的城市更新实践和社会资本问题。正像中国其他许多历史街区一样,这里的物质环境和原有基础设施都非常陈旧。随着西安的快速城市化进程和突进的城市发展政策,该历史住区开展城市更新具有很大的必要性,这些必要性既有来自物质环境的要求也有来自当前城市社会经济现状的需求。随着西安第四次总体规划的提出,旧城区已经由以前的住宅区规划重建为如今的旅游商贸区。相应的,历史住区很多地段的肌理也发生了改变。这些都有力地促进了西安的城市发展,三学街的城市更新规划需求也呼之欲出,尤其是1992年西安被列入内陆对外开放城市的名单,以及1999年作为中国西部大开发项目的桥头堡城市以来。

20世纪80年代末90年代初时,三学街的城市更新政策可以概括为是地方政府主导,社区参与的城市修复过程。20世纪90年代以后,城市更新的特点可概括为地方政府主导、倾向于经济发展的再开发过程。2001年以来,三学街的城市更新政策逐渐演变为地方政府主导的保护与再开发并进的城市空间重组过程。本研究重点探讨了三学街历史街区更新实践中的两个突出特点,即,居民重置和传统城市景观的再开发。

本研究进一步探讨了三学街的低水平社会资本问题。随着西安的快速城市化进程,多种因素对三学街中原本脆弱的社区关系和低水平的社会资本产生影响。主要因素涉及外来务工者的涌入、当地破败的居住条件以

6.4 Conclusion

This chapter explores urban regeneration practices and the issue of social capital in the historic district of SHD. Like many historic residential areas in China, the physical living environment and basic infrastructures within this area are very dilapidated. Due to the rapid urbanization and aggressive urban development policies in Xi'an, this historic residential area has many genuine needs for urban regeneration. These needs come from both the physical environment and the current urban socio-economic situation. With the proposal of Xi'an's fourth masterplan, CWA has been restructured from previous residential to current tourist-cum-commercial uses. Accordingly, many parts of the historic residential areas have also been restructured. These opportunities forcefully speed urban development in Xi'an and call forth the needs of urban regeneration programs in SHD, especially since Xi'an was assigned as an inland open city (*nei lu dui wai kai fang cheng shi*) in 1992 and given the leading role in China's Western Development Program in 1999,

From the late 1980s to the early 1990s, urban regeneration policies in Xi'an SHD can be generalized as local district government-led and community-based urban revitalization. During the 1990s, these urban policies were characterized as local district government-led pro-growth urban redevelopment. Since 2001, urban regeneration policies in SHD have been characterized by government-led conservation-cum-redevelopment urban restructuring. This book thus focuses on a discussion of urban regeneration practices in two historic sites within SHD and the outstanding characteristics of these regeneration policies, namely, residential relocation and the redevelopment of traditional urban scenes.

This study further probes into the issue of low-level social capital within SHD. With the rapid urbanization in Xi'an, various factors

及脆弱的地方依附感。本文认为低水平的社会资本对当地综合化的城市更新目标的实现有很大影响。有关的历史城市环境，以前传统的庭院生活模式已经受到了挑战，并被当今的年轻人所丢弃，尽管庭院生活中良好的邻里关系，尤其是对老年居民非常重要。同时，随着历史街区原住居民的搬离，这对当地非物质文化遗产——如原住居民的集体活动的保护，也造成很大影响，而这些应该是城市更新规划中不可或缺的一部分。

本章进一步讨论了基于当地脆弱的社区关系的低水平社会资本不利于当地社区参与到城市更新项目中。鉴于当前以地方政府主导的倾向于经济发展的更新政策，以及碎片化的城市更新管治模式，地方政府主导的再开发策略受到很多居民的支持，不过这或许是居民要改善当地居住环境的唯一选择。本章最后强调以下观点：基于脆弱的社区关系的城市更新政策和策略以及低水平的社会资本，影响当地城市更新过程中的社区参程度，这些进一步会影响综合式城市更新与保护的最终结果。

have caused impacts on the weak community relationships and low level of social capital in Xi'an SHD. These factors include the influx of migrant workers, local dilapidated living conditions, and weak communities' place attachment. The book argues that low levels of social capital have many impacts on local comprehensive urban regeneration objectives. With regard to historic urban environments, previous traditional courtyard living styles have been challenged and abandoned by many young workers, even though good neighborhood relationships in courtyard living are very important, especially to senior residents. In addition, with the loss of the original residents in this historic district, the conservation of local intangible heritage in terms of the group activities by original residents, which should be an indispensible component of local regeneration plans, is severely undermined.

This chapter further argues that low levels of social capital based on local weak community relationships do not promote a community-based involvement process in local regeneration programs. Considering current government-led, pro-growth urban policies and the local quasi-PPP mode of regeneration governance, the local government-led redevelopment strategy, preferred by many local residents, could also be their only choice for improving their living environments. This chapter concludes by underlining the view that local urban regeneration policies and strategies relate to weak community relationships, and that the lack of community involvement in local regeneration plans due to low levels of social capital further undermines the comprehensive urban regeneration and integrated urban conservation outcomes.

第七章　结论
CHAPTER 7　Conclusion

7.1 综述

本研究探讨了社会资本在中国历史城市的更新过程中所发挥的作用。快速城市化进程使历史城市面临很多城市问题和挑战。快速城市化也呼求制定相应的更新政策和实践。受当地倾向经济发展的城市发展政策以及经济增长目标的驱使，城市物质环境的发展和改善往往是在以妥协社区发展和社会文化保护的代价上进行的。但是，综合化的城市更新和可持续发展所关注的不仅仅是物质环境的改善和经济的发展，还应该包括社会凝聚力的提高和城市历史文化的保护。尤其是，保护历史居住区中原住民生活应该是城市保护和更新规划中的重要组成部分。相关文献表明，社会资本能够促进社区参与到城市再开发过程中，而社区参与能极大地促进历史街区的综合式保护，不过，已有研究表明，在探讨社会资本在促进社区参与到转型期中国历史城市的更新过程中的作用方面还是一个研究空白。

本研究提出如下总的论点：

> 通过促进居民参与到城市更新过程中，社会资本的力量使政府主导的城市更新规划进一步完善，从而有助于中国

7.1 Summary

This study explores the role that social capital plays in the regeneration of Chinese historic cities. Historic cities have generally faced many urban problems and challenges because of rapid urbanization. This urbanization calls forth regeneration policies and practices. Spurred by local pro-growth urban development policies and economic-growth aims, the development and improvement of the urban physical environment have often been conducted at the cost of community development and socio-cultural conservation. Nevertheless, comprehensive urban regeneration and sustainable development should concern not only the improvement of physical environment and economic growth but also the enhancement of social cohesion and urban conservation. In particular, the conservation of indigenous lives in historic residential areas should be a significant part of local urban conservation and regeneration plans. Literature shows that social capital facilitates community involvement in urban development processes, and that community involvement plays a significant role in the integrated conservation of historic urban areas. However, current studies also show that there is a gap that has failed to explore the role of social capital in facilitating community involvement in the regeneration of historic cities of transitional China. This book endeavors to fill this gap.

The overall argument in the study is set out as follows.

> By facilitating local residents' involvement in the urban regeneration process, social capital moderates the government-led regeneration

历史城市的综合式保护，以及全面城市更新目标的实现。

本研究综合利用各种研究资料，包括政府文件和当地的出版物来验证上述论点。这种多重数据论证的方法增加了数据来源，也降低了研究的局限性。同时，通过多方位考察以及多途径资料来核实并论证该论点。

本研究的主要研究问题可概括为"社会资本在中国历史城市更新中发挥怎样的作用？"为解决这个问题，本研究课题主要涉及三方面的研究内容，即历史城市的城市更新、社会资本和城市保护。通过探讨地方城市更新过程中的社区关系和社区参与，研究了社会资本问题。在转型期中国历史城市的背景下，本研究明确了影响城市更新过程中的社区参与以及更新结果的主要因素。这些因素包括宗教和种族、当地居民对历史环境的场所依附感、当地更新政策和实施机制以及私人开发商的参与。通过研究几个主要概念之间的关系以及关键影响因素，本研究建立了一个理论分析架构。

研究选择西安作案例研究，分析了破败的历史街区中的城市更新政策与策略。通过大量的文献研究，探讨了西安市的更新规划和实践以及社会资本问题。为了更好地理解社会资本在西安城市更新过程中的作用，本研究重点分析了西安两个特定历史街区的更新实践，即鼓楼回民区历史街区和三学街历史街区。在对两个历史街区的更新实践进行探讨时，研究也确定了影响社会资本水平和城市更新结果的其他因素。这些因素包括当地倾向于经济发展的城市更新政策、破败的

plan and contributes to the integrated urban conservation and the comprehensive urban regeneration goals in Chinese historic cities.

The study used various data sources, including government documents and local academic publications, to validate the argument. These multiple data add alternative perspectives and reduce the limitations of the study. Triangulating observations and various data sources qualify and validate this argument.

The main research question in the study is "*What is the role of social capital in the regeneration of Chinese historic cities?*". To address this question, the study hinges on three areas of knowledge, namely, urban regeneration, social capital, and urban conservation policies in historic cities. The issue of social capital has been examined through the exploration of community relationships and community involvement in local urban regeneration processes. In the context of historic cities of transitional China, some factors which have substantive impacts on community involvement in local urban regeneration process and regeneration outcome have been identified. These factors include religion and ethnicity, local residents' spatial attachment to the historic environment, local regeneration policies and implementation mechanism, and involvement of private developers. Through examining the relationships among the key concepts and the important influencing factors, the book develops a tentative analytical framework.

Xi'an City of China is selected as the case study site, and the urban regeneration policies and strategies in its dilapidated historic districts are examined. The regeneration plans and practices, as well as the issue of social capital in Xi'an City, are explored mainly through extensive documentary studies. To better understand the role of social capital in the regeneration process of Xi'an, the study concentrates on the regeneration practices in two specific historic districts in Xi'an City, namely, the Drum-Tower Muslim District and the Sanxuejie Historic District. Other factors that affect the level of local social capital and urban regeneration outcomes have been identified while exploring

居住环境品质、外来务工者的大量涌入以及其他因素。这些有助于进一步完善最初绘制的分析框架，并有助于更好地理解社会资本在中国历史城市更新过程中的作用。

基于对所建立的理论框架的分析探讨，本章将解决如下问题："可以改善西安城市更新实践的建议有哪些？"以及"研究成果对中国其他历史城市的更新实践有何普适意义？"

7.2 主要结论与讨论

首先，本章将讨论西安两个案例研究中的主要研究发现；其次，将探讨促进西安实际意义上城市更新过程的影响因素，并提出几个政策建议；最后，本章将讨论研究对中国历史城市更新实践的意义和价值，以及对当前关于该研究领域的理论贡献。

7.2.1 对比讨论西安两个历史街区的研究发现

历史街区进行城市更新的必要性

研究旨在探讨社会资本在中国历史城市更新过程中的作用，以西安两个历史街区为例，即鼓楼回民区历史街区和三学街历史街区。随着西安的快速城市化进程，两个历史街区发生了巨大变化，从而对城市更新提出了许多新的要求。必要性既包括外部因素也包括内部因素。外部因素涉及西安城市空间的重组规划以及当地商业的发展要求；内部因素主要指当地破败的城市物质环境和杂乱

the urban regeneration practices in these two historic districts. These factors include local pro-growth urban regeneration policies, dilapidated residential environments, and the massive influx of migrant workers, among others. This helps refine the initial analytical framework proposed at the beginning and contributes to a better understanding of the role of social capital in the regeneration process of Chinese historic cities.

Based on discussions of the theoretical framework developed in the book, this chapter will address the questions: "What recommendations can be made in order to improve the regeneration practices of Xi'an?" and "What implications do this book' findings have with regard to the regeneration practices in other Chinese historic cities?"

7.2 Major Findings and Discussions

This chapter first discusses the findings from two case studies about Xi'an. Then it explores the implications for promoting genuine regeneration processes in Xi'an. The case studies discussed also propose several policy recommendations. Finally, the chapter discusses this study's contribution to the urban regeneration practices in Chinese historic cities and its theoretical contributions to the literature currently available for this topic.

7.2.1 A comparison and discussion of the findings from case studies of two Xi'an City districts

The need for genuine urban regeneration in two historic districts

This book aims to explore the role of social capital in urban regeneration processes in Chinese historic cities in two Xi'an districts, namely, the Drum-Tower Muslim District (DTMD) and the Sanxuejie Historic District (SHD). With the rapid urbanization in Xi'an, both districts have undergone tremendous urban changes which have led to many genuine needs for urban regeneration practices. These needs cover both external and internal factors. External factors include the urban spatial restructuring plans in Xi'an and the requirements and

的居民自建活动。

除了两个历史街区存在的共同因素之外，各街区还有其更新的特殊需求。例如，由于快速城市化，大量外来务工者涌入西安城区，许多人都想住在三学街区。同时，由于对当地历史城区环境的脆弱依附性，很多原住居民选择搬离该地区，转而把他们的房屋租赁给外来务工者。增加的人口对原本就已破败的城市基础设施与环境造成进一步破坏。留下来的原住居民中不少人都抱怨，由于缺乏公共居住环境的行为规范，外来住户常常破坏他们当地的居住环境。同样，由于生活和工作习惯的差异，外来居民和当地居民之间也产生很多冲突。因此，可以说，除了破败的城市物质环境之外，三学街还存在一定程度的社会问题。

相比之下，鼓楼回民区中快速增加的人口为这里注入了新的血液。由于具有相同的宗教信仰，居民邻里之间的关系更加趋向于亲密。外来移民要么和当地的回民通婚，要么本身就是穆斯林。这种现象与穆斯林团体具有排他的特点也密切相关。在其特定的宗教团体内，他们往往建立非常密切的社区关系和聚合型关系。

地方政府主导的、倾向于经济发展的再开发和保护为特色的城市更新策略

根据不少地方的实际情况，以及第四次总体规划中明确的西安城市空间重组规划，当地区政府在回民区和三学街提出了相应的更新规划。本研究发现两个历史街区的更新政策与策略有很多相似点，尤其是在提出的规划、采取的更新策略、更新机制以及参与的阶层方面。为了促进当地经济发展，吸

development of local commercial business. Internal factors, on the other hand, concern local dilapidated urban physical environments and messy self-construction activities.

In addition to these factors that exist in both districts, there are also specific needs for urban regeneration in each individual district. For instance, because of rapid urbanization, a large number of migrant workers have flooded into Xi'an's urban areas, and many of them went to and resided in the SHD area. At the same time, considering their weak place attachment to the local historic urban environment, many of the original residents chose to move out and rent their houses to migrant workers. This increase in population has further challenged the already-dilapidated basic infrastructure and urban environments. Many of the remaining original residents also complained that migrant residents often destroy their local living environments, following the lack of common residential norms. Likewise, there have been many conflicts between migrants and residents because of the incongruence in their living and working habits. It can thus be said that, aside from the destruction of urban physical environments, the SHD also experiences some social problems.

In comparison, the rapid increase in population in the DTMD area has infused new blood in the area and has resulted in even closer relationships attributable to a common religion. This means that either the new migrants inter-marry with local residents or they are also Muslim migrants. This observation has much to do with the excluding features of Muslim residential groups. Muslim residents tend to build very strong community relationships and bonding ties within their particular groups.

The enactment of local government-led pro-growth urban regeneration policies through redevelopment and conservation strategies

The local district governments proposed regeneration plans in both DMTD and SHD as a response to the many local district realities and Xi'an's urban spatial restructuring plans as indicated in the fourth master plan. This study found many similarities in the regeneration policies and strategies of these two historic

引商业投资，当地区政府提出了以房地产开发主导的发展政策。房地产公司在当地的发展中起到了非常重要的作用。事实上，通过和私人开发商的合作，地方政府也积极参与到大量的再开发活动中。在很大程度上，与西方文献中纯粹的"公私合营"模式相比，中国地方城市的再开发项目是以公司合作融资的模式进行的。但是，城市保护规章对两个历史街区的再开发活动都提出许多的限制和要求。地方城市更新规划以保护和再开发策略为特色，突出保护两个历史街区中有限的历史建筑。回民区的大清真寺和几个院落，以及三学街的碑林博物馆和关中学院都是受保护建筑的范例。大约在同一时间，根据当地的更新规划，西安市政当局又提出重建传统历史城市景观的愿景。

举例说明，1997年，在西安回民区中，当地社区都参与到试点工程"回民历史街区保护项目"当中来。在该项目中，地方社区在与专家和政府部门的合作中发挥出了积极的作用。而区政府则起到了积极的协助作用。该试点项目是由西安房地产第二分局管理下的开发公司来进行的，这明显不同于那些追求最大经济回报的私人开发商所进行的再开发项目。该试点项目强调了对当地遗产古迹的保护和对城市物质环境的改善。同时，该项目也旨在吸引商业投资，促进当地旅游业的发展。其另外一个目标就是探索一个保护和重建破败历史城区的理想模式。

西安市政府、挪威科技大学以及相关住户通过各种方式为该项目共同筹集到了必要的资金。这些资金减轻了其中任何一个部门的压力，对该项目的顺利完成起到非常关键

districts, especially with regard to the proposed regeneration plans, adopted regeneration strategies, regeneration mechanisms, and involved actors. To promote local economic development and attract business investment, property-led urban redevelopment policies were promoted by local district governments. Real estate industries played a significant role in Xi'an's local urban development. In fact, through their close cooperation with private developers, local district governments also became actively involved in substantial local redevelopment activities. To a large extent, local redevelopment projects had been conducted in a kind of quasi-public-private-partnership compared to the pure "Public-Private-Partnership" mode in western literature. Nevertheless, urban conservation regulations posed many constraints and requirements to the redevelopment activities in these two districts. Local urban regeneration policies were featured by conservation-cum-redevelopment strategies, highlighting the conservation of limited historical buildings in these two districts. The Great Mosque and several courtyards in the DTMD area and the Beilin Museum and Guanzhong School in the SHD area are examples of conserved buildings. The redevelopment of traditional urban scenes was proposed according to local regeneration plans at about the same time.

As an example, in 1997, local communities were involved in the pilot project dubbed "Muslim Historic District Protection Project (MHDPP)" in the DTMD area. In this project, local communities played an active role in cooperating with local professionals and governments. The district government, on the other hand, played a facilitative role. The project was conducted by the development company that belonged to the Second Branch of the Xi'an Real Estate Bureau, and it was significantly different from the redevelopment projects undertaken by purely private developers whose sole aim was often the pursuit of maximum economic returns. It was a pilot project, one that underlined the conservation of local heritage buildings and the improvement of the local urban physical environment. It likewise aimed at attracting business investments

作用。结果，当地优秀的历史建筑得到了保存，居住环境得到了改善，原住民的生活，包括宗教生活模式和传统商业活动内容也都得到了很好的保护。

不过，与之前章节中谈到的西方城市更新模式相比，这种工作机制在更新过程中融合了多个部门的共同参与，应该是一种"理想化"的更新模式。图5-16显示了该项目的机构设置。正如联合国教科文组织对该保护项目的评估中所指出的，在整个前期策划、规划、决策和实施阶段，公众参与在项目更新过程中起到了非常关键的作用（UNESCO，2007：235）。当地回民社区和区政府通过成立的项目办公室进行了密切合作，这极大地促成了这种参与式更新机制和实践的成功（XMHDPPO，2003：63）。当地社区在更新过程中的参与也有助于改善居民与地方政府之间的关系，促进当地社会资本的生成。不过，值得注意的是，试点项目的"理想化"的更新机制是研究者"根据中挪协议"提出和实施的（XMHDPPO，2003：63），而并没有在其他项目中得到推广，比如在大麦市和洒金桥地区的项目。

不同水平的社会资本以及影响因素

地方的社会资本在促进社区居民在更新过程中的参与起着突出的作用。本研究表明，之前讨论的两个历史街区中的社会资本水平有着较大的不同。回民区中的大部分穆斯林居民，约80.4%的居民之间保持着良好的邻里关系。而在三学街区中，只有57.5%的受访者和邻里之间保持着良好的关系。两个街区中具体的社会经济发展状况也极大地影响了社区居民间的相互关系和信任度。尤其是

and promoting local tourism industries. Another objective was to explore an ideal mode for the conservation and redevelopment of dilapidated historic environments.

The Xi'an municipal government, the NORAD, and the concerned families collectively provided the funds necessary for this project in various ways. These funds helped mitigate the burden of any one particular sector and were critical to the accomplishment of this project. As a result, excellent heritage buildings were conserved, physical living conditions were improved, and indigenous lives, including religious living styles and traditional business activities, were protected.

This kind of mechanism of integrating various actors in local regeneration processes seems like an "ideal" regenerative mode, compared to the regeneration mode studied in western literature and discussed in previous chapters. The institutional setup for this project is shown in Figure 5-16. As indicated by the UNESCO in its evaluation of this regeneration project, public involvement and participation play a substantial role in the regeneration process throughout the pre-planning, planning, decision-making and implementation stages (UNESCO, 2007: 235). This kind of participatory regeneration mechanism and practice has also been significantly promoted through the frequent connections between the Muslim communities and local district governments via the established Project Office (XMHDPPO, 2003: 63). The local community's involvement in regeneration processes also contributes to the relationships between local residents and government agencies and the generation of local social capital. It was observed, however, that the "ideal" regeneration mechanism of this pilot project was only proposed and implemented by research academics "according to the agreement between China and Norway" (XMHDPPO, 2003: 63) and was not even expanded to other local projects, such as those in the Damaishi and Sajinqiao areas.

The different levels of social capital and various influencing factors

Local social capital plays an outstanding

地方依附感、宗教和种族等因素对社会资本存量的影响度很突出。

回民区的社区居民由于受宗教、种族和其他社会经济因素的影响，有着较强的地方依附感。当地穆斯林居民之间的关系不仅基于邻里关系，而且与通过血缘关系和亲属关系建立的亲情有关。很多穆斯林居民在回民区从事零售商业，作为主要经济来源。该街区的繁华很大程度上依赖于当地繁荣的旅游业市场。而且，由于当地的特有商业类型，街区里的穆斯林零售商业活动与当地传统居住模式密切相关。这种商业背景极大地刺激了社区居民对该历史住区的场所依附感。另外，宗教对当地社会资本的生成也起到了极大的促进作用。相关宗教的教义对穆斯林居民的日常生活规定了很多要求，并形成了一定的社会行为规范，这些规范有助于居民形成共同的观念和生活价值观。这些都有助于在居民之间建立相互信任。宗教活动还为居民进行交流提供了很多机会，主要与他们日常的祈祷活动有关。所有这些都有利于居民之间形成密切的社区关系，从而促进当地社会资本的积累。

在三学街的案例中，我国快速的城市化进程引起了很多社会和政治变化，这给当地居民之间的关系也带来不少冲击。居民之间的紧密关系也受到挑战，很多当地居民表示社区邻里间的关系往往很表面化，尤其是20世纪90年代末以来，由于政府倾向于经济发展的城市发展政策，以及大量的外来人口向该区的迁移，更加重了邻里关系的疏远。当地居民倾向于把他们的关系维持在一个比较"表面"的层次。在相互交流中，人们关注更

role in facilitating the community involvement in local regeneration processes. This book shows that the levels of social capital in the two districts previously discussed differ significantly from each other. The field survey showed that most local Muslim residents, about 80.4%, in the DTMD kept good relationships with their neighborhoods. This finding was observed in only 57.5% of the respondents in the SHD. Different levels of community relationships and mutual trust were closely influenced by specific socio-economic situations in these two districts. Issues of place attachment, religion, and ethnicity especially stand out.

Local communities in the DTMD seem to have a strong place attachment, based on religious, ethnic, and other socio-economic factors. The relationships among local Muslim residents are based on not only neighborhoods but also on relatives through blood ties and kinships. Many local Muslim residents also undertake retail businesses as sources of income within the Muslim area. This area relies heavily on a healthy tourism industry. Moreover, because of local business types, the Muslim retail business activities found in the area often closely relate to traditional living patterns. This business setting strongly promotes the place attachment of the local community to the historic urban area. In addition, religion plays a substantive and central role in promoting the generation of local social capital. Religious doctrines set many requirements and social norms with regard to how Muslim residents conduct their daily lives, and these norms contribute in producing the same viewpoints and life values among local residents. This helps build mutual trust among community members. Many opportunities for local residents to meet and interact through religious activities also exist, partly because of their daily prayer activities. All these promote cohesive community relationships among local residents and the accumulation of local social capital.

In the case of the SHD, rapid urbanization in China has induced many social and political changes, and these have imposed many challenges to the relationships among local

多的也是自身利益。而且，考虑到很多居民只居住在三学街区中，而选择在其他地方工作，因此，当地居民和该历史街区间也较难建立一种经济或商业联系。这使得当地居民对该街区物质环境的场所依附感也较小。所有这些都很大程度上阻碍了当地社区民众之间形成凝聚性较强的社区关系，因而也会影响当地社会资本的积累。

社会资本对更新项目的影响及其结果

由于回民区和三学街区中的社会资本水平的不同，当地居民对地方政府主导的更新规划反应也不一样。面对大规模的居民重置政策，回民区中的穆斯林居民强烈表示希望保护原有的社区关系网络。他们同时希望在新的居住环境中，他们还能继续从事他们的传统商业和零售业活动。和回民区相比，三学街区的再开发活动进行的更为有效。在三学街区，由于居民把政府主导的再开发项目看做改善他们现有居住环境的唯一出路，所以他们基本上都完全赞成地方政府的更新和重置方案。不过，本研究也发现，随着三学街原住居民的重置，也带来不少严重问题，主要有三方面：(1) 随着传统院落生活和老社区的消失，当地年老居民的生活受到不少挑战；(2) 地方政府主导的再开发规划使继续保持传统的庭院生活也受到了挑战；(3) 随着当地原住居民的搬迁，非物质文化遗产的保护也受到挑战。

在回民区，传统的穆斯林生活和商业活动使当地经济一直保持一种繁荣状态。本研究认为，如果地方政府能够对破旧的基础设施给予适当维修，再加上当地紧密的社区邻里关系，回民区实际上可以维持一种自我更

residents. Intimate relationships among local residents have been further challenged and many local people comment that there are few deep interactions among the neighborhoods, especially after the implementation of pro-growth national and local urban development policies and massive population migration in the late 1990s. Community members tended to maintain their relationships on a "faces" level, and people focused on self-interest in their interactions. Moreover, considering the reality that many residents only lived in the SHD but worked in other places in the city, local residents and this historic district share very few economic relations and business settings. This results in the minimal place attachment of local residents to the physical environment. All these substantially hinder both the building of cohesive community relationships among local neighborhoods and the accumulation of local social capital.

The impacts of social capital in local urban regeneration processes and the possible outcomes of local urban regeneration plans

Given the very different levels of social capital in the DTMD and the SHD, local residents also expressed very distinct reactions toward local government-led regeneration plans. Facing massive residential relocation policies, the Muslim communities in the DTMD area expressed strong wishes to keep their original neighborhoods' networks. They also wished that the new residential environments could provide them with the requirements of continuing their commercial and retail businesses as a means of income.

In comparison, the redevelopment activities in the SHD area proceeded much smoother than in the DTMD area. Local residents in the SHD strongly encouraged the local government to carry out the renewal and relocation projects because they considered the government-led redevelopment projects as their only chance at improving their existing living conditions. Nevertheless, this book found that the relocation of the original SHD residents had severe impacts in three aspects: (1) The lives of local senior residents were challenged with vanishing courtyard lives

新过程。目前主要问题是缺少一个促进这种更新过程的制度化机制。在三学街区，整体社会资本水平偏低，居民邻里关系比较薄弱，不过由地方政府主导的更新项目则比回民区显得更为"有效"。然而，从对当地原社区住民生活内容保护方面讲，回民区的更新实践则似乎比三学街区更成功。

表7-1显示出回民区和三学街区，在城市更新政策与策略、社会资本水平与影响以及可能的更新结果方面的比较。

and old neighborhoods; (2) the continuity of traditional courtyard lives was challenged with local government-led redevelopment plans; and (3) the conservation of intangible cultural heritages was challenged with the relocation of local original residents.

In the DTMD, traditional Muslim lives and businesses have sustained a vibrant local economy. Together with cohesive community relationships, this book argues that the historic Muslim district can actually sustain a self-regeneration process, given proper maintenance of the dilapidated urban infrastructures by local governments. The question is an institutionalized mechanism to facilitate this kind of regeneration. The implementation of the government-led regeneration projects within the SHD, where community relationships are weak with low level of social capital, was more "efficient" than that in the DTMD. However, in terms of the conservation of local original neighborhoods and indigenous lives, the DTMD practices were more successful than those in the SHD.

A comparison of the urban regeneration policies and practices, levels of social capital, related influencing factors, and possible regeneration outcomes between the DTMD and the SHD areas are shown in Table 7-1.

回民区和三学街区，在城市更新政策与策略、社会资本水平与影响以及可能的更新结果比较　表7-1
A comparison of the urban regeneration policies and strategies, levels of social capital, related factors, and possible regeneration outcomes between the DTMD and the SHD areas　Table 7-1

	回民区 Drum Tower Muslim District (DTMD)	三学街区 Sanxuejie Historic District (SHD)	案例比较 Comparisons
城市更新的必要性（城市变化、挑战等原因） Needs for Urban Regeneration (Urban Changes, Challenges, And Other Reasons)	外部因素： ● 西安城市重建方案； ● 当地商业的发展。 内部因素： ● 破败的城市基础设施； ● 快速人口增加所导致的当地混乱的自建行为。 Two external factors: ● Urban re-fabrication schemes in Xi'an ● Local business development Two internal factors: ● Local rundown urban infrastructures ● Local messy self-constructions out of rapidly increasing population	● 快速城市化：大量的外来务工者和租户； ● 破败的居住环境； ● 缺乏基础设施建设； ● 地方居民对该区脆弱的地方依附感。 ● Rapid urbanization: Tremendous migrant workers and tenants ● Dilapidated physical living environment ● Lack of infrastructure constructions ● Local residents' weak place attachment to this area	相比之下，三学街区中存在更多的社会问题和挑战 Comparatively, there are more social problems and challenges in the development of the SHD area.

续表
Continued

		回民区 Drum Tower Muslim District (DTMD)	三学街区 Sanxuejie Historic District (SHD)	案例比较 Comparisons
更新政策和策略 Regeneration Policies And Strategies	关键参与者 Key actors	●地方政府； ●私营部门； ●当地规划学院； ●参与到试点项目中的社区居民。 ● Local governments ● Private sectors ● Local planning institutes ● Local communities only involved in pilot project	●地方政府； ●私营部门； ●当地规划学院。 ● Local governments ● Private sectors ● Local planning institutes	两个历史街区中的更新规划、关键参与者以及更新机制很相似。 The proposed regeneration plans, the involved key actors and local regeneration mechanisms in these two historic districts are very similar.
	城市更新政策的特点 Characteristics of urban regeneration policies	转型期的更新政策： ●零星的自助整修和建设（1978年~20世纪80年代）； ●大量的以地方区政府主导的，倾向经济发展的开发政策（20世纪90年代~现在）。 Transitional urban policies: ● Sporadic self-help refurbishment and constructions (from 1978 to the late 1980s) ● Massive local district government-led pro-growth urban development policies (from the early 1990s to present)	转型期的更新政策： ●地方区政府主导的城市修复期（20世纪80年代末~20世纪90年代初）； ●地方区政府主导的，倾向经济发展的再开发（20世纪90年代~21世纪初）； ●地方政府主导的保护和再开发并进的实践（2001年~现在）。 Transitional urban policies: ● Local government-led urban rehabilitation period (from late 1980s to early 1990s) ● Local government-led pro-growth urban redevelopment (from early 1990s to early 2000s) ● Local government-led urban conservation-cum-redevelopment practices (from 2001 to present)	
	更新机制和城市管治模式 Regeneration mechanisms and modes of urban governance	●地方区政府和私营部门之间的合作； ●更新实践促进了经济增长，传统城市景观的重建。 ● Cooperation between local district government and "private" sectors ● Regeneration practices promote economic growth and the re-fabrication of traditional urban scenes	●地方区政府和私营部门之间的合作； ●更新实践促进了经济增长，以及传统城市景观的重建。 ● Cooperation between local district government and "private" sectors ● Regeneration practices promote economic growth and the re-fabrication of traditional urban scenes	
地方社会资本 Local Social Capital	社区信任 Community trust	建立在亲属和血缘关系之上的深层社区信任和关系 Deep community trust and relationships built on kinships and blood ties	邻居之间很少的社区关系和联系 Little community trust and few relationships among neighbors	基于不同的社会经济和社会文化因素，两个街区的社会资本水平也不同。 Based on very different socio-economic and socio-cultural factors, the levels of social capital in these two districts are very different.
	社会规范（种族和宗教） Social norms (ethnicity and religion)	基于宗教规范和要求 Based on religious norms and requirements	商品化和物质利益化 Commoditized and material-benefits-oriented	
	社区关系 Community relationships	在日常的宗教活动中频繁的社区交流 Frequent community interactions in daily religious activities	保持薄弱的邻里关系，维持较"表面"的交流关系 Keeping shallow neighborhoods' relationships and interactions on a "faces" level	

续表
Continued

		回民区 Drum Tower Muslim District (DTMD)	三学街区 Sanxuejie Historic District (SHD)	案例比较 Comparisons
影响社会资本的因素 Factors Affecting Local Social Capital	地方依附感 Place attachment	在当地宗教、亲属和经济方面对回民区有很强的场所依附感 Strong place attachment to the DTMD in relation to local religious, kinship, and economic aspects	由于当地的社会经济因素，对三学街区有较薄弱的场所依附感 Weak place attachment to the SHD based on local socio-economic factors	两个历史街区中，不同的场所依附感 Very different conditions of public attachment in these two districts
	当地的更新政策 Local regeneration policies	当地更新政策倾向对当地原住居民进行重置 Local regeneration policies tend to resettle local original residents	当地更新政策倾向对当地原住居民进行重置 Local regeneration policies tend to resettle local original residents	两个历史街区中影响社会资本的外部因素非常相似 The external factors affecting the level of social capital in these two cases are very similar.
	缺少制度化的更新机制 Lack of an institutionalized regeneration mechanism	这往往破坏原住区的邻里关系和穆斯林社区 It tends to disrupt local original neighborhoods and Muslim communities	这往往破坏原住区的邻里关系和当地传统生活方式 It tends to disrupt local original neighborhoods and their living styles	
	私人开发商的参与 Involvement of private developers	追求最大的经济回报，忽视当地社会文化和宗教问题 Pursuing the maximum economic returns and ignoring local socio-cultural and religious issues	通过重建传统的城市景观，追求最大的经济回报 Pursuing the maximum economic returns through redeveloping traditional urban scenes	
	缺少一个参与式的城市管治模式 Lack of a participatory mode of urban governance	破坏地方居民和地方政府之间的政治信任 Undermining the political trust between local residents and local governments	地方居民较少选择，只能遵循地方政府的更新规划 Local residents follow local governments' regeneration plans without alternative choices	两个历史街区中，当地社区和地方政府之间的联合型社会资本不同 The bridging social capital between local communities and local governments is different in these two cases.
社会资本对城市更新过程的影响 Impacts of Social Capital to Urban Regeneration Processes	更新过程中的社区参与 Community involvement in regeneration processes	当地自下而上的有组织的社区参与 Local self-construction activities	当地社区一般通过地方政府对规划的宣传来了解当地更新项目 Local community gets to know local projects only through the promulgation of the plan by local government	两个历史街区中，在当地更新过程的社区参与程度不同 The levels of communities' involvement in local regeneration processes are very different in these two cases.
	对社会文化发展的影响 Impacts to socio-cultural development	增强了地方社区凝聚力，促进了在地方发展中的参与 Promoting local community cohesion and participation in local development	●影响老年居民的生活； ●影响传统的庭院生活模式； ●影响对地方原住居民的集体活动模式的保护。 ● Impacting the lives of senior residents ● Impacting the traditional courtyard living pattern ● Impacting the conservation of local original residents' group activities	

续表
Continued

		回民区 Drum Tower Muslim District (DTMD)	三学街区 Sanxuejie Historic District (SHD)	案例比较 Comparisons
城市更新规划的可能结果 Possible Outcomes of Urban Regeneration Plans	对当地城市更新规划实施的影响 On the implementation of local urban regeneration plans	修正、完善地方政府倾向经济发展的更新规划 Postponing the implementation of local official pro-growth regeneration plans	促进地方政府倾向经济发展的更新规划 Smooth implementation of local official pro-growth regeneration plans	两个历史街区中，当地更新项目可能产生的结果不同 The possible outcomes of local regeneration projects in these two cases are very different.
	对当地经济发展的影响 On local economic development	促进了基于社区的经济增长 Promoting the community-based economic growth	当地更新项目极大地促进了地区和街道经济的增长 Substantial growth of the district and the street economy from local regeneration projects	
	对保护地方社区和原住民生活的影响 On the conservation of local communities and indigenous lives	●保留了原社区的邻里关系和原住民生活方式； ●保护了当地传统的社会文化。 ●Retaining original neighborhoods and indigenous lives ●Conserving local traditional social culture	影响了原社区和原住民的生活方式与内容 Disrupting original neighborhoods and indigenous lives	

7.2.2 对西安古城实施综合式城市更新的意义与建议

前文分析表明，尽管两个历史街区的区政府所制定的更新政策比较相似，但是两个历史街区的保护与再开发策略却表现出迥异的社会经济特点。同时，由于两个历史街区的社会资本水平的不同，当地居民对有关更新规划的反映也有较大差别，因此，两个街区的更新规划结果也非常不同。地方政府意识到了基于社区的更新过程的重要性，尤其是在回民区的试点项目中，但遗憾是，这种多方参与的更新过程没有在其他类似更新案例中进一步扩展。我国目前的城市发展过程，还未完全实现政府规划过程中的公众参与。由于缺乏一个制度化的多方参与式的更新机制，当地社区也就很难在更新过程中发表意见。尽管旧的《城市规划法》（SCXPC，1990）和新的《城乡规划法》（SCNPC，2008）要求

7.2.2 Implications and policy recommendations to promote genuine urban regeneration practices in Xi'an City

The analysis above shows that, although the local district governments of each district took very similar urban regeneration policies, their urban conservation-cum-redevelopment strategies featured very different socio-economic characteristics. In addition, based on the distinct levels of social capital between the two, local residents responded to the regeneration plans very differently and the results of the regeneration plans also varied from each other significantly. The significance of community-based regeneration processes was realized by the local governments, especially in the pilot project for the DTMD. Unfortunately, this kind of multi-participatory regeneration process did not continue to other similar regeneration cases within the DTMD. China's current urban development process does not include public participation in official planning processes yet. This lack of an institutionalized multi-participatory regeneration mechanism makes it difficult for local communities to have a voice in the regeneration processes. While the old City

地方政府一定要将更新规划进行公示，但是，根据安斯汀（Arnstein）《市民参与的阶梯》一文所主张的，这与实现真正的社区参与尚有较远差距（Arnstein，1969）。

西安的历史城区确实存在进行更新的必要。这种综合式的更新需要不仅涉及促进地方经济的增长、改善破败的物质环境、保护当地优秀的文化遗产，而且包括提高社会的公平性和增强社区的凝聚力。西安当地的城市更新政策与策略很大程度上促进了当地的经济增长，改善了当地的物质环境，保护了个别传统历史建筑，不过，这些成果往往没有较好兼顾到当地社区原住民的利益。鉴于此，研究认为当地政府为了实现历史城区的综合更新目标，需要在其城市更新规划和政策中思考如下内容：

1）完善当前的城市保护和更新策略以及评估标准

当前西安城市更新的目标比较注重经济的发展以及传统城市景观的重建，以此吸引更多的商业投资，并促进当地旅游业的发展。这在地方市政当局和决策管理者出台的政策措施中，也多有反映。不过，这种倾向于经济发展的更新政策和策略，对实现综合的城市保护与更新目标可能尚不充分，尤其是当原住民在再开发项目中被重置后，会对综合的保护与更新目标造成影响。没有原住民的生活内容，当地保护规划的目标也常常会被打折扣。除了对当地珍贵文物古迹和考古遗址的保护，当地综合的城市保护政策还应考虑保护原住居民的传统生活内容，并改善当地居民的居住环境。西安城市更新议程中应包括保护当地社会文化的因素。另外，地方

Planning Act (SCXPC, 1990) and the new Urban and Rural Planning Law (SCNPC, 2008) require that regeneration plans be first promulgated by local governments, this is a far cry from the realities of genuine community participation processes, as contended by Arnstein in the Ladder of Citizen Participation (Arnstein, 1969).

In the context of Xi'an's historic cities, this book shows that there are genuine needs for urban regeneration plans. The actual regeneration needs involve not only the growth of the local economy, the improvement of dilapidated physical environments, and the conservation of local excellent cultural heritages, but also the improvement of social equity and community cohesion. However, the urban regeneration plans and strategies in Xi'an saliently promote the growth of the local economy and the improvement of local physical environments while conserving only a limited number of heritage buildings, often at the expense of the local communities' interest. In this This book therefore argues that the Xi'an municipal government needs to consider the following in its urban regeneration plans and policies within its historic urban areas in order to achieve its objectives:

1) The amendment of current urban conservation and regeneration strategies and evaluation criteria

The objectives of Xi'an's urban regeneration plans currently stress economic development and the re-fabrication of traditional urban scenes to attract more business investments and local tourism industries. This is encouraged by the local decision-makers or authoritarians. However, these pro-growth urban regeneration policies and strategies contribute very little to the comprehensive regeneration or integrated conservation goals, especially when indigenous residents are resettled to give way to urban redevelopment projects. Without the influence of indigenous lives, the outcomes of local conservation plans often fail. In addition to the conservation of valuable heritage monuments and archaeological ruins, the integrated urban conservation policies in Xi'an should also consider the conservation of the traditional lives of indigenous residents and

城市发展的评估标准可以考虑由以经济增长为标准转向更加综合的标准，例如包括改善原住居民的生活和提高社区居民的满意度。为实现这些目标，本研究认为以社区为基础的、以保护为主导的更新方式很关键，同时，制度化的、社区参与式的城市管治模式也尤为重要。

2) 基于更高水平的社会资本，建立制度化的、参与式的城市管治模式

回民区的试点项目是多方参与式的更新机制下的一个案例，它要求各方利益相关者从前期策划到最终规划实施的全方位参与。因此，在进行合理的再开发活动的同时，也能够保护当地物质环境和原住民的生活内容。联合国教科文组织认为该更新项目说明，"享受现代工业文明设施和保护传统建筑可以兼得"（UNESCO, 2007：235）。这个试点研究也证明，制度化的、参与式的城市管治模式对实现当地全面的城市更新结果有很大影响。同时，受到中央政府政策的支持，当地专家和市政当局针对更新规划制定、实施了相应的规章制度，并监管这些制度的具体实施。这些必要法规和机构的设置，在第五章中曾作了阐释，为地方居民在更新过程中的参与注入了活力。制度化参与方式的设置也为不同的参与者赋予了特定的权力和义务。

该试点项目的更新政策强调社区居民的作用，社区居民在更新规划中的参与汇集了当地原住民的集体认识和想法，这样，更新结果也能更好地满足当地实际需求。鉴于回民区的宗教特色，当地穆斯林居民主动参与到更新项目中来，更显得很有必要，因为专业工作人员未必完全了解当地的宗教和文化

the improvement of their living environments. The conservation of local socio-cultural elements must be included in Xi'an's urban regeneration agenda. In addition, the evaluation standards for local development should shift from economic growth-based criteria to more integrated criteria, such as the improvement of original residents' lives and the communities' level of satisfaction. To achieve these, this book argues that a community-based, conservation-led regeneration approach is critical and that an institutionalized community participatory mode of urban governance is essential.

2) An institutionalized participatory mode of urban governance based on higher level of social capital

The pilot project within the DTMD is a good example of a multi-participatory regeneration mechanism requiring the engagement of multi-stakeholders from pre-planning to implementation. Therefore, it is possible to conserve the physical environment and indigenous lives while simultaneously conducting reasonable redevelopment activities. On the outcomes of this regeneration project, the UNESCO stated that, "the enjoyment of modern amenities can be compatible with preserving heritage buildings" (UNESCO, 2007: 235). This pilot study confirms that an institutionalized participatory mode of urban governance is significant to local comprehensive regeneration outcomes. Support from a central government policy, local professionals, and an actively formed district/municipal government could implement corresponding regulations on local regeneration's policy making, implementation, and management. The necessary regulations and institutional setup, elaborated in Chapter 5, provide vigor for local residents' participation in the regeneration process. The institutional setup provides specific rights and obligations for different actors.

The regeneration policy in the pilot project specifically addresses local residents. By involving them in the regeneration plan, local indigenous knowledge is integrated. Thus, regeneration outcomes could better meet local practical needs. Given the religious characteristics of this pilot study in the Muslim

需求。为了确保更新项目的顺利开展，该项目成立了一个中间机构——项目办公室，以便根据既定的法规来进行公众协商与交流。项目办公室的工作人员由当地专家和政府工作人员共同组成。这个中间机构对更新项目中的各种人群的参与起到了很好的协调作用。考虑到当地旅游业将带来的经济回报，私人开发商为更新规划也提供了必要的资金支持，地方市政当局则通过对开发商减免税收的优惠政策支持其在项目中的参与。地方政府通过制定专门政策，发挥了其积极的协调作用，也有效地保障了更新项目的顺利完工和多维保护与更新目标的实现。该项目中，居民和其他机构团体，如地方政府和项目办公室之间的协调关系，即联合型的社会资本，有利地保证了更新规划的顺利实施。考虑到当地社区可以从中直接获益，很多居民也非常支持更新项目，并与项目办公室的工作人员开展了频繁的互动。这些都有助于更新项目取得最后的成功，保护了当地珍贵的历史建筑、改善了破败的居住环境和基础设施、并促进了当地旅游业的发展和经济增长。总之，试点项目的突出了基于社区的决策机制、制度化的机制以及自上而下的政策支持是该项目成功的关键。

本研究认为，在城市更新过程中，必要的法律法规能使各阶层的参与合法化甚至"官方化"。相应的法律法规也为原住居民提供了重要的法律杠杆。在实践中，法律法规可能需要首先在地方政府层面进行制定，然后再由中央政府在国家层面进行颁布。在更新规划中，制度化的或者合法化的原住居民的参与对特定群体来说非常重要，

area, the involvement of local Muslim residents in the regeneration project is necessary because local religious and cultural traditions are often overlooked by the professionals. To conduct the regeneration project smoothly, an intermediate agency was set up as the project office for public consultation and communication according to the established regulation. Staff in the project office consist of professionals and local government members. This intermediate organization coordinates the involvement of various actors in the regeneration process. With the expectation of future economic returns from the local tourism industry, private developers provide the necessary funds to implement the community-based regeneration plan. The local district/municipal government supports the pilot study by offering tax-free privileges to concerned private developers. This facilitating role of the local government through special urban policy effectively guarantees the accomplishment of local regeneration project toward the proposed multi-dimensional conservation and regeneration goals. The harmonious relationships among local residents and other agencies, such as the local government and the intermediate working office, in terms of bridging social capital, are conducive to the implementation of local regeneration plan. Recognizing that they themselves would benefit from the outcome of the local government-led regeneration project, and based on their trust of the regeneration policy, many residents strongly support the plan and have had frequent dialogues with the staff in the project office. All these contribute to the final successful regeneration outcome, featured by the conservation of several valuable historical heritages, the redevelopment of derelict living conditions and infrastructures, and the improvement of local tourism industry and economic growth. The discussion on the pilot project stresses that the community-based regeneration project has much to do with the ad hoc institutionalized setup and with the top-down political and governmental support.

The book argues that necessary regulation and law could help legitimize and

比如本研究中的穆斯林居民。他们独特的宗教与种族需求往往超出了专业工作者的知识范围。他们以家庭为基础的商业活动和传统生活方式，很大程度上依赖于当地繁荣的旅游业，这些都应该在更新规划的决策范畴内进行考虑。同时，他们的宗教生活习惯，已经成为当地非物质文化遗产的重要象征，因此，也应该在当地更新规划的制定当中进行考虑。在这个过程中，地方政府的作用尤为显著，其协助、促进的作用贯穿整个更新过程，从协调各方利益相关者之间的关系——如居民、开发商和专业工作者，到为更新项目的筹集必要资金。政府通常是通过调整当地的更新政策来达成上述目的（Zhai & Ng, 2009：81）。受到回民区试点项目的启发，本研究认为立场更加中立的中间机构似乎更能为当前的实际困境提供许多解决办法。这种中立机构和管理机构中的成员也应来自于不同领域，包括居民代表、专业工作者以及政府工作人员。该中立机构的任务就是提供一个平台，以便协助各方更加有效地参与到更新过程中。

7.2.3 对中国历史城市更新实践的启示

本书中两个历史街区的案例研究，为中国其他历史城市的更新实践提供了许多启示，主要涉及以下三个方面：(1) 保护主导的城市政策和策略，更加适合中国历史城市的更新实践；(2) 社会资本有助于促进参与式的城市更新过程；(3) 城市更新中的制度化的、参与式的城市管治模式，对实现更加全面的城市更新目标至关重要。

1) 保护主导的城市政策和策略，更加适

even "politicize" the role of various actors in the urban regeneration process. These corresponding regulations and laws provide indigenous residents crucial legal leverage in the urban regeneration process. In practice, such regulations may be first enacted at the local municipal government level, and then promulgated by the central government at the national level. The institutionalized or legitimized involvement of indigenous residents in the regeneration plan is important for particular groups, such as the Muslim groups in this study. Their special religious and ethnic needs are often beyond the knowledge of decision-makers. Their family-based business activities and their traditional ways of making a living, which rely much on the tourism industry, should be considered in the regeneration plan. Their religious lives and customs, which serve as significant symbols of an intangible heritage, should also play a part in the forming a local regeneration plan. The local district/municipal government plays a noteworthy role in this process. Its facilitating role runs throughout the regeneration process, from coordinating the relationships of various stakeholders, such as local residents, private developers, and professionals, to raising the necessary funds to carry out the regeneration projects, usually by adjusting local regeneration policies (Zhai & Ng, 2009: 81). Inspired by the pilot project in the DTMD, a more neutral organization or an intermediate agency could suggest solutions to current practical dilemmas when the local government tends to play dual roles as both the public developer and the city development manager. This neutral organization and management unit must consist of members from different domains, including the residents' representatives and the professionals, as well as the government staff. It is intended to provide a platform and to facilitate the involvement of various actors in the regeneration process.

7.2.3 Implications for the urban regeneration practices in Chinese historic cities

The case studies of the two districts

合中国历史城市的更新实践

从西安的城市更新实践可以看出，在快速城市化进程中，城市保护在城市更新过程中发挥着至关重要的作用，并且有助于保护当地优秀的文化遗产。但是，需要强调的是城市保护政策的内容，应该既包括物质环境也包括当地原住民的生活等非物质环境。正如前面第三章所讨论的，在我国，非物质文化遗产被广泛地理解为高层次的文化——如工艺、艺术、歌剧等，所以，历史城市中的原住民生活往往会被忽视掉。事实上，正是当地原住民的生活，尤其是居民传统的生活方式和习俗，与当地非物质文化遗产的保护息息相关。同时，在居民区的更新或再开发当中，原住居民的生活和经验也可以为更新实践提供多方面的知识与智慧。这也突出了社区参与式城市更新过程的重要性。

2）社会资本有助于促进参与式的城市更新过程

城市更新中各阶层之间的关系或者社会资本的水平影响着参与式更新实践的进展。书中两个街区案例从两方面证实了这个论点。第一，当地社区居民之间的关系影响着社区居民在更新过程的参与。一般来说，如果当地社区居民之间的社会资本处于高水平状态，居民会渴望参与到更新项目中，并且保护或加强他们原有的社区关系与联系。这也有助于当地聚合型社会资本的积累。而当社区居民之间的社会资本处于低水平时，居民在更新过程中，往往表现出非常关心自己的利益。另外，低水平的社会资本也使传统的庭院生活和集体活动面临挑战，这反过来进一步破坏了当地社会资本的积累。一些弱

in Xi'an City can provide many theoretical implications for the regeneration practices in other Chinese historic cities. Three implications are included as follows: (1) Conservation-led urban policies and strategies fit the goals of urban regeneration practices in Chinese historic cities better; (2) social capital contributes to the participatory regeneration processes; and (3) the institutionalized participatory mode of urban governance in urban regeneration is essential for the more comprehensive urban regeneration outcomes.

1) Conservation-led urban policies and strategies fit the goals of urban regeneration practices in Chinese historic cities better.

As inferred by the experiences in Xi'an urban regeneration practices, urban conservation plays a critical role in urban regeneration processes and contributes to the conservation of excellent cultural heritages during rapid urbanization. However, it must be stressed that urban conservation policies should cover both the physical environments and local indigenous lives. As discussed in Chapter 3, intangible cultural heritages in China are widely understood as high levels of culture, such as craftsmanship, art, opera, and the like. Thus, indigenous lives in many Chinese historic cities have been overlooked. Nevertheless, it is local indigenous lives, specifically, the local residents' traditional living styles and customs, which often closely relate to the conservation of local intangible heritages. In the regeneration or redevelopment of local residents' residential areas, the opinions and experiences of indigenous residents contribute significant knowledge and wisdom to urban regeneration practices. This underlines the importance of a community-participatory regeneration process.

2) Social capital contributes to the participatory regeneration processes.

Relationships between various actors or the levels of social capital have an impact on the participatory regeneration processes. The studies based on two districts of Xi'an City show this from two aspects. First, the relationships between local communities affect these communities' participation in local regeneration processes.

势群体，包括老年居民，他们经常受益于或依赖于邻里关系与邻里交流，会因此在更新过程中感受到利益受损。第二，社区居民和其他利益团体之间的关系，或称之为联合型社会资本，也影响社区居民在当地更新过程中的参与程度。这一点在西安回民区的试点项目当中得到了很好的阐释。总体而言，更新过程中的社区参与和不同水平的聚合型与联合型社会资本之间的关系，可以分为如下四类进行讨论：

a）高水平的联合型社会资本、低水平的聚合型社会资本（我国 20 世纪 50～70 年代的社会关系特征）

当社区成员和市政当局之间保持一种良好的关系时，这些社区成员非常渴望能够参与到由政府主导的城市发展或更新实践中。这种情况在 20 世纪 50 年代到 20 世纪 70 年代时，以中央政府主导的城市建设和以项目为导向的城市复兴中尤其突出。在意识形态为导向的城市发展过程普及期间，很多人也热衷于参与到政府主导的活动中来，与此同时，人们之间的人际关系也由先前的朋友关系逐渐转变为工作上的同志关系（Vogel，1965）。

b）低水平的联合型社会资本、低水平的聚合型社会资本（我国 20 世纪 80～90 年代末的社会关系特征）

在社区居民和其他利益团体之间的社会资本处于低水平状态时，社区居民表现为愿意遵循地方政府提出的更新规划方案。尤其是当地的聚合型社会资本或当地社区成员之间的关系也处于低水平时，这种情况更为突出，地方政府在更新规划实施中会起到主心骨的作用。通常，为了改善当地居住条件，

Normally, when the level of social capital among local communities is high, residents are eager to get involved in regeneration projects and conserve or strengthen their original neighborhood relationships and connections. This contributes to the accumulation of local bonding social capital. When the existing level of social capital among local communities is low, local communities tend to care more about their individual interests resulting from local regeneration processes. Low levels of social capital have also challenged traditional courtyard lives and group activities, which in turn further undermine the accumulation of local social capital. As a result, disadvantaged groups, including senior residents who usually benefit from or rely on the relationships and mutual interactions among neighbors, would feel the benefits of local regeneration less. Second, relationships between the local community and other actors, or the bridging social capital, also affect the level of community participation in local regeneration process. This was well illustrated through the pilot study in Xi'an DTMD case. Generally, the relationships of community participation in urban regeneration and various levels of bonding and bridging social capital in urban China are discussed in the following four categorizations:

a) High bridging social capital with low bonding social capital (from the 1950s to the 1970s)

When local communities keep good relationships with local authorities, these communities are very keen to be involved in government-led urban development or regeneration practices. This situation was observed during the period of central government-led and project-oriented urban revitalization in socialist China, especially from the 1950s to the 1970s. During the period when ideology-oriented urban development processes were popular, many people were keen to participate in central government-led activities, and human relationships in China transformed from friends to comrades (Vogel, 1965).

b) Low bridging social capital with low bonding social capital (from the late 1980s to the late 1990s)

When the level of social capital between local communities and other actors is low,

居民除了遵循政府主导的更新规划之外，也较难再有其他更多选择。这种情况普遍存在于 20 世纪 80 年代末到 90 年代末，中国快速城市化和片面追求经济发展阶段的城市人际关系当中。

c）高水平的聚合型社会资本、低水平的联合型社会资本（我国 20 世纪 80～90 年代末的社会关系特征）

在有些历史城区，虽然社区居民间有着较强的社区关系，但他们和其他利益团体间的关系却较薄弱。城市更新过程中这种类型的社区参与有助于保护当地传统的生活方式与习俗。但是，却会进一步破坏了社区居民和其他利益团体之间的关系，或者说阻碍了联合型社会资本的积累。而这种问题也会阻碍城市更新项目的顺利开展。

d）高水平的聚合型社会资本、高水平的联合型社会资本

理论上来讲，这种情况代表了一种以社区为基础的、多方参与式的城市更新过程。在这个过程中，地方社区居民会积极地发起并投入到整个过程当中，而地方政府会促成主要利益相关者在项目中的参与（例如，当地的专业工作者和私营部门）。在更新规划实施方面，这种情况也有助于保护原住民的生活方式，并促进社区居民之间聚合型社会资本的积累。同时也有利于社区居民和其他利益团体之间建立相互信任和合作关系，促进联合型社会资本的积累。这种情况往往也有助于城市物质环境的改善和社会文化的保护与发展，例如保护与发展当地传统的生活方式和活动内容，并实现多维的更新目标。

communities continue to follow local government-led regeneration plans. This is true especially when the local bonding social capital, or relationship between local communities, is also low, and local authorities are the dominant power when implementing urban regeneration plans. Often, many residents do not see alternatives to improving living conditions other than following government-led development plans. This situation existed widely during the rapid economic-biased development of many Chinese cities from the late 1980s to the late 1990s.

c) High bonding social capital with low bridging social capital (from the late 1980s to the late 1990s)

In some Chinese historic urban areas, local communities have very strong community ties but their relationships with other sectors are weak. This kind of community involvement in local regeneration processes contributed to the conservation of local traditional living styles and custom. However, it also further undermined the relationships between the local community and other invested interest groups, or the accumulation of bridging social capital. The problem may also become obstacles to many of the local urban regeneration processes.

d) High bonding social capital with high bridging social capital

Theoretically speaking, this indicates a condition in the community-based multi-participatory urban regeneration process where the initiatives of local communities play a significant role and local governments tend to facilitate the involvement of key actors (e.g., local professionals and the private sector). With regard to the implementation of local regeneration plans, this condition contributes to the conservation of indigenous lives and the accumulation of bonding social capital among original neighborhoods. It also contributes to the mutual trust and cooperation between the community and local governments and the accumulation of bridging social capital. Moreover, this condition tends to improve both urban physical environments and socio-cultural development, such as the conservation and development of local traditional lives and activities, towards multi-dimensional

高 High ↑ 联合型社会资本 Bridging social capital	1. 高水平的联合型社会资本、低水平的聚合型社会资本（1950年代–1970年代）：计划经济下，中央政府和政治意识形态为主导的城市发展阶段。 1. High bridging social capital with low bonding social capital (from the 1950s to the 1970s): Under planned economy, central government and political ideology-led urban development.	4. 高水平的聚合型社会资本、高水平的联合型社会资本（理论上"理想"）：以社区为基础、多方参与式的更新过程，其结果既改善城市物质环境，又促进社会文化的发展，尤其是促进当地传统生活内容和活动方式的保护与发展。 4. High bonding social capital with high bridging social capital (theoretically "ideal"): Community-based and multi-participatory regeneration processes, the outcome of which improves both physical urban environments and socio-cultural development, specifically the conservation and development of local traditional lives and activities.
低 Low	2. 低水平的联合型社会资本、低水平的聚合型社会资本（1980年代末–1990年代末）：经济转型期，地方政府主导的、大规模的城市再开发阶段。 2. Low bridging social capital with low bonding social capital (from the late 1980s to the late 1990s): Local government-led massive urban redevelopment in a transitional economy.	3. 高水平的聚合型社会资本、低水平的联合型社会资本（1980年代末–1990年代末）：经济转型期，地方政府主导的大规模城市再开发和以社区为基础的城市更新提议并存。 3. High bonding social capital with low bridging social capital (from the late 1980s to the late 1990s): In a transitional economy, local government-led massive urban redevelopment and community-based proposals coexist.
	低 Low ← 聚合型社会资本 Bonding social capital → **高 High**	

图 7-1：社会资本（聚合型和联合型社会资本）对中国城市更新中社区参与度的影响
Figure 7-1: Impacts of social capital (bonding and bridging social capital) on the level of community participation in the urban regeneration of China

图 7-1 显示出中国城市中，与不同程度的社区参与相关的四种聚合型和联合型社会资本。以社区为基础的、多方参与式的"理想的"更新过程，往往以高水平的聚合型社会资本和联合型社会资本为特点。为了形成这样一种局面，本文认为，在中国历史城市中，建立一种制度化的、参与式的城市更新管治模式至关重要。

以上四种情况中，第三种的特点是高水平的聚合型社会资本和低水平的联合型社会资本，这代表了居民基于紧密的社区邻里关系而参与到城市更新过程中。正如在回民区案例中所观察到的，而这种参与可以对当地更新规划的实施与结果都产生实质性的影响。同时，当地很多居民非法的自建活动会 regeneration goals.

The four categorizations of bonding and bridging social capital related to the various levels of community participation in urban China are revealed in Figure 7-1. In order to transit toward the "ideal" situation of community-based and multi-participatory regeneration processes, characterized by high bonding and high bridging social capital in urban regeneration processes, this book argues that the institutionalized participatory mode of urban regeneration governance in Chinese historic cities is essential.

Among the four situations, the third featuring high bonding and low bridging social capital represents the circumstance in which local residents become involved in the regeneration process based on cohesive community ties. These activities, as revealed in the DTMD case, could bring significant impact on the implementation and outcome of the local government-led regeneration

加剧原本就已破败和拥挤的历史环境。这对历史环境的保护和当地居住环境的改善都会产生很大影响。促进社区居民间形成强大的聚合型社会资本的因素有很多，包括历史城区建筑结构、宗教和种族问题。第四、五章集中讨论了这些因素。基于社区居民间密切关系的、强大的聚合型社会资本的公众参与，有助于实现城市更新设定的全方位更新目标。社区居民在参与到更新规划中时，他们的集体意见往往能够弥补专业工作者，或地方政府提出的更新规划中许多不足的地方。在更新规划中，综合地方居民的意见也能够使更新结果更有利于公共利益，而不仅是地方政府或开发商的利益。另外，原住居民在更新规划中的参与更能保证地方原著居民的生活得到保护，而避免简单的被重置或被分散的情况出现。保护原住居民的生活内容有助于实现综合的城市保护目标。在制度化的以社区为基础的机制下，社区参与体现在整个更新过程中，从制定规划、执行规划到项目管理。为了建立一个制度化的参与式城市更新管治模式，在更新规划中应该明确各利益相关者的权力和义务，并通过特定的法律法规来实施。

3) 制度化的、参与式城市管治模式对实现当地全面的更新目标至关重要

我国目前的社区参与还仅仅停留在规划方案的公示层面，这是地方政府采取的一种比较"家长式"的管理参与方式。根据安斯汀（Arnstein）的理论，"家长式"的参与是介于不参与和象征式的社区参与之间的一种形式（Arnstein，1969：217）。在"参与"过程中，如果没有考量地方居民的初始

plan. At the same time, many local self-built illegal constructions exacerbate the already dilapidated and congested historical environments. This has substantial effects on the conservation of historical environments and the improvement of local living conditions. Various factors contribute to stronger social capital among community members, including the historical urban fabric, religion, and ethnicity issues. These factors have been extensively discussed in Chapter 4 and Chapter 5. Community involvement in the local regeneration process, based on close relationships among residents or stronger bonding social capital, contributes to the comprehensive dimensions in the regeneration plan. The involvement of local residents in the regeneration plan means that their collective opinions make up for the missing aspects in the local government-led regeneration plan, usually proposed by local professionals. Integrating local residents' opinions in the regeneration plan helps the regeneration outcome become in favor of the public interest, rather than the interest of only the local government or developers. In addition, the involvement of indigenous residents in the regeneration plan more likely ensures that local indigenous lives are protected, rather than relocated or dispersed, as is the case in many government-led plans. Conserving indigenous lives contributes to the aim of integrated urban conservation. In an institutionalized community-based regeneration mechanism, community participation in the regeneration process exists in the entire process from planning, to implementation, to project management. To achieve an institutionalized participatory mode of urban regeneration governance, the rights and obligations of various stakeholders in the regeneration plan need to be specified and carried out through special regulations and laws.

3) The institutionalized participatory mode of urban governance is essential for local comprehensive regeneration outcomes.

In China, community participation currently exists only at the level of planning

意见，而仅通过公示更新规划的方式来使居民"参与"更新过程当中，收效会非常有限。为了在更新规划中综合考虑原住居民的意见和见解，建立一个制度化的、参与式城市管治模式至关重要。正如书中案例研究所指明的，在地方更新规划中需要制定关于社区参与的有关法律法规。而且，在地方更新过程中，有必要建立并遵循一种制度化的社区参与机制。应该鼓励各方利益相关者发挥出各自的作用。当地专业工作者需要负责对社区开展教育活动，使居民认识到城市中物质文化遗产和非物质文化遗产的价值，同时意识到实现更加全面的和可持续的发展目标的重要性。当中最为重要的是，地方政府应该发挥其协调作用，促进当地更新规划的实施和管理，同时鼓励私人开发商和当地社区居民参与到城市更新过程中来（Zhai & Ng, 2013）。

promulgation, which is a very paternalistic way of participation adopted by local authorities. According to Arnstein, the paternalistic participation should locate between non-participation and tokenism of community participation (Arnstein, 1969: 217). When local residents' opinions are not considered in the "participation" process, it actually becomes meaningless for them to go through the "participation" process, by means of the promulgation of local regeneration plans. In order to integrate indigenous residents' opinions and knowledge in the regeneration plan, an institutionalized participatory mode of urban governance in local regeneration processes is essential. As inferred in the case study of the book, necessary laws and regulations regarding community participation in local urban regeneration plans need to be enacted. Furthermore, the institutionalized community participation mechanism needs to be set up and followed in local urban regeneration processes. Stakeholders should be encouraged to play their roles. Local professionals should take on the responsibility of educating local communities of the value of tangible and intangible heritages, as well as the importance of more comprehensive and sustainable development aims. Local governments should also take the lead role in facilitating the implementation and management of local regeneration plans, as well as the involvement of private developers and local communities in local regeneration processes (Zhai & Ng, 2013).

参考文献／BIBLIOGRAPHY

[1] AHMAD Y. Scope and Definitions of Heritage: From Tangible to Intangible [J]. International Journal of Heritage Studies, 2006, 12(3):292–300.

[2] ANTHIAS F. Ethnic ties: social capital and the question of mobilisability [J]. The Sociological Review, 2007, 55(4):788–805.

[3] ARNSTEIN S R. Ladder of Citizen Participation [J]. Journal of the American Institute of Planners, 1969, 35(4):216–224.

[4] BAI N. Reservation and Conservation-the Conservation Countermeasure for the Historical Reidential District in Xi'an [D]. Unpublished Master Thesis, Xi'an University of Architecture and Technology, Xi'an, 2000. (白宁 . 保留与保护——西安历史住区保护对策研究 . 西安 : 西安建筑科技大学 , 2000.)

[5] BIAN Y J. Studies on Social Capital [J]. Study and Exploration, 2006, 2:39–40. (边燕杰 . 社会资本研究 . 学习与探索 , 2006, 2:39–40.)

[6] BIAN Y J. The Formation of Social Capital among Chinese Urbanites: Theoretical Explanation and Empirical Evidence [M] // N LIN and B. H. ERICKSON. Social Capital: an international research program. pp., Oxford; New York: Oxford University Press, 2008: 81–104.

[7] BIAN Y J and QIU H X. The Social Capital of Enterprises and Its Efficiency [J]. Social Science in China, 2000, 2:87–99. (边燕杰 , 丘海雄 . 企业的社会资本及其功效 . 中国社会科学 , 2000, 2:87–99.)

[8] BIAN Y J and ZHANG W H. Economic Systems, Social Networks and Occupational Mobility [J]. Social Sciences in China, 2001, 2:77–89. (边燕杰 , 张文宏 . 经济体制、社会网络与职业流动 . 中国社会科学 , 2001, 2:77–89)

[9] BING Z. Cultural Restructuring in Social Transition [J]. Jilin University Journal Social Sciences Edition, 1997, 4:39–44. (邴正 . 社会转型时期的文化建构 . 吉林大学社会科学学报 , 1997, 4: 39–44)

[10] BLAKE J. Developing a new standard-setting instrument for the safeguarding of intangible cultural heritage [M]. Paris: UNESCO, 2002.

[11] BO C L. Social capital and social harmony [M]. Beijing: Social Science Document Press, 2005. (卜长莉 . 社会资本与社会和谐 . 北京 : 社会科学文献出版社 , 2005.)

[12] BOSCHMA R, LAMBOOY J and SCHUTJENS V. Embeddedness and Innovation [M] // M TAYLOR and S LEONARD. Embedded enterprise and social capital: international perspectives. Aldershot: Ashgate, 2002: 19–35.

[13] BOURDIEU P. Forms of Capital [M] // J G RICHARDSON. Handbook of Theory and Research for the Sociology of Education. New York: Greenwood Press, 1986: 241–258.

[14] BOX R C. Citizen Governance: leading American communities into the 21st century [M]. Thousand Oaks, Calif.: Sage Publications, 1998.

[15] BRAY D. Social space and governance in urban China: the danwei system from origins to reform [M]. Stanford, Calif.: Stanford University Press, 2005.

[16] BROWN B and PERKINS D D. Disruptions in place attachment [M] // I ALTMAN and S M LOW. Place attachment. New York: Plenum Press, 1992: 279–304.

[17] BROWN B, PERKINS D D and BROWN G. Place attachment in a revitalizing neighborhood: Individual and block levels of analysis [J]. Journal of Environmental Psychology, 2003, 23(3):259–271.

[18] BRYANT C A and NORRIS D. Measurement of Social Capital – the canadian experience [M]. Paper presented at the International Conference on Social Capital Measursment in London, 2002.

[19] BUCK D D. Directions in Chinese Urban Planning [J]. Urbanism Past and Present, 1975, 1(1):24–35.

[20] BURNS D, HAMBLETON R and HOGGETT P. Politics of decentralisation: revitalising local democracy [M]. London: Macmillan, 1994.

[21] BURT R S. Structural holes: the social structure of competition [M]. Cambridge, Mass: Harvard University Press, 1992.

[22] BURT R S. Contingent Value of Social Capital [M]. Administrative Science Quarterly, 1997a, 42(2):339–365.

[23] BURT R S. A note on social capital and network content [M]. Social Networks, 1997b, 19(4):355–373.

[24] CAI Y S. Managed Participation in China [J]. Political Science Quarterly, 2004, 119(3):425–451.

[25] CALLIES D L. Land Use Planning and Control in Japan [M] // P SHAPIRA, I MASSER and D W EDGINGTON. Planning for cities and regions in Japan. Liverpool, England: Liverpool University Press, 1994: 59–69.

[26] CALLIES D L. Urban Land Use and Control in the Japanese City: A Case Study of Hiroshima, Osaka, and Kyoto [M] // P P KARAN and K E STAPLETON. The Japanese city. Lexington: University Press of Kentucky, 1997: 134–155.

[27] CAMPBELL H and MARSHALL R. Public Involvement and Planning: Looking beyond the One to the Many [J]. International Planning Studies, 2000, 5(3):321–344.

[28] CAMPBELL K E and LEE B A. Name generators in surveys of personal networks [J]. Social Networks, 1991, 13(3):203–221.

[29] CAO W. Study on the Renewal Building in the Historic Drum-tower District of Xi'an [D]. Unpublished Master Thesis, Xi'an University of Architecture and Technology, Xi'an, 2005. (曹苇. 西安鼓楼历史街区更新建筑研究. 西安：西安建筑科技大学, 2005.)

[30] CARLEY M and KIRK K. Sustainable by 2020? : a strategic approach to urban regeneration for Britain's cities [M]. Bristol, England: Policy Press, 1998.

[31] CARMON N. Neighborhood Regeneration: the state of the art [J]. Journal of planning education and research, 1997, 17:131–144.

[32] CAUPD (China Academy of Urban Planning and Design) and DTCPMC (Department of Town and Country Planning Ministry of Construction of China). Urban Planning Resource Book: VIII Urban Historic Conservation and Regeneration [M]. Beijing: China Architecture and Building Press, 2008. (中国城市规划设计研究院，建设部城乡规划司. 城市规划资料集第 8 分册：城市历史保护与城市更新. 北京：中国建筑工业出版社, 2008.)

[33] CEFLAD (Center for Ethnic and Folk Literature and Art Development Ministry of Culture of China). Studies on intangible cultural heritage conservation in China [M]. Beijing: Beijing Normal University Press, 2007. (文化部民族民间文艺发展中心 , 中国非物质文化遗产保护研究 . 北京 , 北京师范大学出版社 , 2007.)

[34] CHAI Y W. Changing social space in transitional urban China from the danwei perspective: Case studies on internal spatial and residential hybridization [C]. Paper presented at the 2009 Annual Meeting of Association of American Geographers, 2009.

[35] CHAN J K. The 99 historic cities of China [M]. Hong Kong: Historic City Books, 1999.

[36] CHANG H Q. Organic Cooperation between "Top–Down" and "Bottom–Up" : Suitable Way for the Development and Renewal of Traditional Urban Housing District [D]. Unpublished Master Thesis, Xi'an University of Architecture and Technology, Xi'an, 1999. (常海青 . "自上而下"与"自下而上"的有机结合——城市传统住区建设和改造的"适宜"途经探寻 . 西安 : 西安建筑科技大学 , 1999.)

[37] CHEN D L, WANG S M and WEN J Q. Conservation Prospectives on the Ming City Area in Xi'an [J]. Architectural Journal, 2003, 2:22–24. (陈道麟，王树茂，温建群 . 西安市明城区保护展望 . 建筑学报 , 2003, 2: 22–24.)

[38] CHEN J F. Approach to the Institution of Public Participation in Urban Planning [J]. City Planning Review, 2000, 24(7):54–57. (陈锦富 . 论公众参与的城市规划制度 . 城市规划 , 2000, 24(7):54–57.)

[39] CHEN W S. Exploring the Real Estate Reality in Xi'an, Problems and Strategies [J]. China Real Estate, 1994, 7:52–55. (陈万生 . 浅谈西安房地产开发市场的现状、问题及对策 . 中国房地产 , 1994, 7:52–55.)

[40] CHEN W Y. Characteristics of Muslim living and relationship with urban regeneration), Economic Geography, 2001, 21(4):431–434. (陈文言 . 回民的生活居住特性及其与旧城改造的关系 . 经济地理 , 2001, 21(4):431–434.)

[41] CHENG A D. Inland–Located City with Outward Mind – Research on Xi'an's Development Strategy [M]. Xi'an: Shanxi People's Publishing House, 1997. (程安东 . 内陆外向型城市——西安发展战略研究 . 西安 : 陕西人民出版社 .)

[42] CHENG F and LIU D X. Xi'an's old streets in modern times [J]. Chinaweek 2003, 10:13–15.

[43] CHENG Q L. Theories and Practices on Xi'an Urban Community Construction [M]. Xi'an: Xi'an Publishing House, 2003. (程群力 . 西安城市社区建设的理论与实践 . 西安 : 西安出版社 .)

[44] CHENG T Q. On Building Socialist Harmonious Society [EB/OL]. (2006) [2007–01–20] http://www.p1001.com/item70180.htm. (程天权 . 论构建社会主义和谐社会 , 2006.)

[45] CHEONG P H, EDWARDS R, GOULBOURNE H and SOLOMOS J. Immigration, social cohesion and social capital: A critical review [J]. Critical Social Policy, 2007, 27(1):24–49.

[46] COCHRANE A. Whatever happened to local government [M]? Buckingham [England]: Open University Press, 1993.

[47] COHEN N. Urban planning conservation and preservation [M]. New York: McGraw–Hill, 2001.

[48] COLEMAN J S. Social Capital in the Creation of Human Capital [J]. American Journal of Sociology, 1988, 94(Supplement):95–120.

[49] COLEMAN J S. Foundations of social theory [M]. Cambridge, Mass.: Harvard University Press, 1990.

[50] COUCH C. Urban renewal: theory and practice [M]. Houndmills, Basingstoke: Macmillan, 1990.

[51] CRESWELL J W. Qualitative inquiry and research design: choosing among five traditions [M]. Thousand Oaks,

Calif.: Sage Publications, 1998.

[52] CRESWELL J W. Research design: qualitative, quantitative, and mixed methods approaches [M]. Thousand Oaks, Calif.: Sage Publications, 2003.

[53] 中国社会科学院人口与劳动经济研究所. 中国人口年鉴 2009. 北京, 中国人口年鉴杂志社, 2009.

[54] 国家统计局人口和就业统计司. 中国人口统计年鉴 2009. 北京市: 中国统计出版社, 2009.

[55] DELLIQUADRI F. Helping the family in urban society [M]. New York: Columbia University Press, 1963.

[56] 邓泉国. 中国城市社区居民自治. 沈阳: 辽宁人民出版社, 2004.

[57] 刁杰成. 人民信访史略. 北京: 北京经济学院出版社, 1996.

[58] DOE (Department of Environment in Great Britain). Five year review of the Birmingham Inner City Partnership [M]. London: HMSO, 1985.

[59] DOE (Department of Environment in Great Britain). Sustainable development: the UK strategy [M]. London: HMSO, 1994a.

[60] DOE (Department of Environment in Great Britain). Community involvement in planning and development processes [M]. London: HMSO, 1994b.

[61] 董鉴泓. 第一个五年计划中关于城市建设工作的若干问题. 建筑学报, 1955, 3:1-12.

[62] 董鉴泓. 中国城市建设史. 北京: 中国建筑工业出版社, 2004.

[63] DONG W. An Ethnic Housing in Transition: Chinese Muslim Housing Architecture in the Framework of Resource Management and Identity of Place [D]. Unpublished PhD Thesis, The Norwegian Institute of Technology, Trondheim, 1995.

[64] DONG W. City Transform under the Free Market Economy: the Study on the Self-renovation of the Hui Nationality Area in Xi'an City [J]. City Planning Review, 1996, 5:42-45. (董卫. 自由市场经济驱动下的城市变革——西安回民区自建更新研究初探. 城市规划, 1996, 5: 42-45.)

[65] DUAN B L. Understanding the connotation of intangible cultural heritage [M] // DEVELOPMENT CENTRE OF ETHNIC AND FOLK LITERATURE AND ART. Research on the Conservation of Intangible Cultural Heritage of China. Beijing: Beijing Normal University Press, 2007: 1-8. (段宝林. 明确非物质文化遗产的主要内涵. 中国非物质文化遗产保护研究. 北京: 北京师范大学出版社, 2007: 1-8.)

[66] DURKHEIM E. The elementary forms of the religious life [M]. New York: Free Press, 1969.

[67] EBUCCC (Editorial Board of Urban Construction of Contemporary China). The Urban Construction of Contemporary China [M]. Beijing: Chinese Social Science Publisher, 1990. (《当代中国的城市建设》编辑委员会. 当代中国的城市建设. 北京: 中国社会科学出版社, 1990.)

[68] EBXY (Editorial Board of Xi'an Yearbook). Xi'an Yearbook 1993~2012 [M]. Xi'an: Xi'an Publishing House, 1993~2012. (西安市地方志办公室. 西安年鉴 1993-2012. 西安: 西安出版社. 1993-2012)

[69] ECUCIMX (Editorial Commission of The Urban Construction in Modern Xi'an). The Urban Construction in Modern Xi'an [M]. Xi'an: Shanxi People's Publishing House, 1988. (《当代西安城市建设》编辑委员会. 当代西安城市建设. 西安: 陕西人民出版社, 1988.)

[70] ECXBDR (Editorial Committee of Xi'an Beilin District Records). Beilin District Book [M]. Xi'an: Sanqin Press, 2003. (西安市碑林区地方志编纂委员会. 碑林区志. 西安: 三秦出版社, 2003.)

[71] ECXLDR (Editorial Committee of Xi'an Lianhu District Records). Lianhu District Book [M]. Xi'an: Sanqin Press, 2001. (西安市莲湖区地方志编纂委员会. 莲湖区志. 西安: 三秦出版社, 2001.)

[72] ECXR (Editorial Committee of Xi'an Records). Xi'an City Records Book [M]. Xi'an: Xi'an Publishing House, 2006. (西安市地方志编纂委员会 . 西安市志 . 西安：西安出版社 , 2006.)

[73] EDITOR. Our Neighborhood [J]. China Reconstructs, 1973, 22(8):2–11.

[74] EGLOFF B and NEWBY P. Towards cultural sustainable tourism at historical places: a critical study of Port Arthur, Tasmania [J]. Conservation and Management of Archaeological Sites, 2005, 7(01):19–33.

[75] ENGELHARDT R. Management of World Heritage Cities: Evolving Concepts, New Strategies [C]. Paper presented at the The Conservation of Urban Heritage: Macao Vision, International Conference, 2002.

[76] FANG L L. Pay attention to the carriers of intangible cultural heritages [M] // W Z WANG, Q S ZHANG and S D MA. A collection of papers of the Forum on the Protection of Chinese Intangible Cultural Heritage. Beijing: Culture and Art Publishing House, 2006: 185–193. (方李莉 . 请关注非物质文化遗产的传承者 // 中国非物质文化遗产保护论坛论文集 . 北京：文化艺术出版社 , 2006: 185–193.)

[77] FEI X T. Chinese Native Soil Society; Birth Institutions [M]. Beijing: Peking University Press, 1998. (费孝通 . 乡土中国；生育制度 . 北京：北京大学出版社 , 1998.)

[78] FILIPPI F D. Sustainable "living" heritage conservation through community-based approaches [C]. Paper presented at the Forum UNESCO University and Heritage 10th International Seminar "Cultural landscape in the 21st century", 2005.

[79] FLAP H. "No man is an island" [M] // E LAZEGA and O FAVEREAU. Conventions and Structures. Oxford: University Press, 2002: 29–59.

[80] FLAP H and VOLKER B. "Social, Cultural, and Economic Capital and Job Attainment: The Position Generator as a Measure of Cultural and Economic Resources" [M] // N LIN and B H ERICKSON. Social capital: an international research program. Oxford ; New York: Oxford University Press, 2008: 65–80.

[81] FLETCHER R, JOHNSON I, BRUCE E and KHUN-NEAY K. Living with heritage: site monitoring and heritage values in Greater Angkor and the Angkor World Heritage Site, Cambodia [J]. World Archaeology, 2007, 39(3):385–405.

[82] FORREST R and KEARNS A. Social Cohesion, Social Capital and the Neighbourhood [J]. Urban Studies, 2001, 38(12):2125–2143.

[83] FUKUYAMA F. Social capital, civil society and development [J]. Third World Quarterly, 2001, 22(1):7–20.

[84] GANS H J. Urban villagers: group and class in the life of Italian-Americans [M]. New York: Free Press of Glencoe, 1962.

[85] GANS H J. Failure of Urban Renewal: a critique and some proposals [M] // J BELLUSH and M HAUSKNECHT. Urban Renewal: people, politics, and planning. New York: Doubleday, Anchor Books, 1967.

[86] GAO J and JIANG M N. Analysis on the Layout Pattern of Xi'an City [J]. Journal of Xi'an University of Architecture and Technology, 2005, 24(3):26–28. (高娟．姜满年 . 西安城市结构布局形态分析 . 西安建筑科技大学学报 (自然科学版), 2005, 24(3):26–28.)

[87] GINSBURG N. Putting the social into urban regeneration policy [M]. Local Economy, 1999, 14(1):55–71.

[88] GLASS R. London: aspects of change [M]. London: MacGibbon and Kee, 1964.

[89] GOLD T B. After Comradeship: Personal Relations in China since the Cultural Revolution [J]. The China Quarterly, 1985, 104:657–675.

[90] GOODWIN M and PAINTER J. Local Governance, the Crises of Fordism and the Changing Geographies of

Regulation [J]. Transactions of the Institute of British Geographers, 1996, 21(4):635–648.

[91] GOULBOURNE H and SOLOMOS J. Families, Ethnicity and Social Capital [J]. Social Policy and Society, 2003, 2(04):329–338.

[92] GRANOVETTER M. Strength of Weak Ties: A Network Theory Revisited [J]. Sociological Theory, 1983, 1:201–233.

[93] GRANOVETTER M. Economic Action and Social Structure: The Problem of Embeddedness [J]. The American Journal of Sociology, 1985, 91(3):481–510.

[94] GROOTAERT C. Social Capital: the missing link [M]. Washington, D.C.: Social Development Department, The World Bank, 1998.

[95] GROOTAERT C, NARAYAN D, JONES V N and WOOLCOCK, M. Measuring Social Capital: an integrated questionnaire [M]. Washington, DC: World Bank, 2004.

[96] HAKANSSON H and JOHANSON J. The Embedded firm: on the socioeconomics of industrial networks [M] // G GRABHER. London: Routledge, 1993: 35–51.

[97] HAMBLETON R. Regeneration of U.S. and British Cities [J]. Local government studies, 1991, 17:53–69.

[98] HAMDI N and MAJALE M. Partnerships in Urban Planning: a guide for municipalities [M]. S.l.: Practical Action Publishing, 2005.

[99] HAN P. Basic characteristics of human relationships of market economy [J]. Behavioural science, 1995, 1:44–45. (韩平. 市场经济条件下人际关系的基本特征. 行为科学, 1995, 1:44–45.)

[100] HANIFAN L J. Rural School Community Center [J]. Annals of the American Academy of Political and Social Science, 1916, 67:130–138.

[101] HANNIGAN J A. Gentrification [J]. Current Sociology, 1995, 43(1):173–182.

[102] HARRIS P B, WERNER C M, BROWN B B and INGEBRITSEN D. Relocation and privacy regulation: a cross-cultural analysis [J]. Journal of Environmental Psychology, 1995, 15(4):311–320.

[103] HAUS M and HEINELT H. How to achieve governability at the local level?: theoretical and conceptual considerations on a complementarity of urban leadership and community involvement [M] // M. HAUS, H. HEINELT and M. STEWART. Urban Governance and Democracy: leadership and community involvement. London: Routledge, 2005: 12–39.

[104] HE H X. Ancient capital Xi'an's integrative environment harmony and analysis [J]. Architectural Journal, 2002, 5:49–51. (和红星. 古城西安整体环境的协调与分析. 建筑学报, 2002, 5:49–51.)

[105] HE H X. Report on the implementation situation of Xi'an 1995–2010 Master Layout Plan [EB/OL]. (2003a) [2007-06-20] http://www.china-xa.gov.cn/ShowArticle.asp?ArticleID=234. (和红星. 关于西安市1995—2010年城市总体规划执行情况的汇报, 2003.)

[106] HE H X. A Few Explanations About Xi'an's 2004–2020 Urban Master Planning [EB/OL]. (2003b) [2007-06-20] http://www.china-xa.gov.cn/printpage.asp?ArticleID=1280. (和红星. 西安市2004—2020年城市总体规划修编若干问题的说明, 2003.)

[107] HE H X. Giving Prominence to the Culture of the Ancient Capital of Xi'an and Creating the Spirit of a Modern City—Major Concern in the Fourth Revision of Xi'an's Master Plan [J]. Planners, 2004, 20(12):13–15. (和红星. 凸现西安古都文化 营造现代城市精神. 规划师, 2004, 20(12):13–15.)

[108] HE H X. Breaking the Uniform Urban Appearances and Creating the Attractive Ancient-capital-city [EB/OL].

(2005) [2007–06–20] http://www.xasohu.com/enews/news/200509/03/0007_0000045190.shtml. (和红星 . 西安模式破千城一面 造魅力古都 , 2005.)

[109] HE H X. The Special Capital City of Xi'an [M]. Beijing: China Architecture and Building Press, 2006. (和红星 . 特色城市：古都西安 . 北京：中国建筑工业出版社 , 2006.)

[110] HEALEY P. Urban Regenenration and the Development Industry [J]. Regional Studies, 1991, 25(2):97–110.

[111] HEALEY P. Collaborative planning: shaping places in fragmented societies [M]. Basingstoke, England; New York: Palgrave Macmillan, 2006.

[112] HIBBITT K, JONES P and MEEGAN R. Tackling social exclusion: The role of social capital in urban regeneration on Merseyside—From mistrust to trust [J]? European Planning Studies, 2001, 9(2):141.

[113] HOLCOMB H B and BEAUREGARD R A. Revitalizing cities [M]. Washington, D.C.: Association of American Geographers, 1981.

[114] HOYEM H. Drum Tower Muslim District Project Report [R]. Xi'an: Xi'an Jiaotong University, The Norwegian Institute of Technology, 1989.

[115] HOYEM H. Epilogue [M] // B. ERRING, H. HØYEM and S. VINSRYGG. The horizontal skyscraper. Trondheim: Tapir Academic Press, 2002: 385–386.

[116] HU A G. Objectives and Political Suggestions for Tenth Five–Year–Plan and Sustainable Development Strategies before 2010 [EB/OL]. (2001) [2007–12–12] http://cssd.acca21.org.cn/news0801d.html#5. (胡鞍钢 . "十五" 计划和 2010 年规划可持续发展战略的目标和政策建议 , 2001.)

[117] HU S. Synopsis of Chinese Phylosophy Histories [M]. Shanghai: Shangwu Press, 1936. (胡适 . 中国哲学史大纲 . 上海：商务出版社， 1936.)

[118] HUA L H. Reconstruire la Chine [M]. Beijing: SDX Joint Publishing Company, 2006. (华揽洪 . 重建中国：城市规划三十年 1949–1979. 北京：生活・读书・新知三联书店 , 2006.)

[119] HUASHANGNEWS. Making Xi'an ancient music be continued [EB/OL]. (2004) [2009–03–04] http://news.hsw.cn/2004–09/13/content_1270927.htm. (华商报 . 不让西安古乐后继无人 老乐手教新人识古谱， 2004.)

[120] HUGHES J and CARMICHAEL P. Building partnerships in urban regeneration: A case study from Belfast [J]. Community Development Journal, 1998, 33(3):205–225.

[121] HUTCHINSON J. Social capital and community building in the inner city [J]. Journal of the American Planning Association, 2004, 70(2):168–175.

[122] IAUS TSINGHUA UNIVERSITY. Old City's Renewal Planning, Design and Research [M]. Beijing: Tsinghua University Press, 1993. (清华大学建筑与城市研究所 . 旧城改造规划、设计、研究 . 北京：清华大学出版社 , 1993.)

[123] ICOMOS. Charter for the Conservation of Historic Towns and Urban Areas [EB/OL]. (1987) [2006–11–23] http://www.international.icomos.org/charters.htm.

[124] ICOMOS. The Nara Document on Authenticity[EB/OL]. (1994) [2008–12–05] http://www.international.icomos.org/naradoc_eng.htm.

[125] ICOMOS. Burra Charter [EB/OL]. (1999) [2007–01–20] http://www.icomos.org/australia/burra.html.

[126] ICOMOS. Principles for the Conservation of Heritage Sites in China [M]. Los Angeles, CA: The Getty Conservation Institute, 2000.

[127] IMON S S. Integrated Conservation: an introduction [C]. Paper presented at the The First Asian Academy Field School "Conserving Asia's Built Heritage: an integrateed management approach", 2003.

[128] IMON S S. Sustainable Urban Conservation: the role of public participation in the conservation of urban heritage in old Dhaka [D]. Unpublished PhD Thesis, The University of Hong Kong, Hong Kong, 2006.

[129] IMRIE R and RACO M. Community and the changing nature of urban policy [M] // M RACO and R IMRIE. Urban Renaissance? : New Labour, community and urban policy. Bristol: Policy Press, 2003: 3–36.

[130] ISHIDA Y. The concept of machi-sodate and urban planning: the case of Tokyu Tama Den'en Toshi [M] // A SORENSEN and C FUNCK. Living cities in Japan: citizens' movements, machizukuri and local environments. London; New York: Routledge, 2007: 115–136.

[131] JESSOP B. The rise of governance and the risks of failure: the case of economic development [J]. International Social Science Journal, 1998, 50:29–45.

[132] JIA H Y. On Chinese Historically and Culturally Famous Cities [M]. Nanjing: Southeast University Press, 2007. (贾鸿雁 . 中国历史文化名城通论 . 南京 : 东南大学出版社 , 2007.)

[133] JIN M J and XU X. Construction of Urban Culture and Conservation of Historic Cities [J]. Urban Studies, 2000, 6:67–69. (金鸣娟 , 徐鑫 . 城市文化建设与历史文化名城保护 . 城市发展研究 , 2000, 6:67–69.)

[134] JOHNSON J H. Urbanisation [M]. London: Macmillan Education, 1990.

[135] JU J P. Study on the planning of conservation and renewal of the historic districts in Xi'an [D]. Unpublished Master Thesis, Xi'an University of Architecture and Technology, Xi'an, 2005. (巨荩蓬 . 西安老城区历史街区保护与更新规划构想 . 西安 : 西安建筑科技大学 , 2005.)

[136] KASARDA J D and JANOWITZ M. Community Attachment in Mass Society [J]. American Sociological Review, 1974, (39)3:328–339.

[137] KEANE M P. Turning Kyoto into kindling [J]. Architecture, 2000, 89(5):75–79, 213.

[138] KEARNS A. Social capital, regeneration and urban policy [M] // M RACO and R IMRIE. Urban Renaissance? : New Labour, community and urban policy. Bristol: Policy Press, 2003: 37–60.

[139] KHOURI-DAGHER N. World Heritage: Living places managed by local people [J]. UNESCO Sources, 1999, 115:10–11.

[140] KING Y C A. From Traditionalism to Modernism [M]. Beijing: China Renmin University Press, 1999. (金耀基 . 从传统到现代 . 北京 : 中国人民大学出版社 , 1999.)

[141] KLEINHANS R, PRIEMUS H and ENGBERSEN G. Understanding Social Capital in Recently Restructured Urban Neighbourhoods: Two Case Studies in Rotterdam [J]. Urban Studies, 2007, 44(5/6):1069–1091.

[142] KOBAYASHI S. Machizukuri ordinances in the era of decentralization (Chiho bunken jidai no machizukuri jorei) [M]. Kyoto: Gakugei Shuppansha, 1999.

[143] KONG L and YEOH B S A. Urban conservation in Singapore: a survey of state policies and popular attitudes [J]. Urban Studies, 1994, 31(2):247–265.

[144] KRISHNA A. Enhancing Political Participation in Democracies—What is the Role of Social Capital [J]? Comparative Political Studies, 2002, 35(4):437–460.

[145] KURODA T. Urban Scenario in Japan in the New Millenium [M] // G D NESS and P P TALWAR. Asian urbanization in the new millennium. Singapore: Marshall Cavendish Academic, 2005: 384–402.

[146] LAI L W C. Hong Kong New Port and Airport Development Strategy [J]. Ekistics, 1990, 57:79–87.

[147] LAUMANN E O. Bonds of pluralism: the form and substance of urban social networks [M]. New York: J. Wiley, 1973.

[148] LEAF M and HOU L. The "Third Spring" of Urban Planning in China: The Resurrection of Professional Planning in the Post-Mao Era [J]. China Information, 2006, 20(3):553-585.

[149] LÉAUTIER F. Cities in a globalizing world : governance, performance, and sustainability [M]. Washington, DC: World Bank, 2006.

[150] LEE S L. Urban conservation policy and the preservation of historical and cultural heritage: the case of Singapore [J]. Cities, 1996, 13(6):399-409.

[151] LEES L. Visions of 'urban renaissance': the Urban Task Force report and the Urban White Paper [M] // M Raco and R Imrie. Urban Renaissance?: New Labour, community and urban policy. Bristol: Policy Press, 2003: 61-82.

[152] LI H B. Decrease of Social Capital [J]. Marxism and Reality, 1999a, 03:65-68. (李惠斌 . 社会资本的衰落 . 马克思主义与现实 , 1999, 3:65-68.)

[153] LI H Y. Impact of Land Tenure Transformation on Physical Development of Drum-Tower Muslim District, Xi'an China [D]. Unpublished Master Thesis, The Norwegian University of Science and Technology, 2002a.

[154] LI H Y. Xi'an Urban Fabric and Layout Pattern [J]. Beijing City Planning and Construction Review, 2005, 6:104-107. (李红艳 . 西安城市结构与布局形态 . 北京规划建设 , 2005, 6:104-107.)

[155] LI J B. A History on Sanqin: Xi'an Muslim and Mosques [M]. Xi'an: Sanqin Press, 2004. (李健彪 . 三秦史話 : 西安回族与清真寺 . 西安 : 三秦出版社 , 2004.)

[156] LI L H. Urban land reform in China [M]. Basingstoke, Hampshire: New York: Macmillan; St. Martin's Press, 1999b.

[157] LI L M. Fundamental Study on the Preservation of Xi'an, the Historic and Cultural City [J]. Human Geography, 2002b, 17(5):25-28. (李骊明 . 关于西安历史文化名城保护的战略思考 . 人文地理 , 2002, 17(5):25-28.)

[158] LI L Q and YIN J. Practices and Considerations on Constructing Harmonious Xi'an [J]. Commercial Times, 2006, 36:81. (李丽琴 , 尹洁 . 构建和谐西安的实践与思考 . 商业时代 , 2006, 36:81.)

[159] LI P L. Again On the Invisible Hand [J]. Sociological Research, 1994, 1:11-18. (李培林 . 再论 "另一只看不见的手" . 社会学研究 , 1994, 1:11-18.)

[160] LI Q R. Urban planning and historic preservation [M]. Nanjing: Southeast China Press, 2003. (李其荣 . 城市规划与历史文化保护 . 南京 : 东南大学出版社 , 2003.)

[161] LI R Q. Methods and Measures to Conserve Intangible Cultural Heritage [M] // W Z WANG, Q S ZHANG and S D MA. A collection of papers of the Forum on the Protection of Chinese Intangible Cultural Heritage. Beijing: Culture and Art Publishing House, 2006: 237-248. (李荣启 . 论非物质文化遗产保护的方法与措施 // 中国非物质文化遗产保护论坛论文集 . 北京 : 文化艺术出版社 , 2006: 237-248.)

[162] LI S D. Definition and conservation of intangible cultural heritage [M] // DEVELOPMENT CENTRE OF ETHNIC AND FOLK LITERATURE AND ART. Research on the Conservation of Intangible Cultural Heritage of China. Beijing: Beijing Normal University Press, 2007a: 42-53. (李顺德 . "非物质文化遗产" 的界定及保护 .《中国非物质文化遗产保护研究》. 北京 : 北京师范大学出版社 , 2007: 42-53.)

[163] LI Z N. The economics of China's unemployment: the inside story [M]. Hong Kong: New Century Publishing House, 2007b. (李志宁 . 中国失业困境的背后 . 香港 : 新世纪出版社 , 2007.)

[164] LIANG S M. Essentials on Chinese Culture [M]. Shanghai: Shanghai People's Publishing House, 2011. (梁漱溟 . 中国文化要义 . 上海：上海人民出版社 , 2011.)

[165] LIN B and WANG H S. Transitional Urban District and Its Economy: the sociological research of a typical urban district [M]. Beijing: China Social Science Publication, 2002. (林彬，王汉生 . 变迁中的城区政府与区街经济：一个典型城区的社会学研究 . 北京：中国社会科学出版社 , 2002.)

[166] LIN N. Building a Network Theory of Social Capital [J]. Connections, 1999a, 22(1):28–51.

[167] LIN N. Social networks and status attainment [J]. Annual Review of Sociology, 1999b, 25:467–487.

[168] LIN N and BIAN Y J. Getting Ahead in Urban China [J]. The American Journal of Sociology, 1991, 97(3):657–688.

[169] LIN N and DUMIN M. Access to occupations through social ties [J]. Social Networks, 1986, 8(4):365–385.

[170] LIN N and ERICKSON B H. Theory, Measurement, and the Research Enterprise on Social Capital [M] // N LIN and B H ERICKSON. Social capital: an international research program. Oxford ; New York: Oxford University Press, 2008: 1–24.

[171] LIN T. Renewal of Modern Cities and Vicissitudes of Social Space: House, Ecology, Harness [M]. Shanghai: Shanghai Guji Publishing House, 2007. (林拓 . 现代城市更新与社会空间变迁：住宅、生态、治理 . 上海：上海古籍出版社 , 2007.)

[172] LIU L A. An experimental mode of the conservation of historic cities and the revival of traditional culture – project practices of Shuyuanmen historic district in Xi'an city [M] // XI'AN URBAN AND RURAL CONSTRUCTION COMMITTEE and RESEARCH INSTITUTE OF XI'AN HISTORIC CITY. On Xi'an Urban Characteristics. Xi'an: Shanxi People's Publishing House, 2006: 78–86. (刘临安 . 历史城市保护与传统文化复兴的实验模式—西安市书院门历史街区的工程实践 // 论西安城市特色 . 西安：陕西人民出版社 , 2006:78–86.)

[173] LIU L P. Social capital in enterprises: conceptual rethinking and assessment methods [J]. Sociological Studies, 2006, 2:204–216. (刘林平 . 企业的社会资本：概念反思和测量途径——兼评边燕杰、丘海雄的《企业的社会资本及其功效》. 社会学研究 , 2:204–216.)

[174] LIU P L, LIU A and WALL G. Eco-Museum Conception and Chinese Application –A Case Study in Miao Villages, Suoga, Guizhou Province [J]. Resources and Environment in the Yangtze Basin, 2005, 14(2):254–257. (刘沛林，Abby Liu, Geoff Wall. 生态博物馆理念及其在少数民族社区景观保护中的作用—以贵州梭嘎生态博物馆为例 . 长江流域资源与环境 , 2005, 14(2):254–257.)

[175] LIU Y. The third climax of unemployment: lay-off, unemployment, re-employment [M]. Beijing: China Book Press, 1998. (刘拥 . 第三次失业高峰：下岗、失业、再就业 . 北京市：中国书籍出版社 , 1998.)

[176] LIU Y W. Analysis on the traditional historical district-Xiyangshi Street in Xi'an [J]. Shanxi Architecture, 2004, 14:13–14. (刘永望 . 谈西安传统历史街区内西羊市街 . 山西建筑 , 2004, 14: 13–14.)

[177] LONDON B, LEE B A and LIPTON S G. Determinants of gentrification in the United State: a city-level analysis [J]. Urban Affairs Quarterly, 1986, 21(3):369–387.

[178] LOWNDES V and WILSON D. Social Capital and Local Governance: Exploring the Institutional Design Variable [J]. Political Studies, 2001, 49(4).

[179] LU F. Workunit: A Type of Special Social Life Scope [J]. Social Science in China, 1989, 1:71–88. (路风. 单位：一种特殊的社会组织形式 . 中国社会科学 , 1989, 1: 71–88.)

[180] LU Y. The History of Beilin: Looking back thousand years of vicissitudes [M]. Xi'an: Xi'an Publishing House, 2000. (路远 . 碑林史话 : 回首千年沧桑 . 西安 : 西安出版社 , 2000.)

[181] LU Y. Protection of the Inferior Group in Plan Making Processes of City Renewal [J]. Modern Urban Research, 2005, 11:22–26. (卢源 . 论旧城改造规划过程中弱势群体的利益保障 . 现代城市研究 , 2005, 11: 22–26.)

[182] LUO J D. Can Guanxi be social capital [M]? // X H ZHOU. Social network and enterprise development. Beijing: Economy and Management Publishing House, 2008: 315–317. (罗家德 . 关系能构成社会资本吗？《社会网络与企业成长》. 北京 : 经济管理出版社 , 2008:315–317.)

[183] LUO Y D. Guanxi and business [M]. Singapore; River Edge, N.J.: World Scientific, 2000.

[184] LUO Z W. Latest Development of the Conservation and Construction of China's Historic Cities [M] // HISTORIC CITIES STUDY SOCIETY IN CHINA. Conservation and Construction of China's Historic Cities. Beijing: Cultural Relics Press, 1987: 58–64. (罗哲文 . 中国历史文化名城保护与建设的新发展 // 中国历史文化名城保护与建设 . 北京 : 文物出版社 , 1987:58–64.)

[185] MA J C L. Economic reforms, urban spatial restructuring, and planning in China [J]. Progress in Planning, 2004, 61(3):237–260.

[186] MA L J C. Chinese Approach to City Planning: Policy, Administration, and Action [J]. Asian Survey, 1979, 19(9):838–855.

[187] MA L J C. Urban transformation in China 1949–2000: a review and research agenda [J]. Environment and Planning A, 2002, 34:1545–1569.

[188] MA P. Regarding the restrictions on Muslim women marriages [M] // XI'AN ISLAMIC CULTURE RESEARCH INSTITUTE. Research on Islamic Culture: proceedings of the second symposium on Xi'an Islamic culture. Yinchuan: Ningxia Renmin Publication, 1998a: 330–342. (马平 . 回族婚姻择偶中的 "妇女外嫁禁忌" // 伊斯兰文化研究 : 第二届西安市伊斯兰文化研讨会论文汇编 . 银川 : 宁夏人民出版社 , 1998: 330–342.)

[189] MA P. Muslim Psychological–consciousness and Behaviour–modes [M]. Yinchuan: Ningxia Renmin Publication, 1998b. (马平 . 回族心理素质与行为方式 . 银川 : 宁夏人民出版社 , 1998.)

[190] MA Z L. Xi'an: the Metropolis of North–West China [M] // F S V SIT. Chinese cities: the growth of the metropolis since 1949. Oxford: Oxford University Press, 1985: 148–166.

[191] MARMON D. Planning and public participation [J]. Urban Planning Overseas, 1995, 1:41–50. (大卫・马门 . 规划与公众参与 . 国外城市规划 . 1995, 1:41–50.)

[192] MASAFUMI Y and WALEY P. Kyoto and the Preservation of Urban Landscapes [M] // N FIÉVÉ and P WALEY. Japanese capitals in historical perspective: place, power and memory in Kyoto, Edo and Tokyo. London: RoutlegeCurzon, 2003: 347–366.

[193] MAXWELL J A. Qualitative research design: an interactive approach [M]. Thousand Oaks, Calif.: Sage Publications, 1996.

[194] MAY N. Challenging assumptions: gender issues in urban regeneration [M]. York, England: YPS for the Joseph Rowntree Foundation, 1997.

[195] MC (Ministry of Construction of China). Code of conservation planning for historic cities [M]. Beijing: China Architecture and Building Press, 2005. (中华人民共和国建设部 . 历史文化名城保护规划规范 . 北京 : 中国建筑工业出版社 , 2005.)

[196] MCCALLISTERL AND FISCHER C S. A Procedure for Surveying Personal Networks [J]. Sociological Methods

and Research, 1978, 7(2):131-148.

[197] MCCARNEY P, HALFANI M and RODRIGUEZ A. Towards an Understanding of Governance [M] // R STREN and J K BELL. Urban research in the developing world. Toronto: Centre for Urban and Community Studies, University of Toronto, 1995, 4: 93-141.

[198] MCKIE R. Cellular Renewal—a policy for the older housing areas [J]. Town Planning Review, 1974, 45(3):274-290.

[199] MIDDLETON A, MURIE A and GROVES R. Social capital and neighbourhoods that work [J]. Urban Studies, 2005, 42(10):1711-1738.

[200] MIMURA H, KANKI K and KOBAYASHI F. Urban Conservation and Landscape Management: The Kyoto Case [M] // G Golany, K Hanaki and O Koide. Japanese urban environment. Oxford: Elsevier, 1998: 39-56.

[201] MISZTAL B and SHUPE A. Fundamentalism and Globalization: Fundamentalist Movements at the Twilight of the Twentieth Century [M] // A D SHUPE and B MISZTAL. Religion, mobilization, and social action. Westport, Conn.: Praeger, 1998: 3-14.

[202] MISZTAL B A. Trust in modern societies: the search for the bases of social order [M]. Cambridge: Polity Press, 1996.

[203] MIURA K. Conservation of a 'living heritage site': a contradiction in terms? A case study of Angkor World Heritage Site [J]. Conservation and Management of Archaeological Sites, 2005, 7(01):3-18.

[204] MLR (Ministry of Land and Resource of China). A Pictorial Collection of Urban Land Prices in Chinese Cities [M]. Beijing: China Maps Press, 2003. (中华人民共和国国土资源部. 中国城市地价图集. 北京：地图出版社, 2003.)

[205] MURAYAMA A. Civic movement for sustainable urban regeneration: downtown Fukaya city, Saitama prefecture [M] // A SORENSEN and C FUNCK. Living cities in Japan: citizens' movements, machizukuri and local environments. London; New York: Routledge, 2007: 206-223.

[206] MUTCHLER S E, MAYS J L and POLLARD J S. Finding Common Ground: Creating Local Governance Structures [M]. Austin, TX.: Southwest Educational Development Lab, 1993.

[207] NDRC (National Development and Reform Commission). Outline of the Eleventh Five-Year Plan [EB/OL]. (2007) [2008-10-30] http://en.ndrc.gov.cn/hot/.

[208] NG M K. Global Overview on Urban Regeneration Practices: Economic Development, Governance, Planning Mode and Regeneration Concerns in Western and Asian Contexts [C]. Paper presented at the Continuing Professional Development Workshop, 1998.

[209] NG M K. Viewpoint: the role of urban planning in China's sustainable development [J]. Town Planning Review, 2004, 75(4):i-v.

[210] NG M K. Quality of Life Perceptions and Directions for Urban Regeneration in Hong Kong [J]. Social Indicators Research, 2005a, 71(1-3):441-465.

[211] NG M K. Sustainable urban regeneration of a decaying urban community facing new uncertainty—the experience of University-led community envisioning in Hong Kong [C]. Paper presented at the International Community Planning Forum, 2005b.

[212] NG M K, COOK A and CHUI E W T. Road Not Travelled: A Sustainable Urban Regeneration Strategy for Hong Kong [J]. Planning Practice and Research, 2001, 16(2):171-183.

[213] NG M K and TANG W S. Politics of urban regeneration in Shenzhen, China: a case study of Shangbu industrial district [M]. Hong Kong: Centre for China Urban and Regional Studies, Hong Kong Baptist University, 2002a.

[214] NG M K and TANG W S. Urban regeneration with Chinese characteristics: a case study of the Shangbu Industrial District, Shenzhen, China [J]. Journal of East Asian Studies, 2002b, 1:29–54.

[215] NG M K and TANG W S. Role of Planning in the Development of Shenzhen, China: Rhetoric and Realities [J]. Eurasian Geography and Economics, 2004a, 45(3):190–211.

[216] NG M K and TANG W S. Theorising urban planning in a transitional economy – the case of Shenzhen, People's Republic of China [J]. Town Planning Review, 2004b, 75(2):173–203.

[217] NG M K and WU F L. A Critique of the 1989 City Planning Act of the People's Republic of China [J]. Third world planning review, 1995, 17:279–293.

[218] NUNOKAWA H. Machizukuri and historical awareness in the old town of Kobe [M] // A SORENSEN and C FUNCK. Living cities in Japan: citizens' movements, machizukuri and local environments. London; New York: Routledge, 2007: 172–186.

[219] OI J C. Role of the Local State in China's Transitional Economy [J]. The China Quarterly, 1995, 144:1132–1149.

[220] PAN G X. A History of Chinese Architecture [M]. Beijing: Chinese Architectural Industrial Press, 2004. (潘谷西. 中国建筑史. 北京：中国建筑工业出版社, 2004.)

[221] PEARCE G. Conservation as a component of urban regeneration [J]. Regional Studies, 1994, 28(1):88–93.

[222] PORTES A. Social capital: its origins and applications in modern sociology [J]. Annual Review of Sociology, 1998, 24:1–24.

[223] PRI (Policy Research Initiative). Measurement of Social Capital: Reference Document for Public Policy Research, Development, and Evaluation [EB/OL]. (2005a) [2007–12–29] http://policyresearch.gc.ca/doclib/Measurement_E.pdf.

[224] PRI (Policy Research Initiative). Social Capital as a Public Policy Tool Project Report [EB/OL]. (2005b) [2007–10–26] http://policyresearch.gc.ca/doclib/SC_Synthesis_E.pdf.

[225] PRI (Policy Research Initiative). Social Capital: a tool for public policy [EB/OL]. (2005c) [2007–10–26] http://policyresearch.gc.ca/doclib/R4_PRI%20SC%20Briefing%20Note_E.pdf.

[226] PUTNAM R D. Prosperous Community: Social Capital and Public Life [J]. The American Prospect 1993, 13:35–42.

[227] PUTNAM R D. Bowling alone: America's declining social capital [J]. Journal of Democracy, 1995a, 6(1):65–78.

[228] PUTNAM R D. Tuning In, Tuning Out: The Strange Disappearance of Social Capital in America [J]. Political Science and Politics, 1995b, 28(4):664–683.

[229] PUTNAM R D. Foreword [J]. Housing Policy Debate, 1998, 9(1):v–viii.

[230] PUTNAM R D. Bowling alone: the collapse and revival of American community [M]. New York: Simon and Schuster, 2000.

[231] PUTNAM R D, LEONARDI R and NANETTI R. Making democracy work: civic traditions in modern Italy [M]. Princeton, N.J.: Princeton University Press, 1993.

[232] QIAN X R, LI Y B and MAO W L. Public Participation in Urban Planning in China [J]. Journal of Zhejiang

Vocational and Technical Institute of Transportation, 2004, 4:65–67. (钱晓如 , 李耀斌 , 毛雯丽 . 我国城市规划中的公众参与 . 浙江交通职业技术学院学报 , 2004, 4: 65–67.)

[233] QIAO J J and MAI Y Y. A scientific way to measure social capital [J]. Journal of Wuhan Textile S. and T. Institute, 2003, 16(5):89–92. (乔俊杰 , 买忆媛 . 社会资本的科学测量方法 . 武汉科技学院学报 , 2003, 16(5): 89–92.)

[234] REID J N. Community Participation—How People Power Brings Sustainable Benefits to Communities [EB/OL]. (2000) [2007–08–27] http://www.rurdev.usda.gov/rbs/ezec/Pubs/commparticrept.pdf.

[235] RENMINNEWS. Intangible cultural heritage: Xi'an ancient music [EB/OL]. (2006) [2009–03–04] http://culture.people.com.cn/BIG5/22226/4210570.html. (人民网 . 非物质文化遗产 : 西安古乐 , 2006.)

[236] RIDLEY N. Local Right: enabling not providing [M]. London: Centre for Policy Studies, 1988, 92.

[237] RIVIÈRE G H. The Ecomuseum: an evolutive definition [J]. Museum, 1985, 37(4):182–183.

[238] ROBERTS P. Evolution, Definition and Purpose of Urban Regeneration [M] // P W ROBERTS and H SYKES. Urban regeneration: a handbook. London: SAGE Publications, 2000: 9–36.

[239] RUAN Y S. Conservation and planning in historic districts [J]. Urban Planning Forum, 2000, 2:46–50. (阮仪三 . 历史街区的保护及规划 . 城市规划汇刊 , 2000, 2: 46–50.)

[240] RUAN Y S. Conseration and Reuse of Historic Cities in China [J]. China Cultural Heritage Scientific Research, 2007, 3:12–18. (阮仪三 . 中国历史古城的保护与利用 . 中国文物科学研究 , 2007, 3:12–18.)

[241] RUAN Y S and SUN M. Conservation and planning in historical streets areas [J]. City Planning Review, 2001, 25(10):25–32. (阮仪三 , 孙萌 . 我国历史街区保护与规划的若干问题研究 . 城市规划 , 25(10): 25–32.)

[242] RUAN Y S and XIANG B J. Preservation and Renewal of Traditional Living Blocks in Suzhou [J]. Urban Planning Review, 1997, 4:45–49. (阮仪三 , 相秉军 . 苏州古城街坊的保护与更新 . 城市规划汇刊 , 1997, 4: 45–49.)

[243] RUSTON D. Social Capital Matrix of Surveys [M]. U.K.: Social Analysis and Reporting Division, Office of National Statistics, 2002.

[244] RYOICHI K. Preservation and Revitalization of machiya in Kyoto [M] // N FIÉVÉ and P WALEY. Japanese capitals in historical perspective: place, power and memory in Kyoto, Edo and Tokyo. London: RoutlegeCurzon, 2003: 367–384.

[245] SACH (State Administration of Cultural Heritage). Cultural Heritage Atlas of China: Shaanxi [M]. Xi'an: Xi'an Map Publishing House, 1998. (国家文物局 . 中国文物地图集 : 陕西分册 . 西安 : 西安地图出版社 , 1998.)

[246] SAEGERT S and WINKEL G. Social Capital and the Revitalization of New York City's Distressed Inner-City Housing [J]. Housing policy debate, 1998, 9(1):17–60.

[247] SAICH T. Governance and politics of China [M]. Basingstoke, Hampshire; New York, N.Y.: Palgrave, 2001.

[248] SALASTIE R R. Living Tradition or Panda's Cage? An Analysis of the Urban Conservation in Kyoto. Case Study: 35 Yamahoko Neighbourhoods [M]. Helsinki: Helsinki University of Technology, 1999.

[249] SBBD (Statistics Bureau of Beilin District). National Economy Statistics of Beilin District of Xi'an in 2004 [M]. Xi'an: Statistics Bureau of Beilin District, 2005. (碑林区统计局 . 西安市碑林区 2004 年国民经济统计资料 . 西安 : 碑林区统计局 , 2005.)

[250] SBLD (Statistics Bureau of Lianhu District). Census Statistics of Lianhu District of Xi'an in 2000 [M]. Xi'an: Statistics Bureau of Lianhu District, 2003. (莲湖区统计局 . 西安市莲湖区 2000 年人口普查资料 . 西安 :

莲湖区统计局, 2003.)

[251] SCC (State Council of China). Agenda 21 of China [M]. Beijing: Environment and Science Press (China), 1994. (国务院 . 中国 21 世纪议程：中国 21 世纪人口、环境与发展白皮书 .)

[252] SCC (State Council of China). Opinions on intensifying the conservation of intangible cultural heritage in China [EB/OL]. (2005a) [2008-11-24] http://www.law-lib.com/law/law_view.asp?id=91926. (国务院 . 国务院办公厅关于加强我国非物质文化遗产保护工作的意见, 2005.)

[253] SCC (State Council of China). Regulation on religious affairs [EB/OL]. (2005b) [2010-03-22] http://www.sara.gov.cn/GB/zcfg/116b855c-0581-11da-adc6-93180af1bb1a.html. (国务院 . 宗教事务条例, 2005.)

[254] SCC (State Council of China). Conservation Regulation for Historic Cities, Towns and Villages [EB/OL]. (2008a) [2008-11-26] http://www.gov.cn/zwgk/2008-04/29/content_957280.htm. (国务院 . 历史文化名城名镇名村保护条例, 2008.)

[255] SCC (State Council of China). Official reply to Xi'an's fourth master plan by the State Council of China [EB/OL]. (2008b) [2008-05-19] http://www.gov.cn/zwgk/2008-05/09/content_965572.htm. (国务院 . 国务院关于西安市城市总体规划的批复, 2008.)

[256] SCFPC (Standing Committee of Fuzhou People's Congress). Conservation Regulations on Fuzhou Historic City [EB/OL]. (1997) [2008-11-26] http://www.cpll.cn/law1406.html. (福州市人大常委会 . 福州市历史文化名城保护条例, 1997.)

[257] SCGPC (Standing Committee of Guangzhou People's Congress). Conservation Regulations on Guangzhou Historic City [EB/OL]. (1999) [2008-11-26] http://www.gzwh.gov.cn/whw/channel/whzwgk/zcfg/wwmsp/5.htm. (广州市人大常委会 . 广州历史文化名城保护条例, 1999.)

[258] SCHMITTER P C. Parcipatory Governance Arrangements: is there any reason to expect it will achieve "sustainable and innovative policies in a multilevel context"？[M] // J GROTE and B GBIKPI. Participatory Governance: political and societal implications. Opladen: Leske+Budrich, 2002: 51-70.

[259] SCNPC (Standing Committee of National People's Congress). Land Administration Act in P.R.China [EB/OL]. (2004a) [2007-06-20] http://news.xinhuanet.com/zhengfu/2004-08/30/content_1925451.htm. (全国人民代表大会常务委员会 . 中华人民共和国土地管理法, 2004.)

[260] SCNPC (Standing Committee of National People's Congress). Ratification of the UNESCO Convention for the Safeguarding of the Intangible Cultural Heritage by the Standing Committee of National People's Congress [EB/OL]. (2004b) [2008-11-24] http://www.66law.cn/channel/lawlib/2005-06-14/89753.aspx?highlight=1andkeyword=%D2%C5%B2%FA. (全国人民代表大会常务委员会 . 全国人民代表大会常务委员会关于批准《保护非物质文化遗产公约》的决定, 2004.)

[261] SCNPC (Standing Committee of National People's Congress). Law of the People's Republic of China on Protection of Cultural Relics [EB/OL]. (2007) [2008-11-25] http://www.gov.cn/flfg/2007-12/29/content_847433.htm. (全国人民代表大会常务委员会 . 中华人民共和国文物保护法, 2007.)

[262] SCNPC (Standing Committee of National People's Congress). The Urban and Rural Planning Law of the People's Republic of China [EB/OL]. (2008) [2009-07-18] http://www.lawinfochina.com/law/display.asp?id=6495andkeyword=. (全国人民代表大会常务委员会 . 中华人民共和国城乡规划法, 2008.)

[263] SCNPC (Standing Committee of National People's Congress). City Planning Act of the Peopl'e Republic of China [J]. Building in China, 1990, 3(3):2-9.

[264] SCXPC (Standing Committee of Xi'an People's Congress). Xi'an Regulations on the Urban Building-Height Control [M] // EDITORIAL BOARD OF XI'AN YEARBOOK. Xi'an Yearbook. Xi'an: Xi'an Publishing House, 1993: 376–377. (西安市人民代表大会常务委员会. 西安市控制市区建筑高度的规定 // 西安年鉴. 西安：西安出版社, 1993: 376–377.)

[265] SCXPC (Standing Committee of Xi'an People's Congress). Conservation Regulations on Xi'an Historic City [EB/OL]. (2002) [2007-06-20] http://www.xafz.gov.cn/ReadNews.asp?NewsID=533. (西安市人民代表大会常务委员会. 西安市历史文化名城保护条例, 2002.)

[266] SCXPC (Standing Committee of Xi'an People's Congress). Xi'an Urban Housing Removal Management Method [EB/OL]. (2003) [2008-06-29] http://www.china-xa.gov.cn/ShowArticle.asp?ArticleID=567. (西安市人民代表大会常务委员会. 西安市城市房屋拆迁管理办法, 2003.)

[267] SCXPC (Standing Committee of Xi'an People's Congress). Xi'an Urban Planning Management Regulations [EB/OL]. (2005) [2007-06-20] http://www.china-xa.gov.cn/ShowArticle.asp?ArticleID=1364. (西安市人民代表大会常务委员会. 西安市城市规划管理条例, 2005.)

[268] SERAGELDIN I and GROOTAERT C. Defining Social Capital: An Integrating View [M] // P DASGUPTA and I SERAGELDIN. Social capital: a multifaceted perspective. Washington, D.C.: World Bank, 2000: 40–58.

[269] SHAH B. Being young, female and Laotian: Ethnicity as social capital at the intersection of gender, generation, 'race' and age [J]. Ethnic and Racial Studies, 2007, 30(1):28–50.

[270] SHAN J X. Urbanization and the Protection of Cultural Heritage [M]. Tianjin: Tianjin University Press, 2006. (单霁翔. 城市化发展与文化遗产保护. 天津：天津大学出版社, 2006.)

[271] SHAN J X. Cultural heritage conservation and urban culture renaissance [M]. Beijing: China Architecture and Building Press, 2009. (单霁翔. 文化遗产保护与城市文化建设. 北京：中国建筑工业出版社, 2009.)

[272] SHEN J F. Scale, state and the state: reorganising urban space in China [M] // L J C MA and F L WU. Restructuring the Cinese City: changing society, economy and space. London; New York: Routledge, 2005: 39–58.

[273] SHI H S. Study of Xi'an Urban Geography in Ming and Qing Dynasties [M]. Beijing: Chinese Social Science Press, 2008. (史红帅. 明清时期西安城市地理研究. 北京：中国社会科学出版社, 2008.)

[274] SHI H S and WU H Q. Xi'an: A strategic important city in northwest China [M]. Xi'an: Xi'an Publishing House, 2007. (史红帅, 吴宏岐. 西北重镇西安. 西安：西安出版社, 2007.)

[275] SHI Y. Research on the Urban Social Problems in Xi'an [M]. Lanzhou: Lanzhou University Press, 2004. (石英. 西安城市社会问题研究. 兰州：兰州大学出版社.)

[276] SHORT J R. Housing in Britain: the post-war experience [M]. London: Methuen, 1982.

[277] SIT F S V. The City in China: review of the city in the PRC in 1949–2009 [C]. Paper presented at the 2009 Annual Meeting of Association of American Geographers, 2009.

[278] SMDPI (Shaanxi Muslim Designing and Planing Institute) and XUAT (Xi'an University of Architecture and Technology). Research on Xi'an Muslim Historical Area's Renewal and Innovation [M]. Xi'an: Xi'an University of Architecture and Technology, 1997. (陕西穆斯林建筑规划设计院, 西安建筑科技大学. 西安市回民区更新改造研究, 1997.)

[279] SMITH A H. Chinese characteristics [M]. New York: Revell, 1894.

[280] SNU (Geography Department of Shanxi Norm University). Records of Xian's geography [M]. Xi'an: Shanxi

[281] SONG J Y and YUAN X L. Studies on the Problems and Strategies of Urban Community Construction in Xi'an [M] // Q L CHENG. Theories and Practices on Xi'an Urban Community Construction. Xi'an: Xi'an Publishing House, 2003: 11–23. (宋景赟，袁晓玲. 社会转型期西安市社区建设存在问题及对策的研究 // 西安城市社区建设的理论与实践. 西安：西安出版社，2003: 11–23.)

[282] SORENSEN A and FUNCK C. Conclusions: a diversity of machizukuri processes and outcomes [M] // A SORENSEN and C FUNCK. Living cities in Japan: citizens' movements, machizukuri and local environments. London; New York: Routledge, 2007a: 269–279.

[283] SORENSEN A and FUNCK C. Living cities in Japan: citizens' movements, machizukuri and local environments [M]. London; New York: Routledge, 2007b.

[284] SSB (Shaanxi Statistics Bureau). Statistical Yearbook of Shaanxi Province [M]. Beijing: China Statistics Press, 2004~2012. (陕西省统计局. 陕西统计年鉴. 北京：中国统计出版社，2006.)

[285] STAKE R E. Art of case study research [M]. Thousand Oaks, Calif.: Sage Publications, 1995.

[286] STEINBERG F. Conservation and rehabilitation of urban heritage in developing countries [J]. Habitat International, 1996, 20(3):463–475.

[287] STEWART M. Collaboration in multi-actor governance [M] // M HAUS, H HEINELT and M STEWART. Urban governance and democracy: leadership and community involvement. London: Routledge, 2005: 149–167.

[288] STOKER G. Urban Development Corporations: a review [J]. Regional studies, 1989, 23(2):156–167.

[289] STOKER G. Redefining Local Democracy [M] // L PRATCHETT and D WILSON. Local democracy and local government. Basingstoke: Macmillan Press, 1996: 188–209.

[290] STORPER M. The regional world: territorial development in a global economy [M]. New York; London: Guilford Press, 1997.

[291] STOVEL H. Approaches to Managing Urban Transformation for Historic Cities [C]. Paper presented at the The Conservation of Urban Heritage: Macao Vision, International Conference, 2002.

[292] SUN L P. Free Floating Resources" and "Free Action Space" —On the Transition of China's Social Structure During China's Reform [J]. Probe, 1993, 1:64–68. (孙立平. "自由流动资源"与"自由活动空间" —论改革过程中中国社会结构的变迁. 探索，1993, 1: 64–68.)

[293] SUN L P. Guanxi, Social Connection and Social Structure [J]. Sociological Studies 1996, 5:20–30. (孙立平. 关系、社会关系与社会结构. 社会学研究，1996, 5: 20–30.)

[294] SUN P. From Historic Cities to Historic Districts [J]. City Planning Review, 1992, 6:36–37. (孙平. 从"名城"到"历史保护地段". 城市规划，1992, 6: 36–37.)

[295] SUN X Y. Xi'an West Street: towards a modern commercial street [EB/OL]. (2007) [2008–08–13] http://www.cb-h.com/2008/shshshow.asp?n_id=31973. (孙孝运. 西安西大街：向现代化商业大街迈进，2007.)

[296] SUSNIK A and GANESAN S. Select Case Study Findings from Comprehensive Urban Renewal in Hong Kong [M] // S GANESAN and D C K HUI. Hong Kong Papers in Design and Development. Hong Kong: Pace Publishing, 1998, 1.

[297] SUSNIK A E. Urban Redevelopment and Displacement Outcomes: case studies of urban renewal in Hong Kong [D]. Unpublished PhD Thesis, University of Hong Kong, Hong Kong, 1997.

[298] TAKAKI A and SHIMOTRUMA K. What is "Living Heritage Site" [C]? Paper presented at the ICCOROM's

Living Heritage Sites Programme First Strategy Meeting, 2003.

[299] TANG B S and LIU S C. Property developers and speculative development in China [M] // C R DING and Y SONG. Emerging land and housing markets in China. Cambridge, Mass.: Lincoln Institute of Land Policy, 2005: 199-231.

[300] TANG Z X. Introduction to community development in China [M]. Tianjin: Tianjin Renmin Press, 2000. (唐忠新 . 中国城市社区建设概论 . 天津 : 天津人民出版社 , 2000.)

[301] TAO D M and CHEN M M. Political Participation in Contemporary China [M]. Hangzhou: Zhejiang People's Press, 1998. (陶东明 , 陈明明 . 当代中国政治参与 . 杭州 : 浙江人民出版社 , 1998.)

[302] TEMKIN K and ROHE W M. Social Capital and Neighborhood Stability: An Empirical Investigation [J]. Housing Policy Debate, 1998, 9(1):61-88.

[303] THOMPSON R. City planning in China [J]. World Development, 1975, 3:595-605.

[304] TUROK I. Property-led urban regeneration: panacea or placebo [J]? Environment and planning A, 1992, 24(3):361-379.

[305] UN-HABITAT. UN-HABITAT for a better urban future [M]. Nanjing: World Urban Forum IV, 2008.

[306] UNESCO. Agenda 21—United Nations Conference on Environemnt and Development [EB/OL]. (1992) [2007-01-20] http://www.sidsnet.org/docshare/other/Agenda21_UNCED.pdf.

[307] UNESCO. Convention for the Safeguarding of the Intangible Cultural Heritage [EB/OL]. (2003) [2007-01-20] http://unesdoc.unesco.org/images/0013/001325/132540e.pdf.

[308] UNESCO. Operational Guidelines for the Implementation of the World Heritage Convention [EB/OL]. (2005) [2007-01-20] http://whc.unesco.org/en/guidelines.

[309] UNESCO. Asia conserved: lessons learned from the UNESCO Asia-Pacific Heritage Awards for culture heritage conservation (2000-2004) [M]. Bangkok: UNESCO, 2007.

[310] USLANER E M and CONLEY R S. Civic Engagement and Particularized—Trust – the Ties that Bind People to their Ethnic Communities [J]. American Politics Research, 2003, 31(4):331-360.

[311] VAN DER GAAG M and SNIJDERS T A B. A comparison of measures for individual social capital [EB/OL]. (2003) [2009-01-06] http://www.xs4all.nl/~gaag/work/comparison_paper.pdf.

[312] VAN DER GAAG M and SNIJDERS T A B. The Resource Generator: Measurement of Individual Social Capital with Concrete Items [C]. Paper presented at the the XXII Sunbelt International Social Networks Conference, 2004.

[313] VAN DER GAAG M, SNIJDERS T A B and FLAP H. Position Generator Measures and Their Relationship to Other Social Capital Measures [M] // N LIN and B H ERICKSON. Social capital: an international research program. Oxford; New York: Oxford University Press, 2008: 27-48.

[314] VOGEL E F. From Friendship to Comradeship: The Change in Personal Relations in Communist China [J]. The China Quarterly, 1965, 21:46-60.

[315] WAGER J. Developing a strategy for the Angkor World Heritage Site [J]. Tourism Management, 1995, 16(7):515-523.

[316] WALDER A G. China's Transitional Economy: Interpreting Its Significance [J]. The China Quarterly, 1995a, 144:963-979.

[317] WALDER A G. Local Governments as Industrial Firms: An Organizational Analysis of China's Transitional

Economy [J]. The American Journal of Sociology, 1995b, 101(2):263–301.

[318] WALEY P and FIÉVÉ N. Introduction: Kyoto and Edo–Tokyo: Urban Histories in Parallels and Tangents [M] // N FIÉVÉ and P WALEY. Japanese capitals in historical perspective: place, power and memory in Kyoto, Edo and Tokyo. London: RoutlegeCurzon, 2003: 1–37.

[319] WALSH M L. Building Citizen Involvement: strategies for local government [M]. Washington: ICMA, 1997.

[320] WANG D G and CHAI Y W. The jobs–housing relationship and commuting in Beijing, China: the legacy of Danwei [J]. Journal of Transport Geography, 2009, 17(1):30–38.

[321] WANG D H. The outline Of city planning history Of China [M]. Nanjing: Southeast University Press, 2005. (汪德华 . 中国城市规划史纲 . 南京：东南大学出版社 , 2005.)

[322] WANG H, HE X Y and BAO H. Sustainable Development of Historical District in Western City [J]. Journal of Architecture and Civil Engineering, 2006, 3:90–94. (王华，贺小宇，宝华 . 西部城市历史街区的可持续发展 . 建筑科学与工程学报 , 2006, 3: 90–94.)

[323] WANG J H. National Policies for Preserving Historical District in China [M] // B ERRING, H HØYEM and S VINSRYGG. The horizontal skyscraper. Trondheim: Tapir Academic Press, 2002a: 1–3.

[324] WANG J H. On the various levels of conservation of historic and cultural heritages [J]. Planners, 2002b, 6:9–13. (王景慧 . 论历史文化遗产保护的层次 . 规划师 , 2002, 6: 9–13.)

[325] WANG J H, RUAN Y S and WANG L. Theory and Planning of the Conservation of Historic Cities [M]. Shanghai: Tongji University Press, 1999. (王景慧，阮仪三，王林 . 历史文化名城保护理论与规划 . 上海：同济大学出版社 , 1999.)

[326] WANG J X, YANG Y Y and CHEN W Q. Investigation report on social mentality of China in 2006 [M] // X RU, X Y LU and P L LI. Blue Book of China's Society 2007. Beijing: Social Sciences Academic Press (China), 2006:63–75. (王俊秀，杨宜音，陈午晴 . 2006 年中国社会心态调查报告 // 社会蓝皮书：2007 年中国社会形势分析与预测 . 北京：社会科学文献出版社 , 2006: 63–75.)

[327] WANG J Y. Participatory governance: a positive study on urban community building in China [M]. Beijing: Chinese Social Science Press, 2006a. (王敬尧 . 参与式治理：中国社区建设实证研究 . 北京市：中国社会科学出版社 , 2006.)

[328] WANG J Z. Human Relationships in China [M]. Taiyuan: Shanxi Renmin Press, 1989. (王举忠 . 中国人际关系 . 太原市：山西人民出版社 , 1989.)

[329] WANG L L and WEI H K. Policies of China's Western Development Program [M]. Beijing: Economy and Management Publishing House, 2003. (王洛林，魏后凯 . 中国西部大开发政策 . 北京：经济管理出版社 , 2003.)

[330] WANG M H. Social Reform and Stratum Changes in China [M]. Taipei: Dao Xiang Publishing House, 2006b. (王明辉 . 中国社会改革与阶层变迁 . 台北：稻乡出版社 , 2006.)

[331] WANG S J and FAN L D. Urbanization in China: 1959–2000 [M] // G D NESS and P P TALWAR. Asian urbanization in the new millennium. Singapore: Marshall Cavendish Academic, 2005: 358–383.

[332] WANG T. Research on the Renewal Method of Xi'an Muslim District [M] // B ERRING, H HØYEM and S VINSRYGG. The horizontal skyscraper. Trondheim: Tapir Academic Press, 2002c: 91–95.

[333] WANG T. Drafting and Approval Check of the Detailing Planning of Historical Areas [J]. Planners, 2004, 3(20):67–68. (王涛 . 关于历史街区详细规划的编制与审批 . 规划师 , 2004, 3(20), 67–68.)

[334] WANG Y N. Methods and principles on the conservation of intangible cultural heritage [M] // W Z WANG. A collection of papers of the Forum on the Protection of Chinese Intangible Cultural Heritage. Beijing: Culture and Art Publishing House, 2006c: 231-236. (王亚南 . 论非物质文化遗产保护的方法和原则 // 中国非物质文化遗产保护论坛论文集 . 北京：文化艺术出版社 , 2006: 231-236.)

[335] WANG Y P. Planning and Conservation in Historic Chinese Cities: The case of Xi'an [J]. Town Planning Review, 2000, 71(3):311-331.

[336] WANG Y P and HAGUE C. The Development and Planning of Xi'an since 1949 [J]. Planning Perspectives, 1992, 7:1-26.

[337] WATANABE H. Kyoto fragments: Divisions and dicontinuities in an ancient city [J]. Japan Quarterly, 1994, 41(4):416-433.

[338] WATANABE S I J. Toshi keikaku vs machizukuri [M] // A SORENSEN and C FUNCK. Living cities in Japan: citizens' movements, machizukuri and local environments. London; New York: Routledge, 2007: 39-55.

[339] WB (World Bank). China: urban land management in an emerging market economy [M]. Washington, D.C.: World Bank, 1993.

[340] WB (World Bank). The Initiative on Defining, Monitering and Measuring Social Capital [EB/OL]. (1998) [2008-04-09] http://siteresources.worldbank.org/INTSOCIALCAPITAL/Resources/SCI-WPS-02.pdf.

[341] WCED (World Commission on Environment and Development). Our common future [M]. Oxford: Oxford University Press, 1987.

[342] WESTERN J S, WELDON P D and HAUNG T T. Poverty, urban renewal, and public housing in Singapore [J]. Environment and planning A, 1973, 5(5):589-599.

[343] WILLIAMS O. Governmental Intervention into Local Economies under Market Conditions: the case of urban renewal [M] // J-E LANE. State and market: the politics of the public and the private. London: Sage Pubn, 1985: 142-157.

[344] WU B A. Defining intangible cultural heritage in Chinese cultural context [M] // DEVELOPMENT CENTRE OF ETHNIC AND FOLK LITERATURE AND ART. Studies on intangible cultural heritage conservation in China. Beijing: Beijing Normal University Press, 2007: 23-41. (乌丙安 . 中国文化语境中的"非物质文化遗产"界定 // 中国非物质文化遗产保护研究 . 北京：北京师范大学出版社 , 2007: 23-41.)

[345] WU B L. A Brief Introduction to Xi'an History [M]. Xi'an: Shanxi People's Publishing House, 1979. (武伯纶 . 西安历史述略 . 西安：陕西人民出版社 , 1979.)

[346] WU F L. Urban restructuring in China's emerging market economy: towards a framework for analysis [J]. International Journal of Urban and Regional Research, 1997, 21(4):640–663.

[347] WU F L. China's changing urban governance in the transition towards a more market-oriented economy [J]. Urban Studies, 2002a, 39(7):1071-1093.

[348] WU F L. Residential relocation under market-oriented redevelopment: the process and outcomes in urban China [J]. Geoforum, 2004, 35(4):453-470.

[349] WU F L, XU J and YEH A G O. Urban development in post-reform China: state, market, and space[M]. London; New York: Routledge, 2007.

[350] WU F L and ZHANG J X. Strategic planning in urban development – responses of Chinese local governments to surrounding fierce competitions [M] // F L WU, L J C MA and J X ZHANG. Transition and Reconstruction:

Multi-perspectives of Urban China Development. Nanjing: Southeast University Press, 2007a: 66-82. (吴缚龙，张京祥. 城市发展战略规划 – 中国地方政府对激烈竞争环境的回应 // 转型与重构：中国城市发展多维透视. 南京：东南大学出版社, 2007: 66-82.)

[351] WU H Q. Study of Xi'an Historic Geography [M]. Xi'an: Xi'an Map Publishing House, 2006. (吴宏岐. 西安历史地理研究. 西安：西安地图出版社, 2006.)

[352] WU J L. Economic Reforms in Contemporary China: Strategies and Implementation [M]. Shanghai: Shanghai Far East Publishing House, 1999. (吴敬琏. 当代中国经济改革：战略与实施. 上海：上海远东出版社, 1999.)

[353] WU L Y. Conserving Beijing with scrutiny [J]. Beijing Planning Review, 1998, 2:1-4. (吴良镛. 北京旧城要审慎保护. 北京规划建设, 1998, 2: 1-4.)

[354] WU L Y. Sober Contemplations on the Third-spring of Urban Planning [J]. City Planning Review, 2002, 2:9-14. (吴良镛. 面对城市规划"第三个春天"的冷静思考. 城市规划, 2002, 2: 9-14.)

[355] WU P S and ZHANG Y D. Construction and management of urban communities [M]. Shanghai: Shanghai People's Publishing House, 2007b. (吴鹏森，章友德. 城市社区建设与管理. 上海市：上海人民出版社, 2007.)

[356] XCPB (Xi'an City Planning Bureau). Concept Plan of the Regeneration of the Tang Imperial City [M]. Xi'an: Xi'an City Planning Bureau, 2005a. (西安城市规划局. 唐皇城复兴规划. 西安：西安城市规划局, 2005.)

[357] XCPB (Xi'an City Planning Bureau). Planning of Xi'an [M]. Xi'an: Xi'an City Planning Bureau, 2005b. (西安城市规划局. 西安. 规划. 西安：西安城市规划局, 2005.)

[358] XCPB (Xi'an City Planning Bureau). Planning of Xi'an City Cultural System [M]. Xi'an: Xi'an City Planning Bureau, 2005c. (西安城市规划局. 西安市城市建设文化体系规划. 西安：西安城市规划局, 2005.)

[359] XCPB (Xi'an City Planning Bureau). Report on Current Xi'an Urban Condition [M]. Xi'an: Xi'an City Planning Bureau, 2007. (西安城市规划局. 西安市城市现状报告. 西安：西安城市规划局, 2007.)

[360] XCPB (Xi'an City Planning Bureau), XACH (Xi'an Administration of Cultural Heritage), and XUPDI (Xi'an Urban Planning and Design Institute). Conservation Master Plan of Xi'an Historic City [M]. Xi'an: Xi'an City Planning Bureau, 1994/2005. (西安市规划局，西安市文物局，西安市城市规划设计研究院. 西安市历史文化名城保护规划. 西安：西安城市规划局, 2005.)

[361] XCPDI (Xi'an Urban Planning and Design Institute). The Preservation and Renewal of Beiyuanmen Historic District [M] // B ERRING, H HØYEM and S VINSRYGG. The horizontal skyscraper. Trondheim: Tapir Academic Press, 2002: 86-90.

[362] XCPDI (Xi'an Urban Planning and Design Institute). Landscape and Planning Project of Beiyuanmen Street in Xi'an [M]. Xi'an: Xi'an City Planning and Design Institute, 2005. (西安市城市规划设计研究院. 北院门回坊文化风情街规划方案. 西安：西安市城市规划设计研究院)

[363] XCUCRC (Xi'an Committee of Urban Construction Records Compilation). Xi'an Urban Construction Records Compilation [M]. Xi'an: Xi'an Map Press, 2000. (西安市城建系统志编纂委员会. 西安市城建系统志. 西安：西安地图出版社, 2000.)

[364] XIA Q. Establishing the Sustainable Development of the Mechanism of Preservation and Renewal of Historic Cultural Cities [J]. Urbanism and Architecture, 2006, 12:15-17. (夏青. 建立可持续发展的历史文化名城保护更新机制. 城市建筑, 2006, 12: 15-17.)

[365] XIAO L. Conservation Projects and Research Studies of Drum-tower Muslim Historical District [J]. Infor Mation of China Construction, 2004, 13:59-62. (肖莉. 西安鼓楼历史街区保护项目与保护研究. 中国建设信息, 2004, 13: 59-62.)

[366] XIE F Z. China: land institutions and housing policy [M]. Beijing: Earth Publishing House (China), 2008. (谢伏瞻. 土地制度与住房政策. 北京市：中国大地出版社, 2008.)

[367] XIE Y C and COSTA F J. Urban planning in socialist China: theory and practice [J]. Cities, 1993, 10(2):103-114.

[368] XINHUANEWS. Xi'an applying its ancient music to the World Intangible Cultural Heritage [EB/OL]. (2004) [2009-03-04] http://news.xinhuanet.com/house/2004-06/09/content_1516898.htm. (新华网. 陕西省西安古乐申报"世界非物质文化遗产", 2004.)

[369] XLDG (Xi'an Lianhu District Government). (), Work-plan of the External Trade and Economy Cooperation Bureau, Lianhu District, Xi'an City in 2006 [EB/OL]. (2005) [2008-08-13] http://www.lianhu.gov.cn/content.asp?leaf_id=2925. (西安市莲湖区政府. 西安市莲湖区对外贸易经济合作局2006年工作计划, 2005.)

[370] XMG (Xi'an Municipal Government). Measures to implement the policy of Xi'an Open City [M] // EDITORIAL BOARD OF XI'AN YEARBOOK. Xi'an Yearbook. Xi'an: Shaanxi People's Publishing House, 1993: 1-6. (西安市人民政府. 西安市实施开放城市政策的若干办法 // 西安年鉴. 西安：陕西人民出版社, 1993: 1-6.)

[371] XMG (Xi'an Municipal Government). Announcement about the implementation of the street widening project in Sajinqiao Street promoted by Xi'an municipal government [J]. Xi'an Municipal Government Gazette, 2005a, 4:4. (西安市人民政府. 关于洒金桥大街拓宽改造工程拆迁工作的通告. 西安市人民政府公报, 2005, 4:4.)

[372] XMG (Xi'an Municipal Government). Perfecting Urban Functions, Improving Living Environment, Building Five-Convenient Urban District [EB/OL]. (2008a) [2009-04-08] http://news.idoican.com.cn/xarb/html/2008-05/07/content_4725309.htm#. (西安市人民政府. 完善城市功能，改善居住环境，建设"五宜"城区——碑林区乐居场地区综合改造项目昨日全面启动. 2008.)

[373] XMG (Xi'an Municipal Government). Renewal and management method of dilapidated urban areas in Xi'an [EB/OL]. (2008b) [2009-04-10] http://www.law-star.com/cacnew/200809/195023241.htm. (西安市人民政府. 西安市棚户区改造管理办法, 2008.)

[374] XMG (Xi'an Municipal Government) and XSB (Xi'an Statistics Bureau). Xi'an Fifty Years [M]. Beijing: China Statistics Press, 1999. (西安市人民政府，西安市统计局. 西安五十年. 北京市：中国统计出版社, 1999.)

[375] XMG (Xi'an Municipal Government). Xi'an Master Plan 2004-2020 [M]. Xi'an: Xi'an People's Municiple Government, 2005b. (西安市人民政府. 西安城市总体规划（2004-2020）. 西安：西安市人民政府, 2005.)

[376] XMHDPPO (Xi'an Muslim Historical District Protection Project Office). Report of Sino-Norwegian Cooperative Xi'an Muslim Historical District Protection Project [M]. Xi'an: Xi'an Municipal Government, 2003. (西安回民区保护项目办公室. 中挪合作西安回民历史街区保护项目研究成果汇编. 西安：西安市人民政府, 2003.)

[377] XSB (Xi'an Statistics Bureau). Xi'an Statistical Yearbook [M]. Beijing: China Statistics Press, 1990-2012. (西

安市统计局 . 西安统计年鉴 <1990–2012>. 北京：中国统计出版社 , 1990–2012.)

[378] XSFM (Xi'an Stele Forest Museum). Xi'an Stele Forest Museum [M]. Xi'an: Shaanxi People's Publishing House, 2008. (西安碑林博物馆 . 西安：西安碑林博物馆 .)

[379] XSSA (Xi'an Social Science Academy). Studies on Xi'an Urban Community Constructions [M] // Q L CHENG. Theories and Practices on Xi'an Urban Community Construction. Xi'an: Xi'an Publishing House, 2003: 50–73. (西安市社会科学院 . 西安城市社区建设研究 // 西安市社区建设的理论与实践 . 西安：西安出版社 , 2003: 50–73.)

[380] XU J and NG M K. Socialist urban planning in transition: The case of Guangzhou, China [J]. International Development Planning Review, 1998, 20(1):35–51.

[381] XU Q W and CHOW J C. Urban community in China: service, participation and development [J]. International Journal of Social Welfare, 2006, 15(3):199–208.

[382] XU X and YUAN H. Meaning and Role of Social Capital, and its Constructions in China [J]. Sociology, 2004, 03:66–72. (徐翔 , 袁虹 . 社会资本的内涵、作用及在我国的建设 . 社会学 , 2004, 3: 66–72.)

[383] XUPDI (Xi'an Urban Planning and Design Institute). Conservation and renewal planning of Sanxuejie historic district in Xi'an city [M]. Xi'an: Xi'an Urban Planning and Design Institute, 2001. (西安市城市规划设计研究院 . 西安市三学街传统地段保护与更新规划 . 西安：西安市城市规划设计研究院 , 2001.)

[384] XUPDI (Xi'an Urban Planning and Design Institute). Landscape and Planning Project of Beiyuan Road in Xi'an [M]. Xi'an: Xi'an Urban Planning and Design Institute, 2007. (西安市城市规划设计研究院 . 北院门回坊文化风情街规划方案 . 西安：西安市城市规划设计研究院 , 2007.)

[385] XUPEMB (Xi'an Urban Planning and Environmental Management Bureau). Xi'an Central Urban District Planning [M]. Xi'an: Xi'an Urban Planning and Environmental Management Bureau, 1989. (西安城市规划与环境管理局 . 西安市市中心区规划 . 西安：西安城市规划与环境管理局 , 1989.)

[386] YANG J Q and WU M W. Modern urban renewal [M]. Nanjing: Southeast University Press, 1999. (阳建强 , 吴明伟 . 现代城市更新 . 南京：东南大学出版社 , 1999.)

[387] YANG J Q. Status Quo, Characteristics and Tendency of Urban Renewal in China [J]. City Planning Review, 2000, 24(4):53–63. (阳建强 . 中国城市更新的现况、特征与趋向 . 城市规划 , 2000, 24(4)：53–63.)

[388] YANG M R. Thoughts on the conservation of Xi'an historic core [M] // H X He. The Special Capital City of Xi'an. Beijing: Chinese Architectural and Industrial Press, 2006: 93-94. (杨明瑞 . 保护西安老城的思考 // 特色城市：古都西安 . 北京：中国建筑工业出版社 , 2006:93–94.)

[389] YANG W J. Interaction adaptation and reconstruction: a study of the northwestern urban Hui Muslim community and its cultural transition [M]. Beijing: Ethnic Nationality Press, 2007. (杨文炯 . 互动调适与重构：西北城市回族社区及其文化变迁研究 . 北京：民族出版社 , 2007.)

[390] YANG W X. Legislation Construction as the Inevitable Way for the Conservation of Historical Cities [M] // XI'AN URBAN–RURAL CONSTRUCTION COMMITTEE and RESEARCH COMMITTEE ON XI'AN HISTORIC AND CULTURAL CITY. A collection on the conservation and development of Xi'an Historic and Cultural City in the new era. Xi'an: Shaanxi People's Publishing House, 2003: 34–44. (杨文晓 . 法制建设是名城保护的必由之路 // 新世纪西安历史文化名城保护与发展论文集 . 西安：陕西人民出版社 , 2003:34–44.)

[391] YANG Y R and CHANG C H. An Urban Regeneration Regime in China: A Case Study of Urban

[392] YEH A G O. The Dual Land Market and Urban Development in China [M] // C R DING and Y SONG. Emerging land and housing markets in China. Cambridge, Mass.: Lincoln Institute of Land Policy, 2005: 39–57.

[393] YIN R K. Case study research: design and methods [M]. Thousand Oaks: Sage Publications, 1994.

[394] YOUNG M and WILLMOTT P. Family and kinship in East London [M]. London: Routledge and Kegan Paul, 1957.

[395] Yu X W and Ju H S. (2006), (Constructing the Harmonious Xi'an), Xi'an: Shaanxi People's Publishing House. (于小文, 巨杭生. 构建和谐西安——公众主观社会指标调查与研究. 西安: 陕西人民出版社, 2006.)

[396] YU Z C. Conservation, Regeneration and Reuse of Xi'an Qixianzhuang District [D]. Unpublished Master Thesis, Xi'an University of Architecture and Technology, Xi'an, 2003. (虞志淳. 西安七贤庄地区保护、更新与利用. 西安: 西安建筑科技大学, 2003.)

[397] YUAN J J. Xi'an Fu Xue: Thousands years of official school [EB/OL]. (2008) [2009-01-10] http://www.xazz.cn/Article_Show.asp?ArticleID=6434. (原建军. 西安府学官办经学教育千年一叹, 2008.)

[398] YUE G A. Research on the Human Relationships of Contemporary China [M]. Tianjin: Nankai University Press, 2002. (乐国安. 当前中国人际关系研究. 天津: 南开大学出版社, 2002.)

[399] ZETTER J. Challenges for Japanese Urban Policy [M] // P SHAPIRA, I MASSER and D W EDGINGTON. Planning for cities and regions in Japan. Liverpool, England: Liverpool University Press, 1994: 25–32.

[400] ZHAI B Q. Community Participation in Xi'an Drum Tower Muslim District [N]. China Jianshe Newspaper, 2009-12-01. (翟斌庆. 西安鼓楼回坊街的公众参与. 中国建设报, 2009-12-01.)

[401] ZHAI B and NG M K. Urban regeneration and social capital in China: A case study of the Drum-tower Muslim District in Xi'an [J]. Cities, 2013, 35:14–25.

[402] ZHAI B Q and NG M K. Urban Regeneration and Its Realities in Urban China [J]. Urban Planning Forum, 2009, 2:75–82. (翟斌庆, 伍美琴. 城市更新理念与中国城市现实. 城市规划汇刊, 2009, 2: 75–82.)

[403] ZHAI B Q and ZHAI B W. Social Capital in China's Urban Regeneration [J]. Urban Planning International, 2010, 25(1):53–59. (翟斌庆, 翟碧舞. 中国城市更新中的社会资本. 国际城市规划, 2010, 25(1):53–59.)

[404] ZHAI X W. Is it Guanxi or social capital [M]? // X H ZHOU. Social network and enterprise development. Beijing: Economy and Management Publishing House, 2008: 306–314. (翟学伟. 是关系, 还是社会资本? // 中国社会网络与社会资本研究报告 2007-2008 社会网络与企业成长. 北京: 经济管理出版社, 2008: 306–314.)

[405] ZHANG D Q. Importance of Civil Organizations Involvement Considering the Characteristics of Intangible Cultural Heritage [M] // DEVELOPMENT CENTRE OF ETHNIC AND FOLK LITERATURE AND ART OF MINISTRY OF CULTURE OF CHINA. Research on the Conservation of Intangible Cultural Heritage of China. Beijing: Beijing Normal University Press, 2007a: 402–406. (张笃勤. 从非物质文化遗产特点看民间文化组织参与的重要性 // 中国非物质文化遗产保护研究. 北京: 北京师范大学出版社, 2007: 402+406.)

[406] ZHANG F. Strategies on the Conservation of Historical Culture in Urban Development [M]. Nanjing: Southeast University Press, 2006a. (张凡. 城市发展中的历史文化保护对策. 南京市: 东南大学出版社.)

[407] ZHANG F C. Experiencing Xi'an Urban Development [M]. Xi'an: Shaanxi People's Publishing House, 2002a. (张富春. 亲历西安的城市建设. 西安: 陕西人民出版社, 2002.)

[408] ZHANG J. Historic Preservation and Refurbishment in Beijing [J]. City Planning Review, 2002b, 26(2):73–75.

(张杰 . 北京城市保护与改造的现状与问题 . 城市规划 , 2002, 26(2), 73-75.)

[409] ZHANG J F and HAN J. Protecting Ancient City and Giving Play to Strong Points [J]. Architectural Journal, 1981, 2:15-17. (张景沸，韩骥 . 保护古城与发挥优势 . 建筑学报 , 1981, 2: 15-17.)

[410] ZHANG J G. Urban Planning System Involved by the Public Participation [J]. City Planning Review, 2000, 24(7):57-58. (张继刚 . 浅谈城市规划中的公众参与 . 城市规划 , 2000, 24(7):57-58.)

[411] ZHANG L. Culture Heritage, Info-carrier and Historic Memory – A Study for Historic Infor and Its Carrier of Shuyuanmen, Beiyuanmen and Defuxiang of Xi'an [D]. Unpublished Master Thesis, Xi'an University of Architecture and Technology, Xi'an, 2002c. (张凌 . 遗产、载体与记忆——西安书院门、北院门和德福巷历史街道保护研究 . 西安：西安建筑科技大学 , 2002.)

[412] ZHANG L and ZHAO J Y. Getting out of misunderstanding about the conservation of historic buildings – Reflections on the current development situations of the South Avenue of Xi'an [J]. Journal of Architecture and Civil Engineering, 2003, 20(3):28-30. (张琳，赵敬源 . 走出古建保护的误区——对西安南大街发展现状的反思 . 长安大学学报 (建筑与环境科学版), 2003, 20(3): 28-30.)

[413] ZHANG M J. Whether Da Tang Xi Shi can support Xi'an's dream of prosperous Tang dynasty or not [EB/OL]. (2006b) [2007-06-20] http://www.xtour.cn/2005-12/2005128150320.htm. (张敏洁 . 大唐西市能否承载西安的盛唐梦想 , 2006.)

[414] ZHANG P Y. Urban Regeneration: theory and practice in China's new urbanization [J]. City Planning Review, 2004, 4:25-30. (张平宇 . 城市再生：我国新型城市化的理论与实践问题 . 城市规划 , 2004, 4: 25-30.)

[415] ZHANG Q, LI Z M and FENG Q. Study on the current living courtyards and environments of historical urban areas in the rapid urbanization – case study of Xi'an Stele Forest district [M] // J Y TSOU, R L XU, D J JIN and W K POON. Sustainability and harmony: development of eco and social sustainable housing and human settlements under rapid urbanization process – proceedings of the sixth China Urban Housing Conference. Beijing: China City Press, 2007: 215-222. (张倩，李志民，冯青 . 城市高速建设时期的历史街区民居建筑院落与环境生存现状研究 – 以西安碑林历史街区为例 // 永续　和谐：快速城镇化背景下的住宅与人居环境建设 – 第六届中国城市住宅研讨会论文集 . 北京市：中国城市出版社 , 2007: 215-222.)

[416] ZHANG S H. Low-lying land renewal project of Wayao village [M] // EDITORIAL BOARD OF XI'AN YEARBOOK. Xi'an Yearbook. Xi'an: Shanxi People's Publishing House, 1993: 307. (张韶辉 . 瓦窑村低改工程进展顺利 // 西安年鉴 . 西安：陕西人民出版社 , 1993: 307.)

[417] ZHANG S S. Role of Civil Society in the Conservation of Intangible Cultural Heritage in China [M] // DEVELOPMENT CENTRE OF ETHNIC AND FOLK LITERATURE AND ART. Research on the Conservation of Intangible Cultural Heritage of China. Beijing: Beijing Normal University Press, 2007b: 407-420. (张士闪 . 论我国非物质文化遗产保护工程中民众主体的作用 // 中国非物质文化遗产保护研究 . 北京：北京师范大学出版社 , 2007: 407-420.)

[418] ZHANG T W. Urban Development and a Socialist Pro-Growth Coalition in Shanghai [J]. Urban Affairs Review, 2002d, 37(4):475-499.

[419] ZHANG W H. Class structure and social networks in urban China [M]. Shanghai: Shanghai People's Publishing House, 2006c. (张文宏 . 中国城市的阶层结构与社会网络 . 上海：上海人民出版社 , 2006c.)

[420] ZHANG Y. Xi'an Stele Forest Museum [M]. Xi'an: Shaanxi Tourism Publishing House, 2002e. (张云 . 西安碑林博物馆 . 西安：陕西旅游出版社 , 2002.)

[421] ZHANG Y W. Urbanization and social network in China: two cases in Daqing city and Pudong of Shanghai city [M]. Beijing: Social Sciences Academic Press (China), 2008. (张云武 . 中国的城市化与社会关系网络：以大庆市和上海浦东新区为例 . 北京：社会科学文献出版社，2008.)

[422] ZHAO L Y. Ancient buildings in Shaanxi province [M]. Xi'an: Shanxi People's Publishing House, 1992. (赵立瀛 . 陕西古建筑 . 西安：陕西人民出版社，1992.)

[423] ZHAO T, LEE Y F and SUN Y S. Analysis of main problems of urban renewal in China [J]. Engineering Journal of Wuhan University, 2006, 5:80–84. (赵涛，李煜绍，孙蕴山 . 当前我国城市更新中的主要问题分析 . 武汉大学学报（工学版），2006, 5: 80–84.)

[424] ZHAO Z R. Value Conception of Preservation for Historic Districts [J]. Urban Planning Review, 1999, 1:75–77. (赵志荣 . 历史地段保护的价值观——追求可持续的资源、环境与效益 . 城市规划汇刊，1999, 1: 75–77.)

[425] ZHENG L J and YANG C M. Public Participation in the Dynamic Conservation of Historic Districts [J]. City Planning Review, 2005, 7:63–65. (郑利军，杨昌鸣 . 历史街区动态保护中的公众参与 . 城市规划，2005, 7: 63–65.)

[426] ZHENG Y. Urbanization and its characteristics in Japan after World War II [J]. World Regional Studies, 2008, 17(2):56–63. (郑宇 . 战后日本城市化过程与主要特征 . 世界地理研究，2008, 17(2): 56–63.)

[427] ZHOU G Z. Memory on the 1st round of Xi'an master plan [M] // Urban Planning Society of China. Urban Planning Face-to-Face – Proceedings of 2005 China Urban Planning Annual Meeting. Beijing: China Water and Power Press. (周干峙 . 西安首轮城市总体规划回忆，2005a: 1–7 // 城市规划面对面——2005 城市规划年会论文集 . 北京：中国水利水电出版社 .)

[428] ZHOU G Z. Meeting the Third Spring in Urban Planning [J]. City Planning Review, 2002, 26(1):9–10. (周干峙 . 迎接城市规划的第三个春天 . 城市规划，2002, 26(1): 9–10.)

[429] ZHOU G Z. For a Better Spring [J]. City Planning Review, 2004a, 11:20–25. (周干峙 . 为了一个更加美好的春天 . 城市规划，2004, 11: 20–25.)

[430] ZHOU H P. Speeding the enactment of conservation law on intangible cultural heritage [EB/OL]. (2005b) [2008-12-09] http://www.china.com.cn/chinese/kuaixun/848032.htm. (周和平 . 加快非物质文化遗产保护法的立法进程，2005.)

[431] ZHOU H Y. Social Capital, its Study and Application in China [J]. Comparative Economic and Social Systems, 2004b, 2:135–144. (周红云 . 社会资本及其在中国的研究与应用 . 经济社会体制比较，2004, 2: 135–144.)

[432] ZHOU L and CHENG M J. On the Transformation of Urban Planning under the Guidance of Scientific View of Development [J]. Planners, 2005, 2:12–13. (周岚，程茂吉 . 论科学发展观指引下的城市规划转变 . 规划师，2005, 2: 12–13.)

[433] ZHOU L and HE L. Challenges and Reform of Urban Planning of China [J]. City Planning Review, 2005, 3:9–14. (周岚，何流 . 中国城市规划的挑战和改革——探索国家规划体系下的地方特色之路 . 城市规划，2005, 3: 9–14.)

[434] ZHOU L and JIANG X. China Social Welfare [M]. Beijing: Peking University Press, 2008. (周良才 . 中国社会福利 . 北京：北京大学出版社，2008.)

[435] ZHOU L and TONG B. Q. Discussion on the Method of Planning for Protection and Renewal of Ancient Cities –

With the Ancient City of Nanjing as an Example [J]. Planners, 2005, 1:40–42. (周岚 , 童本勤 . 老城保护与更新规划编制办法探讨——以南京老城为例 . 规划师 , 2005, 1: 40–42.)

[436] ZHOU L Y and LIN C. Public Participation in Shenzhen. Planners, 2000, 5:66–69. (周丽亚 , 林晨 . 深圳城市规划中的公众参与 . 规划师 , 2000, 5: 66–69.)

[437] ZHOU M. Ethnicity as social capital: community-based institutions and embedded networks of social relations [M] // G C LOURY, T MODOOD and S M TELES. Ethnicity, social mobility and public policy: comparing the USA and UK. Cambridge: Cambridge University Press, 2005c: 131–159.

[438] ZHOU Y H and YANG X M. Danwei system in China [M]. Beijing: China Economic Publishing House, 1999. (周翼虎 , 杨晓民 . 中国单位制度 . 北京 : 中国经济出版社 , 1999.)

[439] ZHU C L. Assumptions on Xi'an Muslim District Renewal [M] // C L ZHU, L S CHEN and Z BAI. Research on Islamic Culture: proceedings of the second symposium on Xi'an Islamic culture. Yinchuan: Ningxia People's Publishing House, 1998: 254–270. (朱崇礼 . 关于西安回坊改建的设想 . 伊斯兰文化研究 : 第二届西安市伊斯兰文化研讨会论文汇编 . 银川 : 宁夏人民出版社 , 1998: 254–270.)

[440] ZHU J M. Local Growth Coalition: The Context and Implications of China's Gradualist Urban Land Reforms [J]. International Journal of Urban and Regional Research, 1999, 23(3):534–548.

[441] ZHU S G and WU H Q. Historical changes and development of Xi'an [M]. Xi'an: Xi'an Publishing House, 2003. (朱士光 , 吴宏岐 . 西安的历史变迁与发展 . 西安 : 西安出版社 , 2003.)

[442] ZHU W L. Study on Conservation and Revival by the Historic Site in Old City of Xi'an [D]. Unpublished Master Thesis, Xi'an University of Architecture and Technology, Xi'an, 2006. (朱文龙 . 西安老城历史街区的保护与更新研究——以大吉昌巷改造为例 . 西安 : 西安建筑科技大学 , 2006.)

[443] ZHUANG L D and ZHANG J X. A History of Urban China Development and Construction [M]. Nanjing: Southeast University Press, 2002. (庄林德 , 张京祥 . 中国城市发展与建设史 . 南京 : 东南大学出版社 , 2002.)

[444] ZUKIN S. Loft living: culture and capital in urban change [M]. Baltimore: Johns Hopkins University Press, 1982.

[445] ZUKIN S. Gentrification: Culture and Capital in the Urban Core [J]. Annual Review of Sociology, 1987, 13:129–147.

[446] ZUKIN S. Landscapes of power: from Detroit to Disney World [M]. Berkeley: University of California Press, 1991.

后记 / POSTSCRIPT

本书是作者在其博士论文的基础上，进一步删减修改而成。原香港大学博士论文最初由英文完成，决定出版时，考虑到阅读对象以内地读者为主，遂将原文全部翻译成中文，并采用中英对照的形式出版。在翻译过程中，原文中的部分内容进行了适当删减，有些内容和数据进行了修正与更新。概括来讲，本书主要探讨了社会资本对地方社区更新的影响与作用。书中有相当部分是对有关内容的理论分析，致力于建立一个理解我国历史城市更新过程中社会资本作用的理论架构。如果本书能对其他研究有所启发，作者将倍感欣慰。

即将付梓之际，作者要从心底感谢其在香港大学城市规划及设计系的博士导师伍美琴教授。伍教授的悉心指导与教诲，使作者逐步掌握并建立起一个坚实的研究理论架构，从而保证了后期研究工作的顺利完成。当然，本研究中尚存的任何谬误或者问题，都完全是作者本人的责任。同时，作者要感谢中国建筑工业出版社编辑们的鼓励与支持。她们认真负责的编辑工作，促成了本书的顺利完成和面世。本书的出版还受到西安交通大学人文社会科学学术著作出版基金的资助，作

This book is based on the author's doctoral thesis at The University of Hong Kong (HKU). The original thesis was finished in English. When it came into publication, considering that potential readers might be from China mainland, the author had the entire writing translated into Chinese and the book is published in bilingual format. During the translation, parts of the original work were left out or revised and some data were updated. Generally, the research work explored the impact and role of social capital on the regeneration of local communities. Plenty of work in the book was devoted to theoretical analysis of related issues, with the aim to build an analytical framework for the better understanding of the role of social capital in the regeneration of Chinese historic cities. If the book could invoke any other valuable thoughts, the author would feel most gratified.

Upon imminent publication, the author would like to thank from the bottom of his heart his doctoral thesis supervisor, Professor Mee Kam Ng, at the Department of Urban Planning and Design in HKU. It was Professor Ng's inspiration and painstaking supervision that made the author acquire the knowledge and establish a solid theoretical framework, which promised the smooth accomplishment of later research work. Of course, if any errors or problems still remain in the research, they are the author's sole responsibility. The author also appreciates the help and support from the editors of China Architecture and Building Press, whose dedicated editing work facilitated

者一并致谢。最后，作者要感谢家人一如既往的理解与支持，使作者得以全身心地投入于研究工作当中。没有家人的大力支持，本书是无法顺利完成和出版的。

the book's smooth completion and high quality delivery. The author would like to thank the generous support of the Humanities and Social Science Academic Publication Fund from Xi'an Jiaotong University. Last but not least, the author's heartfelt thanks go to his beloved family for their best understanding and support of the author's long-term dedication to the research work. Without their relentless love and support, it would be unimaginable for the research work to go this far.